Progress in Molecular and Subcellular Biology

Volume 62

This series gives an insight into the most current, cutting edge topics in molecular biology, including applications in biotechnology and molecular medicine. In the recent years, the progress of research in the frontier area of molecular and cell biology has resulted in an overwhelming amount of data on the structural components and molecular machineries of the cell and its organelles and the complexity of intra- and intercellular communication. The molecular basis of hereditary and acquired diseases is beginning to be unravelled, and profound new insights into development and evolutionary biology, as well as the genetically driven formation of 3D biological architectures, have been gained from molecular approaches. Topical volumes, written and edited by acknowledged experts in the field, present the most recent findings and their implications for future research. This series is indexed in PubMed.

Mahmut Çalışkan · Muzaffer Arıkan
Editors

Microbial World: Bacteria and Archaea

Springer

Editors
Mahmut Çalışkan
Istanbul University
Istanbul, Türkiye

Muzaffer Arıkan
Istanbul University
Istanbul, Türkiye

ISSN 0079-6484 ISSN 2197-8484 (electronic)
Progress in Molecular and Subcellular Biology
ISBN 978-3-032-11437-2 ISBN 978-3-032-11438-9 (eBook)
https://doi.org/10.1007/978-3-032-11438-9

This Springer imprint is published by the registered company Springer Nature Switzerland AG
The registered company address is: Gewerbestrasse 11, 6330 Cham, Switzerland

Dedicated to my son Ilgaz Çalışkan

Contents

Contributors

Ahmet Arıhan Erözden Division of Biotechnology, Biology Department, Faculty of Science, Istanbul University, Vezneciler, Istanbul, Türkiye
Program of Biotechnology, Biology Department, Institute of Graduate Studies in Sciences, Istanbul University, Vezneciler, Istanbul, Türkiye

Alexa Gannon Strathclyde Institute of Pharmacy and Biomedical Sciences, University of Strathclyde, Glasgow, UK

Andreas C. Lawaetz Strathclyde Institute of Pharmacy and Biomedical Sciences, University of Strathclyde, Glasgow, UK

Ayten Kimiran Department of Biology, Section of Fundamental and Industrial Microbiology, Istanbul University, Faculty of Science, Istanbul, Türkiye

Ayten Yazgan-Karataş Faculty of Science and Letters, Department of Molecular Biology and Genetics, Istanbul Technical University, Maslak, Istanbul, Türkiye
Dr. Orhan Öcalgiray Molecular Biology, Biotechnology, and Genetics Research Center (ITU-MOBGAM), Istanbul Technical University, Maslak, Istanbul, Türkiye

Daniel E. Larcombe Strathclyde Institute of Pharmacy and Biomedical Sciences, University of Strathclyde, Glasgow, UK

Dilan Ergün Faculty of Science and Letters, Department of Molecular Biology and Genetics, Istanbul Technical University, Maslak, Istanbul, Türkiye
Dr. Orhan Öcalgiray Molecular Biology, Biotechnology, and Genetics Research Center (ITU-MOBGAM), Istanbul Technical University, Maslak, Istanbul, Türkiye

Elif Özlem Arslan Aydoğdu Istanbul University, Faculty of Science, Department of Biology, Section of Fundamental and Industrial Microbiology, Vezneciler, Istanbul, Türkiye

Emily Addington School of Infection and Immunity, University of Glasgow, Glasgow, UK

Ezgi Koman Faculty of Science and Letters, Department of Molecular Biology and Genetics, Istanbul Technical University, Maslak, Istanbul, Türkiye
Dr. Orhan Öcalgiray Molecular Biology, Biotechnology, and Genetics Research Center (ITU-MOBGAM), Istanbul Technical University, Maslak, Istanbul, Türkiye

Göksu Tartar Faculty of Science and Letters, Department of Molecular Biology and Genetics, Istanbul Technical University, Maslak, Istanbul, Türkiye
Dr. Orhan Öcalgiray Molecular Biology, Biotechnology, and Genetics Research Center (ITU-MOBGAM), Istanbul Technical University, Maslak, Istanbul, Türkiye

Hilal Ay Department of Molecular Biology and Genetics, Yildiz Technical University, Faculty of Arts and Science, Istanbul, Türkiye

İbrahim Halil Kılıç Gaziantep Universıty, Science and Art Faculty, Biology Department, Gaziantep, Turkey

John T. Munnoch Strathclyde Institute of Pharmacy and Biomedical Sciences, University of Strathclyde, Glasgow, UK

Mahmut Çalışkan Division of Biotechnology, Biology Department, Faculty of Science, Istanbul University, Vezneciler, Istanbul, Türkiye

Merve Mücahide Çağ Department of Biology, Istanbul University, Institute of Graduate Studies in Science, Istanbul, Türkiye

Muzaffer Arıkan Biotechnology Division, Department of Biology, Faculty of Science, Istanbul University, Vezneciler, Istanbul, Türkiye

Mustafa Çilkiz Transition Zone Agricultural Research Institute Directorate, Eskişehir, Türkiye

Nalan Tavşanlı Division of Biotechnology, Biology Department, Faculty of Science, Istanbul University, Vezneciler, Istanbul, Türkiye
Program of Biotechnology, Biology Department, Institute of Graduate Studies in Sciences, Istanbul University, Vezneciler, Istanbul, Türkiye

Nazmiye Ozlem Sanli Istanbul University, Faculty of Science, Department of Biology, Section of Biotechnology, Vezneciler, Istanbul, Türkiye

Nihal Doğruöz Güngör Department of Biology, Section of Fundamental and Industrial Microbiology, Istanbul University, Faculty of Science, Istanbul, Türkiye

Ozgenur Hacioglu Kirklareli University, Faculty of Health Sciences, Kirklareli, Türkiye

Öznur Pehlivan-Günaydın Faculty of Science and Letters, Department of Molecular Biology and Genetics, Istanbul Technical University, Maslak, Istanbul, Türkiye
Dr. Orhan Öcalgiray Molecular Biology, Biotechnology, and Genetics Research Center (ITU-MOBGAM), Istanbul Technical University, Maslak, Istanbul, Türkiye

Rebecca E. McHugh School of Infection and Immunity, University of Glasgow, Glasgow, UK

Silja Vahtokari Strathclyde Institute of Pharmacy and Biomedical Sciences, University of Strathclyde, Glasgow, UK

Chapter 1
Microbial Evolution and Systematics: Archaea and Bacteria

Hilal Ay

Abstract The origin of life on Earth is a profound biological question, with *Bacteria* and *Archaea*—the two principal prokaryotic lineages—central to the inquiry. Together, they represent microbial diversity and offer insights into Earth's earliest biosphere and evolutionary history. Microorganisms are of significant relevance to humanity, not only as disease agents for some infections but also due to their indispensable contributions to ecosystem functioning, primarily because of their involvement in biogeochemical cycling in various habitats. They influence soil fertility, plant growth, and the overall stability of biological communities across different habitats by mediating the turnover of energy and matter through processes such as decomposition, nutrient cycling, and regulating atmospheric gases. The fields of microbial evolution and systematics are mainly concerned with elucidating the origins, diversification, and classification of these two domains of life. These disciplines are fundamental for comprehending the extensive diversity of life on Earth and the evolutionary mechanisms that have shaped it. Notably, horizontal gene transfer, recombination, mutation, and selection are key evolutionary mechanisms driving genetic innovation and ecological differentiation in microbial populations, influencing phylogeny, function, and ecosystem dynamics. Advances in genomics and bioinformatics have transformed microbial systematics by enhancing polyphasic taxonomy through the integration of phenotypic and phylogenetic data, and have also provided valuable tools to gain deep insight into microbial evolution. This chapter examines the evolutionary history of microorganisms in the context of *Bacteria* and *Archaea*, the mechanisms underlying their evolution, the modern methodologies employed in microbial systematics, and the broader implications of these studies for science and society.

H. Ay (✉)
Department of Molecular Biology and Genetics, Yildiz Technical University, Faculty of Arts and Science, Istanbul, Türkiye
e-mail: hilal.ay@yildiz.edu.tr

M. Çalışkan and M. Arıkan (eds.), *Microbial World: Bacteria and Archaea*, Progress in Molecular and Subcellular Biology 62, https://doi.org/10.1007/978-3-032-11438-9_1

Keywords Horizontal gene transfer · Microbial diversity · Polyphasic taxonomy · Tree of life

1.1 Introduction

The origin of life on Earth represents a complex and interesting subject in biology, particularly in the context of the two principal domains of prokaryotic organisms, i.e., *Bacteria* and *Archaea*. Earth is estimated to be approximately 4.6 billion years old, and the earliest life forms are considered to have emerged between 3.7 and 4.1 billion years ago, during the transition from the late Hadean to the early Archean aeons (Rosing 1999; Bell et al. 2015). These earliest life forms have been considered likely simple, single-celled microorganisms like extant bacteria and archaea (Westall 2011).

Members of the domain *Bacteria* are characterized by their remarkable metabolic diversity and ecological adaptability and have played a pivotal role in shaping the early biosphere by contributing to global biogeochemical cycles. Notably, bacteria contributed to the gradual oxygenation of Earth's atmosphere through photosynthesis, and so the rise of many eukaryotic lineages, including plants. On the other hand, members of the domain *Archaea*, despite their morphological similarity to bacteria, possess distinct genetic and biochemical characteristics, including unique membrane lipids and specialized metabolic pathways, and are considered the first domain of diversified cellular life (Caetano-Anollés et al. 2014). Furthermore, recent discoveries, particularly relating to the Asgard archaea, have revealed possible evolutionary relations between prokaryotes and eukaryotes by revealing a more intricate evolutionary history than previously recognized (Zaremba-Niedzwiedzka et al. 2017).

Both bacteria and archaea exhibit a wide array of metabolic strategies from oligotrophy to copiotrophy, which enable them to exploit diverse energy sources, including sunlight, organic carbon, and inorganic minerals, and to switch flexibly between these metabolic modes (Westoby et al. 2021). The emergence of microbial life on early Earth was closely linked to dynamic changes in the planet's oceans, landmasses, and atmosphere, particularly as global oxygen levels rose. Fossil evidence, such as stromatolites—laminated structures created by ancient microbial mats—reveals that anoxygenic photosynthetic bacteria were among the earliest to colonize the planet. These bacteria thrived in anoxic environments, utilizing sunlight as an energy source and H_2, H_2S, or Fe^{2+} as reductants without producing oxygen as a metabolic byproduct (Olson 2006). The advent of oxygenic photosynthesis by cyanobacteria marked a transformative event by providing the biologically available oxygen required for aerobic respiration, facilitating the evolution of a vast array of ecological niches and promoting the diversification and proliferation of both aerobic and anaerobic microorganisms. Conversely, microorganisms challenged or excluded by oxygen were compelled to adapt to an evolving global redox landscape marked by protracted oxygenation, as oxygenation extended from surface waters to the atmosphere and, eventually, the deep sea over billions of years (Lyons et al. 2024).

Microorganisms are renowned for their capacity to inhabit a vast range of physicochemical environments, from boiling thermal springs to acidic mine drainages. Their ability to produce and metabolize a multitude of compounds—including complex carbohydrates, antibiotics, peptides, pigments, and lipids—enables extensive metabolic cross-feeding and niche specialization across various habitats. Moreover, the rapid evolution rates and speciation characteristics of microorganisms contribute significantly to ecological diversity within relatively short timescales (Gibbons and Gilbert 2015).

Microbial evolution and systematics are mainly concerned with elucidating the origins, diversification, and classification of microorganisms to comprehend the extensive diversity of life on Earth and the evolutionary mechanisms that have shaped it. In this chapter, the evolutionary history of microbes in the context of *Bacteria* and *Archaea*, the mechanisms underlying their evolution, the modern methodologies employed in microbial systematics, and the broader implications of these studies for science and society will be examined.

1.2 Microbial Evolution

Microbial evolution constitutes a fundamental biological process that has profoundly shaped the history of life on Earth by driving the emergence, adaptation, and diversification of microorganisms across an extraordinary array of environments. Through various genetic mechanisms, including mutation, gene duplication, recombination, and, notably, horizontal gene transfer, microbial populations acquire novel traits and rapidly adapt to fluctuating environmental conditions. These evolutionary dynamics enable microbes to colonize a wide spectrum of ecological niches, ranging from the sunlit surface waters of the oceans to the extreme environments of deep-sea hydrothermal vents, hot springs, and polar ice (Okafor 2011; Maier and Neilson 2015; Magnabosco et al. 2024).

Microorganisms are of significant relevance to humanity, not only as disease agents for some infections but also due to their indispensable contributions to ecosystem functioning, primarily because of their involvement in biogeochemical cycling in a wide range of habitats. They influence soil fertility, plant growth, and the overall stability of biological communities across different habitats by mediating the turnover of energy and matter through processes such as decomposition, nutrient cycling, and the regulation of atmospheric gases. Therefore, understanding microbial evolution is of considerable practical importance, as it informs our knowledge of contemporary challenges such as the emergence and dissemination of antibiotic resistance affected by the interactions within microbial communities, the evolution of virulence factors in pathogenic bacteria, and the resilience of microbial communities to environmental disturbances, which is a key part of the complex processes determining the fate of ecosystem functioning (Philippot et al. 2021).

The evolutionary innovations pioneered by microorganisms, such as the advent of oxygenic photosynthesis in cyanobacteria, methanogenesis in certain archaea, and nitrogen fixation by diazotrophs, have played pivotal roles in shaping Earth's atmosphere and biogeochemical cycles. These processes established the environmental conditions necessary for the evolution of complex multicellular life (Lyons et al. 2024).

A primary distinction between microbial and macrobial (i.e., plant and animal) evolution lies not merely in organismal size or generational turnover, but rather in the predominance of asexual reproduction among microbes, in contrast to the primarily sexual reproduction observed in most plants and animals. It is crucial to recognize the immense diversity among microorganisms and to avoid treating them as a homogenous group revealed by genome-based analyses (Rosenberg and Zilber-Rosenberg 2016; Hörandl et al. 2020). Notably, recent advances in high-throughput sequencing, metagenomics, and computational biology have dramatically enhanced our ability to investigate microbial genetic and functional diversity at an unprecedented scale, revealing a vast and largely unexplored microbial world (Wani et al. 2021). Consequently, the study of microbial evolution is essential not only for reconstructing the deep history of life but also for predicting future evolutionary trajectories and harnessing microbial capabilities for applications in medicine, industry, agriculture, and environmental sustainability.

1.2.1 Microbial Diversity

The tree of life is a foundational concept in biology, which provides a comprehensive and integrative framework to understand the vast diversity and evolutionary relationships of all life forms on Earth. Central to this framework is the recognition that microorganisms represent the most significant reservoir of biological diversity, which is estimated to number more than 4×10^{30} individuals globally. Despite their ubiquity, abundance, and fundamental roles in ecosystems, the microbial communities remain largely unexplored and enigmatic, with much of their diversity and functional potential yet to be uncovered (Horner-Devine et al. 2004; Hug et al. 2016a).

Microbial diversity is primarily shaped and sustained by divergent natural selection, whereby different taxa evolve to exploit distinct ecological niches. The availability and variety of these niches are intimately linked to the physical structure and heterogeneity of the environment. Complex and heterogeneous habitats, such as soils, sediments, and aquatic systems with varying chemical gradients, provide a multitude of ecological opportunities, thereby promoting and maintaining high levels of microbial diversity (Bach et al. 2018). In contrast, more homogeneous environments may limit ecological specialization and diversity (Zhao et al. 2022). The adaptive radiation of microorganisms into diverse environments by occupying space and consuming nutrient resources, altering the chemical composition resulting from metabolic and catabolic pathways, releasing bioactive macromolecules, causing horizontal gene transfer events, or providing structural surfaces exemplifies the dynamic interplay between environmental heterogeneity and evolutionary processes (Horner-Devine et al. 2004; Baquero et al. 2021).

Over the course of billions of years, microorganisms have evolved to colonize virtually every conceivable ecological niche due to their extraordinary metabolic versatility resulted from their capability to utilize a wide range of energy sources, including sunlight (phototrophy), organic carbon (heterotrophy), and inorganic minerals (chemolithotrophy) (Matheron and Caumette 2014); thus, establishing themselves as the foundational pillars of life on Earth. Through a remarkable array of metabolic innovations, bacteria and archaea have shaped the biosphere and facilitated the evolution of more complex life forms, including multicellular organisms (Baquero et al. 2021). For instance, the advent of oxygenic photosynthesis by cyanobacteria fundamentally transformed Earth's atmosphere, which paved the way for the evolution of aerobic life (Soo et al. 2017). Similarly, processes such as methanogenesis by certain archaea under anoxic conditions (Whiticar 2020) and nitrogen fixation by diazotrophic bacteria in various environments, including oceans, estuaries, lakes, and forests (Liu et al. 2023), have profoundly impacted global biogeochemical cycles, which influence the availability of key elements and the regulation of greenhouse gases.

The metabolic versatility of *Bacteria* and *Archaea* leads to remarkable adaptations to thrive in environments with extreme physicochemical conditions, such as high or low temperatures, acidic or alkaline pH, high salinity, and varying levels of radiation or pressure, as exemplified by the most inhospitable habitats on Earth, including hot springs, deep-sea hydrothermal vents, glacier ice, the atmosphere, terrestrial subsurface, and groundwater ecosystems (Miteva et al. 2004; Griebler and Lueders 2009; Pearce et al. 2009; Dick 2019; Kochetkova et al. 2022; Beaver and Neufeld 2024). Although both bacterial and archaeal cells are diminutive in size (approximately 10^{-6} meters), the global abundance and capacity of these microorganisms to adapt to diverse environments underscore their evolutionary success and ecological importance.

Extensive research has demonstrated that taxonomic distributions among microorganisms are highly differentiated across all domains of microbial life and within a wide variety of ecosystems, which supports the concept that individual microbial taxa possess distinct environmental niches (Bajic and Sanchez 2020). A particularly illustrative case is the niche differentiation observed between ammonia-oxidizing archaea (AOA) and ammonia-oxidizing bacteria (AOB) (Prosser et al. 2020). Although these groups belong to separate domains—*Archaea* and *Bacteria*—they both play crucial roles in the biogeochemical process of ammonia oxidation. AOA are mostly associated with marine and terrestrial ecosystems, but their presence has also been documented in a range of other environments, including thermal springs and the digestive tracts of animals, indicating their ecological versatility. In contrast, AOB are predominantly found in freshwater habitats and engineered environments such as wastewater treatment facilities, where they contribute to nitrogen cycling (Malard and Guisan 2023).

Traditionally, the study of microbial diversity was constrained by cultivation-based techniques, which limited early ecophysiological research to a small subset of "model" organisms that could be readily isolated and grown in the laboratory. This methodological limitation led to a substantial underestimation of microbial diversity, as most bacteria and archaea are recalcitrant to cultivation under standard laboratory conditions (Prakash et al. 2021). The subsequent development and adoption of cultivation-independent molecular methods, such as 16S ribosomal RNA (rRNA) gene clone library analysis and metagenomics, revolutionized the field by revealing a previously hidden wealth of microbial taxa not represented among cultured isolates (Dunbar et al. 1999; Handelsman 2004). These approaches demonstrated that cultured microorganisms constitute only a small fraction of the total microbial diversity in natural environments. Moreover, advances in genomic sequencing technologies and the development of sophisticated computational tools have provided holistic insights into the metabolic potential, ecological dynamics, and evolutionary trajectories of microbial communities across spatial and temporal scales (Madison et al. 2023). These innovations have facilitated the recovery of genomes from previously uncultivated, phylum-level lineages—termed "candidate phyla"—especially within the complex microbial communities, thereby expanding our understanding of the tree of life and revealing novel branches of microbial evolution (Fig. 1.1) (Hu et al. 2016; Hug et al. 2016b; Bickhart et al. 2022). Genomic data derived from these efforts, as well as technological advancements and interdisciplinary approaches, have shed light on the metabolic versatility and ecological significance of *Bacteria* and *Archaea*, providing insight into the evolutionary origins of early life and the emergence of eukaryotes.

1.2.2 Evolutionary Mechanisms and Genetic Variation

Studies on microbial genetics have elucidated three principal qualitative mechanisms that collectively drive the creation of genetic variation within prokaryotic populations, each contributing uniquely to evolutionary adaptability and diversification. The first mechanism causing genetic variation involves localized alterations in DNA sequences, such as nucleotide substitutions, small-scale insertions or deletions, and limited rearrangements of nucleotides. These mutations typically affect single genes or short genomic regions and constitute a primary source of genetic diversity. In prokaryotes, such sequence-level variation is frequently detected as single-nucleotide polymorphisms (SNPs), small indels, or point mutations that can give rise to phenotypic differences within microbial populations. These changes enable rapid microevolutionary adaptation in microorganisms, allowing microbial populations to respond to selective pressures such as antibiotics, nutrient availability, or other environmental stressors (Arber 2000).

The second mechanism encompasses segmental rearrangements of genomic DNA, primarily mediated by homologous or site-specific recombinational processes. These rearrangements can involve the duplication, inversion, translocation, or deletion of larger segments of DNA, often spanning multiple genes or even entire operons. In prokaryotes, segmental rearrangements are responsible for generating structural diversity within and between genomes, resulting in phenomena such as gene duplications that fuel gene family expansions, the shuffling of metabolic gene clusters, and the

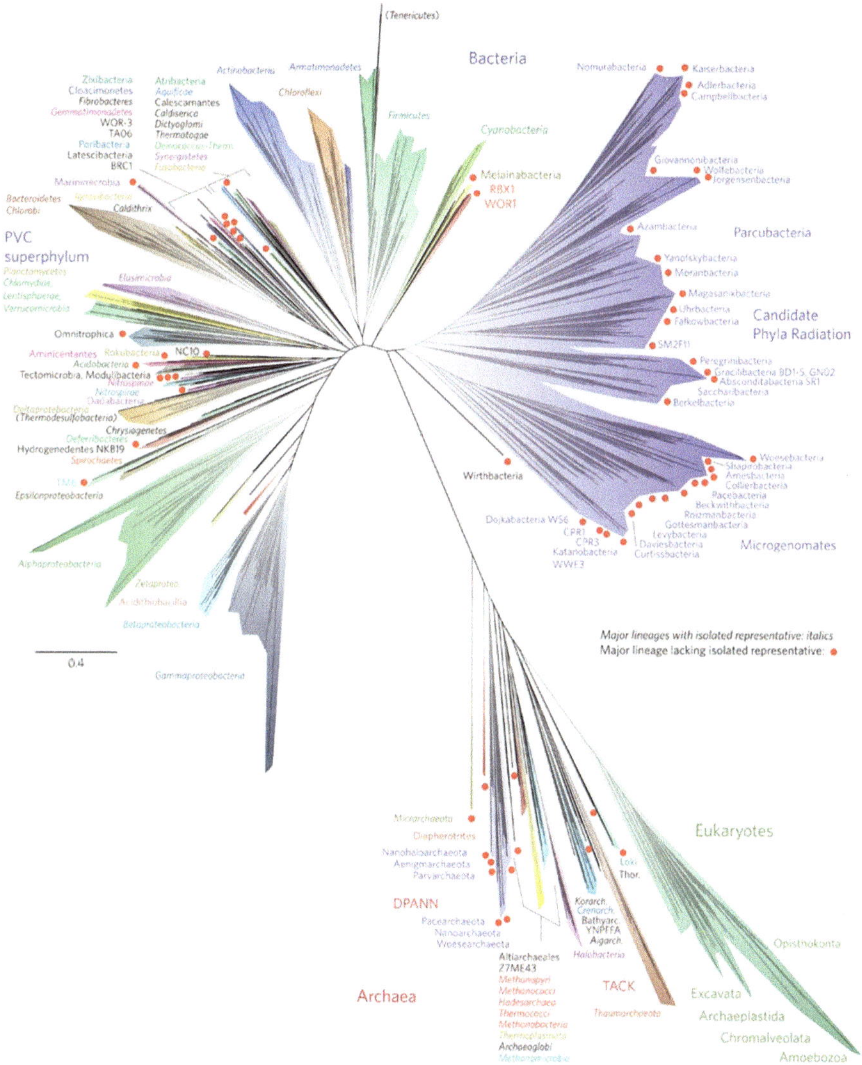

Fig. 1.1 Phylogenetic tree illustrating the evolutionary relationships among various microbial lineages, including *Bacteria*, *Archaea*, and *Eukarya*. The tree is color-coded to differentiate between major groups, with labels for specific lineages such as *Actinomycetota* (formerly *Actinobacteria*), *Bacillota* (formerly *Firmicutes*), and *Methanobacteriota* (formerly *Euryarchaeota*). Red dots indicate major lineages lacking isolated representatives, while italics denote those with isolated representatives. The scale bar represents evolutionary distance (from Hug et al. (2016b))

formation of genomic islands with adaptive significance for medical or environmental interests. These structural changes might lead to the modular reorganization of genetic material, the evolution of novel metabolic pathways, and the fine-tuning of regulatory networks in response to environmental complexity or fluctuating conditions and thus are considered as an adaptive nature of microorganisms (Arber 2000; West et al. 2006).

The third mechanism for the genetic variation among microorganisms operates through horizontal gene transfer (HGT), which is a hallmark of prokaryotic evolution that facilitates the acquisition and integration of exogenous DNA sequences from other, often distantly related, organisms. HGT occurs via multiple mobile genetic elements, including plasmids, transposons, bacteriophages, integrative conjugative elements, and natural transformation (Tokuda and Shintani 2024). This process allows both bacteria and archaea to rapidly acquire new traits, such as antibiotic resistance, virulence factors, or novel metabolic capabilities, with profound implications for their ecological success and adaptability (Arnold et al. 2022). HGT has been shown to be especially significant in the dissemination of antibiotic resistance genes among pathogenic bacteria, the spread of toxin-production genes, and the transfer of entire metabolic modules that enable colonization of new ecological niches (Moran et al. 2012; Von Wintersdorff et al. 2016). Comparative genomic studies have revealed that a substantial proportion of prokaryotic genes, especially those associated with accessory functions, have been acquired through HGT, contributing to the mosaic nature of prokaryotic genomes (Sengupta and Azad 2023).

These three strategies for generating genetic diversity—mutation, recombination, and horizontal gene transfer—are qualitatively distinct in their scope and evolutionary consequences (Ares-Arroyo et al. 2024). While mechanisms within each category may offer some functional redundancy, such substitution is generally not observed between categories, underscoring their complementary roles in microbial evolution. Accordingly, highly adaptable bacteria and archaea are expected to possess at least one, and preferably multiple, systems corresponding to each of these mechanisms, thereby maximizing their potential for genetic innovation and environmental adaptation. The cumulative effect of these processes is evident in the extraordinary genetic variability and plasticity observed among prokaryotes, manifesting as differences in genome size, gene content, operon organization, regulatory elements, and mobile genetic elements across species and even among strains of the same species (Arnold et al. 2022; Wani et al. 2022; Drebes Dörr et al. 2025). Genetic variation among prokaryotes as a result of these evolutionary mechanisms is further amplified by differences in mutation rates, genome repair systems, and ecological strategies (Li and Wang 2021). Some prokaryotes, for example, exhibit hypermutability as an adaptive response to the stress resulting from the presence of antibiotics, accelerating the rate at which beneficial mutations arise (Ridderberg et al. 2020). Others may rely heavily on HGT to supplement their genetic repertoire, especially in dynamic or competitive environments such as soil, marine ecosystems, or the human microbiome (de Sousa et al. 2023). These processes facilitate the rapid emergence of new ecotypes, pathotypes, and symbiotic associations, which contribute to the remarkable versatility and resilience of prokaryotic microorganisms (Swiecicka et al. 2013).

It can be deduced that the evolutionary trajectories of microbial populations are mainly shaped by a complex interplay between vertical inheritance and extensive HGT events. The prevalence of HGT has fundamentally challenged the traditional view of evolution as a strictly branching, tree-like process, instead suggesting a reticulate or network-like model where gene flow occurs both within and across taxonomic boundaries. This paradigm shift has led to the proposal of alternative evolutionary frameworks, including composite evolutionary units and overlapping taxonomic groups, to better capture the dynamic nature of prokaryotic genome evolution (Koonin and Wolf 2012). Beyond HGT, other mechanisms, such as Lamarckian-type inheritance and stress-induced mutagenesis, have been implicated in microbial evolution (Osmanovic et al. 2018). Lamarckian-like processes involve the environmentally induced, directed acquisition of genetic changes—such as the activation of specific mutational pathways under stress—while stress-induced mutagenesis refers to an increased mutation rate in response to adverse conditions (Koonin and Wolf 2012). These mechanisms further enhance the evolvability of microbial populations by generating new genetic variants that can be subject to natural selection.

The ecological strategies adopted by prokaryotic lineages also play a crucial role in shaping the patterns of genetic variation and evolutionary dynamics. Generalist microbes, which are capable of exploiting a wide range of environmental conditions and resources, tend to exhibit higher speciation rates and greater evolutionary persistence than specialist lineages (Sriswasdi et al. 2017; Xu et al. 2022). However, evolutionary transitions from generalist to specialist are observed to occur more frequently, which reflects a dynamic process wherein generalist microorganisms act as reservoirs for new species and contribute disproportionately to the maintenance of taxonomic and genetic diversity within communities (Sriswasdi et al. 2017). This dynamic is evident in environments such as soil and aquatic systems, where rapid niche differentiation and adaptation are more common (Yan et al. 2022).

The scope of microbial evolution has been further broadened by the hologenome concept, which posits the holobiont—the host organism together with its associated microbiota—as a fundamental unit of selection (Rosenberg 2021). Symbiotic microorganisms play integral roles in host development, physiological homeostasis, immune function, behavior, and genetic variability. Through both cooperative and competitive interactions, microbial symbionts can drive the evolution of complexity in multicellular hosts by facilitating adaptation to new environments and the emergence of novel traits (Rosenberg and Zilber-Rosenberg 2016). The hologenome theory emphasizes the relationship between microbial and macroorganismal evolution and highlights the importance of genetic variation among prokaryotes in shaping broader evolutionary outcomes (Moran et al. 2012; Husnik and McCutcheon 2018).

In addition to these intrinsic genetic and evolutionary mechanisms, advances in molecular biology, genomics, and systems biology have greatly expanded our understanding of prokaryotic genetic variation (Koonin et al. 2021; Zhong et al. 2021; Wani et al. 2022). High-throughput sequencing technologies, comparative genomics, and shotgun metagenomics have revealed vast, previously unappreciated diversity within prokaryotic populations, uncovering extensive pan-genomes, accessory gene pools, and the widespread presence of mobile genetic elements (Zhong et al. 2021; Huang et al. 2023). Population genomics studies in microbial communities have shown that even within a single species or environment, prokaryotic genomes can vary markedly in gene content, regulatory networks, and adaptive traits—a phenomenon known as genome fluidity. These findings underscore the importance of ongoing gene gain, loss, and rearrangement in maintaining prokaryotic diversity and adaptability (Andreani et al. 2017; Li et al. 2024).

Collectively, these insights illustrate the multifaceted and dynamic nature of genetic variation and evolutionary processes among prokaryotes. The interplay between localized genetic changes, large-scale genomic rearrangements, and HGT, coupled with ecological strategies, symbiotic relationships, and environmental pressures, underpins the extraordinary adaptability and evolutionary potential of prokaryotic life. Continued research, leveraging advanced molecular and computational tools, promises to further illuminate the complexity of prokaryotic evolution, with far-reaching implications for evolutionary biology, ecology, biotechnology, and our broader understanding of life on Earth.

1.2.3 Environmental Influences

Microbial evolution is a continuous, multifaceted, and dynamic process profoundly shaped by a diverse range of environmental conditions. Both natural factors—such as temperature, pH, nutrient availability, and habitat heterogeneity—and anthropogenic influences contribute to the complexity of evolutionary trajectories in microbial populations. Over the past century, human activities have introduced unprecedented and often abrupt changes to ecosystems, which have intensified selective pressures and accelerated the pace of microbial adaptation and diversification (Nguyen et al. 2021).

The environment acts simultaneously as a selective agent and a catalyst for microbial evolution, and thus a significant factor in creating new opportunities for adaptation and metabolic versatility in microbial populations. Microbial communities respond to the biotic and abiotic drivers through genetic, physiological, and ecological mechanisms, which may result in the emergence of novel traits, lineages, and interactions. These evolutionary processes are fundamental to sustaining ecosystem functions, as microorganisms play critical roles in shaping biodiversity, driving nutrient cycling, facilitating energy flow, and enhancing ecosystem resilience (van den Berg et al. 2022).

Fundamental environmental parameters—including temperature, pH, salinity, moisture content, and nutrient availability—are primary determinants of microbial diversity and evolutionary change (Morrissey et al. 2019). Alterations in any of these variables can create new ecological niches and drive the selection of adaptive traits within microbial populations. For example, microorganisms exhibit remarkable physiological specialization to extreme environments: psychrophilic bacteria are adapted to thrive in polar and glacial habitats (Ay et al. 2025; Sahin et al. 2025), while thermophilic and hyperthermophilic organisms are commonly found in hot springs, hydrothermal vents, and geothermal soils (Alain et al. 2021; Inskeep et al. 2025). Acidophiles and alkaliphiles have evolved mechanisms to survive at the pH extremes of acid mine drainage sites and soda lakes, respectively. These adaptations are underpinned by specific genetic, biochemical, and structural modifications, such as the production of cold-adapted enzymes or heat-stable proteins and membranes (Goyal et al. 2021; Zhang et al. 2023).

The ability of prokaryotes to adapt to a broad spectrum of environmental conditions has resulted in the vast diversification and speciation of microbial lineages, with new species frequently emerging to exploit novel or ephemeral environments. Notably, microbial populations display considerable phenotypic plasticity, which allows them to generate a variety of phenotypes in response to environmental stressors. This plasticity is a key driver of evolutionary potential, which facilitates both short-term survival and long-term adaptation to rapidly changing or unpredictable environments (Wang et al. 2022; Gubry-Rangin et al. 2024).

1.2.3.1 Anthropogenic Influences: Pollution, Antibiotics, and Land Use Change

Human activities have imposed dramatic and unprecedented selective pressures on microbial communities, which profoundly alter their evolutionary dynamics (Zhou et al. 2023). The introduction of pollutants—including pharmaceuticals, heavy metals, pesticides, and industrial chemicals—into soils, water bodies, and the atmosphere has created complex chemical landscapes. Microorganisms exposed to these stressors often experience strong selection for resistance and tolerance traits (Rzymski et al. 2024). Notably, the most well-documented example is the emergence and global spread of antibiotic resistance among pathogenic and environmental bacteria. The widespread use and misuse of antibiotics in medicine, agriculture, and aquaculture have driven the rapid evolution of multidrug-resistant strains, facilitated by both genetic mutations and HGT mechanisms. The proliferation of antibiotic resistance genes in clinical and natural environments is now recognized as a major public health threat, which demonstrates the capacity of human-induced selective pressures to shape microbial evolution on a global scale (Baquero et al. 2021; Mutuku et al. 2022). Similar evolutionary responses are observed in microbes exposed to heavy metals, where the acquisition or upregulation of metal resistance genes enables survival in contaminated sites. These adaptive processes may also confer cross-resistance to antibiotics, which further complicates efforts to manage resistance worldwide. Moreover, agricultural intensification, urbanization, and deforestation have led to widespread habitat modification and fragmentation, disrupting native microbial communities and selecting for lineages capable of rapid adaptation to altered or degraded environments (Kumari et al. 2024). Consequently, microbial diversity may decline, and community composition may become homogenized, which may result in reduced ecosystem resilience and functionality.

The destruction and fragmentation of habitats through urbanization, deforestation, mining, and agricultural expansion cause the loss of native biodiversity, including not only that of animals and plants but also that of microorganisms. The remaining microbial populations in such disturbed habitats often experience intense selective pressures, which might lead to rapid evolutionary responses as they adapt to new environmental realities. In some cases, this results in the dominance of a few generalist or opportunistic species, while more specialized taxa are lost, leading to community homogenization and reduced ecosystem resilience (Chen et al. 2021).

The introduction of non-native plant and animal species can also disrupt microbial ecosystems by triggering evolutionary changes in indigenous microorganisms. Invasive species might alter nutrient cycling, resource availability, or competitive dynamics, forcing resident microorganisms to adapt to new ecological pressures or to form novel associations with the invaders. These interactions can have far-reaching implications for ecosystem function and stability (Lear et al. 2024).

As anthropogenic environmental changes continue to accelerate, understanding the evolutionary processes that strengthen microbial adaptation has become increasingly important. Such knowledge is essential for predicting ecosystem responses to global change to inform conservation strategies and ensure the sustainable management of natural resources. Recent advances, especially in genomics and ecological modeling, have further illuminated the exceptional adaptability and ecological significance of microorganisms within the Earth's biosphere (van den Berg et al. 2022; Shu and Huang 2022). Thus, recognizing and integrating the pivotal role of microbial evolution is not only crucial for maintaining ecosystem health and promoting sustainability but also for mitigating the impacts of rapid environmental change in the Anthropocene.

1.2.3.2 Climate Change and Microbial Evolution

Climate change represents another major driver of microbial evolutionary processes since rising global temperatures, altered precipitation patterns, increased frequency of extreme weather events, and shifts in atmospheric composition (such as elevated CO_2 levels) are transforming microbial habitats at regional and global scales. These changes have direct and indirect effects on microbial communities (Wiens and Zelinka 2024). For instance, rising temperatures can accelerate microbial metabolic rates, alter community structure, and influence the rates of decomposition, nutrient cycling, and greenhouse gas production. Shifts in moisture regimes, including droughts and increased rainfall, can modify soil microbial diversity, which may favor certain functional groups while suppressing others (Sheik et al. 2011; Cregger et al. 2012).

Due to shifted climate conditions, the altered microbial composition and activity may have cascading effects on ecosystem functions such as carbon sequestration, nitrogen fixation, and methane emissions (Waoo et al. 2025). For instance, permafrost thawing exposes previously frozen organic matter to microbial decomposition, releasing significant quantities of CO_2 and methane—greenhouse gases that further exacerbate climate change (Ernakovich et al. 2022). Anaerobic methanogens, for example, contribute to methane emissions in wetlands and rice paddies (Malyan et al. 2021), while other microbes facilitate the sequestration of carbon in soils and sediments (Martin-Pozas et al. 2022). These activities directly influence climate feedback mechanisms and the Earth's capacity to buffer against atmospheric changes, underscoring the importance of microbial evolution and adaptation in mediating biogeochemical responses to a changing climate.

1.2.3.3 Microbe–Mineral Interactions and Co-evolution

Minerals are fundamental components of Earth's crust, and their physical and chemical properties reflect the environmental conditions present during their formation. Microorganisms, which constitute a significant portion of the tree of life, have played a pivotal role in shaping the planet's mineral diversity throughout geological time. In both subsurface and surface environments, minerals and microbes interact through energy flow and matter exchange. Minerals provide essential nutrients and electron donors or acceptors necessary for microbial metabolism, while microbial activities drive the dissolution, transformation, and formation of minerals (Pradhan et al. 2025).

These mineral–microbe interactions are central to Earth's habitability and have contributed to major geological and environmental transitions. For example, during the early history of the planet, microbial metabolism—such as oxygenic photosynthesis by cyanobacteria—catalyzed the Great Oxidation Event (GOE), which dramatically increased atmospheric oxygen levels and led to the diversification of mineral species from approximately 1500 to over 4000 (Sánchez-Baracaldo et al. 2022). The subsequent evolution of eukaryotes and the biosynthesis of organo-minerals further expanded mineral diversity, culminating in the current total of more than 5000 mineral species (Schirrmeister et al. 2013). Throughout Earth's history, these interactions have typically occurred at microscopic scales, but their cumulative effects are evident on a global scale, influencing events such as the formation of ore deposits and major biogeochemical shifts (Davín et al. 2025). However, current understanding remains limited by challenges in identifying biosignatures in the geological record and a lack of comprehensive laboratory simulation studies.

1.2.3.4 Microbial Evolution and Ecosystem Function

Microbial evolution has profound implications for ecosystem structure and function. Microorganisms are central players in global biogeochemical cycles, including those of carbon, nitrogen, sulfur, and phosphorus, due to their evolutionary adaptations that enable the utilization of diverse energy sources and substrates. For instance, nitrogen-fixing bacteria such as *Rhizobium* spp. convert atmospheric nitrogen into bioavailable ammonia, enhancing soil fertility and supporting plant productivity. Fungi and bacteria are also crucial for decomposing complex organic matter, recycling nutrients, and facilitating the flow of energy through food webs (Rousk and Frey 2015).

Microbial evolution also influences interactions with other organisms, affecting community dynamics and ecosystem processes. Evolutionary processes in microorganisms foster a wide range of symbiotic relationships that enhance ecosystem stability and diversity. Microbial communities inhabiting animal hosts, such as the human gut microbiome, co-evolve with their hosts, influencing digestion, immunity, and behavior (Shapira 2016). These interactions can cascade through trophic levels, shaping broader ecological networks and impacting community structure. Pathogenic microbes, for instance, can evolve rapidly, impacting host populations and triggering co-evolutionary responses both in animals and plants. These interactions drive evolutionary arms races that shape the trajectories of both microbial and host lineages (Perreau and Moran 2022). Microbial communities may also influence primary productivity by mediating nutrient availability, modulating disease pressures, and competing with other organisms, thereby impacting energy flow and biomass production within aquatic and terrestrial ecosystems (Glibert and Mitra 2022).

The rapid evolutionary potential of microorganisms allows them to respond to environmental stresses, such as increased salinity, temperature fluctuations, or pollution, by metabolic and biosynthetic versatility at rates far exceeding those of many macroorganisms. This adaptive capacity can help stabilize ecosystems during periods of turbulence, providing a buffer against environmental perturbations. Moreover, microbial communities often exhibit functional redundancy together with taxonomic diversity, where multiple species can perform similar ecological roles. Such redundancy is crucial for ecosystem resilience: if one species is lost, others can compensate, maintaining ecosystem processes and stability in the face of disturbance or environmental change (Louca et al. 2018).

1.2.4 Experimental Systems to Study Microbial Evolution

The study of microbial evolution has become especially critical in the context of contemporary global challenges, such as the emergence of antibiotic-resistant pathogens (Card et al. 2021), the effects of climate change on microbial communities (Santos et al. 2023), and the necessity for sustainable ecosystem management (Rillig and Antonovics 2019). In this context, experimental systems play a pivotal role in advancing our understanding of microbial evolution by providing powerful platforms to unravel the mechanisms underpinning adaptation, diversification, and the influence of environmental factors. Therefore, the multifaceted dynamics of microbial evolutionary processes might be dissected with increasing precision and realism by leveraging a diversity of experimental systems, ranging from controlled laboratory cultures to high-throughput sequencing technologies and complex environmental mesocosms.

1.2.4.1 Laboratory-Based Experimental Evolution

Laboratory-based experimental evolution is a cornerstone of microbial evolutionary research, which enables direct observation and quantification of evolutionary changes in microbial populations across defined timescales, often spanning hundreds or thousands of generations. In these controlled settings, environmental variables such as nutrient availability, temperature, or the presence of specific stressors can be manipulated, thereby imposing selective pressures on microorganisms that drive adaptation. One of the most well-known examples of the long-term evolution experiments (LTEEs) is Richard Lenski's long-lasting experiment with *Escherichia coli*, in which 12 populations were founded from the same ancestral strain and have evolved for almost

30 years, creating more than 65,000 generations under the same conditions (Lenski 2017). This experiment has yielded profound insights into the rates and repeatability of adaptation, the emergence of novel traits, the dynamics of clonal interference, and the roles of mutation, selection, and genetic drift over tens of thousands of generations. Through regular sampling and whole-genome sequencing integrated with LTEEs, evolutionary trajectories, adaptive mutations, and the genetic basis of key phenotypic innovations can be elucidated effectively.

1.2.4.2 High-Throughput and Automated Systems

Recent advances in high-throughput sequencing and automated culturing systems have further transformed experimental evolution studies. These technologies allow for parallel evolution experiments across diverse microbial strains, environmental conditions, and selection regimes. High-throughput sequencing provides unprecedented resolution for real-time monitoring of genomic, transcriptomic, and epigenetic changes, enabling fine-scale mapping of evolutionary dynamics, while automated culturing systems, such as chemostats and morbidostats, facilitate the maintenance of constant or dynamic environmental conditions, supporting continuous culture and the study of evolutionary responses to fluctuating or stressful environments (Leyn et al. 2021). These platforms are invaluable for investigating the evolution of antibiotic resistance, metabolic specialization, and responses to environmental perturbations. Moreover, the integration of robotics with microbial experimental evolution marks an innovative frontier in the field. Evolutionary robotics involves the use of robotic systems to simulate complex ecological interactions and environmental fluctuations. Robotic mappers and automated environmental controllers can precisely modulate spatial and temporal parameters, such as resource gradients, physical barriers, or predation pressures (Theodosiou et al. 2017). These systems are suitable platforms to study microbial community assembly, adaptive behaviors, and interspecies interactions in environments that more closely mimic the heterogeneity of natural ecosystems. Such interdisciplinary approaches are also valuable for exploring questions related to microbial cooperation, competition, dispersal, and the evolution of collective behaviors (Brancato et al. 2024).

1.2.4.3 Mesocosms and Field-Based Studies

Experimental systems that bridge the laboratory and the field provide essential insights into the ecological realism of microbial evolution. Mesocosms—controlled, enclosed experimental ecosystems that replicate key aspects of natural environments—are employed to investigate evolutionary processes in mixed microbial communities. These ecosystems can be established in terrestrial, freshwater, or marine settings, thus allowing researchers to manipulate variables such as nutrient inputs, temperature, or species composition (Land et al. 2023). Mesocosms facilitate the study of competition, cooperation, HGT, as well as the effects of environmental change on evolutionary outcomes. For instance, nutrient enrichment experiments in freshwater mesocosms have revealed how changes in resource availability can drive shifts in microbial community structure, metabolic function, and evolutionary trajectories (Lever et al. 2015).

Field-based experiments extend experimental evolution to natural ecosystems by allowing the observation of microbial adaptation and evolutionary change in situ. These studies may involve the introduction of specific microbial strains or consortia into native communities, manipulating land management practices, or monitoring microbial responses to environmental disturbances such as pollution or habitat fragmentation. Field experiments are particularly crucial for understanding the context-dependence of evolutionary processes, the role of local adaptation, and the resilience of microbial communities to environmental stress such as introduction or enrichment of opportunistic bacteria in a host-associated environment (Ezzat et al. 2020).

1.2.4.4 Synthetic Biology and Adaptive Laboratory Evolution

The intersection of synthetic biology and microbial evolution has ushered in a new era of experimental systems for probing evolutionary dynamics (Bassalo et al. 2016), which is enabled by designing and constructing genetically modified microorganisms with defined traits, synthetic pathways, or regulatory circuits. The engineered strains can be subjected to adaptive laboratory evolution (ALE), exposing them to specific selective pressures—such as novel substrates, toxins, or fluctuating environmental conditions—to observe evolutionary responses (Godara and Kao 2020). ALE experiments provide valuable data on the principles governing adaptation, the evolution of metabolic networks, and the stability of synthetic traits in variable environments. By integrating high-throughput sequencing, phenotypic characterization, and modeling in an ALE system, the genetic and systems-level determinants of microbial fitness, robustness, and evolvability can be uncovered (Van den Bergh et al. 2018).

1.2.4.5 Applications and Future Directions

Experimental systems for studying microbial evolution are not only central to fundamental research but also have wide-ranging practical applications. For instance, insights derived from these approaches inform the development of novel antimicrobial strategies, the optimization of industrial microbial strains, the remediation of contaminated environments, and the engineering of stable and resilient microbiomes for agriculture and human health (Kaminsky et al. 2025). As environmental challenges and biotechnological demands grow, the continued refinement and integration of experimental systems—including microfluidics, real-time imaging, and multi-omics technologies—will be essential for advancing our ability to predict, manipulate, and harness microbial evolution.

In conclusion, experimental systems are indispensable for elucidating the mechanisms of microbial evolution across scales—from genes to communities and from controlled environments to natural ecosystems. The complex interplay between genetic variation, selective pressures, and environmental context can be unraveled by integrating laboratory, mesocosm, field, and synthetic biology approaches. The expanding toolkit of experimental systems advances evolutionary theory and underpins efforts to address pressing challenges in medicine, biotechnology, and ecosystem sustainability. As our understanding of microbial evolution deepens, experimental systems will remain at the forefront of innovation, thus enabling transformative discoveries with profound implications for science and society.

1.3 Microbial Systematics

Microbial systematics provides a rigorous and standardized framework for organizing the extraordinary diversity within the prokaryotic domains of *Bacteria* and *Archaea*, based on their evolutionary history, phenotypic traits, and genotypic characteristics (Vandamme et al. 1996; Hugenholtz et al. 2021). Systematics has utmost importance for the study of prokaryotes, as it enables delineating species, genera, and higher taxonomic ranks to facilitate clear communication and reproducibility across the scientific community. In the context of microbial systematics, accurate classification is crucial not only for naming and describing novel taxa but also for predicting the physiological, ecological, and metabolic properties of newly discovered organisms based on their phylogenetic relationships to well-characterized lineages (Pallen et al. 2021; Wang and Cole 2024).

In recent decades, significant advances in molecular techniques have profoundly transformed bacterial and archaeal systematics since traditional methods—such as morphological observation, metabolic profiling, and biochemical tests—have been complemented and often superseded by molecular approaches. Sequencing of conserved genes, particularly the 16S rRNA gene, has become the cornerstone of prokaryotic phylogeny and enabled the reconstruction of evolutionary relationships and the identification of novel lineages. More recently, whole-genome sequencing and comparative genomics have provided unprecedented resolution for distinguishing closely related species, uncovering cryptic diversity, and revising taxonomic boundaries within both domains (Goodfellow 2000; Moore et al. 2010; Riesco and Trujillo 2024).

A systematic framework for bacteria and archaea is essential for a wide range of scientific and practical purposes. For instance, in clinical microbiology, accurate identification is vital for diagnosing infectious diseases and tracking antibiotic resistance (Munson et al. 2023). In environmental microbiology, systematics aids in elucidating the roles of bacteria and archaea in nutrient cycling, biogeochemical processes, and ecosystem functioning (Oren 2010). Furthermore, industrial and biotechnological applications—such as wastewater treatment, bioremediation, and the production of pharmaceuticals—heavily rely on the correct identification and classification of prokaryotic strains with desired properties (Harirchi et al. 2022).

1.3.1 Principles of Microbial Classification

Microbial systematics is underpinned by a set of guiding principles that facilitate the classification, identification, and nomenclature of microorganisms. These principles are fundamental for organizing microbial diversity, elucidating evolutionary relationships, and ensuring consistency in scientific communication. The core areas of microbial systematics are informed by the integration of phenotypic and genotypic data, as well as phylogenetic analysis (Klenk and Göker 2010; Oren 2010; Gao and Gupta 2012).

Classification refers to the orderly arrangement of microorganisms into hierarchical groups, or taxa, based on shared characteristics and inferred evolutionary relationships. The primary aim is to construct a natural classification system that reflects phylogeny, thereby representing the evolutionary history and relatedness of organisms. The taxonomic hierarchy comprises several ranks, including domain, phylum, class, order, family, genus, and species. Modern microbial classification increasingly relies on molecular data, particularly the comparative analysis of conserved genetic loci such as the 16S rRNA gene, to infer evolutionary relationships and delineate taxa. Nevertheless, phenotypic properties—such as cellular morphology, metabolic capabilities, and ecological niches—are still important, especially for practical identification and functional characterization (Logan 2009; Vandamme 2011).

Identification is the process of determining the taxonomic placement of an unknown microorganism, which is achieved by comparing the organism's characteristics to those of established taxa by utilizing a combination of phenotypic assays (e.g., staining techniques, biochemical tests), molecular techniques (e.g., PCR amplification and sequencing of marker genes), and increasingly, genome-based approaches (e.g., calculation of overall genome-relatedness indices) (Vandamme 2011; Uvarova et al. 2020). Precise identification of microorganisms is critical in diverse fields such as clinical microbiology, environmental monitoring, and biotechnology, where accurate recognition of microbial strains underpins diagnosis, surveillance, and exploitation for industrial purposes.

Nomenclature involves the assignment of standardized scientific names to microbial taxa. This process is governed by internationally recognized codes, including the International Code of Nomenclature of Prokaryotes (ICNP) and the International Code of Nomenclature for algae, fungi, and plants (ICN). These codes provide rules for valid publication, priority, and typification, ensuring the use of type strains or specimens as reference points for taxa. Microbial names are typically binomial—comprising a genus and species epithet—and must be unique, descriptive, and stable to promote unambiguous communication throughout the global scientific community (Sneath 2015; Oren 2024).

The diversity of microorganisms is vast and complex, encompassing millions of species with highly varied ecological, physiological, and genetic attributes. Thus, accurate identification and classification are essential for progress in medicine, agriculture, biotechnology, and environmental science. Since traditional taxonomy, which relied primarily on observable characteristics, often failed to capture true evolutionary and genetic relationships among microorganisms (Thompson et al. 2015), the advent of molecular and chemotaxonomic methods has greatly enhanced the resolution of microbial classification, yet these approaches also have limitations when applied in isolation (Lawson and Patel 2021).

1.3.2 Methods in Microbial Systematics

The discipline of microbial systematics employs a diverse and evolving array of methodologies to classify, identify, and elucidate the evolutionary relationships among microorganisms (Vandamme et al. 1996). Over the past century, these methods have tremendously advanced in parallel with technological developments, shifting from traditional, phenotype-based analyses to the incorporation of sophisticated molecular, genomic, and computational approaches (Chun and Rainey 2014). The integration of these diverse techniques has culminated in the adoption of a holistic or polyphasic approach, which is now widely regarded as the gold standard in microbial systematics. The methodology adopting a polyphasic approach synthesizes phenotypic, genotypic, and phylogenetic data to provide a comprehensive and robust framework for microbial classification and identification (Vandamme et al. 1996; Gillis et al. 2005). By leveraging multiple, independent lines of evidence, polyphasic taxonomy enhances the reliability and precision of taxonomic assignments and is particularly well-suited to accommodate the complex, mosaic nature of microbial genomes (Vandamme and Peeters 2014). As advances in genomics, bioinformatics, and systems biology continue to accelerate, the polyphasic approach is expected to further refine our understanding of microbial diversity and the critical roles that microorganisms play across natural ecosystems and industrial applications, as well as their effects on human health (Mahato et al. 2017).

Historically, microbial taxonomy was rooted in the analysis of observable phenotypic traits, including cell morphology (e.g., shape, size, motility), staining reactions (such as the Gram stain), metabolic activities (e.g., carbon source utilization, enzymatic reactions), and growth characteristics under varying environmental conditions (Moore et al. 2010; Godbole and Sachdev 2024). These methods formed the basis for early schemes of microbial classification and remain valuable for practical identification in clinical and environmental microbiology (Gillis et al. 2005). However, phenotypic approaches are constrained by several fundamental limitations. Phenotypic plasticity—the ability of a single genotype to produce multiple phenotypes in response to environmental factors—can obscure true taxonomic relationships. Convergent evolution may result in unrelated organisms exhibiting similar phenotypes due to adaptation to comparable ecological niches. Furthermore, the vast majority of microbes cannot be cultured using standard laboratory methods, leaving much of microbial diversity inaccessible to phenotype-based study (Konstantinidis et al. 2017; Overmann et al. 2019).

The emergence of molecular techniques in the late twentieth century marked a paradigm shift in microbial systematics. Notably, DNA–DNA hybridization became an important tool for quantifying genome-wide similarity between strains, providing a more objective and reproducible measure than phenotype-based methods

(Brenner 1973; Stackebrandt et al. 2002). The introduction of 16S rRNA gene sequencing further revolutionized the field, as this highly conserved genetic marker enabled the reconstruction of deep evolutionary relationships among prokaryotes and facilitated the discovery of novel lineages. The use of 16S rRNA gene sequences as a universal phylogenetic marker allowed for the establishment of a standardized framework for bacterial and archaeal taxonomy, leading to major revisions in the classification of many groups (Woese 1987; Stackebrandt and Goebel 1994; Yarza et al. 2014).

Despite their transformative impact on microbial systematics, molecular approaches also have inherent limitations. For instance, 16S rRNA gene sequencing, while invaluable for broad phylogenetic analysis, often lacks the high resolution required for differentiating closely related species or strains and provides limited information about functional capabilities. Additionally, horizontal gene transfer, genome reduction, and the presence of multiple copies of rRNA genes can complicate phylogenetic inference and taxonomic assignments (Bartoš et al. 2024). By recognizing these challenges, microbial taxonomists in the 1990s formalized the polyphasic approach, which integrates data from phenotypic, genotypic, and phylogenetic sources to achieve a more holistic and accurate classification (Vandamme et al. 1996; Goodfellow 2000). In practice, polyphasic taxonomy may encompass classical phenotypic assays (morphological, physiological, biochemical tests), molecular techniques (PCR amplification and sequencing of marker genes, such as the 16S rRNA gene), and increasingly, whole-genome sequencing and comparative genomics (overall genome-relatedness indices). Phylogenetic analyses, based on multiple genetic loci or genome-scale data, are used to reconstruct evolutionary relationships and to validate the coherence of taxonomic groups (Gillis et al. 2005; Kämpfer and Glaeser 2012; Vandamme and Peeters 2014).

The polyphasic approach has proven particularly valuable in resolving complex taxonomic questions, delineating cryptic species, and accommodating the genetic and ecological diversity inherent in microbial populations (Nuñez et al. 2017; Venter et al. 2017). In environmental microbiology, polyphasic analyses have facilitated the discovery and description of previously uncultivated bacteria and archaea through metagenomic sequencing, single-cell genomics, and targeted cultivation efforts (Hedlund et al. 2015; Zhang and Zhang 2023). Furthermore, contemporary microbial systematics is increasingly driven by advances in high-throughput sequencing, metagenomics, and computational biology. Whole-genome sequencing now enables the calculation of genome-based metrics such as average nucleotide identity (ANI) and digital DNA–DNA hybridization (dDDH), which offer high-resolution tools for species delineation and phylogenetic analysis. Comparative genomics allows for the identification of core and accessory genomes, shedding light on the evolutionary dynamics, ecological adaptation, and functional potential of microbial taxa (Mahato et al. 2017; Meier-Kolthoff and Göker 2019; Nouioui and Sangal 2022). Moreover, the integration of chemotaxonomic data—such as lipid composition, cell wall structure, and secondary metabolite profiles—further enriches the polyphasic framework (Lawson and Patel 2021).

The ongoing refinement of the polyphasic approach, supported by continued technological innovation, is crucial for unraveling the true extent of microbial diversity and for understanding the evolutionary processes that shape microbial communities. As new techniques emerge and datasets expand, microbial systematics will play an increasingly central role in advancing our knowledge of microbial life, thereby informing applications in medicine, agriculture, biotechnology, and environmental management. The evolution of methods in microbial systematics—from traditional phenotypic analyses to the polyphasic integration of molecular, genomic, and ecological data—has transformed the classification and understanding of microorganisms. Nevertheless, the polyphasic approach, by synthesizing multiple lines of evidence, remains indispensable for achieving robust, meaningful, and universally accepted taxonomic frameworks in the ever-expanding field of microbial diversity research.

1.3.2.1 Phenotypic Characterization

Among the foundational pillars of polyphasic taxonomy, phenotypic characterization remains indispensable since it provides critical information about the observable traits and biochemical capabilities of bacteria and archaea, offering essential insights into their functionality, adaptability, and ecological significance (Vandamme et al. 1996). Despite the increasing dominance of molecular methodologies, phenotypic data continue to play a crucial role in the comprehensive and practical taxonomy of prokaryotes. By integrating observable traits with molecular and chemotaxonomic data (Vandamme and Peeters 2014), phenotypic characterization ensures robust and meaningful classification within the polyphasic framework.

Morphological Characterization

Morphological analysis is traditionally the starting point for taxonomic investigations of prokaryotic microorganisms, forming the basis upon which further physiological, biochemical, and molecular studies are constructed. Morphological characterization involves the meticulous examination of cellular features, typically using light microscopy for routine analyses and electron microscopy for high-resolution ultrastructural details. Key attributes assessed include cell shape—such as cocci (spherical), bacilli (rod-shaped), spirilla (spiral), vibrio (comma-shaped), and filamentous forms—which often correlate with specific taxonomic groups or ecological strategies. The spatial arrangement of cells (e.g., diplococci, streptococci, staphylococci, tetrads) also provides diagnostic value and may reflect underlying genetic or environmental influences (On and Holmes 1991; Tindall et al. 2010). Moreover, external appendages, such as flagella, pili, and fimbriae, are evaluated for their number, arrangement, and function. Flagella arrangements (monotrichous, lophotrichous, peritrichous, amphitrichous) are directly associated with motility mechanisms, while pili and fimbriae facilitate adhesion, biofilm formation, and conjugation. The detection of specialized structures—such as endospores in *Bacillus* and *Clostridium*, or cysts in certain archaea and bacteria—is vital for distinguishing taxa with unique survival strategies (Logan et al. 2009; Galperin 2016). Similarly, extracellular features like capsules and sheaths, which can be visualized using specific stains, signal adaptations for environmental resilience or evasion of host defenses. Staining techniques are a critical component of morphological characterization. The Gram stain differentiates bacteria based on cell wall architecture into Gram-positive and Gram-negative groups; each group is associated with distinct phylogenetic and physiological traits (Wiegel 1981; Tindall et al. 2010). Moreover, acid-fast staining, which highlights taxa with high mycolic acid content (e.g., *Mycobacterium, Nocardia*), provides further taxonomic refinement (Reynold 2009). Morphological observations extend to colony characteristics on solid media, such as color, shape, margin, elevation, and texture, which, although influenced by growth conditions, supplement the differentiation of closely related taxa (Tindall et al. 2007; Sousa et al. 2015).

Despite these strengths, it is important to recognize the limitations of morphological analysis. Phenotypic plasticity, convergent evolution, and environmental variability can diminish the reliability of morphological traits as sole taxonomic markers (Lahr et al. 2014). Nevertheless, when integrated with physiological, biochemical, and molecular data, morphological characterization enriches our understanding of microbial diversity, ecology, and evolutionary relationships.

Physiological and Biochemical Characterization

Physiological and biochemical properties represent core dimensions of prokaryotic identification and classification, remaining highly relevant in both traditional and modern systematics. These properties encompass metabolic capacities and environmental tolerances, which serve as indicators of ecological adaptation and evolutionary divergence (Vandamme et al. 1996). Key physiological traits include growth requirements such as optimal temperature (psychrophilic, mesophilic, thermophilic), pH range (acidophilic, neutrophilic, alkaliphilic), salinity tolerance (halophilic, halotolerant), and oxygen needs (aerobic, anaerobic, microaerophilic, facultatively anaerobic) (Gillis et al. 2005; Sharma et al. 2015).

Biochemical characterization involves an array of standardized assays to determine substrate utilization (e.g., carbohydrates, amino acids, lipids), energy production pathways (fermentation, respiration, photosynthesis, chemolithotrophy), and the presence of characteristic enzymes (e.g., catalase, oxidase, urease, amylase, nitrate reductase) (Vandamme et al. 1996). The detection of unique metabolic end-products—such as acids, gases, alcohols, pigments, or volatile compounds—further aids in the differentiation of taxa (Oren 2011; Lawson and Patel 2021). In addition, antibiotic susceptibility profiles, as well as resistance mechanisms, are routinely assessed in clinical and environmental monitoring (Watson et al. 2022).

The integration of physiological and biochemical data enables the construction of detailed metabolic fingerprints, invaluable for distinguishing closely related species and strains. For example, the differential ability to ferment sugars, reduce nitrates, or hydrolyze complex polymers is often used to characterize genera and species. These datasets are critical for practical identification in diagnostic laboratories, biotechnological exploitation, and ecological studies, while also enhancing our understanding of microbial functional potential in natural environments (Fusco et al. 2015).

It is essential to acknowledge, however, that physiological and biochemical traits can also exhibit significant variability depending on environmental factors, culture conditions, and gene regulation. This necessitates rigorous standardization and replication in experimental protocols for the context of microbial systematics. Within the polyphasic taxonomy framework, physiological and biochemical characterizations complement molecular and phylogenetic analyses, thus ensuring that taxonomic decisions reflect both genetic identity and functional phenotype (Moore et al. 2010).

Chemotaxonomic Characterization

Chemotaxonomy represents an important and increasingly sophisticated component of contemporary microbial systematics, functioning as a bridge between traditional phenotypic methods and modern molecular approaches. By focusing on the analysis of stable, species- or group-specific chemical constituents within microbial cells, chemotaxonomy may provide powerful supplementary and, in some cases, decisive criteria for classification, identification, and the elucidation of evolutionary relationships among bacteria and archaea. The markers employed in chemotaxonomy are often conserved at deep phylogenetic levels, which serve as reliable indicators of evolutionary divergence, ecological adaptation, and physiological specialization (Vandamme et al. 1996; Vandamme and Sutcliffe 2021).

The impetus for chemotaxonomy arose from the recognition that many phenotypic traits—such as morphology, metabolic activity, or growth requirements—can be subject to plasticity under environmental influences or be convergent among unrelated taxa (Logan 2009). In contrast, many chemical constituents of microbial cells—particularly those involved in fundamental cellular architecture or metabolism—are less susceptible to environmental variation and tend to be phylogenetically informative (Logan 2009; Oren 2010). Consistently, chemotaxonomy emerged as a means to resolve ambiguities in traditional taxonomy, supplement molecular data, and provide insights into microbial physiology and evolution that may not be accessible through other approaches (Vandamme and Sutcliffe 2021).

A cornerstone of chemotaxonomic differentiation is the analysis of cell wall polymers. The peptidoglycan (murein) layer is of particular taxonomic significance in bacteria. The amino acid composition of the peptide subunits—such as the presence of meso-diaminopimelic acid, L-lysine, or other diamino acids—delineates major bacterial lineages and supports the distinction between broad groups like the *Actinomycetota* and *Bacillota*. Gram-positive bacteria are further characterized by the presence of teichoic and lipoteichoic acids, which are absent in Gram-negative bacteria. Conversely, Gram-negative bacteria possess a distinctive outer membrane rich in lipopolysaccharides, a feature that correlates with phylogenetic relationships and underpins the diagnostic Gram stain reaction (Komagata and Suzuki 1988; Potekhina et al. 2011; Lawson and Patel 2021).

In archaea, the absence of peptidoglycan and the presence of alternative cell wall polymers—such as pseudopeptidoglycan, polysaccharides, glycoproteins, or exclusive S-layer proteins—are diagnostic features. The unique ether-linked isoprenoid lipids that constitute archaeal membranes serve as robust chemotaxonomic markers, distinguishing archaea from bacteria and providing granularity for resolving relationships within the archaeal domain. These structural adaptations not only facilitate survival in extreme environments (e.g., high temperature, salinity, or acidity) but also reflect deep evolutionary divergence among archaeal lineages (König et al. 2007).

Membrane lipid composition is another core aspect of chemotaxonomic analysis. Fatty acid methyl ester (FAME) analysis is widely employed to generate reproducible profiles of cellular fatty acids, identifying the presence and proportions of saturated, unsaturated, branched, cyclopropane, or hydroxy fatty acids (Sasser 2006). These profiles are often considered diagnostic at the genus level before the genomics era and have been instrumental in clarifying the taxonomy of diverse bacterial groups, such as the methylotrophic *Pseudomonadota* and *Actinomycetota*. The relative abundance of specific fatty acids, such as cyclopropane fatty acids or branched-chain fatty acids, often correlates with phylogenetic groupings and ecological specialization (De Carvalho and Caramujo 2018).

The analysis of respiratory quinones—ubiquinones (Q) and menaquinones (MK)—adds another layer of chemotaxonomic resolution. Quinones are key components of the electron transport chain, and their types, including chain length and degree of saturation, are frequently conserved within phylogenetic groups. For example, the distribution of ubiquinones and menaquinones can distinguish between major bacterial families and can also indicate metabolic adaptations, such as aerobic versus anaerobic respiration (Collins and Jones 1981; Komagata and Suzuki 1988).

Polar lipid analysis, typically conducted using two-dimensional thin-layer chromatography (TLC) or advanced mass spectrometry techniques, reveals complex patterns of phospholipids, glycolipids, and other polar components. These patterns are highly informative, particularly in actinobacteria and archaea, where specific lipid signatures are often genus-specific (Minnikin et al. 1984).

Polyamines—organic compounds such as putrescine, spermidine, and cadaverine—play roles in cellular stabilization and stress response. Their presence, absence, and relative proportions serve as additional taxonomic markers, especially when combined with other chemotaxonomic and molecular data (Hamana and Matsuzaki 1992). Whole-cell sugar analysis, which involves profiling di- and polysaccharides in the cell wall or capsule, can further assist in distinguishing closely related taxa (Lechevalier and Lechevalier 1970; Saddler et al. 1991). Diagnostic secondary metabolites and pigments (e.g., carotenoids, phenazines, bacteriochlorophylls, or antibiotics) may also serve as important chemotaxonomic markers, particularly in environmental and industrial microbiology (Oren 2011; Venil et al. 2013).

The reliability of chemotaxonomic characterization depends on rigorous standardization of cultivation conditions, extraction protocols, and analytical methodologies. Variability in growth phase, medium composition, or environmental parameters can influence chemical phenotypes and potentially confound taxonomic assignments. As such, best practices dictate the use of defined growth conditions and replicable extraction techniques to ensure comparability across studies (Lawson and Patel 2021; Vandamme and Sutcliffe 2021).

Recent advances in analytical chemistry have substantially enhanced the power and accessibility of chemotaxonomic approaches. High-performance liquid chromatography (HPLC), gas chromatography-mass spectrometry (GC-MS), and tandem mass spectrometry (MS/MS), including matrix-assisted laser desorption ionization time-of-flight mass spectrometry (MALDI-TOF MS), now allow for highly sensitive, high-throughput, and detailed chemical profiling of microbial cells. Moreover, the integration of bioinformatics tools facilitates the interpretation of chemotaxonomic data and its correlation with genomic and phylogenetic information, thereby strengthening the polyphasic taxonomy framework (Chudzik et al. 2024; Alaribe et al. 2025).

Chemotaxonomic characterization also has numerous practical applications. In clinical microbiology, chemotaxonomic markers assist in the rapid identification of pathogens, especially when molecular or phenotypic data are ambiguous (Gupta 2022). In industrial and environmental microbiology, these markers guide the selection and monitoring of microbial strains for biotechnological processes, bioremediation, or quality control (Harirchi et al. 2022). Furthermore, chemotaxonomic analysis is essential for the formal validation and description of new taxa, especially above species level, as mandated by international codes of microbial nomenclature, and plays a critical role in the ongoing re-evaluation and reclassification of misidentified or polyphyletic groups (Vandamme and Sutcliffe 2021).

1.3.2.2 Phylogenetic Analysis

Phylogenetic characterization refers to the study of evolutionary relationships among organisms, typically by comparing gene sequences that are conserved yet variable enough to reflect divergence over time. In bacteria and archaea, phylogenetic analysis provides insights into the lineage, speciation, and genetic diversity of microbial populations. The central principle is that homologous gene sequences, inherited from a common ancestor, accumulate mutations over generations. By analyzing these mutations in conserved genes or genomes, microbial systematists can infer evolutionary relationships and construct phylogenetic trees that depict the relatedness of different taxa (Ludwig and Klenk 2005; Tindall et al. 2010; Vandamme et al. 2014).

The choice of molecular markers is pivotal in determining the resolution, reliability, and interpretive power of phylogenetic studies in bacteria and archaea. Ideally, a molecular marker must be universally present across the taxa under investigation and possess an optimal balance between conserved sequences—allowing for reliable alignment—and variable regions that enable the discrimination of evolutionary relationships at different taxonomic levels. The 16S ribosomal RNA (rRNA) gene epitomizes these qualities and has long been the gold standard for prokaryotic phylogenetic analysis. The 16S ribosomal RNA (rRNA) gene is universally regarded as the cornerstone of phylogenetic characterization in bacteria and archaea, and its pivotal role in microbial systematics cannot be overstated (Tindall et al. 2010). The broad utility of the 16S rRNA genes arises from their ubiquitous presence in all prokaryotic organisms, as well as their unique structure, which intersperses highly conserved regions with nine hypervariable segments (V1–V9) (Yang et al. 2016). This architecture enables the development of universal primers for PCR amplification, allowing researchers to investigate the evolutionary relationships among an extensive range of microorganisms, including those that are recalcitrant to cultivation or entirely novel to science (Chun and Goodfellow 1995). The conserved regions facilitate robust sequence alignment and comparative analyses across distantly related taxa, providing a universal framework for organizing prokaryotic diversity, while the variable regions offer the necessary resolution for distinguishing between genera and, in many cases, between closely related species. In addition, the slow evolutionary rate of the 16S rRNA gene ensures that phylogenetic signals remain intact over deep evolutionary timescales, making it an essential tool for genus-level identification and elucidating higher-order relationships and reconstructing the prokaryotic tree of life (Ludwig and Schleifer 1994).

The profound impact of the 16S rRNA gene is further reflected in the creation and continual expansion of comprehensive sequence databases such as SILVA (Yilmaz et al. 2014), the Ribosomal Database Project (RDP) (Cole et al. 2014), and Greengenes (DeSantis et al. 2006), all of which serve as vital resources for microbial taxonomy, ecology, and evolution. These databases have standardized the identification and classification of bacteria and archaea, thus enabling both rapid taxonomic assignment and the discovery of novel lineages through comparative sequence analysis. In environmental microbiology, the 16S rRNA gene has opened new horizons, as it underpins culture-independent, high-throughput sequencing approaches (e.g., amplicon sequencing, metagenomics) by allowing the characterization of complex microbial communities directly from environmental samples such as soil, water, and the human gut. This has led to the recognition of a vast "microbial dark matter"—diverse lineages that would otherwise remain undetected by cultivation-based methods (Barak et al. 2023). Despite these advantages, the 16S rRNA gene also exhibits notable limitations. Its relatively slow evolutionary rate can result in insufficient resolution among closely related species or strains, and the frequent presence of multiple 16S rRNA operons with intragenomic heterogeneity can complicate sequence interpretation. To address these challenges, phylogenetic studies have increasingly incorporated additional single-copy housekeeping genes—such as *gyrB* (DNA gyrase subunit B), *rpoB* (RNA polymerase β subunit), *recA* (recombinase A), *atpD* (ATP synthase subunit D), and *infB* (translation initiation factor 2)—as alternative or complementary markers. These protein-coding genes generally evolve more rapidly than 16S rRNA, providing enhanced phylogenetic resolution, particularly in multilocus sequence analysis (MLSA), which integrates sequence data from multiple loci to generate robust and finely resolved phylogenies (Glaeser and Kämpfer 2015; Alanjary et al. 2019).

Consequently, the selection of molecular markers must be carefully aligned with the specific objectives and scope of the phylogenetic study, considering factors such as the breadth of taxonomic coverage, the evolutionary depth of interest, and the practical constraints of data generation and analysis. The thoughtful integration of multiple markers—ranging from single genes to whole genomes—not only enhances the robustness of phylogenetic inferences but also mitigates the biases and limitations inherent to any individual locus.

1.3.2.3 Genotypic Characterization

The advent of genotypic approaches, particularly whole-genome sequencing (WGS) and the application of overall genome-relatedness indices (OGRIs), has redefined the epistemological and technical standards of microbial systematics (Chun and Rainey 2014; Riesco and Trujillo 2024). These molecular strategies afford unprecedented resolution, objectivity, and reproducibility in the delineation of microbial species and higher taxa, thereby overcoming many of the constraints associated with phenotype-based classification.

Early molecular approaches, such as DNA–DNA hybridization (DDH), represented a significant methodological advance over phenotypic methods. DDH quantifies genomic similarity between organisms, with a 70% relatedness threshold historically demarcating species boundaries (Wayne et al. 1987). Despite its foundational role in microbial systematics, wet-lab DDH is labor-intensive, difficult to standardize across laboratories, and necessitates high-quality DNA from both strains under comparison, which compromises reproducibility and reliable cross-study comparisons. Subsequent technological innovations, particularly the proliferation of next-generation sequencing platforms, have rendered whole-genome sequencing (WGS) both routine and economically feasible (Mahato et al. 2017). Therefore, the increasing availability of public genome sequences has enabled the transition to *in silico* methodologies, hereby facilitating the calculation of OGRIs, which have become the gold standard for species delineation (Goris et al. 2007; Auch et al. 2010). Consistently, digital DNA–DNA hybridization (dDDH) has emerged as a pivotal and highly reliable tool in microbial systematics, fundamentally reshaping the criteria for species boundaries among prokaryotes (Auch et al. 2010). The dDDH simulates DNA reassociation processes under standardized and reproducible conditions by calculating the proportion of homologous genomic regions shared between two genomes. This approach yields values that closely parallel those obtained via wet-lab DDH, with a 70% dDDH threshold now serving as the contemporary species boundary (Meier-Kolthoff et al. 2014). The principal advantages of dDDH include its scalability, rapid processing of large genomic datasets, and resistance to technical artifacts and operator-dependent variability, thereby providing a universally applicable and reproducible criterion for species delineation, also exemplified by computational platforms such as the Genome-to-Genome Distance Calculator (GGDC) (Meier-Kolthoff et al. 2014; Meier-Kolthoff and Göker 2019).

Another OGRI commonly used to differentiate prokaryotic taxa at species-level taxonomic resolution is the average nucleotide identity (ANI), which quantifies the mean nucleotide-level similarity across all orthologous genomic segments shared between two bacterial or archaeal genomes, typically via pairwise genome alignments or algorithmically identified orthologous gene clusters (Arahal 2014; Yoon et al. 2017). Extensive empirical research has established that an ANI threshold of approximately

95–96% aligns closely with the classical 70% DDH species cutoff, rendering ANI a scientifically rigorous and objective proxy for species delineation (Kim et al. 2014; Palmer et al. 2020). This threshold is now widely accepted within the microbiological community and underpins the reclassification, validation, and discovery of numerous bacterial and archaeal species. The computational efficiency and scalability of ANI, facilitated by bioinformatics tools such as OrthoANI (Lee et al. 2016), JSpecies (Richter and Rosselló-Móra 2009), and FastANI (Jain et al. 2018), enable the rapid comparison of thousands of genomes, thereby supporting large-scale taxonomic surveys and high-throughput annotation pipelines. Unlike single-gene approaches, ANI analysis captures the full spectrum of genomic diversity—including both core and accessory genes—providing a comprehensive perspective on genomic cohesion and evolutionary divergence. However, utmost attention must be warranted since the different approaches to calculate ANI might give different results; thus, the appropriate interpretation of ANIs is heavily dependent on the method used to determine them as well as the evolutionary history of the taxa being investigated (Palmer et al. 2020). In addition to its central function in species definition, ANI elucidates patterns of microbial evolution, biogeography, and ecology, as genomic similarity patterns can reveal recent speciation events, genomic flux, and adaptation to environmental niches. As the accessibility and affordability of whole genome sequencing continue to increase, the utility of ANI analysis is further amplified by enabling the inclusion of uncultured and rare taxa from metagenomic assemblies and single-cell genomics into the broader taxonomic landscape. Its compatibility with automated bioinformatics pipelines and expanding genome databases ensures that ANI will remain a cornerstone of prokaryotic systematics, fostering the development of a precise, scalable, and evolutionarily informed taxonomy commensurate with the rapidly expanding scope of microbial diversity (Gosselin et al. 2022).

Similar to ANI, average amino acid identity (AAI) has emerged as a powerful and indispensable metric in microbial systematics, particularly for resolving evolutionary relationships and delineating taxonomic boundaries above the species level. AAI measures the mean percentage of amino acid identity across all orthologous protein-coding genes shared between two genomes, offering a genome-wide and functionally relevant estimate of genetic relatedness. Unlike nucleotide-based indices such as ANI, which excel at distinguishing closely related species, AAI operates at a broader evolutionary scale, as amino acid sequences evolve more slowly and are less susceptible to the confounding effects of synonymous substitutions and codon usage bias. This property makes AAI especially effective for classifying and comparing genera, families, and orders, where nucleotide variation may have reached saturation and lost its discriminatory power (Konstantinidis and Tiedje 2005). The standard workflow for calculating AAI involves the identification of sets of bidirectional best-hit orthologs, alignment of their protein sequences, and computation of the mean identity across these alignments, a process that is now facilitated by advanced automated algorithms and can be applied to increasingly large genomic datasets (Medlar et al. 2018). It is considered that AAI thresholds in the range of approximately 65–72% reliably correspond to genus boundaries, providing a quantitative and objective framework for genus-level taxonomy and enabling the reassessment and redefinition of polyphyletic or ambiguously assigned taxa (Konstantinidis and Tiedje 2005). In addition, AAI has proven especially useful for clarifying the relationships among highly diverse or deeply branching lineages, where traditional phenotypic or single-gene approaches often fail to provide sufficient resolution (Nicholson et al. 2020). Its robustness to horizontal gene transfer and resilience against sequence saturation further enhance the reliability and applicability of AAI across the breadth of bacterial and archaeal diversity (Feng et al. 2025). In addition, AAI is less influenced by genome size, gene content variability, and the presence of non-homologous regions compared to some other genome-wide measures, which makes it particularly suitable for comparative analyses that span broad taxonomic distances.

Another genome-based metric utilized in microbial systematics is the percentage of conserved proteins (POCP), which is calculated by identifying the number of proteins encoded in one genome that have homologs in a second genome, based on strict bidirectional best-hit criteria and typically using a preset amino acid sequence identity cutoff (often 40% identity over at least 50% of the length of the protein) (Hölzer 2024). The resulting value, expressed as a percentage, reflects the extent to which two organisms share their protein-coding gene content—an attribute that is particularly meaningful when considering the evolutionary and ecological coherence of microbial genera (Sbissi et al. 2025). Generally, a POCP threshold of 50% is widely accepted as a genus boundary, which reflects that two genomes share more than half of their protein-coding genes and so they are likely to belong to the same genus, whereas lower values point to more distant taxonomic relationships (Qin et al. 2014). However, this operational cutoff might not provide an objective and reproducible standard for all bacterial or archaeal genera since POCP-relatedness values are influenced by extreme differences in genome sizes and may show widespread distribution and deviation for some genera (Riesco and Trujillo 2024). Nevertheless, the integration of POCP with other genome-based indices and comprehensive phylogenomic analyses may enhance the rigor and multidimensionality of polyphasic taxonomy, notably in evaluating the monophyly and taxonomic cohesion of diverse or complex genera, helping to identify polyphyletic groupings and prompting the redefinition or subdivision of genera when warranted by genomic evidence. Consistently, the POCP calculation has been instrumental in clarifying the boundaries of notoriously heterogeneous genera, such as *Streptomyces* (Madhaiyan et al. 2022), *Bacillus* (Liu et al. 2020), and *Pseudomonas* (Lalucat et al. 2020) and has been widely applied to the reclassification of misassigned or ambiguously placed taxa (Sangal et al. 2022).

In order to construct a robust polyphasic framework for microbial taxonomy, it is a common practice to integrate dDDH with other genome-wide metrics—such as ANI, AAI, and phylogenomic analyses based on conserved proteins (Segata et al. 2013). This integrative approach ensures that taxonomic assignments are grounded in comprehensive, genome-wide evidence that reflects both evolutionary relationships and genetic cohesion. As the scope of available genome sequences continues to expand, including those derived from metagenome-assembled genomes (MAGs) and single-cell genomics, the genome-wide metrics will play a critical role in the systematic placement of uncultured or previously uncharacterized taxa within the microbial tree of life, as well as the identification of numerous novel genera, the rectification of erroneous classifications, and the refinement of taxonomic hierarchies to more accurately reflect evolutionary relationships.

1.3.3 Modern Taxonomy and Genomics

Core and pan-genome analysis, in conjunction with genome-based phylogenomics and the application of OGRIs, has fundamentally transformed the practice and philosophy of modern microbial taxonomy and genomics (Caputo et al. 2019). At its foundation, core genome analysis identifies the set of genes that are universally conserved across all members of a defined group including broad microbial clades such as *Archaea* (Gribaldo and Brochier-Armanet 2006), *Cyanobacteria* (Shi and Falkowski 2008), and *Gammaproteobacteria* (Susko et al. 2006); thus, offering a molecular blueprint that anchors taxonomic cohesion and reflects essential biological functions and evolutionary heritage. Conversely, the pan-genome encompasses the full complement of genes found among all strains or species within the group, including both the core and the highly variable accessory genome, which is often enriched in genes related to environmental adaptation, niche specialization, pathogenicity, antimicrobial resistance, and metabolic versatility (Mira et al. 2010; Suresh et al. 2025). This dual perspective provides a nuanced understanding of genetic diversity, enabling researchers to discern the evolutionary forces shaping microbial populations, such as gene gain and loss, horizontal gene transfer, and genome reduction or expansion. The comparative analysis of core and accessory genomes not only reveals the degree of genetic cohesion within taxa but also informs species concepts by highlighting the continuum between genetic unity and divergence across various habitats (Mesa et al. 2022; Piper et al. 2024).

When integrated, core and pan-genome analyses, genome-based phylogenomics, and OGRIs create a comprehensive, data-driven, and evolutionarily coherent taxonomy that reflects not only genomic similarity but also functional potential and ecological relevance (Carro et al. 2021). This integrative approach underpins the polyphasic taxonomy paradigm by supporting the re-evaluation of established taxa, the discovery and accurate placement of novel or uncultured lineages, and the ongoing refinement of microbial systematics in the face of rapid technological advancement. Moreover, the comprehensive use of genomics-based phylogeny approach facilitates the detection of polyphyletic or paraphyletic groups, informs the revision of genus and species concepts, and provides a framework for investigating the evolutionary mechanisms that drive adaptation and diversification in the microbial world (de la Haba et al. 2021; Carpi et al. 2022; Morreale et al. 2023). As genome sequencing technologies become more affordable and accessible, and as bioinformatics tools for pan-genome and phylogenomic analysis, such as BPGA (Chaudhari et al. 2016) and PanACoTA (Perrin and Rocha 2021), continue to mature, this multidimensional strategy is poised to remain at the forefront of microbial taxonomy, ensuring that classifications are robust, reproducible, and reflective of the true genetic and ecological complexity of prokaryotic life.

1.3.4 Contemporary Classification of Bacteria and Archaea

The classification of *Bacteria* and *Archaea* has evolved significantly over the past several decades. Contemporary classification systems for *Bacteria* and *Archaea* are codified in recognized taxonomic frameworks, such as the List of Prokaryotic names with Standing in Nomenclature (LPSN) (Parte et al. 2020; Meier-Kolthoff et al. 2022) and the Genome Taxonomy Database (GTDB) (Chaumeil et al. 2020), which integrate both phylogenetic and phenotypic information. These databases are continuously updated to reflect discoveries and methodological advances, ensuring that nomenclature remains consistent with current scientific understanding. Furthermore, using standardized criteria for genome quality and completeness helps maintain the robustness and reliability of taxonomic assignments.

Currently, the domain *Bacteria* is subdivided into the kingdoms *Bacillati*, *Fusobacteriati*, *Pseudomonadati*, and *Thermotogati*, reflecting the contemporary phylogenetic hypotheses as well as previous taxonomic proposals based on cell wall structure (Göker and Oren 2024). As an elevation in rank of the division *Firmicutes*, the kingdom *Bacillati* harbors seven phyla *Actinomycetota*, *Armatimonadota*, *Bacillota*, *Cyanobacteriota*, *Chloroflexota*, *Mycoplasmatota*, and recently discovered *Minisyncoccota* (Göker and Oren 2024; Nakajima et al. 2025). Having a relatively isolated phylogenetic position, *Fusobacteria* had been considered a close relative of *Firmicutes* (Mira et al. 2004) while their close association with hyperthermophiles was shown in large phylogenetic studies (Battistuzzi and Hedges 2009). In the current classification system for higher ranks in domain *Bacteria*, the kingdom *Fusobacteriati* harbors the phylum *Fusobacteriota*, which includes members with extensive HGT history and isolated phylogeny (Battistuzzi and Hedges 2009). Contrary to fusobacteria, the kingdom *Pseudomonadati* includes the phyla forming a relatively monophyletic group mostly formed by the former division of *Gracilicutes* (Moody et al. 2022; Göker and Oren 2024). The kingdom *Thermotogati* harbors bacterial phyla of *Thermotogota* and *Synergistota*, mostly isolated from anaerobic environments (Sexton et al. 2023; Bhandari and Gupta 2012), and also *Deinococcota*, famous for the members with extreme resistance to ionizing radiation and other stress-related conditions (Lim et al. 2018).

The domain *Archaea* has been subdivided into the kingdoms *Methanobacteriati*, *Nanobdellati*, and *Thermoproteati* (Göker and Oren 2024), compatible with the previous division into *"Euryarchaeota,"* "DPANN superphylum," and "TACK superphylum" based on comprehensive analyses of genomic features as well as metabolic and ecological preferences (Williams et al. 2017). The *Methanobacteriati* include the organisms formerly classified as *"Euryarchaeota,"* which contain physiologically and morphologically heterogeneous microorganisms, including some hyperthermophiles and thermoacidophilic archaea (Oren 2019). The kingdom *Nanobdellati* includes the microorganisms previously classified in the "DPANN superphylum" (*"Diapherotrites," "Parvarchaeota," "Aenigmarchaeota", "Nanoarchaeota,"* and *"Nanohalarchaeota"*) (Rinke et al. 2013; Göker and Oren 2024), most of which are characterized by sparse metabolisms with limited anabolic and catabolic activities (Dombrowski et al. 2019). The kingdom *Thermoproteati* was designed to include the microorganisms previously classified in the "TACK superphylum" (*"Thaumarchaeota," "Aigarchaeota," "Crenarchaeota,"* and *"Korarchaeota"*) (Guy and Ettema 2011; Göker and Oren 2024), of which eukaryotic cells might have originated based on the evidence obtained through phylogenomics and functional genomics as outlined by Guy and Ettema (2011).

1.4 Conclusion and Future Perspectives

The exploration of microbial evolution and systematics underlines the pivotal role that bacteria and archaea have played—not only in the context of Earth's deep history, but also in the ongoing processes that shape our planet's biosphere and influence its future. Prokaryotic domains of life have demonstrated extraordinary resilience, adaptability, and innovation by evolving a remarkable array of metabolic capabilities that have driven the major transitions in Earth's geochemical and ecological landscapes from their origins in the primordial environments of the early Earth. The profound metabolic diversity of *Bacteria* and *Archaea* has enabled these microorganisms to colonize environments ranging from the most benign to the most extreme—from oceanic hydrothermal vents and acidic hot springs to the coldest polar ice and the deepest sediments. Metabolic capabilities of prokaryotic microorganisms, especially the development of oxygenic photosynthesis by cyanobacteria, methanogenesis by certain archaea, and nitrogen fixation by diazotrophs, have fundamentally transformed the planet. These innovations established the environmental preconditions necessary for the evolution and radiation of complex multicellular life and continue to underpin the cycling of essential elements such as carbon, nitrogen, and sulfur.

Evolution of prokaryotic microorganisms is distinct from the evolution of larger, multicellular organisms since the prevalence of asexual reproduction, the capacity for horizontal gene transfer, and the ability to rapidly acquire and disseminate new genetic material through plasmids and mobile genetic elements have resulted in a rapid pace of adaptation and diversification among bacteria and archaea. The metabolic versatility of prokaryotes resulting from their ability for genomic innovation has enabled microbial populations to thrive in various habitats, and it is a significant factor in the persistence and resilience of microbial communities across geological time.

Microbial systematics is of utmost importance not only for a comprehensive insight into the evolution of prokaryotes but also for their robust identification for various fields, from clinical microbiology to microbial biotechnology. Scientific and technological advances in molecular systematics, utilizing particularly high-throughput sequencing and metagenomics, have revolutionized our understanding of the evolutionary relationships among prokaryotic microorganisms. These technologies have unveiled a vast and previously hidden world of microbial diversity by revealing that most microbial lineages remain uncultured and poorly characterized under standard laboratory conditions. The emerging picture is a dynamic, interconnected web of life, in which gene exchange, metabolic cooperation, and ecological interactions drive evolutionary processes at both micro and macro scales.

The study of microbial systematics—classifying organisms based on evolutionary relationships—has become an indispensable tool for deciphering the complex patterns of microbial diversity. Phylogenetic approaches based on comparative genomics have clarified the deep branches of the tree of life and uncovered the relationships among the major taxonomic groups by providing insights into the evolutionary innovations that have enabled microorganisms to adapt to new niches and tackle environmental challenges.

Microorganisms are also central to human well-being, driving the processes that sustain agriculture, purify water, and regulate the climate. At the same time, they present challenges, such as the emergence of antibiotic resistance and the evolution of new pathogens, which demand a deep understanding of evolutionary mechanisms and microbial ecology. Consistently, harnessing microbial capabilities for biotechnology, medicine, and environmental sustainability relies on our continued efforts to unravel the complexity of microbial evolution and systematics.

In conclusion, the domains of *Bacteria* and *Archaea* represent not only our planet's earliest inhabitants but also remain the most abundant, diverse, and resilient forms of life as their evolutionary stories are inextricably linked to the history of Earth itself. As we refine our methods and expand our knowledge, the study of microbial evolution and systematics will remain at the forefront of efforts to understand the diversity of life, the workings of our planet, and the prospects for a sustainable future. Ultimately, the microbial world stands as both a testament to the power of evolution and a rich resource for addressing the scientific, environmental, and societal challenges of the coming century.

Acknowledgments I would like to acknowledge Laura A. Hug for providing the figure reproduced from "A new view of the tree of life," published in Nature Microbiology (https://doi.org/10.1038/nmicrobiol.2016.48) for this chapter. I am grateful for her permission and valuable contribution to this study.

References

Alain K, Vince E, Courtine D, Maignien L, Zeng X, Shao Z, Jebbar M (2021) Thermococcus henrietii sp. nov., a novel extreme thermophilic and piezophilic sulfur-reducing archaeon isolated from a deep-sea hydrothermal chimney. Int J Syst Evolut Microbiol 71(7):004895

Alanjary M, Steinke K, Ziemert N (2019) AutoMLST: an automated web server for generating multi-locus species trees highlighting natural product potential. Nucleic Acids Res 47(W1):W276–W282

Alaribe SC, Titilayo BE, Ikpatt UN, Oladipupo AR, Alabi FA, Ikwugbado RI (2025) Computational methods for advanced mass spectrometry–a review. Phys Sci Rev 10:589–620

Andreani NA, Hesse E, Vos M (2017) Prokaryote genome fluidity is dependent on effective population size. ISME J 11(7):1719–1721

Arahal DR (2014) Whole-genome analyses: average nucleotide identity. Method Microbiol 41:103–122
Arber W (2000) Genetic variation: molecular mechanisms and impact on microbial evolution. FEMS Microbiol Rev 24(1):1–7
Ares-Arroyo M, Coluzzi C, Moura de Sousa JA, Rocha EP (2024) Hijackers, hitchhikers, or co-drivers? The mysteries of mobilizable genetic elements. PLoS Biol 22(8):e3002796
Arnold BJ, Huang IT, Hanage WP (2022) Horizontal gene transfer and adaptive evolution in bacteria. Nat Rev Microbiol 20(4):206–218
Auch AF, von Jan M, Klenk HP, Göker M (2010) Digital DNA-DNA hybridization for microbial species delineation by means of genome-to-genome sequence comparison. Stand Genomic Sci 2(1):117–134
Ay H, Sahin SM, Duman M, Ajmi N, Saticioglu IB (2025) Comprehensive genome analysis of Planococcus sp. S3-L1 isolated from horseshoe island in Antarctica reveals its biotechnological potential. Turk J Fish Aquat Sci 25(6):TRJFAS27193
Bach EM, Williams RJ, Hargreaves SK, Yang F, Hofmockel KS (2018) Greatest soil microbial diversity found in micro-habitats. Soil Biol Biochem 118:217–226
Bajic D, Sanchez A (2020) The ecology and evolution of microbial metabolic strategies. Curr Op Biotechnol 62:123–128
Baquero F, Coque TM, Galán JC, Martinez JL (2021) The origin of niches and species in the bacterial world. Front Microbiol 12:657986
Barak H, Fuchs N, Liddor-Naim M, Nir I, Sivan A, Kushmaro A (2023) Microbial dark matter sequences verification in amplicon sequencing and environmental metagenomics data. Front Microbiol 14:1247119
Bartoš O, Chmel M, Swierczková I (2024) The overlooked evolutionary dynamics of 16S rRNA revises its role as the "gold standard" for bacterial species identification. Sci Rep 14(1):9067
Bassalo MC, Liu R, Gill RT (2016) Directed evolution and synthetic biology applications to microbial systems. Curr Opin Biotechnol 39:126–133
Battistuzzi FU, Hedges SB (2009) A major clade of prokaryotes with ancient adaptations to life on land. Mol Biol Evol 26(2):335–343
Beaver RC, Neufeld JD (2024) Microbial ecology of the deep terrestrial subsurface. ISME J 18(1):wrae091
Bell EA, Boehnke P, Harrison TM, Mao WL (2015) Potentially biogenic carbon preserved in a 4.1 billion-year-old zircon. Proc Natl Acad Sci USA 112(47):14518–14521
Bhandari V, Gupta RS (2012) Molecular signatures for the phylum Synergistetes and some of its subclades. Antonie Van Leeuwenhoek 102(4):517–540
Bickhart DM, Kolmogorov M, Tseng E, Portik DM, Korobeynikov A, Tolstoganov I, Uritskiy G, Liachko I, Sullivan ST, Shin SB, Zorea A (2022) Generating lineage-resolved, complete metagenome-assembled genomes from complex microbial communities. Nat Biotechnol 40(5):711–719
Brancato SM, Salzano D, Fiore D, Russo G, Di Bernardo M (2024) Ratiometric control of two microbial populations via a dual chamber bioreactor. IEEE Control Syst Lett 8:1301–1306
Brenner DJ (1973) Deoxyribonucleic acid reassociation in the taxonomy of enteric bacteria. Int J Syst Evolut Microbiol 23(4):298–307
Caetano-Anollés G, Nasir A, Zhou K, Caetano-Anollés D, Mittenthal JE, Sun FJ, Kim KM (2014) Archaea: first domain diversified life Archaea 2014(1):590214
Caputo A, Fournier PE, Raoult D (2019) Genome and pan-genome analysis to classify emerging bacteria. Biol direct 14(1):5
Card KJ, Thomas MD, Graves JrJL, Barrick JE, Lenski RE (2021) Genomic evolution of antibiotic resistance is contingent on genetic background following a long-term experiment with *Escherichia coli*. Proc Natl Acad Sci USA 118(5):e2016886118
Carpi FM, Coman MM, Silvi S, Picciolini M, Verdenelli MC, Napolioni V (2022) Comprehensive pan-genome analysis of *Lactiplantibacillus plantarum* complete genomes. J Appl Microbiol 132(1):592–604

Carro L, Peix Á, Velázquez E (2021) The taxonomy of bacteria in the genomic era. In: Developmental biology in prokaryotes and lower eukaryotes. Springer International Publishing, Cham, pp 289–309

Chaudhari NM, Gupta VK, Dutta C (2016) BPGA-an ultra-fast pan-genome analysis pipeline. Sci Rep 6(1):24373

Chaumeil PA, Mussig AJ, Hugenholtz P, Parks DH (2020) GTDB-Tk: a toolkit to classify genomes with the Genome Taxonomy Database. Bioinformatics. 36(6):1925–1927

Chen YJ, Leung PM, Wood JL, Bay SK, Hugenholtz P, Kessler AJ, Shelley G, Waite DW, Franks AE, Cook PL, Greening C (2021) Metabolic flexibility allows bacterial habitat generalists to become dominant in a frequently disturbed ecosystem. ISME J 15(10):2986–3004

Chudzik A, Jalkanen K, Täubel M, Szponar B, Paściak M (2024) Identification of environmental Actinobacteria in buildings by means of chemotaxonomy, 16S rRNA sequencing, and MALDI-TOF MS. Microbiol Spectrum 12(3):e03596–23

Chun J, Goodfellow M (1995) A phylogenetic analysis of the genus Nocardia with 16S rRNA gene sequences. Int J Syst Evolut Microbiol 45(2):240–245

Chun J, Rainey FA (2014) Integrating genomics into the taxonomy and systematics of the Bacteria and Archaea. Int J Syst Evolut Microbiol 64(Pt_2):316–324

Cole JR, Wang Q, Fish JA, Chai B, McGarrell DM, Sun Y, Brown CT, Porras-Alfaro A, Kuske CR, Tiedje JM (2014) Ribosomal Database Project: data and tools for high throughput rRNA analysis. Nucleic Acids Res 42(D1):D633–D642

Collins MD, Jones D (1981) Distribution of isoprenoid quinone structural types in bacteria and their taxonomic implication. Microbiol Rev 45(2):316–354

Cregger MA, Schadt CW, McDowell NG, Pockman WT, Classen AT (2012) Response of the soil microbial community to changes in precipitation in a semiarid ecosystem. Appl Environ Microbiol 78(24):8587–8594

Davín AA, Woodcroft BJ, Soo RM, Morel B, Murali R, Schrempf D, Clark JW, Álvarez-Carretero S, Boussau B, Moody ER, Szánthó LL (2025) A geological timescale for bacterial evolution and oxygen adaptation. Science 388(6742):eadp1853

De Carvalho CC, Caramujo MJ (2018) The various roles of fatty acids. Molecules 23(10):2583

de la Haba RR, Minegishi H, Kamekura M, Shimane Y, Ventosa A (2021) Phylogenomics of haloarchaea: the controversy of the genera Natrinema-Haloterrigena. Front Microbiol 12:740909

de Sousa JM, Lourenço M, Gordo I (2023) Horizontal gene transfer among host-associated microbes. Cell Host Microbe 31(4):513–527

DeSantis TZ, Hugenholtz P, Larsen N, Rojas M, Brodie EL, Keller K, Huber T, Dalevi D, Hu P, Andersen GL (2006) Greengenes, a chimera-checked 16S rRNA gene database and workbench compatible with ARB. Appl Environ Microbiol 72(7):5069–5072

Dick GJ (2019) The microbiomes of deep-sea hydrothermal vents: distributed globally, shaped locally. Nat Rev Microbiol 17(5):271–283

Dombrowski N, Lee JH, Williams TA, Offre P, Spang A (2019) Genomic diversity, lifestyles and evolutionary origins of DPANN archaea. FEMS Microbiol Lett 366(2):fnz008

Drebes Dörr NC, Lemopoulos A, Blokesch M (2025) Exploring mobile genetic elements in Vibrio cholerae. Genome Biol Evol 17(5):evaf079

Dunbar J, Takala S, Barns SM, Davis JA, Kuske CR (1999) Levels of bacterial community diversity in four arid soils compared by cultivation and 16S rRNA gene cloning. Appl Environ Microbiol 65(4):1662–1669

Ernakovich JG, Barbato RA, Rich VI, Schädel C, Hewitt RE, Doherty SJ, Whalen ED, Abbott BW, Barta J, Biasi C, Chabot CL (2022) Microbiome assembly in thawing permafrost and its feedbacks to climate. Glob Change Biol 28(17):5007–5026

Ezzat L, Lamy T, Maher RL, Munsterman KS, Landfield KM, Schmeltzer ER, Clements CS, Vega Thurber RL, Burkepile DE (2020) Parrotfish predation drives distinct microbial communities in reef-building corals. Anim Microbiome 2(1):5

Feng W, Wan X, Zhang Y, Quensen J, Williams TA, Thompson M, Streeter M, Zhang Y, Jiao S, Wei G, Zhu Y (2025) Diversification, niche adaptation, and evolution of a candidate phylum thriving in the deep Critical Zone. Proc Natl Acad Sci USA 122(12):e2424463122

Fusco V, Quero GM, Cho GS, Kabisch J, Meske D, Neve H, Bockelmann W, Franz CM (2015) The genus Weissella: taxonomy, ecology and biotechnological potential. Front Microbiol 6:155

Galperin MY (2016) Genome diversity of spore-forming *Firmicutes*. In: The bacterial spore: from molecules to systems. ASM Press, Washington DC, pp 1–18

Gao B, Gupta RS (2012) Microbial systematics in the post-genomics era. Antonie Van Leeuwenhoek 101(1):45–54

Gibbons SM, Gilbert JA (2015) Microbial diversity—exploration of natural ecosystems and microbiomes. Curr Opin Genet Dev 35:66–72

Gillis M, Vandamme P, DeVos P, Swings J, Kersters K (2005) Polyphasic taxonomy. In: Bergey's Manual® of Systematic Bacteriology. Springer, Boston, MA, pp 43–48

Glaeser SP, Kämpfer P (2015) Multilocus sequence analysis (MLSA) in prokaryotic taxonomy. Syst Appl Microbiol 38(4):237–245

Glibert PM, Mitra A (2022) From webs, loops, shunts, and pumps to microbial multitasking: evolving concepts of marine microbial ecology, the mixoplankton paradigm, and implications for a future ocean. Limnol Oceanography 67(3):585–597

Godara A, Kao KC (2020) Adaptive laboratory evolution for growth coupled microbial production. World J Microbiol Biotechnol 36(11):175

Godbole S, Sachdev DP (2024) Basic concepts and recent advances in microbial diversity, taxonomy, speciation and evolution. Cambridge Scholar Publishing, Newcastle upon Tyne, UK, pp 79–106

Göker M, Oren A (2024) Valid publication of names of two domains and seven kingdoms of prokaryotes. Int J Syst Evolut Microbiol 74(1):006242

Goodfellow M (2000) Microbial systematics: background and uses. In: Applied microbial systematics. Springer Netherlands, Dordrecht, pp 1–18

Goris J, Konstantinidis KT, Klappenbach JA, Coenye T, Vandamme P, Tiedje JM (2007) DNA–DNA hybridization values and their relationship to whole-genome sequence similarities. Int J Syst Evol Microbiol 57(1):81–91

Gosselin S, Fullmer MS, Feng Y, Gogarten JP (2022) Improving phylogenies based on average nucleotide identity, incorporating saturation correction and nonparametric bootstrap support. Syst Biol 71(2):396–409

Goyal D, Swaroop S, Prakash O, Pandey J (2021) Survival strategies in cold-adapted microorganisms. In: Survival strategies in cold-adapted microorganisms. Springer Singapore, Singapore, pp 173–186

Gribaldo S, Brochier-Armanet C (2006) The origin and evolution of Archaea: a state of the art. Philos Trans R Soc B Biol Sci 361(1470):1007–1022

Griebler C, Lueders T (2009) Microbial biodiversity in groundwater ecosystems. Freshw Biol 54(4):649–677

Gubry-Rangin C, Aigle A, Herrera-Alsina L, Lancaster LT, Prosser JI (2024) Niche breadth specialization impacts ecological and evolutionary adaptation following environmental change. ISME J 18(1):wrae183

Gupta RS (2022) Microbial taxonomy: how and why name changes occur and their significance for (clinical) microbiology. Clin Chem 68(1):134–137

Guy L, Ettema TJ (2011) The archaeal 'TACK' superphylum and the origin of eukaryotes. Trends Microbiol 19(12):580–587

Hamana K, Matsuzaki S (1992) Polyamines as a chemotaxonomic marker in bacterial systematics. Crit Rev Microbiol 18(4):261–283

Handelsman J (2004) Metagenomics: application of genomics to uncultured microorganisms. Microbiol Mol Biol Rev 68(4):669–685

Harirchi S, Sar T, Ramezani M, Aliyu H, Etemadifar Z, Nojoumi SA, Yazdian F, Awasthi MK, Taherzadeh MJ (2022) Bacillales: from taxonomy to biotechnological and industrial perspectives. Microorganisms 10(12):2355

Hedlund BP, Dodsworth JA, Staley JT (2015) The changing landscape of microbial biodiversity exploration and its implications for systematics. Syst Appl Microbiol 38(4):231–236
Hölzer M (2024) POCP-nf: an automatic Nextflow pipeline for calculating the percentage of conserved proteins in bacterial taxonomy. Bioinformatics 40(4):btae175
Hörandl E, Bast J, Brandt A, Scheu S, Bleidorn C, Cordellier M, Nowrousian M, Begerow D, Sturm A, Verhoeven K and Boenigk J (2020) Genome evolution of asexual organisms and the paradox of sex in eukaryotes. In: Evolutionary biology—a transdisciplinary approach. Springer, Basel, pp 133–167
Horner-Devine MC, Carney KM, Bohannan BJ (2004) An ecological perspective on bacterial biodiversity. Proc R Soc Lond Ser B Biol Sci 271(1535):113–122
Hu P, Tom L, Singh A, Thomas BC, Baker BJ, Piceno YM, Andersen GL, Banfield JF (2016) Genome-resolved metagenomic analysis reveals roles for candidate phyla and other microbial community members in biogeochemical transformations in oil reservoirs. MBio 7(1):10–1128
Huang G, Shi W, Wang L, Qu Q, Zuo Z, Wang J, Zhao F, Wei F (2023) PandaGUT provides new insights into bacterial diversity, function, and resistome landscapes with implications for conservation. Microbiome 11(1):221
Hug LA, Thomas BC, Sharon I, Brown CT, Sharma R, Hettich RL, Wilkins MJ, Williams KH, Singh A, Banfield JF (2016) Critical biogeochemical functions in the subsurface are associated with bacteria from new phyla and little studied lineages. Environ Microbiol 18(1):159–173
Hug LA, Baker BJ, Anantharaman K, Brown CT, Probst AJ, Castelle CJ, Butterfield CN, Hernsdorf AW, Amano Y, Ise K, Suzuki Y (2016) A new view of the tree of life. Nat Microbiol 1(5):1–6
Hugenholtz P, Chuvochina M, Oren A, Parks DH, Soo RM (2021) Prokaryotic taxonomy and nomenclature in the age of big sequence data. ISME J 15(7):1879–1892
Husnik F, McCutcheon JP (2018) Functional horizontal gene transfer from bacteria to eukaryotes. Nat Rev Microbiol 16(2):67–79
Inskeep WP, Jay ZJ, McKay LJ, Dlakić M (2025) Respiratory processes of early-evolved hyperthermophiles in sulfidic and low-oxygen geothermal microbial communities. Nat Commun 16(1):277
Jain C, Rodriguez-R LM, Phillippy AM, Konstantinidis KT, Aluru S (2018) High throughput ANI analysis of 90K prokaryotic genomes reveals clear species boundaries. Nat Commun 9(1):5114
Kaminsky LM, Burghardt L, Bell TH (2025) Evolving a plant-beneficial bacterium in soil vs. nutrient-rich liquid culture has contrasting effects on in-soil fitness. Appl Environ Microbiol 91(4):e02085–24
Kämpfer P, Glaeser SP (2012) Prokaryotic taxonomy in the sequencing era–the polyphasic approach revisited. Environ Microbiol 14(2):291–317
Kim M, Oh HS, Park SC, Chun J (2014) Towards a taxonomic coherence between average nucleotide identity and 16S rRNA gene sequence similarity for species demarcation of prokaryotes. Int J Syst Evolut Microbiol 64(Pt_2):346–351
Klenk HP, Göker M (2010) En route to a genome-based classification of Archaea and Bacteria? Syst Appl Microbiol 33(4):175–182
Kochetkova TV, Podosokorskaya OA, Elcheninov AG, Kublanov IV (2022) Diversity of thermophilic prokaryotes inhabiting Russian natural hot springs. Microbiology 91(1):1–27
Komagata K, Suzuki KI (1988) 4 Lipid and cell-wall analysis in bacterial systematics. Method Microbiol. 19:161–207
König H, Rachel R, Claus H (2007) Proteinaceous surface layers of Archaea: ultrastructure and biochemistry. In: Archaea: Molecular and Cellular Biology. ASM Press, Washington, pp 315–340
Konstantinidis KT, Tiedje JM (2005) Towards a genome-based taxonomy for prokaryotes. J Bacteriol 187(18):6258–6264
Konstantinidis KT, Rosselló-Móra R, Amann R (2017) Uncultivated microbes in need of their own taxonomy. ISME J 11(11):2399–2406
Koonin EV, Wolf YI (2012) Evolution of microbes and viruses: a paradigm shift in evolutionary biology? Front Cell Infect Microbiol 2:119

Koonin EV, Makarova KS, Wolf YI (2021) Evolution of microbial genomics: conceptual shifts over a quarter century. Trends Microbiol 29(7):582–592

Kumari S, Devi M, Thakur K, Minhas B, Bhatt AK, Kaushik N (2024) Impact of anthropogenic activities on microbial diversity and soil health. In: Advancements in microbial biotechnology for soil health. Springer Nature Singapore, Singapore, pp 227–248

Lahr DJ, Laughinghouse IVHD, Oliverio AM, Gao F, Katz LA (2014) How discordant morphological and molecular evolution among microorganisms can revise our notions of biodiversity on Earth. Bioessays 36(10):950–959

Lalucat J, Mulet M, Gomila M, García-Valdés E (2020) Genomics in bacterial taxonomy: impact on the genus Pseudomonas. Genes 11(2):139

Land M, Bundschuh M, Hopkins RJ, Poulin B, McKie BG (2023) Effects of mosquito control using the microbial agent *Bacillus thuringiensis israelensis* (Bti) on aquatic and terrestrial ecosystems: a systematic review. Environ Evid 12(1):26

Lawson PA, Patel NB (2021) The strength of chemotaxonomy. In: Trends in the systematics of bacteria and fungi. CABI Publishing, Wallingford, pp 141–167

Lear L, Hesse E, Buckling A (2024) Disturbances can facilitate prior invasions more than subsequent invasions in microbial communities. Ecol Lett 27(8):e14493

Lechevalier MP, Lechevalier H (1970) Chemical composition as a criterion in the classification of aerobic actinomycetes. Int J Syst Evol Microbiol 20(4):435–443

Lee I, Ouk Kim Y, Park SC, Chun J (2016) OrthoANI: an improved algorithm and software for calculating average nucleotide identity. Int J Syst Evol Microbiol 66(2):1100–1103

Lenski RE (2017) Experimental evolution and the dynamics of adaptation and genome evolution in microbial populations. ISME J 11(10):2181–2194

Lever MA, Rogers KL, Lloyd KG, Overmann J, Schink B, Thauer RK, Hoehler TM, Jørgensen BB (2015) Life under extreme energy limitation: a synthesis of laboratory-and field-based investigations. FEMS Microbiol Rev 39(5):688–728

Leyn SA, Zlamal JE, Kurnasov OV, Li X, Elane M, Myjak L, Godzik M, de Crecy A, Garcia-Alcalde F, Ebeling M, Osterman AL (2021) Experimental evolution in morbidostat reveals converging genomic trajectories on the path to triclosan resistance. Microb Genomics 7(5):000553

Li W, Wang A (2021) Genomic islands mediate environmental adaptation and the spread of antibiotic resistance in multiresistant Enterococci-evidence from genomic sequences. BMC Microbiol 21(1):55

Li W, Wu Q, Kwok LY, Zhang H, Gan R, Sun Z (2024) Population and functional genomics of lactic acid bacteria, an important group of food microorganism: current knowledge, challenges, and perspectives. Food Front 5(1):3–23

Lim ML, Brooks MD, Boothe MA, Krzmarzick MJ (2018) Novel bacterial diversity is enriched with chloroperoxidase-reacted organic matter under anaerobic conditions. FEMS Microbiol Ecol 94(5):fiy050

Liu GH, Liu B, Wang JP, Che JM, Li PF (2020) Reclassification of Brevibacterium frigoritolerans DSM 8801T as *Bacillus frigoritolerans* comb. nov. based on genome analysis. Curr Microbiol 77(8):1916–1923

Liu Y, Wang Y, Song X, Hou X, Cao X, Wang Y (2023) The evolution of nitrogen transformation microorganism consortium under continued manganese domestication conditions. Sci Total Environ 899:165656

Logan NA (2009) Bacterial systematics. Hoboken: John Wiley & Sons

Logan NA, Berge O, Bishop AH, Busse HJ, De Vos P, Fritze D, Heyndrickx M, Kampfer P, Rabinovitch L, Salkinoja-Salonen MS, Seldin L (2009) Proposed minimal standards for describing new taxa of aerobic, endospore-forming bacteria. Int J Syst Evolut Microbiol 59(8):2114–2121

Louca S, Polz MF, Mazel F, Albright MB, Huber JA, O'Connor MI, Ackermann M, Hahn AS, Srivastava DS, Crowe SA, Doebeli M (2018) Function and functional redundancy in microbial systems. Nat Ecol Evol 2(6):936–943

Ludwig W, Klenk HP (2005) Overview: a phylogenetic backbone and taxonomic framework for procaryotic systematics. Bergey's Manual® of Systematic Bacteriology. Springer, Boston, pp 49–66

Lyons TW, Tino CJ, Fournier GP, Anderson RE, Leavitt WD, Konhauser KO, Stüeken EE (2024) Co-evolution of early Earth environments and microbial life. Nat Rev Microbiol 22(9):572–586

Madhaiyan M, Saravanan VS, See-Too WS, Volpiano CG, Sant'Anna FH, Faria da Mota F, Sutcliffe I, Sangal V, Passaglia LMP, Rosado AS (2022) Genomic and phylogenomic insights into the family Streptomycetaceae lead to the proposal of six novel genera. Int J Syst Evolut Microbiol 72(10):005570

Madison JD, LaBumbard BC, Woodhams DC (2023) Shotgun metagenomics captures more microbial diversity than targeted 16S rRNA gene sequencing for field specimens and preserved museum specimens. PLoS One 18(9):e0291540

Magnabosco C, Husain F, Paoletti MM, Parsons C, Payette JG, Schwartz SL, Tamre E, Fournier GP (2024) Toward a natural history of microbial life. Annu Rev Earth Planet Sci 52(1):85–108

Mahato NK, Gupta V, Singh P, Kumari R, Verma H, Tripathi C, Rani P, Sharma A, Singhvi N, Sood U, Hira P (2017) Microbial taxonomy in the era of OMICS: application of DNA sequences, computational tools and techniques. Antonie Van Leeuwenhoek 110(10):1357–1371

Maier RM, Neilson JW (2015) Extreme environments. In: Environmental microbiology. Academic Press, San Diego CA, pp 139–153

Malard LA, Guisan A (2023) Into the microbial niche. Trends Ecol Evol 38(10):936–945

Malyan SK, Kumar SS, Singh A, Kumar O, Gupta DK, Yadav AN, Fagodiya RK, Khan SA, Kumar A (2021) Understanding methanogens, methanotrophs, and methane emission in rice ecosystem. In: Microbiomes and the global climate change. Springer Singapore, Singapore, pp 205–224

Martin-Pozas T, Cuezva S, Fernandez-Cortes A, Cañaveras JC, Benavente D, Jurado V, Saiz-Jimenez C, Janssens I, Seijas N, Sanchez-Moral S (2022) Role of subterranean microbiota in the carbon cycle and greenhouse gas dynamics. Sci Total Environ 831:154921

Matheron R, Caumette P (2014) Structure and functions of microorganisms: production and use of material and energy. In: Environmental microbiology: fundamentals and applications: microbial ecology. Springer Netherlands, Dordrecht, pp 25–71

Medlar AJ, Törönen P, Holm L (2018) AAI-profiler: fast proteome-wide exploratory analysis reveals taxonomic identity, misclassification and contamination. Nucleic Acids Res 46(W1):W479–W485

Meier-Kolthoff JP, Göker M (2019) TYGS is an automated high-throughput platform for state-of-the-art genome-based taxonomy. Nat Commun 10(1):2182

Meier-Kolthoff JP, Carbasse JS, Peinado-Olarte RL, Göker M (2022) TYGS and LPSN: a database tandem for fast and reliable genome-based classification and nomenclature of prokaryotes. Nucleic Acids Res 50(D1):D801–D807

Meier-Kolthoff JP, Klenk HP, Göker M (2014) Taxonomic use of DNA G+ C content and DNA–DNA hybridization in the genomic age. Int J Syst Evolut Microbiol 64(Pt_2):352–356

Mesa V, Monot M, Ferraris L, Popoff M, Mazuet C, Barbut F, Delannoy J, Dupuy B, Butel MJ, Aires J (2022) Core-, pan-and accessory genome analyses of *Clostridium neonatale*: insights into genetic diversity. Microb Genomics 8(5):000813

Minnikin DE, O'Donnell AG, Goodfellow M, Alderson G, Athalye M, Schaal A, Parlett JH (1984) An integrated procedure for the extraction of bacterial isoprenoid quinones and polar lipids. J Microbiol Methods 2(5):233–241

Mira A, Martín-Cuadrado AB, D'Auria G, Rodríguez-Valera F (2010) The bacterial pan-genome: a new paradigm in microbiology. Int Microbiol 13(2):45–57

Mira A, Pushker R, Legault BA, Moreira D, Rodríguez-Valera F (2004) Evolutionary relationships of *Fusobacterium nucleatum* based on phylogenetic analysis and comparative genomics. BMC Evol Biol 4(1):50

Miteva VI, Sheridan PP, Brenchley JE (2004) Phylogenetic and physiological diversity of microorganisms isolated from a deep Greenland glacier ice core. Appl Environ Microbiol 70(1):202–213

Moody ER, Mahendrarajah TA, Dombrowski N, Clark JW, Petitjean C, Offre P, Szöllősi GJ, Spang A, Williams TA (2022) An estimate of the deepest branches of the tree of life from ancient vertically evolving genes. elife 11:e66695
Moore ER, Mihaylova SA, Vandamme P, Krichevsky MI, Dijkshoorn L (2010) Microbial systematics and taxonomy: relevance for a microbial commons. Res Microbiol 161(6):430–438
Moran Y, Fredman D, Szczesny P, Grynberg M, Technau U (2012) Recurrent horizontal transfer of bacterial toxin genes to eukaryotes. Mol Biol Evol 29(9):2223–2230
Morreale DP, St Geme III JW, Planet PJ (2023) Phylogenomic analysis of the understudied Neisseriaceae species reveals a poly-and paraphyletic Kingella genus. Microbiol Spectrum 11(6):e03123–23
Morrissey EM, Mau RL, Hayer M, Liu XJA, Schwartz E, Dijkstra P, Koch BJ, Allen K, Blazewicz SJ, Hofmockel K, Pett-Ridge J (2019) Evolutionary history constrains microbial traits across environmental variation. Nat Ecol Evol 3(7):1064–1069
Munson E, Lawhon SD, Burbick CR, Zapp A, Villaflor M, Thelen E (2023) An update on novel taxa and revised taxonomic status of bacteria isolated from domestic animals described in 2018 to 2021. J Clin Microbiol 61(2):e00281–22
Mutuku C, Gazdag Z, Melegh S (2022) Occurrence of antibiotics and bacterial resistance genes in wastewater: resistance mechanisms and antimicrobial resistance control approaches. World J Microbiol Biotechnol 38(9):152
Nakajima M, Nakai R, Hirakata Y, Kubota K, Satoh H, Nobu MK, Narihiro T, Kuroda K (2025) Minisyncoccus archaeiphilus gen. nov., sp. nov., a mesophilic, obligate parasitic bacterium and proposal of Minisyncoccaceae fam. nov., Minisyncoccales ord. nov., Minisyncoccia class. nov. and Minisyncoccota phyl. nov. formerly referred to as Candidatus Patescibacteria or candidate phyla radiation. Int J Syst Evolut Microbiol 75(2):006668
Nguyen J, Lara-Gutiérrez J, Stocker R (2021) Environmental fluctuations and their effects on microbial communities, populations and individuals. FEMS Microbiol Rev 45(4):fuaa068
Nicholson AC, Christopher AG, Anne MW, Ben WH, Melissa EB, Barry H, Arnie GS et al (2020) Division of the genus Chryseobacterium: Observation of discontinuities in amino acid identity values, a possible consequence of major extinction events, guides transfer of nine species to the genus Epilithonimonas, eleven species to the genus Kaistella, and three species to the genus Halpernia gen. nov., with description of Kaistella daneshvariae sp. nov. and Epilithonimonas vandammei sp. nov. derived from clinical specimens. Int J Syst Evolut Microbiol 70(8):4432–4450
Nouioui I, Sangal V (2022) Advanced prokaryotic systematics: the modern face of an ancient science. N Microbes N Infect 49:101036
Nuñez H, Moya-Beltrán A, Covarrubias PC, Issotta F, Cárdenas JP, González M, Atavales J, Acuña LG, Johnson DB, Quatrini R (2017) Molecular systematics of the genus Acidithiobacillus: insights into the phylogenetic structure and diversification of the taxon. Front Microbiol 8:30
Okafor N (2011) Taxonomy, physiology, and ecology of aquatic microorganisms. In: Environmental microbiology of aquatic and waste systems. Springer Netherlands, Dordrecht, pp 47–107
On SL, Holmes B (1991) Effect of inoculum size on the phenotypic characterization of Campylobacter species. J Clin Microbiol 29(5):923–926
Oren A (2010) Microbial systematics. In: Environmental biotechnology. Humana Press, Totowa, NJ, pp 81–120
Olson JM (2006) Photosynthesis in the Archean era. Photosynthesis Res 88(2):109–117
Oren A (2011) Characterization of pigments of prokaryotes and their use in taxonomy and classification. Method Microbiol 38:261–282
Oren A (2019) Extremophiles and their natural niches on Earth. In: Handbook of astrobiology. CRC Press (Taylor & Francis Group), Boca Raton, Florida, USA, pp 635–660
Oren A (2024) Naming new taxa of prokaryotes: rules and recommendations. In: Modern taxonomy of bacteria and archaea: new methods, technology and advances. Springer Nature Singapore, Singapore, pp 13–67

Osmanovic D, Kessler DA, Rabin Y, Soen Y (2018) Darwinian selection of host and bacteria supports emergence of Lamarckian-like adaptation of the system as a whole. Biol Direct 13(1):24
Overmann J, Huang S, Nübel U, Hahnke RL, Tindall BJ (2019) Relevance of phenotypic information for the taxonomy of not-yet-cultured microorganisms. Syst Appl Microbiol 42(1):22–29
Pallen MJ, Telatin A, Oren A (2021) The next million names for Archaea and Bacteria. Trends Microbiol 29(4):289–298
Palmer M, Steenkamp ET, Blom J, Hedlund BP, Venter SN (2020) All ANIs are not created equal: implications for prokaryotic species boundaries and integration of ANIs into polyphasic taxonomy. Int J Syst Evolut Microbiol 70(4):2937–2948
Parte AC, Sardà Carbasse J, Meier-Kolthoff JP, Reimer LC, Göker M (2020) List of prokaryotic names with Standing in Nomenclature (LPSN) moves to the DSMZ. Int J Syst Evolut Microbiol 70(11):5607–5612
Pearce DA, Bridge PD, Hughes KA, Sattler B, Psenner R, Russell NJ (2009) Microorganisms in the atmosphere over Antarctica. FEMS Microbiol Ecol 69(2):143–157
Perreau J, Moran NA (2022) Genetic innovations in animal–microbe symbioses. Nat Rev Genet 23(1):23–39
Perrin A, Rocha EP (2021) PanACoTA: a modular tool for massive microbial comparative genomics. NAR Genom Bioinform 3(1):lqaa106
Philippot L, Griffiths BS, Langenheder S (2021) Microbial community resilience across ecosystems and multiple disturbances. Microbiol Mol Biol Rev 85(2):10–1128
Piper KR, Ikhimiukor OO, Souza SS, Garcia-Aroca T, Andam CP (2024) Evolutionary dynamics of the accessory genomes of *Staphylococcus aureus*. Msphere 9(4):e00751–23
Pradhan N, Singh S, Saxena G, Pradhan N, Koul M, Kharkwal AC, Sayyed R (2025) A review on microbe-mineral transformations and their impact on plant growth. Front Microbiol 16:1549022. p
Prakash O, Parmar M, Vaijanapurkar M, Rale V, Shouche YS (2021) Recent trend, biases and limitations of cultivation-based diversity studies of microbes. FEMS Microbiol Lett 368(17):fnab118
Prosser JI, Hink L, Gubry-Rangin C, Nicol GW (2020) Nitrous oxide production by ammonia oxidizers: physiological diversity, niche differentiation and potential mitigation strategies. Glob Change Biol 26(1):103–118
Qin QL, Xie BB, Zhang XY, Chen XL, Zhou BC, Zhou J, Oren A, Zhang YZ (2014) A proposed genus boundary for the prokaryotes based on genomic insights. J Bacteriol 196(12):2210–2215
Reynolds TB (2009) Strategies for acquiring the phospholipid metabolite inositol in pathogenic bacteria, fungi and protozoa: making it and taking it. Microbiology 155(5):1386–1396
Richter M, Rosselló-Móra R (2009) Shifting the genomic gold standard for the prokaryotic species definition. Proc Natl Acad Sci USA 106(45):19126–19131
Ridderberg W, Handberg KJ, Nørskov-Lauritsen N (2020) Prevalence of hypermutator isolates of Achromobacter spp. from cystic fibrosis patients. Int J Med Microbiol 310(2):151393
Riesco R, Trujillo ME (2024) Update on the proposed minimal standards for the use of genome data for the taxonomy of prokaryotes. Int J Syst Evolut Microbiol 74(3):006300
Rillig MC, Antonovics J (2019) Microbial biospherics: the experimental study of ecosystem function and evolution. Proc Natl Acad Sci USA 116(23):11093–11098
Rinke C, Schwientek P, Sczyrba A, Ivanova NN, Anderson IJ, Cheng JF, Darling A, Malfatti S, Swan BK, Gies EA, Dodsworth JA (2013) Insights into the phylogeny and coding potential of microbial dark matter. Nature 499(7459):431–437
Rosenberg E, Zilber-Rosenberg I (2016) Microbes drive evolution of animals and plants: the hologenome concept. MBio 7(2):10–1128
Rosenberg, E. (2021). Evolution of Holobionts: The Hologenome Concept. In: Microbiomes. The Microbiomes of Humans, Animals, Plants, and the Environment, Springer, Cham, pp 317–352
Rosing MT (1999) 13C-depleted carbon microparticles in> 3700-Ma sea-floor sedimentary rocks from West Greenland. Science 283(5402):674–676
Rousk J, Frey SD (2015) Revisiting the hypothesis that fungal-to-bacterial dominance characterizes turnover of soil organic matter and nutrients. Ecol Monogr 85(3):457–472

Rzymski P, Gwenzi W, Poniedziałek B, Mangul S, Fal A (2024) Climate warming, environmental degradation and pollution as drivers of antibiotic resistance. Environ Pollut 346:123649

Saddler GS, Tavecchia P, Lociuro S, Zanol M, Colombo L, Selva E (1991) Analysis of madurose and other actinomycete whole cell sugars by gas chromatography. J Microbiol Methods 14(3):185–191

Sahin SM, Saticioglu IB, Duman M, Ay H (2025) Streptomyces antarcticus sp. nov., isolated from Horseshoe Island, Antarctica. Int J Syst Evolut Microbiol 75(7):006856

Sánchez-Baracaldo P, Bianchini G, Wilson JD, Knoll AH (2022) Cyanobacteria and biogeochemical cycles through Earth history. Trends Microbiol 30(2):143–157

Sangal V, Goodfellow M, Jones AL, Sutcliffe IC (2022) A stable home for an equine pathogen: valid publication of the binomial Prescottella equi gen. nov., comb. nov., and reclassification of four rhodococcal species into the genus Prescottella. Int J Syst Evolut Microbiol 72(9):005551

Santos ES, Lima ACM, Breda GL, Tomaz APDO, Nabhan SK, Funke VAM, Loth G, Nogueira KDS (2023) Colonization by multidrug-resistant bacteria in hematological patients undergoing hematopoietic stem cell transplantation and clinical outcomes: a single-center retrospective cohort study. Transpl Infect Dis 25(5):e14119

Sasser M (2006) Bacterial identification by gas chromatographic analysis of fatty acids methyl esters (GC-FAME). Microbial ID, Newark, NY

Sbissi I, Hana Dali RH, Emeni Boujemaa MAHN, Faten Ghodhbane-Gtari, Maher G (2025) Taxonomic revision of the family Aurantimonadaceae: proposal of Dennerimonas gen. nov., Mesocryomonas gen. nov., Rathsackimonas gen. nov. and Plantimonas gen. nov., along with the reclassification of Jeongeupella Jiang et al. 2024 as a later heterotypic synonym of Antarcticirhabdus Du et al. 2023. Int J Syst Evolut Microbiol 75(3):006733

Schirrmeister BE, de Vos JM, Antonelli A, Bagheri HC (2013) Evolution of multicellularity coincided with increased diversification of cyanobacteria and the great oxidation event. Proc Natl Acad Sci USA 110(5):1791–1796

Schleifer KH, Ludwig W (1994) Molecular taxonomy: classification and identification. In: Bacterial diversity and systematics. Springer US, Boston, MA, pp 1–15

Segata N, Börnigen D, Morgan XC, Huttenhower C (2013) PhyloPhlAn is a new method for improved phylogenetic and taxonomic placement of microbes. Nat Commun 4(1):2304

Sengupta S, Azad RK (2023) Leveraging comparative genomics to uncover alien genes in bacterial genomes. Microb Genomics 9(1):000939

Sexton DL, Hashimi A, Beskrovnaya P, Sibanda L, Huan T, Tocheva EI (2023) The cell envelope of Thermotogae suggests a mechanism for outer membrane biogenesis. Proc Natl Acad Sci USA 120(18):e2303275120

Shapira M (2016) Gut microbiotas and host evolution: scaling up symbiosis. Trends Ecol Evol 31(7):539–549

Sharma R, Polkade AV, Shouche YS (2015) 'Species concept'in microbial taxonomy and systematics. Curr Sci 108:1804–1814

Sheik CS, Beasley WH, Elshahed MS, Zhou X, Luo Y, Krumholz LR (2011) Effect of warming and drought on grassland microbial communities. ISME J 5(10):1692–1700

Shi T, Falkowski PG (2008) Genome evolution in cyanobacteria: the stable core and the variable shell. Proc Natl Acad Sci USA 105(7):2510–2515

Shu WS, Huang LN (2022) Microbial diversity in extreme environments. Nat Rev Microbiol 20(4):219–235

Sneath PH (2015) Bacterial nomenclature. In: Bergey's manual of systematics of Archaea and bacteria. John Wiley & Sons, Hoboken NJ, pp 1–8

Soo RM, Hemp J, Parks DH, Fischer WW, Hugenholtz P (2017) On the origins of oxygenic photosynthesis and aerobic respiration in Cyanobacteria. Science 355(6332):1436–1440

Sousa AM, Pereira MO, Lourenço A (2015) MorphoCol: an ontology-based knowledgebase for the characterisation of clinically significant bacterial colony morphologies. J Biomed Inform 55:55–63

Sriswasdi S, Yang CC, Iwasaki W (2017) Generalist species drive microbial dispersion and evolution. Nat Commun 8(1):1162
Stackebrandt E, Goebel BM (1994) Taxonomic note: a place for DNA-DNA reassociation and 16S rRNA sequence analysis in the present species definition in bacteriology. Int J Syst Evol Microbiol 44(4):846–849
Stackebrandt E, Frederiksen W, Garrity GM, Grimont PA, Kämpfer P, Maiden MC, Nesme X, Rosselló-Mora R, Swings J, Trüper HG, Vauterin L (2002) Report of the ad hoc committee for the re-evaluation of the species definition in bacteriology. Int J Syst Evol Microbiol 52(3):1043–1047
Potekhina NV, Streshinskaya GM, Tul'skaya EM, Shashkov AS (2011) Cell wall teichoic acids in the taxonomy and characterization of gram-positive bacteria, vol 38. In: Rainey FA, Oren A (eds) Methods in Microbiology. Academic Press, Amsterdam, pp 131–164
Suresh R, Jayachandiran S, Balu P, Ramasamy D (2025) Comparative genomics reveals genetic diversity and differential metabolic potentials of the species of Arachnia and suggests reclassification of Arachnia propionica E10012 (= NBRC_14587) as novel species. Arch Microbiol 207(4):93
Susko E, Leigh J, Doolittle WF, Bapteste E (2006) Visualizing and assessing phylogenetic congruence of core gene sets: a case study of the γ-Proteobacteria. Mol Biol Evol 23(5):1019–1030
Swiecicka I, Bartoszewicz M, Kasulyte-Creasey D, Drewnowska JM, Murawska E, Yernazarova A, Lukaszuk E, Mahillon J (2013) Diversity of thermal ecotypes and potential pathotypes of *Bacillus thuringiensis* soil isolates. FEMS Microbiol Ecol 85(2):262–272
Theodosiou P, Faina A, Nejatimoharrami F, Stoy K, Greenman J, Melhuish C, Ieropoulos I (2017) EvoBot: towards a robot-chemostat for culturing and maintaining microbial fuel cells (MFCs). In: Conference on Biomimetic and Biohybrid Systems. Springer International Publishing, Cham, pp 453–464
Thompson CC, Amaral GR, Campeão M, Edwards RA, Polz MF, Dutilh BE, Ussery DW, Sawabe T, Swings J, Thompson FL (2015) Microbial taxonomy in the post-genomic era: rebuilding from scratch?. Arch Microbiol 197(3):359–370
Tindall BJ, Rosselló-Móra R, Busse HJ, Ludwig W, Kämpfer P (2010) Notes on the characterization of prokaryote strains for taxonomic purposes. Int J Syst Evol Microbiol 60(1):249–266
Tindall BJ, Sikorski J, Smibert RA and Krieg NR (2007) Phenotypic characterization and the principles of comparative systematics. In: Methods for general and molecular microbiology. ASM Press, Washington DC, pp 330–393
Tokuda M, Shintani M (2024) Microbial evolution through horizontal gene transfer by mobile genetic elements. Microb Biotechnol 17(1):e14408
Uvarova Y, Bryanskaya AV, Rozanov AS, Shlyakhtun VN, Demidov EA, Starostin KV, Goryachkovskaya TN, Shekhovtsov SV, Slynko NM, Peltek SE (2020) An integrated method for taxonomic identification of microorganisms. Vavilov J Genet Breed 24(4):376
van den Berg NI, Machado D, Santos S, Rocha I, Chacón J, Harcombe W, Mitri S, Patil KR (2022) Ecological modelling approaches for predicting emergent properties in microbial communities. Nat Ecol Evol 6(7):855–865
Van den Bergh B, Swings T, Fauvart M, Michiels J (2018) Experimental design, population dynamics, and diversity in microbial experimental evolution. Microbiol Mol Biol Rev 82(3):10–1128
Vandamme P, Peeters C (2014) Time to revisit polyphasic taxonomy. Antonie Van Leeuwenhoek 106(1):57–65
Vandamme P, Sutcliffe I (2021) Out with the old and in with the new: time to rethink twentieth century chemotaxonomic practices in bacterial taxonomy. Int J Syst Evolut Microbiol 71(11):005127
Vandamme P, De Bruyne K, Pot B (2014) Phylogenetics and systematics. In: Lactic acid bacteria: biodiversity and taxonomy. John Wiley & Sons, Chichester, UK, pp 31–44
Vandamme P, Pot B, Gillis M, De Vos P, Kersters K, Swings J (1996) Polyphasic taxonomy, a consensus approach to bacterial systematics. Microbiol Rev 60(2):407–438

Vandamme PA (2011) Taxonomy and classification of bacteria. In: Manual of clinical microbiology. ASM Press , Washington DC, USA, pp 210–227
Venil CK, Zakaria ZA, Ahmad WA (2013) Bacterial pigments and their applications. Process Biochem 48(7):1065–1079
Venter SN, Palmer M, Beukes CW, Chan WY, Shin G, Van Zyl E, Seale T, Coutinho TA, Steenkamp ET (2017) Practically delineating bacterial species with genealogical concordance. Antonie Van Leeuwenhoek 110(10):1311–1325
Von Wintersdorff CJ, Penders J, Van Niekerk JM, Mills ND, Majumder S, Van Alphen LB, Savelkoul PH, Wolffs PF (2016) Dissemination of antimicrobial resistance in microbial ecosystems through horizontal gene transfer. Front Microbiol 7:173
Wang P, Zhao Y, Wang W, Lin S, Tang K, Liu T, Wood TK, Wang X (2022) Mobile genetic elements used by competing coral microbial populations increase genomic plasticity. ISME J 16(9):2220–2229
Wang Q, Cole JR (2024) Updated RDP taxonomy and RDP Classifier for more accurate taxonomic classification. Microbiol Resour Announc 13(4):e01063–23
Wani AK, Akhtar N, Sher F, Navarrete AA, Américo-Pinheiro JHP (2022) Microbial adaptation to different environmental conditions: molecular perspective of evolved genetic and cellular systems. Arch Microbiol 204(2):144
Wani GA, Khan MA, Dar MA, Shah MA, Reshi ZA (2021) Next generation high throughput sequencing to assess microbial communities: an application based on water quality. Bull Environ Contamination Toxicol 106(5):727–733
Waoo AA, Vyas C, Chopade BA (2025) Soil bacteria: an invisible army for climate resilience and mitigation. In: Climate change and soil microorganisms for environmental sustainability. Springer Nature Singapore, Singapore, pp 429–449
Watson AK, Kepplinger B, Bakhiet SM, Mhmoud NA, Chapman J, Allenby NE, Mickiewicz K, Goodfellow M, Fahal AH, Errington J (2022) Systematic whole-genome sequencing reveals an unexpected diversity among actinomycetoma pathogens and provides insights into their antibacterial susceptibilities. PLoS Negl Trop Dis 16(7):e0010128
Wayne LG, Brenner DJ, Colwell RR, Grimont PAD, Kandler O, Krichevsky MI, Moore LH, Moore WEC, Murray R, Stackebrandt ESMP, Starr MP (1987) Report of the ad hoc committee on reconciliation of approaches to bacterial systematics. Int J Syst Evol Microbiol 37(4):463–464
West SA, Griffin AS, Gardner A, Diggle SP (2006) Social evolution theory for microorganisms. Nat Rev Microbiol 4(8):597–607
Westall F (2011) Chapter 25: Early life: nature, distribution and evolution. In: Origins and evolution of life: an astrobiological perspective. Cambridge University Press, Cambridge, pp 391–413
Westoby M, Nielsen DA, Gillings MR, Gumerov VM, Madin JS, Paulsen IT, Tetu SG (2021) Strategic traits of bacteria and archaea vary widely within substrate-use groups. FEMS Microbiol Ecol 97(11):fiab142
Whiticar MJ (2020) The biogeochemical methane cycle. Hydrocarbons, oils and lipids: diversity, origin, chemistry and fate. Springer, Cham, pp 1–78
Wiegel J (1981) Distinction between the Gram reaction and the Gram type of bacteria. Int J Syst Evolut Microbiol 31(1):88–88
Wiens JJ, Zelinka J (2024) How many species will Earth lose to climate change?. Glob Change Biol 30(1):e17125
Williams TA, Szöllősi GJ, Spang A, Foster PG, Heaps SE, Boussau B, Ettema TJ, Embley TM (2017) Integrative modeling of gene and genome evolution roots the archaeal tree of life. Proc Natl Acad Sci USA 114(23):E4602–E4611
Woese CR (1987) Bacterial evolution. Microbiol Rev 51(2):221–271
Xu Q, Vandenkoornhuyse P, Li L, Guo J, Zhu C, Guo S, Ling N, Shen Q (2022) Microbial generalists and specialists differently contribute to the community diversity in farmland soils. J Adv Res 40:17–27
Yan Q, Liu Y, Hu A, Wan W, Zhang Z, Liu K (2022) Distinct strategies of the habitat generalists and specialists in sediment of Tibetan lakes. Environ Microbiol2 4(9):4153–4166

Yang B, Wang Y, Qian PY (2016) Sensitivity and correlation of hypervariable regions in 16S rRNA genes in phylogenetic analysis. BMC Bioinforma 17(1):135

Yarza P, Yilmaz P, Pruesse E, Glöckner FO, Ludwig W, Schleifer KH, Whitman WB, Euzéby J, Amann R, Rosselló-Móra R (2014) Uniting the classification of cultured and uncultured bacteria and archaea using 16S rRNA gene sequences. Nat Rev Microbiol 12(9):635–645

Yilmaz P, Parfrey LW, Yarza P, Gerken J, Pruesse E, Quast C, Schweer T, Peplies J, Ludwig W, Glöckner FO (2014) The SILVA and "all-species living tree project (LTP)" taxonomic frameworks. Nucleic Acids Res 42(D1):D643–D648

Yoon SH, Ha SM, Kwon S, Lim J, Kim Y, Seo H, Chun J (2017) Introducing EzBioCloud: a taxonomically united database of 16S rRNA gene sequences and whole-genome assemblies. Int J Syst Evol Microbiol 67(5):1613–1617

Zaremba-Niedzwiedzka K, Caceres EF, Saw JH, Bäckström D, Juzokaite L, Vancaester E, Seitz KW, Anantharaman K, Starnawski P, Kjeldsen KU, Stott MB (2017) Asgard archaea illuminate the origin of eukaryotic cellular complexity. Nature 541(7637):353–358

Zhang Y, Zhang T (2023) Culturing the uncultured microbial majority in activated sludge: a critical review. Crit Rev Environ Sci Technol 53(5):601–624

Zhang Y, Liu T, Li MM, Hua ZS, Evans P, Qu Y, Tan S, Zheng M, Lu H, Jiao JY, Lücker S (2023) Hot spring distribution and survival mechanisms of *Thermophilic comammox* Nitrospira. ISME J 17(7):993–1003

Zhao Z, Li H, Sun Y, Shao K, Wang X, Ma X, Hu A, Zhang H, Fan J (2022) How habitat heterogeneity shapes bacterial and protistan communities in temperate coastal areas near estuaries. Environ Microbiol 24(4):1775–1789

Zhong C, Wang L, Ning K (2021) Pan-genome study of Thermococcales reveals extensive genetic diversity and genetic evidence of thermophilic adaption. Environ Microbiol 23(7):3599–3613

Zhou X, Lennon JT, Lu X, Ruan A (2023) Anthropogenic activities mediate stratification and stability of microbial communities in freshwater sediments. Microbiome 11(1):191

Chapter 2
Diversity of Bacteria

Öznur Pehlivan-Günaydın, Ezgi Koman, Dilan Ergün, Göksu Tartar and Ayten Yazgan-Karataş

Abstract Bacteria exhibit extraordinary evolutionary and ecological diversity. They range from dominant, well-characterized phyla to rare lineages that are known only through environmental sequencing. This chapter reviews four key bacterial phyla, including *Pseudomonadota*, *Bacillota*, *Actinomycetota*, and *Bacteroidota*. These phyla are widely distributed, metabolically versatile, and play a central role in ecosystem functioning and human health. We discuss unique phyla within the PVC superphylum (*Planctomycetota*, *Verrucomicrobiota*, *Chlamydiota*) for their unusual cell biology, compartmentalization, and host associations. We also highlight hyperthermophilic phyla, such as *Thermotogota, Aquificota*, and *Thermodesulfobacteriota*, that thrive in geothermal ecosystems and drive sulfur and carbon cycling. We consider less-cultivated lineages, including *Deinococcota, Acidobacteriota, Nitrospirota, Fusobacteriota, Fibrobacterota, Synergistota, Deferribacterota*, and *Chrysiogenota*, in terms of their ecological niches, metabolic specializations, and roles in biogeochemical cycles, symbiosis, and disease. Collectively, these examples demonstrate the remarkable metabolic flexibility and ecological impact of bacteria, ranging from host-associated commensals and pathogens to free-living autotrophs in extreme environments. Despite advances in genomics and cultivation-independent methods, vast portions of bacterial

Ö. Pehlivan-Günaydın · E. Koman · D. Ergün · G. Tartar · A. Yazgan-Karataş (✉)
Department of Molecular Biology and Genetics, Faculty of Science and Letters, Istanbul Technical University, Maslak, Istanbul, Türkiye

Dr. Orhan Öcalgiray Molecular Biology, Biotechnology, and Genetics Research Center (ITU-MOBGAM), Istanbul Technical University, Maslak, Istanbul, Türkiye

A. Yazgan-Karataş
e-mail: karatasay@itu.edu.tr

M. Çalışkan and M. Arıkan (eds.), *Microbial World: Bacteria and Archaea*, Progress in Molecular and Subcellular Biology 62, https://doi.org/10.1007/978-3-032-11438-9_2

diversity remain uncultured and poorly understood. Continued exploration of both dominant phyla and rare lineages promises to refine bacterial taxonomy, expand our understanding of microbial evolution, and reveal novel metabolic pathways with implications for ecology, medicine, and biotechnology.

Keywords Bacterial phyla · Microbial taxonomy · Metabolic versatility · Phylogeny · Prokaryotic evolution

2.1 Introduction

Bacterial diversity encompasses not only species richness but also significant variation in morphology, metabolism, genomic architecture, and ecological function. Advances in molecular biology, particularly genome-resolved metagenomics and high-throughput sequencing, have greatly expanded our understanding of this diversity by revealing numerous uncultured and previously unknown lineages. However, a significant portion of bacterial diversity remains unexplored, particularly in extreme or complex habitats.

A comprehensive understanding of bacterial diversity is essential not only for microbial systematics but also for practical applications in biotechnology, agriculture, environmental management, and medicine. Accurate phylogenetic classification enables better identification, functional prediction, and utilization of bacterial lineages in both fundamental and applied sciences.

The early classification of bacteria was primarily based on observable phenotypic characteristics, such as cell morphology, staining properties, and growth behavior under laboratory conditions. This classical approach lacked genetic or molecular insights, relying instead on practical, descriptive traits. In the late nineteenth century, pioneers such as Ferdinand Cohn proposed one of the first systematic frameworks for bacterial classification by grouping organisms into categories such as rods, cocci, spirilla, and spore-formers (Drews 2000). A significant paradigm shift occurred in the 1970s and 1980s with the advent of molecular phylogenetics, particularly through the comparative analysis of 16S and 18S rRNA genes (Pace et al. 2012). These molecular markers enabled a phylogenetically informed classification system and led to the formal recognition of the three domains of life: Bacteria, Archaea, and Eukarya into distinct domains.

Beginning in the late 2000s, the transition into the genomic era introduced high-throughput sequencing technologies and large-scale comparative genome analysis. These advances enabled the development of genome-based metrics such as Average Nucleotide Identity (ANI) and digital DNA–DNA hybridization (dDDH), which have become central to modern species delineation. Genomic approaches now form the backbone of bacterial taxonomy, providing high-resolution tools for evaluating evolutionary relationships across diverse taxa (Ferraz Helene et al. 2022).

Contemporary bacterial classification integrates genomic data with ecological and phenotypic traits to construct a more comprehensive view of microbial diversity. Importantly, metagenomic and single-cell genome sequencing have revealed an enormous array of novel bacterial lineages, many of which are currently uncultivable yet can be taxonomically characterized based on genomic evidence. This shift underscores a growing recognition of uncultured microbial dark matter while also reinforcing the importance of culture-based taxa as reference points within the taxonomic framework.

The nomenclature of bacterial taxa is regulated by the International Code of Nomenclature of Prokaryotes (ICNP), which sets the rules for valid publication, type strain designation, and name priority. Recent revisions have expanded the framework to include taxa defined solely by genomic data and to formalize "*Candidatus*" names for uncultured organisms. These processes are managed by the International Committee on Systematics of Prokaryotes (ICSP). The List of Prokaryotic Names with Standing in Nomenclature (LPSN) serves as the official reference for all validly published bacterial names. It also incorporates genome-resolved taxa and applies standardized suffixes such as *-ota* for phyla and *-ales* for orders to ensure consistency (Parte et al. 2020).

As the discovery of microbial lineages accelerates, particularly through metagenomic approaches, automated naming strategies are being developed to manage the increasing number of uncultured taxa while ensuring clarity and stability in nomenclature. LPSN continues to play a central role in integrating these advances, providing a coherent and accessible framework for prokaryotic taxonomy. Unless otherwise stated, taxonomic nomenclature in this chapter follows the currently valid names recorded in the LPSN, which serves as the authoritative reference for names published under the International Code of Nomenclature of Prokaryotes (ICNP) (Parte et al. 2020; Parks et al. 2022; Oren et al. 2025).

As of July 2025, LPSN recognizes four major kingdoms: *Bacillati, Fusobacteriati, Pseudomonadati*, and *Thermotogati*, which together encompass 39 validly published bacterial phyla, each supported by formal nomenclatural standards (Parte et al. 2020) (Fig. 2.1). In contrast, the Genome Taxonomy Database (GTDB), which employs a genome-based classification system independent of the rules of ICNP, currently recognizes 169 bacterial phyla in its latest release (R10-RS226 2025), many of which consist of uncultured or metagenome-only lineages.

This chapter emphasizes phyla of particular ecological, metabolic, medical, and biotechnological relevance. Attention is given to their phylogenetic placement, defining traits, and roles in natural ecosystems and applied fields. The discussion covers dominant phyla, such as *Pseudomonadota, Bacillota, Actinomycetota*, and *Bacteroidota*; specialized lineages within the PVC superphylum; and hyperthermophilic groups, including *Thermotogota*, *Thermodesulfobacteriota*, and *Aquificota*, as well as less-cultured yet environmentally significant phyla, including *Deinococcota, Acidobacteriota, Nitrospirota, Fusobacteriota, Fibrobacterota, Synergistota, Deferribacterota*, and *Chrysiogenota*. Together, these lineages demonstrate the extraordinary range of bacterial adaptations and their critical contributions to biogeochemical cycles, symbiosis, pathogenesis, and biotechnology. They also emphasize the significance of environmental genomics in uncovering the diversity of uncultured microorganisms.

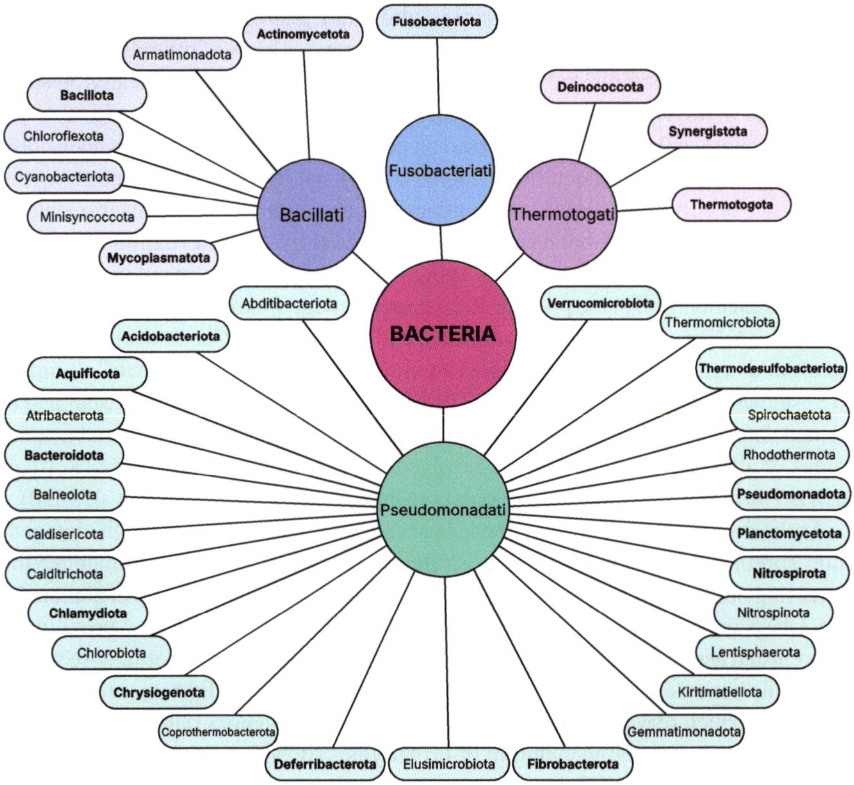

Fig. 2.1 Taxonomic framework of the domain *Bacteria*, showing the four major kingdoms (*Bacillati*, *Fusobacteriati*, *Pseudomonadati*, and *Thermotogati*) and their constituent phyla as recognized by the List of Prokaryotic Names with Standing in Nomenclature (LPSN) as of July 2025

2.2 Pseudomonadota

Pseudomonadota (formerly *Proteobacteria*) is one of the most ecologically and metabolically diverse phyla within the bacterial domain. *Pseudomonadota* represents a diverse group of Gram-negative bacteria. They are notable for their wide range of cell architectures, including simple rods and cocci, spirals, filaments, appendaged and sheathed forms. They also exhibit various modes of movement, including flagellar swimming and gliding. Metabolically, they encompass chemoorganotrophs, chemolithotrophs, or phototrophs that harvest energy. They also tolerate diverse oxygen conditions, which span from strict aerobes to obligate anaerobes, facultative aerobes, and microaerophiles. Ecologically, many are free-living; others form mutualistic partnerships, such as nitrogen-fixing rhizobia in legume nodules, or persist as intracellular associates of protozoa and invertebrates. Some lineages act as obligate pathogens. Their morphological, physiological, and ecological diversity is fundamental to their pivotal roles in global biogeochemical cycles and interactions with higher organisms (Kersters et al. 2006; Madigan et al. 2022).

This group was first described in 1987 by the American microbiologist Carl Woese as "the purple bacteria and their relatives" (Woese 1987) and was later formally named *Proteobacteria*. Recent phylogenomic analyses have led to substantial taxonomic revisions, resulting in the renaming of the phylum as *Pseudomonadota* (Oren and Garrity 2021; Oren 2024). Although several classes are currently recognized under the phylum *Pseudomonadota*, as listed in the LPSN database, including *Alphaproteobacteria*, *Betaproteobacteria*, *Gammaproteobacteria*, *Deltaproteobacteria*, *Epsilonproteobacteria*, as well as *Acidithiobacillia*, *Hydrogenophilia*, and *Oligoflexia*, we focused on five major and widely studied classes comprising *Alphaproteobacteria*, *Betaproteobacteria*, *Gammaproteobacteria*, *Deltaproteobacteria*, and *Epsilonproteobacteria* in this phylum. The first three were discussed in detail, while the subsequent two were discussed through an examination of their general characteristics. The purpose of this overview of recent genomic reclassifications is to elucidate the rationale behind the selection of lineages that are emphasized in this section.

2.2.1 Alphaproteobacteria

The class *Alphaproteobacteria* comprises a diverse group of phylogenetically and metabolically distinct Gram-negative bacteria. These microorganisms can be found in a variety of habitats, from aquatic to terrestrial ecosystems. While some exist as free-living organisms, others form symbiotic relationships with eukaryotic hosts. Their metabolic capabilities are remarkable, encompassing essential biological processes such as photosynthesis, nitrogen fixation, methylotrophy, and ammonia oxidation. They exhibit considerable morphological diversity with various cellular structures and reproductive strategies such as binary fission, budding, and the formation of specialized cell types (Batut et al. 2004; Williams et al. 2007). The major orders within the class *Alphaproteobacteria* are *Hyphomicrobiales*, *Rhodobacterales*, *Rhodospirillales*, *Rickettsiales*, *Sphingomonadales*, and *Caulobacterales*.

The order *Hyphomicrobiales* (heterotypic synonym: *Rhizobiales*) is characterized by remarkable ecological diversity, encompassing free-living soil bacteria, intracellular pathogens, and plant symbionts. This broad range of lifestyles, from mutualism to parasitism, highlights their importance in agriculture and medicine. The group is believed to have originated as free-living soil bacteria, which then co-evolved with plants and animals to form symbiotic and pathogenic lineages (Wang et al. 2020a). These bacteria are typically rod-shaped, and many are motile via flagella. Some also exhibit polar growth or budding cell division. This unipolar growth appears to be an ancestral trait in *Hyphomicrobiales* (Brown et al. 2011; Brown et al. 2012). Important families within the order include *Rhizobiaceae*, *Phyllobacteriaceae*, and *Nitrobacteraceae* (heterotypic synonym: *Bradyrhizobiaceae*) for their nitrogen-fixing symbioses with plants; *Brucellaceae* and *Bartonellaceae* as animal pathogens;

Methylobacteriaceae, *Methylocystaceae*, and *Beijerinckiaceae* as methylotrophs or methanotrophs; and *Hyphomicrobiaceae* and *Rhodobiaceae* for their budding or photosynthetic capabilities, respectively.

Hyphomicrobiales play a key role in the nitrogen cycle, as many of them are nitrogen-fixing bacteria that convert atmospheric nitrogen (N2) into bioavailable ammonia. *Rhizobia*, including the genera *Rhizobium*, *Bradyrhizobium*, and *Mesorhizobium*, form symbiotic relationships with leguminous plants to induce root nodule formation and facilitate N2 fixation, thereby naturally enriching the soil (Poole and Terpolilli 2018). Other species (e.g., *Beijerinckia* spp.) are free-living nitrogen fixers (Brenner et al. 2005). *Hyphomicrobiales* also contribute to nitrification: Facultative chemolithoautotrophs such as *Nitrobacter* species obtain energy from the oxidation of nitrite to nitrate, thereby linking ammonium oxidation to the production of plant-available nitrate in soils (Brenner et al. 2005; Starkenburg et al. 2006). *Rhodopseudomonas palustris*, a purple non-sulfur photosynthetic bacterium within this order, is one of the most metabolically versatile microorganisms known. It can grow phototrophically or chemoheterotrophically, fix nitrogen, degrade a variety of aromatic compounds anaerobically, and even produce hydrogen gas as a biofuel byproduct (Pechter et al. 2016). The order *Hyphomicrobiales* also includes methanotrophs, which are bacteria that use methane as a carbon and energy source. Type II methanotrophs, such as *Methylocella* spp., are facultative methanotrophs that can grow on methane as well as multi-carbon compounds, such as acetate or ethanol. Similarly, certain *Methylocystis* spp. can exhibit facultative and mixotrophic lifestyles, utilizing methane alongside substrates such as hydrogen (Hakobyan and Liesack 2020). Pink-pigmented facultative methylotrophs (PPFMs), notably the genus *Methylobacterium*, utilize one-carbon compounds such as methanol and methylamine. Certain species can synthesize phytohormones that promote plant growth. Some species, most notably *M. nodulans*, are associated with nitrogen fixation and nodule formation (Dourado et al. 2015). *Hyphomicrobiales* are prevalent in plant microbiomes, such as the rhizosphere and endosphere, as well as in bryophytes and lichens (Erlacher et al. 2015). As mentioned earlier, this order also contains pathogenic species, including plant pathogens such as *Agrobacterium tumefaciens,* which causes crown gall disease by transferring T-DNA into plant cells (Chilton et al. 1977). Notable animal pathogens include *Brucella*, which causes brucellosis in livestock and can be transmitted to humans (Alton and Forsyth 1996), and *Bartonella henselae*, which causes cat-scratch disease and is transmitted through cat scratches or bites (Baranowski and Huang 2023).

The order *Rickettsiales* comprises obligate intracellular bacteria with reduced genomes and host-dependent physiology. This monophyletic group includes the families the *Rickettsiaceae* and *Ehrlichiaceae* (syn. *Anaplasmataceae*), as well as various "*Candidatus*" lineages originating from uncultivated symbionts. Members are generally small, rod-shaped, or coccoid cells, and most lack flagella, rendering them non-motile. Extensive reductive evolution has resulted in metabolic auxotrophy, making them strictly dependent on eukaryotic host cells (e.g., arthropods, vertebrates, and protists) (Brenner et al. 2005; Salje 2021; Schön et al. 2022). A thin peptidoglycan layer is variably present between the inner and outer membranes; its absence or incomplete synthesis in some species may facilitate evasion of host immune recognition (Guillotte et al. 2018; Salje 2021). Classic members of the order, such as *Rickettsia* and *Ehrlichia*, are well-known arthropod-borne pathogens that infect vertebrates. They typically circulate between mammalian hosts and blood-feeding arthropods, such as ticks, mites, fleas, or lice, which act as vectors for transmission (Salje 2021). Spotted fever group *Rickettsia* species, for example, infect tick vectors and can be transmitted to vertebrate hosts, including humans. Once inside the host, they invade and replicate within the cytosol of endothelial cells (Parola et al. 2013; Salje 2021). Similarly, *Ehrlichia* and *Anaplasma* species are transmitted by ticks and infect vertebrates. These infections can cause diseases like human monocytic ehrlichiosis (*E. chaffeensis*) and human granulocytic anaplasmosis (*A. phagocytophilum*) (Salje 2021). Some *Rickettsiales*, such as "*Ca. Xenohaliotis californiensis*," infect aquatic animals (Cicala et al. 2017). Several *Rickettsiales* members function as endosymbionts; for example, *Wolbachia* infects a wide range of arthropods and nematode worms. In arthropod hosts, *Wolbachia* induces reproductive manipulations, such as cytoplasmic incompatibility, male-killing, or confers mutualistic benefits, such as pathogen protection. These effects can significantly influence host population dynamics and potentially speciation (Darby et al. 2012; Hilgenboecker et al. 2008). In contrast, in filarial nematodes, *Wolbachia* acts as an obligate mutualist, providing essential metabolites (e.g., heme, riboflavin) that the worm cannot synthesize itself. Elimination of *Wolbachia* with antibiotics disrupts nematode development or viability, a mechanism exploited in anti-filarial therapies (Darby et al. 2012). Beyond their role as modern symbionts, *Rickettsiales* may have contributed to eukaryotic evolution. Phylogenomic evidence suggests that mitochondria evolved from an *Alphaproteobacteria* ancestor closely related to *Rickettsiales*, suggesting a shared lineage (Martijn et al. 2018).

The order *Rhodobacterales* is characterized by a significant number of photoautotrophic and photoheterotrophic species that possess bacteriochlorophyll. Members of this order can perform various core metabolic processes, such as aerobic respiration, anaerobic fermentation, sulfur compound oxidation, autotrophic carbon fixation, nitrogen fixation, and hydrogen production. These processes are often performed in diverse combinations depending on the species. Notable genera include *Rhodobacter*, *Paracoccus* and members of the marine "*Roseobacter*" clade (e.g., *Ruegeria*). These bacteria are predominantly found in marine and estuarine ecosystems, where they play key roles in biogeochemical cycles, such as those involving sulfur, nitrogen, and carbon. Many species degrade complex organic compounds and form close ecological associations with phytoplankton, often colonizing algal blooms and suspended particles (Brenner et al. 2005; Kopejtka et al. 2017; Simon et al. 2017; Ahmed and Campbell 2025).

The order *Rhodospirillales* includes two main families: *Rhodospirillaceae* and *Acetobacteraceae*. This highly eclectic order comprises phototrophic purple non-sulfur bacteria (e.g., *Rhodospirillum rubrum*), nitrogen-fixing soil bacteria (e.g., *Azospirillum* spp.), magnetotactic bacteria (*Magnetospirillum*), and acetic-acid bacteria (*Acetobacter*, *Gluconobacter*) (Brenner et al. 2005).

The order *Sphingomonadales* predominantly comprises aerobic bacteria, though some members exhibit facultative anaerobic or phototrophic adaptations. A defining characteristic of its primary family, *Sphingomonadaceae*, is the incorporation of glycosphingolipids rather than lipopolysaccharides in their outer membrane. Members of the genus *Sphingomonas* are known for their ability to degrade recalcitrant compounds, such as aromatic hydrocarbons (Brenner et al. 2005; Glaeser and Kämpfer 2014; García-Romero et al. 2024).

Members of the order *Caulobacterales* are typically oligotrophic and chemoorganotrophic organisms that rely on strictly aerobic respiratory metabolism. *Caulobacter crescentus* is a model organism for cell cycle studies and is characterized by its dimorphic life cycle consisting of motile swarmer and sessile stalked cells (Brenner et al. 2005; Jenal and Stephens 2002).

2.2.2 *Betaproteobacteria*

Betaproteobacteria is a diverse class that exhibits significant metabolic versatility, including aerobic, facultative anaerobic, chemolithotrophic, and phototrophic lifestyles. This versatility enables the class to occupy a wide range of ecological niches. Members of this class also exhibit considerable morphological diversity, ranging from cocci and bacilli to filamentous forms with specialized structures. Several genera within this class are notable human and animal pathogens. (Brenner et al. 2005). The orders within the class *Betaproteobacteria* are *Burkholderiales*, *Neisseriales*, *Rhodocyclales*, and *Spirillales*.

The order *Burkholderiales* exhibits extensive metabolic and ecological versatility. Its species encompass strictly aerobic, facultatively aerobic, and obligately anaerobic chemoorganotrophs; anoxygenic phototrophs; obligate and facultative chemolithotrophs; free-living nitrogen fixers; and pathogens that affect plants, animals, and humans (Brenner et al. 2005; Madigan et al. 2022). Members of *Burkholderiales* are widespread in soil, water, and host-associated habitats, reflecting their functional roles. Many species degrade complex organics. For instance, some *Burkholderia* and *Comamonas* spp. break down phenols and aromatic hydrocarbons, contributing to carbon turnover and bioremediation (Morya and Thakur 2020; Mortazavi et al. 2023). Several genera carry nitrogenase genes and perform nitrogen fixation, notably *Cupriavidus* and *Burkholderia vietnamiensis* (Estrada-De Los Santos et al. 2001; da Silva et al. 2012). In addition to their beneficial roles in nutrient cycling and bioremediation, certain members of *Burkholderiales* have evolved pathogenic traits that enable them to colonize and harm a wide range of hosts, including plants, animals, and humans. For instance, the *Burkholderia cepacia* complex causes severe respiratory infections in cystic fibrosis patients, *B. pseudomallei* leads to melioidosis, and *B. mallei* causes glanders, a zoonotic and occupational disease (Van Zandt et al. 2013; Gutiérrez Santana and Coria Jiménez 2024; Meumann et al. 2024).

The order *Neisseriales* comprises non-spore-forming, aerobic, and mesophilic bacteria that typically occur as nonmotile cocci or distinct rods (Brenner et al. 2005). *Neisseriales* includes two validly published families: *Neisseriaceae* (the type family) and *Chromobacteriaceae*. Most strains within the *Neisseriaceae* family are indigenous to the mucosal membranes of humans and animals. Phylogenetically, this family forms a host-associated clade: genera such as *Neisseria*, *Eikenella*, *Kingella*, and *Simonsiella* are grouped together as obligate mucosal symbionts. Important clinical members of the *Neisseriaceae* include *Neisseria gonorrhoeae* and *N. meningitidis*, which cause gonorrhea and meningitis, respectively (Brenner et al. 2005; Adeolu and Gupta 2013). In contrast, *Chromobacteriaceae* is composed of free-living, Gram-negative rods that are typically motile by polar flagella. The type genus, *Chromobacterium*, includes *C. violaceum*, a purple-pigmented rod commonly found in tropical soil and water. Most strains produce violacein, a bioactive compound responsible for the characteristic violet pigmentation. While the genus exhibits a saprophytic, free-living lifestyle, *C. violaceum* has been associated with opportunistic infections in humans and animals (Adeolu and Gupta 2013; Batista and da Silva Neto 2017).

The order *Rhodocyclales* is characterized by its diverse metabolic lifestyles. Originally classified as a single family (*Rhodocyclaceae*), the order *Rhodocyclales* has been reclassified into four families: *Rhodocyclaceae*, *Zoogloeaceae*, *Azonexaceae*, and *Fluviibacteraceae*. This reclassification was based on 16S rRNA and genomic data. Its members are typically curved or spiral rods that are often motile. There is great physiological diversity within the order. For example, *Rhodocyclus spp.* are anoxygenic photoheterotrophs, while most other *Rhodocyclales* encode catalase and superoxide dismutase, suggesting aerobic or aerotolerant metabolism. Many genera fix nitrogen (e.g., *Azospira*, *Azonexus*, *Azoarcus*, *Azovibrio*), and some respire unusual electron acceptors (e.g., *Azospira*, *Dechloromonas*, *Propionivibrio*), reducing chlorate/perchlorate. They play key roles in nutrient cycling and biodegradation. For example, *Candidatus Accumulibacter phosphatis* removes and stores phosphate under alternating anaerobic/aerobic conditions in enhanced biological phosphorus removal (EBPR) reactors. *Zoogloea spp.* form stable flocs in activated sludge, and *Thauera* and *Azoarcus* strains degrade aromatic pollutants under denitrifying conditions. Many *Rhodocyclales* are active denitrifiers that contribute to nitrogen removal in wastewater and soils (Hesselmann et al. 1999; Brenner et al. 2005; Shao et al. 2009; Wang et al. 2020b).

The order *Spirillales* exhibits metabolic diversity, including families such as *Nitrosomonadaceae* (ammonia-oxidizers like *Nitrosomonas*), Methylophilaceae (C_1-methylotrophs like *Methylophilus* and *Methylotenera*), *Gallionellaceae* (e.g., iron-oxidizers like *Gallionella* and *Ferriphaselus*), and *Spirillaceae* (e.g., *Spirillum*) (Vorobev et al. 2013; Prosser et al. 2014; Hoover et al. 2023). *Spirillales* members play major ecological roles in biogeochemical cycles. For example, the autotrophic ammonia-oxidizing bacteria *Nitrosomonas* regulate the nitrogen cycle in terrestrial and aquatic environments and in wastewater treatment processes (Vorobev et al. 2013). Methylotrophs from the family *Methylophilaceae* (e.g., *Methylophilus* and *Methylotenera*) utilize one-carbon compounds, such as methanol and methylamines, and are key contributors to aquatic carbon cycling (Prosser et al. 2014; Salcher et al. 2019).

2.2.3 *Gammaproteobacteria*

The class *Gammaproteobacteria* is highly diverse and significant within the *Pseudomonadota* phylum. It is one of the most genus-rich bacterial taxa, comprising approximately 250 genera. The members of this class exhibit a wide range of metabolic capabilities, including aerobic and anaerobic respiration, chemolithotrophy, and heterotrophy. They occupy diverse ecological niches, including soil, freshwater, marine, and host-associated environments. They also exhibit different morphologies comprising rods, curved rods, cocci, spirilla, and filaments (Brenner et al. 2005; Williams et al. 2010). Members of this class range from well-known model organisms, such as *Escherichia coli*, to major human pathogens, including *Salmonella*, *Vibrio*, and *Pseudomonas*, as well as endosymbionts and free-living environmental microorganisms. The members of this class play crucial roles in various environmental processes, including bioremediation of pollutants, nitrogen fixation, which enhances soil fertility, and potential applications in energy production through microbial fuel cells (Williams et al. 2010; Qian and Morse 2011; Baker et al. 2015; Al-Awthan et al. 2024). Reflecting their metabolic versatility, *Gammaproteobacteria* exhibit unique physiological traits, including the ability to incorporate exogenous fatty acids into membrane phospholipids, influencing membrane properties, antibiotic resistance, and virulence-related behaviors (Saksena et al. 2023).

The current taxonomy of *Gammaproteobacteria* encompasses several orders. This section, however, focuses on *Enterobacterales*, *Pseudomonadales*, and *Vibrionales* due to their medical and ecological significance. *Vibrionales* is not a validly published order under ICNP and is excluded from LPSN, but it is widely used in applied literature to group *Vibrio*, *Aliivibrio*, and related genera. For this reason, we retain this terminology, emphasizing its functional utility while acknowledging its nomenclatural limitation.

The order *Enterobacterales* (type genus *Enterobacter*) comprises a diverse group of Gram-negative, facultatively anaerobic, non-spore-forming, rod-shaped bacteria. Most are catalase-positive and oxidase-negative. Members of this group occupy diverse ecological niches, including soil and water, as well as association with living organisms, such as plants, insects, animals, and humans. Many members of this order are recognized as human and animal pathogens, including *Escherichia coli*, *Salmonella enterica*, *Yersinia pestis*, *Shig*ella, *Klebsiella*, *Proteus*, *Serratia*, and *Citrobacter*, as well as genera that cause plant diseases such as *Dickeya*, *Pectobacterium*, *Brenneria*, *Erwinia*, and *Pantoea* (Brenner et al. 2005; Adeolu et al. 2016; Doern 2023). Eight families are now formally recognized within the order *Enterobacterales*; however, the family *Enterobacteriaceae* continues to encompass most of the described genera within this taxonomic group.

Most members of the order *Enterobacterales* ferment sugars into various end products and reduce nitrate to nitrite. However, under certain conditions, some can further reduce nitrite to ammonia or other products (Stewart 1994; Brenner et al. 2005; Madigan et al. 2022). *Enterobacterales* are functionally categorized into two main groups based on the end products resulting from glucose fermentation. The first group, known as mixed-acid fermenters, includes genera such as *Escherichia*, *Shigella*,

Salmonella, and *Proteus*. These organisms predominantly yield acidic end products, including acetic, formic, lactic, and succinic acids, as well as small amounts of ethanol, carbon dioxide (CO2), and hydrogen gas (H2). Conversely, the second group, known as 2,3-butanediol fermenters, includes most species in the genera *Klebsiella*, *Serratia*, and *Erwinia*. This group produces significantly lower levels of acids. Instead, their primary fermentation end products are neutral compounds, notably ethanol, acetoin, and substantial amounts of 2,3-butanediol. These fundamental metabolic distinctions underpin the classification of these enteric bacteria (Vivijs et al. 2014; Madigan et al. 2022).

The genus *Escherichia* is recognized for its clinical and ecological importance. The most well-known species, *E. coli*, is a model organism in molecular biology and genetics. It is also a common inhabitant of the human gastrointestinal tract and other warm-blooded animals. As a commensal, *E. coli* plays beneficial roles in the gut, such as synthesizing essential vitamins, such as vitamin K_2 and several B vitamins, and preventing the colonization of pathogenic microbes. However, not all *E. coli* strains are harmless. Certain pathogenic strains are associated with diseases ranging from urinary tract infections and meningitis to different forms of diarrhea. One such strain is Shiga toxin-producing *E. coli* O157:H7, a well-characterized enterohemorrhagic (EHEC) serotype associated with severe gastrointestinal disease, including hemorrhagic colitis and hemolytic uremic syndrome. Human infection typically occurs through consuming contaminated food products, such as undercooked ground beef, raw vegetables, or unpasteurized milk or juice. It can also occur via direct contact with infected animals or fecal-contaminated environments (Nataro and Kaper 1998; Kaper et al. 2004; Gatsios et al. 2021). Although they are taxonomically distinct, *Shigella* and *E. coli* share a high degree of genetic similarity. Comparative studies have shown that *E. coli* is more closely related to *Shigella* than to other species within its genus. However, *Shigella* is differentiated from *E. coli* by key physiological traits. For example, *Shigella* is unable to ferment lactose or decarboxylate lysine. Additionally, *Shigella* has distinct pathogenic mechanisms and serological profiles (Hale and Keusch 1996; Doern 2023). Clinically, *Shigella* causes shigellosis, a diarrheal disease often marked by fever, cramps, and blood in the stool. It is typically transmitted via the fecal–oral route (Hale and Keusch 1996; Kotloff et al. 2018). *Salmonella* species are well-known foodborne pathogens that can infect humans and animals, including mammals, birds, reptiles, and amphibians. In humans, *Salmonella* infections, collectively termed salmonellosis, can lead to various clinical conditions, including enteric fever, gastroenteritis, bloodstream infections (septicemia), localized infections, and asymptomatic carriage, particularly with typhoidal strains (Agbaje et al. 2011; Billah and Rahmen 2024; Lamichhane et al. 2024).

The genus *Enterobacter* is composed of motile organisms that typically possess peritrichous flagella. Members of this genus are widely distributed in nature and are often found in soil, water, and plant surfaces, where they may act as endophytes or phytopathogens. Additionally, *Enterobacter spp.* are part of the normal gut microbiota in both humans and animals (Davin-Regli et al. 2019). While most species are considered commensal or environmental, certain taxa, particularly those within the *Enterobacter cloacae* complex (ECC), which comprises at least 22 phylogenetically defined groups based on average nucleotide identity, have been associated with hospital-acquired infections and nosocomial outbreaks (Sutton et al. 2018). The closely related genus *Klebsiella* within the *Enterobacteriaceae* family comprises clinically significant species and is characterized by distinct phenotypic traits, including a prominent capsule and lack of motility. *Klebsiella pneumoniae* is a significant opportunistic pathogen among its members. It commonly colonizes the human gastrointestinal tract and oropharyngeal mucosa, leading to hospital-acquired infections worldwide, such as pneumonia and urinary tract infections. The increasing resistance of *K. pneumoniae* to multiple antibiotics makes this species a critical priority pathogen (Holt et al. 2015; Dong et al. 2022). The genus *Serratia,* belonging to the family *Yersiniaceae,* shares clinical characteristics with other *Enterobacterales*. It is particularly noted for its opportunistic behavior in healthcare settings. *S. marcescens* is the most clinically significant species in the genus and is responsible for nosocomial infections, which are frequently linked to contaminated respiratory equipment and inadequate catheter management (Williamson et al. 2006). *S. marcescens* strains often produce prodigiosin, a distinctive red pigment which serves as a phenotypic identification marker (Khanna et al. 2013).

The order *Pseudomonadales* predominantly consists of aerobic chemoorganotrophs, most of which are motile by one or more polar flagella. Many species can degrade a variety of organic compounds, facilitating nutrient cycling and bioremediation; others are opportunistic pathogens. These metabolically versatile bacteria inhabit diverse environments such as soil, water, plants, and animal hosts (Brenner et al. 2005; Flores-Félix et al. 2025). There are two major families within the order: *Pseudomonadaceae* (heterotypic synonym *Azotobacteraceae*) and *Moraxellaceae*. The type genus, *Pseudomonas*, comprises species of ecological, agricultural, and clinical relevance, including *Pseudomonas aeruginosa*, an opportunistic human pathogen; *P. syringae*, a plant pathogen responsible for frost damage; *P. stutzeri*, a denitrifying soil bacterium; and *P. putida* and *P. fluorescens*, which are common inhabitants of the rhizosphere known for promoting plant growth and for their application in bioremediation (Lamichhane et al. 2014; Tuon et al. 2022; Alsukaibi et al. 2023; Mehmood et al. 2023; Flores-Félix et al. 2025). Additionally, related genera such as *Azotobacter* contribute to nitrogen cycling by fixing atmospheric nitrogen (Young and Park 2007). *P. aeruginosa*, the type species of its genus, is a model organism for studying fundamental bacterial processes, including gene regulation, quorum sensing, antimicrobial resistance, and biofilm formation. This metabolically versatile bacterium exhibits characteristic green-blue pigmentation due to its production of both pyocyanin (blue) and pyoverdine (yellow-green). Its genomic plasticity enables it to adapt to diverse environments and transition from a saprophytic environmental organism to an opportunistic pathogen capable of infecting multiple hosts (Letizia et al. 2025). *P. aeruginosa* causes complex infections in clinical settings, including ventilator-associated pneumonia, urinary tract infections, bloodstream infections, and chronic conditions such as cystic fibrosis. These clinical challenges result from its combined traits of adaptive virulence mechanisms, multidrug resistance, and robust biofilm formation (De Sousa et al. 2021; Tuon et al. 2022; Letizia et al. 2025).

The *Vibrionales* are straight or curved, facultatively aerobic rods that are fermentative and mostly oxidase-positive. This distinguishes them from the oxidase-negative *Enterobacterales* and the non-fermentative, oxidase-positive *Pseudomonas* species (Brenner et al. 2005). The type genus of the order, Vibrio, naturally inhabits freshwater, estuarine, and marine ecosystems, particularly favoring warm, brackish conditions. The species in this genus share a characteristic two-chromosome genome architecture, shaped by recombination and horizontal gene transfer. Clinically relevant species, such as *V. cholerae*, *V. parahaemolyticus*, and *V. vulnificus*, cause a significant proportion of human infections linked to aquatic environments and seafood. Case numbers typically peak during warmer seasons. *V. cholerae* is the causative agent of cholera outbreaks by transmitting through contaminated water or food, resulting in severe watery diarrhea and acute dehydration (Baker-Austin et al. 2018).

2.2.4 Deltaproteobacteria and Epsilonproteobacteria

The class *Deltaproteobacteria* includes a diverse group of bacteria that are predominantly mesophilic and anaerobic, with considerable metabolic and ecological diversity. This class comprises several orders: *Myxococcales*, *Desulfovibrionales*, *Desulfuromonadales*, *Desulfobacterales*, and *Syntrophobacterales*. A defining feature of this class is its role in dissimilatory sulfate reduction. Beyond this, members of the *Geobacteraceae* and *Desulfuromonadaceae* families are capable of extracellular electron transfer. They reduce metals, such as iron, by using external solid surfaces as terminal electron acceptors. This process can generate an electric current. Unlike the strictly anaerobic lifestyle of many members, members of the order *Myxococcales* display distinctive ecological strategies, demonstrating gliding motility, sporulation, and a complex form of social behavior associated with aerobic predation. During their active phase, they swarm around prey and release enzymes to break down target cells in a coordinated, pack-like manner. Interestingly, the order *Syntrophobacterales* is predicted to have reduced metabolic capabilities compared to other orders. However, some members of the order are involved in syntrophic metabolism, degrading aromatic compounds such as benzoate and potentially benzene in cooperation with hydrogen-consuming microorganisms. These organisms play a key role in anaerobic environments by producing intermediate compounds, such as fermentation products, hydrogen, and formate, that support methanogenesis (Waite et al. 2020; Langwig et al. 2022).

Epsilonproteobacteria is a metabolically versatile class and recognized for its pathogenic genera, *Campylobacter* and *Helicobacter*, which are commonly associated with both animal and human hosts. *Campylobacter jejuni* is a leading cause of human diarrhea worldwide, though it exists harmlessly in animals, particularly poultry. *Helicobacter pylori* colonizes the human stomach in approximately 50% of the population and can induce serious gastric pathologies, including cancer. Both pathogens primarily inhabit the mucosal layers of their hosts, where flagellar motility is critical for colonization, enabling movement to microniches conducive to growth

(Lertsethtakarn et al. 2011; Waite et al. 2017). This class is renowned not only for its pathogenic members but also for its ecological significance in sulfur-rich environments. Many species have unique motility adaptations, such as polar flagella, which enable them to navigate dynamic habitats, such as hydrothermal vents and sulfide-rich sediments. These bacteria predominantly oxidize reduced sulfur compounds (e.g., thiosulfate, sulfide) or hydrogen. They couple these reactions with nitrate/nitrite reduction to sustain chemolithoautotrophic growth (Waite et al. 2017).

2.3 Bacillota, Mycoplasmatota, and Actinomycetota

2.3.1 Bacillota

The *Bacillota* phylum, formerly known as *Firmicutes*, comprises mainly Gram-positive bacterial groups characterized by low G+C content and containing a monophyletic lineage (Argentino 2025; Oren and Garrity 2021). Phylogenetic and evolutionary studies suggest that the monoderm cell wall of Gram-positive members in this phylum evolved from the diderm cell envelope of Gram-negative ancestors through lipopolysaccharide outer membrane loss events (Choi et al. 2024). Members of *Bacillota* exhibit diverse morphologies, including rod-shaped and coccoid forms, and many are capable of forming endospores, highly resistant dormant structures that enable survival under unfavorable environmental conditions (Argentino 2025; Madigan et al. 2022). *Bacillota* includes a wide range of chemoorganotrophic bacteria, many of which produce lactic acid, ethanol, and butyrate as end products of carbohydrate fermentation. Their metabolic versatility enables members to be obligate anaerobes, facultative anaerobes, or strict aerobes, allowing them to inhabit diverse environments, such as soil, sediments, and the gastrointestinal tracts of mammals (Vos et al. 2009; Madigan et al. 2022). Ecologically, *Bacillota* members play critical roles in host-microbe interactions and nutrient cycling; for example, *Bacillus* species contribute to organic matter decomposition and nitrogen cycling (Madigan et al. 2022).

Beyond their ecological importance, *Bacillota* species hold substantial industrial relevance due to their ability to produce enzymes, antibiotics, and fermented food products. Conversely, some members act as opportunistic or obligate pathogens, causing diseases such as botulism and food poisoning (Vos et al. 2009; Hashmi et al. 2020).

The *Bacillota* phylum comprises nine recognized classes: *Bacilli, Clostridia, Culicoidibacteria, Erysipelotrichia, Negativicutes, Limnochordia, Syntrophomonadia, Thermolithobacteria*, and *Thermaerobacteria*, reflecting its considerable phylogenetic and phenotypic diversity. Genomic analyses identified conserved features across the phylum, most notably the presence of sporulation gene clusters, which serve as a defining signature. In particular, the classes *Bacilli, Clostridia, Erysipelotrichia*, and *Negativicutes* share a core sporulation gene cluster. However, many species that inhabit nutrient-rich environments have lost the ability to sporulate, evolving into non-spore-forming taxa such as *Staphylococcus, Listeria, Lactobacillus, and Ruminococcus*.

This pattern suggests that sporulation ability in *Bacillota* is shaped by ecological lifestyle and that the phylum descends from a common ancestor whose diversification was driven by genetic specializations, including both gene loss and acquisition (Argentino et al. 2025; Galperin 2013). Among *Bacillota*, the classes *Bacilli* (typically aerobic or facultative aerobes) and *Clostridia* (strict anaerobes) are the most extensively studied. Comparative genomic analyses using *Bacillus subtilis* and *Clostridioides difficile* as model organisms revealed approximately 60 genes conserved among all spore-forming members of *Bacillota* (Galperin et al. 2022). Notably, the composition of sporulation gene clusters varies among classes. For example, *Clostridia* and certain other groups lack many sporulation genes present in *Bacillus* lineages, suggesting that evolutionary restructuring has occurred in response to adaptation to specific ecological niches (Galperin et al. 2022; Ramos-Silva et al. 2019).

The class *Bacilli* comprises two phylogenetically and phenotypically distinct orders: *Caryophanales (formerly Bacillales)* and *Lactobacillales*, each defined by characteristic genomic and morphological traits (Li et al. 2024). *Caryophanales* exhibits extensive phenotypic diversity, with members displaying rod-shaped, coccoid, filamentous, or trichome-forming morphologies. Many are motile via flagella or gliding, while others are nonmotile. These bacteria are globally distributed and inhabit a broad range of environments, including soil, water, plants, animals, and extreme environments. Ecologically, *Caryophanales* members contribute to organic matter decomposition and nutrient cycling, often acting as saprophytes (Harirchi et al. 2022). The order *Caryophanales* comprises five families: the spore-forming *Bacillaceae, Paenibacillaceae*, and *Caryophanaceae*, and the non-spore-forming *Listeriaceae* and *Staphylococcaceae* (Harirchi et al. 2022). Sporulation follows the canonical pathway described in *Bacillus subtilis* and is regulated by phosphorylation of Spo0A and sequential activation of division-specific sigma factors σ^F, σ^E, σ^G, and σ^K. Families that have lost sporulation ability have adapted alternative ecological strategies such as rapid growth, biofilm formation, and host association (Galperin et al. 2012; Harirchi et al. 2022; Li et al. 2024).

Lactobacillales are Gram-positive, non-sporulating, and catalase-negative bacteria that primarily produce lactic acid via carbohydrate fermentation (De Angelis and Gobbetti 2016). Homofermentative species produce lactic acid as the sole end product, whereas heterofermentative species also generate ethanol and/or CO2 (Ludfiani et al. 2021). This order includes five families: *Aerococcaceae, Carnobacteriaceae, Enterococcaceae, Lactobacillaceae*, and *Streptococcaceae*. Members play key roles in food fermentation, human and animal health, and probiotic applications (Madigan et al. 2022). Morphologically, *Lactobacillus* genera are rod-shaped, while *Streptococcus* and *Enterococcus* genera appear as cocci or chain-shaped (Ludfiani et al. 2021). Lactic acid bacteria lack porphyrins and cytochromes, relying solely on substrate-level phosphorylation for ATP production. Despite being anaerobes, they are aerotolerant, capable of growing in the presence of oxygen (Madigan et al. 2022). Ecologically, they are adapted to nutrient-rich environments and depend on external sources for essential amino acids, vitamins, and growth factors due to their reduced biosynthetic capacity (Madigan et al. 2022; Duar et al. 2017).

The class *Clostridia* comprises strictly anaerobic bacteria that inhabit diverse environments, including soil, river sediments, and the intestinal tracts of animals and humans (Candeliere et al. 2023). Morphologically, members are typically Gram-positive, spore-forming, rod-shaped (bacilli) bacteria; however, some pleomorphic species exhibit bacilli of varying shapes and lengths even within a single strain (Tiffany et al. 2021; Johnson 2009). Owing to their obligate anaerobic metabolism, oxygen is toxic to *Clostridia*, and they can proliferate under anaerobic conditions (Johnson 2009). Ecologically, *Clostridia* play essential roles in both natural ecosystems and host-associated microbiomes. In the human gut, they constitute a substantial fraction of the phylum *Bacillota*, where they contribute to the fermentation of undigested proteins and carbohydrates. This process results in the production of short-chain fatty acids such as propionate and butyrate, which are known to promote intestinal health and support metabolic regulation. Conversely, some species are capable of generating potentially harmful metabolites through protein catabolism, which may negatively impact host health (Candeliere et al. 2023; Tiffany et al. 2021).

The phylum *Bacillota* comprises multiple classes with distinct physiological, morphological, and ecological characteristics. While phylogenetically affiliated with Gram-positive bacteria, the class *Negativicutes* is unique in retaining a Gram-negative-type outer membrane. This diderm cell membrane, characteristic of classical Gram-negative taxa such as *Proteobacteria*, is thought to have been acquired through lateral gene transfer events (Choi et al. 2024).

Thermophilic lineages within *Bacillota* are represented by the classes *Thermolithobacteria* and *Thermaerobacteria*, which inhabit strictly anaerobic, high-temperature environments (Sokolova et al. 2007; Whitman et al. 2015; Vos et al. 2009). Other lineages include *Erysipelotrichia, Culicoidibacteria, Limnochordia*, and *Syntrophomonadia*, each with varying ecological roles and degrees of characterization. *Erysipelotrichia* are Gram-positive bacteria primarily associated with the intestinal tracts of animals, where they contribute to fermentation processes and modulate host immune functions. *Culicoidibacteria* represent a more recently described class whose genomic and physiological traits remain incompletely characterized. *Limnochordia* consists of poorly studied anaerobes inhabiting sedimentary environments. In contrast, *Syntrophomonadia* comprises obligate anaerobes engaged in syntrophic fatty acid degradation, playing key roles in carbon turnover within methanogenic ecosystems (Whitman et al. 2015; Vos et al. 2009; Tiffany et al. 2021; Santinello et al. 2024).

2.3.2 *Mycoplasmatota*

The *Mycoplasmatota* phylum contains a single class, *Mollicutes*, which was formerly classified under the phylum *Tenericutes*. Advances in genome-wide phylogenetic analyses have led to its reclassification as a distinct lineage within *Mycoplasmatota* (Gupta et al. 2018; Oren and Garrity 2021; Zheng et al. 2021). Members of this unusual class are characterized by the absence of a cell wall, extreme genome reduction, and pleomorphic morphology. Their cell membranes are enriched with sterols obtained from their host or the environment, which confer and enable adaptation to a wide range of ecological niches as obligate parasites, commensals, or, in rare cases, free-living saprophytes (May et al. 2014).

The *Mollicutes* possess small cell sizes, low genomic G+C content (typically 25–30% mol%), and highly reduced genomes ranging from 0.5 to 1.5 Mb. Phylogenetic evidence indicates that they evolved from the *Bacillota* through a deep reductive evolutionary process, giving rise to their unique taxonomic position (Faburay and McVey 2022; Gupta et al. 2018). Due to their lack of a cell wall, *Mycoplasmatota* are inherently resistant to cell wall-targeting antibiotics and exhibit high structural flexibility. As metabolically limited chemoorganotrophs, they depend heavily on host-derived nutrients. Members include facultative anaerobes and microaerophiles capable of colonizing mucosal surfaces, such as the respiratory and urogenital tracts. Many are significant pathogens of humans and animals, causing diseases of clinical and veterinary importance (Madigan et al. 2022).

Based on shared genome sets, distinct metabolic profiles, and ecological specializations revealed through molecular analyses, the class *Mollicutes* is subdivided into five orders: *Acholeplasmatales*, *Anaeroplasmatales*, *Haloplasmatales*, *Mycoplasmatales*, and *Mycoplasmoidales* (Gupta et al. 2018; Yan et al. 2024). Species with diverse ecological niches and physiological traits are grouped within *Acholeplasmatales* and *Anaeroplasmatales*, whereas halophilic taxa adapted to saline environments are classified under *Haloplasmatales*, which contains numerous medically important species. The order *Mycoplasmoidales* was established to accommodate species formerly assigned to *Mycoplasma*, such as *Mycoplasmoides pneumoniae* (formerly *Mycoplasma pneumoniae)*, *Mycoplasmoides pirum* (formerly *Mycoplasma pirum)*, and *Mycoplasmoides genitalium* (formerly *Mycoplasma genitalium)*, that were shown through molecular phylogenetic analyses to form distinct and coherent clades (Gupta et al. 2018).

The order *Mycoplasmatales* includes two families: *Mycoplasmataceae* (synonym *Entomoplasmataceae*) and *Spiroplasmataceae*. *Mycoplasmataceae* includes five genera: *Mycoplasma*, *Haemobartonella*, *Mesoplasma* (synonym *Edwardiiplasma*), *Entomoplasma*, and *Williamsoniiplasma*. Members of the genus *Mycoplasma* are notable for their extremely small cell size, strict requirement for exogenous sterols, and characteristic "fried egg" colony morphology on solid media (Madigan et al. 2022).

The family *Spiroplasmataceae* is composed of helical or spiral-shaped *Mollicutes* that lack flagella but display motility via contractile intracellular fibrils. This distinctive morphology and unique genome signature set them apart from symbionts or pathogens, inhabiting ticks, the hemolymph and gut of insects, vascular fluids of plants, and the digestive tracts of plant-feeding insects. Notable plant pathogens include *Spiroplasma citri*, the causative agent of citrus stubborn disease, and *Spiroplasma kunkelii*, responsible for maize stunting, both of which lead to significant agricultural losses worldwide (Madigan et al. 2022; Harne et al. 2020; May et al. 2014).

The order *Acholeplasmatale*s comprises a single family, *Acholeplasmataceae*, which includes the genera *Acholeplasma, Mariniplasma, Peloplasma*, and *Candidatus Phytoplasma*. *Phytoplasma* species are wall-less obligate symbionts associated exclusively with plants and insects, and they cannot be cultivated under axenic laboratory conditions. These bacteria are significant plant pathogens, causing economically important diseases such as aster yellows, coconut lethal yellowing, and pear decline (Bertaccini et al. 2022; Firrao et al. 2004). Transmission typically occurs via phloem-feeding insects, primarily leafhoppers and planthoppers, which serve as both hosts and vectors.

The genus *Acholeplasma* contains free-living species, including *Acholeplasma laidlawii*, which is notable for its lack of sterol requirement. Phylogenomic analyses indicate that *Acholeplasma* is paraphyletically positioned within the core of the *Mycoplasmatales* clade, a placement consistent with its fundamental physiology and broad environmental adaptability (Lazarev et al. 2011).

The order *Anaeroplasmatales* constitutes a component of the gut microbiota. Their presence has been associated with various immune responses, including elevated levels of Immunoglobulin A (IgA) in both feces and the intestinal lumen, suggesting a potential role in modulating intestinal immune functions. These strictly anaerobic microorganisms thrive in oxygen-deprived environments, consistent with their integration into the broader anaerobic community of the gut microbiome. Although the precise mechanistic contributions of *Anaeroplasma* species remain to be fully elucidated, emerging evidence links their occurrence to shifts in microbiota composition that may be associated with both health and disease states (von Goetze 2024; Whitman et al. 2015; Aivelo and Norber 2018).

2.3.3 *Actinomycetota*

Actinomycetota, formerly known as *Actinobacteria*, are a diverse phylum of Gram-positive, high G+C bacteria recognized for their phylogenetic breadth and ecological significance. Initially classified through 16S rRNA gene analyses, they represent one of the largest bacterial lineages (Barka et al. 2016; Embley and Stackebrandt 1994; Mohammadipanah and Dehhaghi 2017). Their morphology spans from unicellular genera such as *Micrococcus*, *Arthrobacter* to filamentous forms like *Streptomyces* and *Frankia*, many of which develop branching substrate mycelia for nutrient uptake and aerial hyphae for spore production (Barka et al. 2016; Hamedi and Poorinmohammad 2017). Spore characteristics, shape, arrangement, and surface texture are key taxonomic features. Non-mycelial taxa such as *Corynebacterium* and *Mycobacterium* may also produce short apical filaments. This morphological plasticity, combined with pigmentation and specialized cellular structures, supports their ecological adaptability and biosynthetic potential (Barka et al. 2016; Hamedi and Poorinmohammad 2017; Uesugi et al. 2024). Ecologically, *Actinomycetota* occupy diverse niches across terrestrial, aquatic, and extreme environments. Soil-dwelling species such as *Streptomyces* function as major saprophytes in alkaline, organic-rich soils, contributing to decomposition, nitrogen fixation, nutrient solubilization, siderophore production, and biocontrol through secondary metabolites (Aguilar et al. 2024; Helmi 2025; Lewin et al. 2016). Their ability to form spores and produce stress-adaptive compounds enables survival in deserts, caves, hot springs, volcanic habitats, and polar ecosystems (Helmi 2025).

These traits highlight their importance in biogeochemical cycling and their biotechnological value as prolific producers of antibiotics and other bioactive compounds (Lewin et al. 2016). The phylum is currently divided into six classes, *Acidimicrobiia, Actinomycetes (syn. Arthrobacteria), Coriobacteriia, Nitriliruptoria, Rubrobacteria*, and *Thermoleophilia*, distinguished by their ecological distribution, metabolic specializations, and molecular traits. Ongoing genomic and ecological research continues to expand understanding of their functional diversity and biotechnological potential.

The class *Actinomycetes* comprises aerobic, spore-forming, Gram-positive bacteria with a high genomic G+C content (57–75%). They characteristically produce both substrate and aerial mycelia, exhibiting filamentous growth reminiscent of fungi, including authentic aerial hyphae. They exhibit filamentous characteristics similar to fungi and possess authentic aerial hyphae. *Actinomycetes* are ubiquitous and form stable, persistent populations in diverse environments, particularly dominating dry, alkaline soils. Their life cycles, highly distinctive among prokaryotes, enable efficient decomposition of organic material and contribute fundamentally to soil nutrient cycling. *Actinomycetes* are biotechnologically and medically significant microorganisms. They produce a vast array of secondary metabolites with pharmaceutical importance, including antibiotics, antifungals, antivirals, antiprotozoals, anticancer agents, and immunosuppressants. Notably, many frontline antibiotics, such as streptomycin, gentamicin, rifamycin, and erythromycin, originate exclusively from actinomycetes. Their contributions extend beyond medicine into agriculture, where their metabolites provide biocontrol and plant health benefits (Bhatti et al. 2017).

Taxonomically, the class *Actinomycetes* comprises of 21 orders, including *Actinomycetales, Bifidobacteriales, Kineosporiales, Acidothermales, Streptosporangiale, Sporichthyales, Pseudonocardiales, Propionibacteriales, Nakamurellales, Mycobacteriales, Motilibacterales, Micromonosporales, Micrococcales, Kitasatosporales, Jiangellales, Jatrophihabitantales, Geodermatophilales, Glycomycetales, Frankiales, Cryptosporangiales*, and *Catenulisporales*. Among these, *Actinomycetales* and *Bifidobacteriale* are particularly prominent. *Actinomycetales*, predominantly soil dwellers, are prolific producers of antibiotics and bioactive metabolites. The genus *Streptomyces* is especially notable for its crucial role in addressing multidrug-resistant pathogens and driving biotechnological innovation. *Bifidobacteriales*, represented chiefly by *Bifidobacterium*, are dominant commensals of mammalian gastrointestinal tracts. They support health through immune modulation, maintenance of gut homeostasis, and disease prevention. Their probiotic applications include reducing inflammation, strengthening epithelial barrier function, and mitigating infectious diseases. Together, these groups exemplify the dual ecological and biomedical importance of Actinomycetes, highlighting their roles in both sustaining ecosystems and supporting human health (Barka et al. 2016; O'Callaghan et al. 2016).

The class *Acidimicrobiia* represents a widely distributed and ecologically abundant lineage of deep-sea *Actinobacteria*. Most members remain uncultivable, and thus, their phylogenetic placement is primarily inferred from 16S rRNA sequence analyses of the few cultivable representatives. Taxonomically, the class comprises a single order, *Acidimicrobiales*, which includes the families *Acidimicrobiaceae*,

Iamiaceae, *Ilumatobacteraceae*, and *Salinilacustritrichaceae*. Members of this lineage predominantly inhabit diverse aquatic environments, ranging from marine systems to acidic mine waters. Species such as *Acidimicrobium ferrooxidans* and *Ferrithrix thermotolerans* (family *Acidimicrobiaceae*) are highly adapted to extreme acidic conditions, including acidic mine drainage and geothermal ecosystems. In contrast, taxa belonging to families *Iamiaceae* and *Ilumatobacteraceae* thrive in neutral to mildly alkaline aquatic environments. Notably, all species within *Iamiaceae*, with the exception of the human skin isolate *Dermatobacter hominis*, were recovered from aquatic habitats, highlighting their strong ecological association with water-based systems (Hu et al. 2018; Huang et al. 2025).

The class *Thermoleophilia* is widely distributed, with members predominantly inhabiting hot springs and soil ecosystems. As the majority of representatives remain uncultivable, phylogenetic and taxonomic insights are largely derived from 16S rRNA gene sequence analyses of a limited number of isolates. This class currently comprises four orders: *Solirubrobacterales, Gaiellales, Miltoncostaeales*, and *Thermoleophilales*. Phylogenetic analyses indicate that *Gaiellales* and *Miltoncostaeales*, primarily associated with marine habitats, form parallel clades that are phylogenetically distinct from the soil- and thermophilic-associated clades represented by *Solirubrobacterales* and *Thermoleophilales*. Members of *Solirubrobacterales* are predominantly recovered from soil-related environments, including activated sludge and rhizosphere soils, and exhibit marked phenotypic diversity and ecological adaptability. In contrast, *Thermoleophilale* encompasses strictly aerobic, thermophilic rod-shaped bacteria with a characteristic metabolic restriction to *n*-alkanes. Due to this distinctive adaptation, they are often described as "heat- and oil-loving" microorganisms, collectively designated as *Thermoleophilum* (Whitman et al. 2015; Hu et al. 2019; Li et al. 2021).

The class *Coriobacteriia* comprises strictly anaerobic, nonmotile, nonsporulating, rod-shaped bacteria that are predominantly associated with the gastrointestinal tract of humans and animals. The class currently includes three orders: *Coriobacteriales, Anaerosomatales*, and *Eggerthellales*. Taxonomic assignments within *Coriobacteriia* remain partly unresolved, posing challenges for the interpretation of 16S rRNA gene-based diversity analyses in both human and rodent microbiome studies (Gupta et al. 2013; Whitman et al. 2015; Hoyles 2019). Members of *Coriobacteriales* and *Atopobiaceae* (within *Coriobacteriales*) primarily colonize the human intestinal tract, where they display saccharolytic activity, contributing to carbohydrate fermentation and host metabolic interactions. In contrast, *Eggerthellaceae* (order *Eggerthellales*) are predominantly found in the rodent gut and are generally *asaccharolytic*, reflecting distinct metabolic profiles within the class. These differences highlight the functional heterogeneity of *Coriobacteriia* in host-associated microbiota. The order *Anaerosomatales* is distinguished from its sibling lineages by its ecological distribution, which spans deep subsurface environments, terrestrial mud volcanoes, and oil-bearing habitats. Its members are characterized as small, obligately anaerobic rods that lack motility but produce numerous pili-like surface extensions, which may contribute to adhesion and environmental persistence (Khomyakova et al. 2022). Collectively, *Coriobacteriia* exhibit a broad ecological spectrum ranging from host-associated intestinal niches to extreme anaerobic environments, underscoring their taxonomic complexity, metabolic diversity, and ecological adaptability.

The class *Nitriliruptoria* represents a distinct lineage within the phylum *Actinomycetota*, characterized by unique biochemical, evolutionary, and phenotypic features. Phylogenetic reconstruction based on 16S rRNA sequence analyses, supported by chemotaxonomic markers, reveals six orders: *Egicoccales, Euzebyales, Nitriliruptorales, Salsipaludibacterales*, and *Stomatohabitantales*. Ecologically, *Nitriliruptoria* thrive in extreme saline and alkaline habitats, particularly haloalkali and soda lakes, where their metabolic flexibility is linked to the ability to use aliphatic nitriles as growth substrates. Chemotaxonomically, their respiratory systems are dominated by menaquinones MK-9(H4), a reliable marker for the lineage. Structural variations in the side-chain saturation and length of these quinones reflect adaptive mechanisms to haloalkaline conditions. These features underscore both the taxonomic distinctiveness and environmental resilience of the class (Chen et al. 2020; Stackebrandt and Otten 2014; Liu et al. 2023).

The class *Rubrobacter* comprises Gram-positive, non-motile, non-spore-forming actinobacteria with morphologies ranging from pleomorphic rods to coccoid forms. It contains a single order, *Rubrobacterales*, whose members are notable for their extremophilic adaptations, including halotolerance, thermotolerance, thermophily, and extraordinary resistance to UV and gamma radiation. Ecologically, *Rubrobacter* species are primarily associated with arid soils and radiation-exposed environments. They synthesize red or pink carotenoid pigments that likely confer photoprotection (Saito et al. 1994; Basile et al. 2025), while intracellular accumulation of polyhydroxyalkanoates (PHAs) enhances drought and thermal resilience (Kouřilová et al. 2021; Egas et al. 2015). Genomic studies further reveal genes encoding for light and hydrogen utilization, oxidative stress defense, and highly efficient DNA repair systems, which collectively support survival under extreme environmental stress (Meier et al. 2021). Collectively, *Rubrobacter* exemplifies the remarkable adaptability of *Actinomycetota* to polyextreme conditions, with ecological and biotechnological significance in understanding stress tolerance and novel metabolic strategies.

2.4 Bacteroidota

The *Bacteroidota* phylum contains six classes, which are *Bacteroidia*, *Chitinophagia*, *Cytophagia*, *Flavobacteriia*, *Saprospiria*, and *Sphingobacteriia* (García-López et al. 2019). Bacteria that are part of this phylum are gram-negative, nonsporulating, and rod-shaped. Metabolically, *Bacteroidota* are typically saccharolytic, fermenting sugars or proteins to produce acetate or succinate as their primary fermentation products. They exhibit diverse respiratory strategies, encompassing obligate aerobes, facultative aerobes, and obligate anaerobes (Krieg et al. 2010; Hahnke et al. 2016). While many taxa display gliding motility, others are nonmotile or motile via flagella. Ecologically, *Bacteroidota* are highly versatile, inhabiting terrestrial and aquatic ecosystems across temperate, tropical, and polar regions, as well as forming a key component of the animal gut microbiota (Hahnke et al. 2016). This ecological breadth, combined with metabolic flexibility, underscores their importance in nutrient cycling, organic matter degradation, and host-associated symbioses.

2.4.1 *Bacteroidia*

Most of the species that constitute the *Bacteroidia* class are obligately anaerobic fermenters. The key genus *Bacteroides* comprises predominantly commensal bacteria that form a substantial proportion of the human intestinal microflora. While usually beneficial, some species may act as opportunistic pathogens, occasionally causing bacteremia and other infections in humans (Krieg et al. 2010). One notable species, *Bacteroides thetaiotaomicron*, is a central component of the human gut microbiome. Its genome encodes an exceptionally large and diverse repertoire of carbohydrate-active enzymes, enabling the degradation of a wide variety of plant-derived polymers that humans cannot digest on their own. This metabolic flexibility enhances nutrient availability and contributes to host health (Krieg et al. 2010). A recent study highlighted *Bacteroides caecimuris* and *Bacteroides muris* as promising next-generation probiotic candidates. These species exhibit quorum-quenching activities, biofilm inhibition, and bile salt hydrolase activity, suggesting potential therapeutic benefits in regulating gut microbiota and preventing pathogen colonization (Ladewig et al. 2025).

A distinctive biochemical feature of *Bacteroides* is their ability to synthesize sphingolipids, molecules common in mammalian tissues but rare among bacteria (Krieg et al. 2010). Evidence suggests that Bacteroides-derived sphingolipids in the gut are inversely associated with inflammation and may modulate the host's sphingolipid metabolism, further underscoring their functional importance in maintaining gut homeostasis (Brown et al. 2025).

2.4.2 *Cytophagia*

The class *Cytophagia* is primarily composed of obligate aerobes, although a few species exhibit limited fermentative capacity. Members are widely distributed in soils and freshwater environments, including habitats with toxic or recalcitrant organic matter. Morphologically, cells are generally long, slender, and pointed rods that move by gliding motility, a common feature across the class. A notable exception is the genus *Sporocytophaga*, which forms resting spherical microcysts, enabling survival under unfavorable conditions. Metabolically, *Cytophagia* species specialize in the degradation of complex polysaccharides such as cellulose, chitin, and agar. Most species secrete extracellular enzymes that hydrolyze insoluble polysaccharides into soluble products for cellular uptake (Krieg et al. 2010). However, *Cytophaga hutchinsonii* employs a unique strategy by expressing cell-surface-associated cellulases, allowing direct degradation of cellulose fibers without relying on secreted enzymes

(Gao et al. 2021). Ecologically, members of this class play a central role in carbon cycling, particularly in the breakdown of plant-derived polymers, thereby contributing significantly to nutrient turnover in soil and aquatic ecosystems.

2.4.3 Flavobacteriia, Sphingobacteriia, Chitinophagia, and Saprospiria

The class *Flavobacteriia* consists of facultatively aerobic chemoorganotrophs, with some species exhibiting limited fermentative capacity. A characteristic feature shared with other related classes (e.g., Sphingobacteriia, Chitinophagia, and Saprospiria) is the production of yellow pigments (Krieg et al. 2010). Ecologically, members of this class are widespread in aquatic environments, including marine waters, polar ecosystems, and freshwater habitats, as well as food and food-processing environments. Morphologically, they are Gram-negative, rod-shaped bacteria, typically nonsporulating. While most are obligate aerobes, some species are capable of nitrate reduction via anaerobic respiration, reflecting metabolic versatility (Krieg et al. 2010). The genus *Flavobacterium* is the key representative of this class. Although generally commensal, pathogenicity has been observed in specific cases. *Flavobacterium meningosepticum* (now *Elizabethkingia meningoseptica*) is associated with infant meningitis (Ratner 1984), and several species are fish pathogens. Notably, *F. columnare* causes columnaris disease, and *F. psychrophilum* causes cold-water disease in fish (Declercq et al. 2013; Park et al. 2023). Other genera, such as *Polaribacter* and *Psychroflexus*, are psychrophilic or psychrotolerant taxa adapted to polar waters and sea ice and are recognized agents of refrigerated food spoilage (Krieg et al. 2010). Overall, Flavobacteriia play important roles in nutrient cycling and food web interactions in aquatic ecosystems, while also including species of medical and veterinary significance.

The class *Sphingobacteriia* is characterized by members containing unusually high concentrations of sphingophospholipids as major cellular lipid components, a distinctive trait among bacteria (Krieg et al. 2010). Ecologically, species of the type genus *Sphingobacterium* are frequently isolated from soil and composted manure, highlighting their role in organic matter decomposition and environmental nutrient cycling. Morphologically, they are Gram-negative, non-sporulating rods, generally facultatively aerobic chemoorganotrophs. From a clinical perspective, some species are recognized as opportunistic pathogens. For example, *Sphingobacterium spiritivorum* has been isolated from blood and urine samples, while *Sphingobacterium mizutaii* was reported from a case of cellulitis in the leg (Nemoto et al. 2019). Metabolically, certain sphingobacteria exhibit the ability to degrade unusual or xenobiotic compounds. For instance, *Sphingobacterium multivorum*, isolated from aerobic sludge, can utilize the herbicide mefenacet and fungicide hexaconazole as sole carbon and energy sources (Ye et al. 2004; An et al. 2020). Overall, *Sphingobacteriia* demonstrate both environmental adaptability and biotechnological potential, while also including species of opportunistic clinical relevance.

The class *Chitinophagia* includes species isolated from diverse environments such as soil, rock surfaces, plants, and freshwater shorelines. Members of the type genus Chitinophaga are chitinolytic bacteria, capable of degrading chitin and forming resting spherical microcysts (Krieg et al. 2010). Ecologically, *Chitinophaga* species contribute to carbon cycling in soils by breaking down fungal glycans, thereby facilitating nutrient turnover and organic matter decomposition. Their chitinolytic and glucanolytic enzyme systems are of particular interest for biomass valorization, bio-refining, and sustainable production of bioactive chito-oligosaccharides (Brinkmann et al. 2022). Clinically, while generally environmental, some members are opportunistic pathogens, as exemplified by *Chitinophaga terrae*, which caused bacteremia in an immunosuppressed patient (Cremet et al. 2009).

The class *Saprospiria* is characterized by members that form long, filamentous cells, including genera such as *Saprospira* and *Aureispira,* which develop helical filaments and exhibit gliding motility. They are typically found in freshwater and marine environments, where they play an ecological role in the degradation of organic matter. Notably, *Saprospira* species are specialized predators, acquiring essential amino acids by preying on other microorganisms such as diatoms, dinoflagellates, and bacteria, breaking down their proteins for nutritional requirements (Krieg et al. 2010). This predatory lifestyle distinguishes Saprospiria from other Bacteroidota classes and highlights their role in microbial community dynamics within aquatic ecosystems

2.5 Chlamydiota, Planctomycetota, and Verrucomicrobiata

The PVC superphylum comprises a monophyletic group of Gram-negative bacteria characterized by distinctive cellular and evolutionary features rarely observed in other prokaryotic lineages (Madigan et al. 2022; Vitorino et al. 2025). Within this clade, *Planctomycetota, Verrucomicrobiota*, and *Chlamydiota* form its phylogenetic and functional core, each of which has notable biotechnological and medical significance (Kallscheuer and Jogler 2021). Members of the PVC superphylum share genetic affinities and exhibit an unusual degree of cellular complexity, including intracellular compartmentalization, atypical modes of cell division, and diverse ecological adaptations (Madigan et al. 2022). This section highlights the biology, phylogeny, and ecological significance of *Chlamydiota, Planctomycetota*, and *Verrucomicrobiata*, with a particular emphasis on the shared features that unify the superphylum as well as the unique traits that distinguish each lineage.

2.5.1 Chlamydiota

Chlamydiaota phylum was historically regarded as a narrow lineage of obligate intracellular human pathogens. However, advances in next-generation sequencing approaches, including amplicon sequencing, single-cell genomics, and metagenomics, have expanded its scope, revealing that *Chlamydiota* represents a phylogenetically and ecologically diverse group with broad biological significance (Collingro et al. 2020; Luu et al. 2023). Members of this phylum have been detected across a wide range of hosts and habitats, including free-living protists, animals, aquatic systems, and other environmental reservoirs (Collingro et al. 2020; Dharamshi et al. 2023). While some species maintain classical chlamydial traits, exemplified by *Chlamydia trachomatis* as a major human pathogen, other lineages exhibit substantial genetic, metabolic, and ecological diversity, reflecting distinct host interactions and survival strategies. Based on molecular phylogenetics, the phylum is currently divided into two major orders: *Chlamydiales,* which includes extensively studied pathogens infecting human and animal cells, with major biomedical relevance, and *Parachlamydiales*, which comprises lineages associated with protists (particularly amoebae) and aquatic ecosystems, underlining their role in environmental microbial networks (Collingro et al. 2020; Dharamshi et al. 2023). Collectively, these findings highlight the evolutionary and ecological breadth of *Chlamydiota*, extending beyond human pathogens to include organisms that play key roles in ecosystem functioning and host–microbe interactions.

Recent research has revised the earlier misconception that the *Chlamydiota* phylum consists solely of obligate parasitic taxa. Several recently described members of the *Parachlamydiales* family demonstrate intracellular symbiosis with their hosts, in which they adopt commensal or even mutualistic lifestyles. Notably, *Parachlamydia acanthamoebae* and *Neochlamydia* spp. protect their amoebal hosts from co-infection by *Legionella pneumophila* and giant viruses, illustrating that some chlamydial lineages can enhance host survival under environmental stress and pathogen pressure (Maita et al. 2018; König et al. 2019; Arthofer et al. 2022).

Chlamydiota species share a biphasic developmental cycle, alternating between two morphologically and functionally distinct forms. The extracellular, infectious elementary body (EB) initiates infection by entering host cells, where it differentiates into a metabolically active reticulate body (RB). RBs replicate via binary fission within membrane-bound inclusions and subsequently revert to EBs, which are released by host cell lysis to infect new cells. The resilience of Ebs to environmental stressors and their ability to disseminate without a vector underpin their highly efficient respiratory transmission strategy (Luu et al. 2023; Madigan et al. 2022). On a cellular level, *Chlamydiota* exhibit several unique features. They lack FtsZ, the tubulin homolog essential for bacterial cytokinesis, and were initially believed to be devoid of peptidoglycan (PG), despite their sensitivity to β-lactam antibiotics, an observation termed the "chlamydial anomaly" (Madigan et al. 2022; Liechti 2021). Recent studies have clarified this paradox, revealing that PG is in fact present but confined to a narrow band at the division septum, synthesized through a minimal MreB/Rod system. In these bacteria, cell shape and division are regulated through coordinated membrane expansion and localized PG remodeling. Despite their reduced PG architecture, *Chlamydiota* sustain efficient intracellular replication via streamlined yet essential division machinery (Liechti 2021). Together, these findings highlight the evolutionary adaptability of *Chlamydiota*, encompassing parasitic, commensal, and mutualistic strategies, while also showcasing their distinctive cell biology and life cycle innovations that distinguish them from other bacterial lineages.

2.5.2 *Planctomycetota*

The *Planctomycetota* phylum, a major lineage of the PVC superphylum, exhibits remarkable cellular and ecological diversity. Members of this phylum are ubiquitous in terrestrial and aquatic environments and play key roles in the global carbon and nitrogen cycles (Kallscheuer and Jogler 2021). Taxonomically, *Planctomycetota* is divided into two established classes, *Planctomycetia* and *Phycisphaerae*, and one candidate class, *Candidatus Brocadiia*. Among these, *Planctomycetia* is the most extensively studied and taxonomically diverse group. Its representatives are found in diverse environments such as aquatic systems, wetlands, soils, and marine sediments. While some members are free-living, others are distinguished by their ability to attach to biotic or abiotic surfaces using stalk-like appendages, facilitating colonization and biofilm formation (Kallscheuer and Jogler 2021). *Physicphaerae*, in contrast, include species adapted to marine, brackish, freshwater sediments, and some terrestrial habitats, notable for their broad repertoire of carbohydrate-active enzymes, highlighting their ecological role in organic matter turnover (Klimek et al. 2024; Lenferink et al. 2024). The candidate class *Candidatus Brocadiia* is particularly unique, comprising obligate anaerobes capable of anaerobic ammonium oxidation (anammox) within specialized compartments (Vitorino et al. 2025). *Planctomycetota* are characterized by unusual modes of reproduction and cell structures. Unlike most bacteria, members of *Planctomycetia* do not divide by binary fission, instead, reproduce through polar budding, where a daughter cell emerges from one end of the mother cell (Fuerst 2017). Distinctive surface structures are also present, including crateriform complexes (crater-like outer membrane depressions associated with pili fibers) and stalks, which are more complex than pili. These features often localize at the opposite pole of cell division and contribute to surface adhesion, rosette formation, and biofilm development (Boedeker et al. 2017; Fuerst 2017). Another hallmark of *Planctomycetota* is their complex internal membrane organization. Cells possess a pirellulosome, a compartment containing the nucleoid and riboplasm. This structure is surrounded by the paryphoplasm, a ribosome-free zone located between the intracytoplasmic membrane (ICM) and the cytoplasmic membrane (CM). In *Candidatus Brocadiia*, the pirellulosome is further specialized to contain the anammoxosome, a membrane-bound compartment where anammox metabolism takes place. The anammoxosome harbors essential enzymes, cytochrome c proteins, and ATP synthase, enabling energy conservation via a proton motive force. Its membrane is enriched in ladderane lipids, which provide tight compartmentalization by preventing the leakage of toxic intermediates into the cytoplasm (Kartal et al. 2008; Fuerst 2017; Madigan et al. 2022). In summary, *Planctomycetota* combine complex cellular architecture, diverse reproductive strategies, and specialized metabolic pathways, particularly the anammox process, positioning them as a functionally and evolutionarily significant group within the PVC superphylum.

2.5.3 *Verrucomicrobiota*

The *Verricomicrobiota* phylum represents the final major branch of the PVC superphylum, with considerable ecological and evolutionary significance. Members of this group have been discovered in a wide range of environments, most commonly in freshwater, marine, and soil ecosystems, though their ecological roles remain poorly understood due to difficulties in cultivation and isolation (Chiang et al. 2018). Current taxonomy divides *Verrucomicrobiota* into three classes: *Verrucomicrobiia, Terrimicrobiia*, and *Opitutia. Verrucomicrobiia* includes species inhabiting soils, aquatic environments, and animal gastrointestinal tracts, where they contribute to mucin degradation, host-associated symbiosis, and carbon cycling (Cabello-Yeves et al. 2017; Clark and Mach 2023; Tan et al. 2025). *Terrimicrobiia* are primarily adapted to soil environments, particularly nutrient-poor subsurface and mineral layers. These species specialize in the decomposition of complex organic matter and play an important role in long-term carbon sequestration (Herlemann et al. 2013; Schulze-Makuch et al. 2018). The class *Opitutia* encompasses diverse species distributed across soil, freshwater, peatlands, rhizospheres, activated sludge, wetlands, and marine habitats. Most members are aerobic bacteria with versatile metabolic capabilities, including the degradation of complex biopolymers such as lignin, xylan, and starch, as well as photoheterotrophic metabolism (Islam et al. 2025). Structurally, *Verrucomicrobiota* show intriguing cellular features. Some members, such as *Verrucomicrobium spinosum, Prosthecobacter dejongeii*, and *Chthoniobacter flavus*, exhibit internal membrane structures analogous to the pirellulosome of *Planctomycetota*, though generally less complex (Lee et al. 2009). Unlike *Planctomycetota*, which display deep internal invaginations, *Verrucomicrobiota* cells often produce prosthecae, projections of cytoplasm and cell wall that extend the surface area (Lee et al. 2009). A remarkable evolutionary feature of *Verrucomicrobiota* is their association with both prokaryotic-type and eukaryotic-type tubulin homologs. In *Verrucomicrobium spinosum*, cell division relies solely on the bacterial FtsZ protein, whereas *Prosthecobacter* species possess both FtsZ and eukaryotic-like tubulins BtubA and BtubB, representing a unique case of cytoskeletal protein coexistence in bacteria (Fuerst 2019). In summary, *Verrucomicrobiota* combine broad habitat diversity, metabolic versatility, and unusual cellular features, making them ecologically important and evolutionarily distinctive within the PVC superphylum.

2.6 Hyperthermophiles

2.6.1 *Thermotogota*

The *Thermotogota* phylum represents a lineage of predominantly thermophilic and hyperthermophilic bacteria, thriving at temperatures between 60 and 90°C, though some members also tolerate more moderate environments (Bhandari and Gupta 2014).

These organisms inhabit organic-rich, anoxic ecosystems such as hydrothermal vents, terrestrial hot springs, and deep petroleum reservoirs. Phylogenomic studies divide the class *Thermotogae* into four main orders: *Thermotogales, Petrotogales, Kosmotogales*, and *Mesoaciditogales*. Within *Thermotogaceae*, highly thermophilic genera such as *Thermotoga* include well-studied representatives like *T. maritima* and *T. neapolitana*. The family *Fervidobacteriaceae* consists of thermophiles with fermentative and sulfur-reducing capabilities, whereas *Kosmotogaceae* displays broad temperature tolerance, ranging from mesophilic to thermophilic conditions, and includes *Kosmotoga* and *Mesotoga*. *Petrotogaceae* are predominantly isolated from petroleum reservoirs and deep subsurface sediments, while *Mesoaciditogaceae* are characterized as moderately thermophilic and acidophilic taxa. Across the phylum, most species are strict anaerobes, relying on chemoorganotrophic metabolism, with many capable of reducing thiosulfate or elemental sulfur to hydrogen sulfide (Peña-Salinas et al. 2024). Morphologically, *Thermotogota* are rod-shaped, Gram-negative staining cells surrounded by a distinctive sheath-like outer envelope, the "toga". This structure consists of β-barrel proteins and unique lipids forming a bipartite system connected to the inner membrane (Boniface et al. 2009; Sexton et al. 2023). Their peptidoglycan composition is unusual, containing D-lysine, a rare feature among bacteria (Boniface et al. 2009). From an evolutionary perspective, *Thermotogota* genomes reveal extensive lateral gene transfer with both *Bacillota* and Archaea, reflecting adaptations to extreme environments and suggesting evolutionary linkages across domains (Haverkamp et al. 2021). Collectively, their thermophily, metabolic diversity, and distinctive cell architecture highlight *Thermotogota* as a unique lineage of hyperthermophilic bacteria with both ecological and evolutionary significance.

2.6.2 Thermodesulfobacteria

The *Thermodesulfobacteriota* phylum is composed predominantly of thermophilic anaerobes that inhabit high-temperature environments such as hydrothermal vents, hot springs, and other geothermal ecosystems. Members of this group play a crucial role in the global sulfur cycle through their versatile metabolism of sulfur compounds (Hashimoto et al. 2022). The main recognized class is *Thermodesulfobacteriia*, which includes strict anaerobes specialized in sulfate reduction. In this process, sulfate functions as the terminal electron acceptor and is reduced to hydrogen sulfide during anaerobic respiration (Madigan et al. 2022). Beyond classical sulfate reduction, some genera are capable of sulfur compound disproportionation, whereby a single sulfur species acts simultaneously as electron donor and acceptor, generating both sulfate and hydrogen sulfide without the need for external electron donors. Representative species include *Thermosulfurimonas dismutan* and *Thermosulfurimonas marina*, which metabolize thiosulfate or elemental sulfur via enzymatic pathways involving tetrathionate reductase (TtrBCA) and proteins encoded by the YTD gene cluster, both of which are widely distributed across thermophilic members of the phylum (Hashimoto et al. 2022). A classic example of chemoorganotrophic sulfate reduction is *Thermodesulfobacterium commune*, which oxidizes organic substrates while reducing sulfate (Zeikus et al. 1983). In addition to sulfur metabolism, several *Thermodesulfobacteriota* species are able to use alternative electron acceptors such as nitrate and nitrite, reducing them to ammonium via dissimilatory nitrate reduction pathways (Mardanov et al. 2016).

2.6.3 Aquificota

The *Aquificota* phylum is composed of thermophilic, chemolithoautotrophic, Gram-negative bacteria that thrive in extreme, high-temperature environments such as hot springs, sulfur pools, and deep-sea hydrothermal vents (Power et al. 2024). Members of this phylum are metabolically adapted to exploit hydrogen and sulfur compounds as primary energy sources, thereby playing a significant role in energy flow and carbon fixation within geothermal ecosystems. Within the class *Aquificia*, the order *Aquificales* includes the families *Aquificaceae* and *Hydrogenothermaceae*. These organisms are typically microaerophilic and capable of oxidizing hydrogen, sulfur, or thiosulfate as electron donors (Gupta 2014). The genus *Aquifex* (*Aquificaceae*) represents extreme thermophiles, with species such as *Aquifex aeolicus* growing optimally between 85°C and 95°C. These bacteria oxidize hydrogen and generate water as a byproduct, using oxygen as a terminal electron acceptor (Madigan et al. 2022). The order *Desulfurobacteriales* includes strictly anaerobic species that rely exclusively on hydrogen oxidation for energy metabolism (Gupta 2014). Another lineage, *Thermosulfidibacterales*, is phylogenetically distinct and strongly associated with sulfur metabolism. For instance, *Thermosulfidibacter takaii* exemplifies chemolithoautotrophic sulfate reduction, oxidizing hydrogen while fixing carbon via a reversible tricarboxylic acid (TCA) cycle (Nunoura et al. 2018). Overall, *Aquificota* contribute substantially to carbon fixation and sulfur cycling in high-temperature ecosystems. Molecular phylogenetic analyses place them close to *Thermotogota* and possibly *Proteobacteria*, while their distinction is further supported by the presence of conserved signature indels (CSIs) that define the phylum and its constituent orders (Keller et al. 2025; Mondal et al. 2024; Rogers et al. 2023).

2.7 Other Bacteria

Up to this point, our discussion has focused on bacterial phyla with relatively large numbers of described species and well-established ecological or biotechnological roles. However, beyond these dominant groups, the bacterial domain also includes a wide array of lineages represented by only one or a few cultured species. In many cases, entire phyla remain known exclusively from environmental 16S rRNA gene surveys, without any laboratory isolates or physiological characterization. These understudied groups often occupy specialized ecological niches and may play key, yet poorly understood, roles in global biogeochemical cycles.

This section highlights several such phyla, including *Deinococcota*, *Acidobacteriota*, *Nitrospirota*, *Fusobacteriota*, *Fibrobacterota*, *Synergistota*, *Deferribacterota*, and *Chrysiogenota*. While their taxonomic breadth and metabolic diversity vary widely, together they exemplify the frontier of bacterial systematics, where molecular phylogenetics continues to uncover deep branches of evolutionary history that remain only sparsely sampled at the organismal level. By summarizing the limited but growing insights into their biology, ecology, and potential applications, this section provides a broader perspective on the diversity of the bacterial domain beyond the more extensively studied lineages.

2.7.1 *Deinococcota*

The *Deinococcota* phylum consists primarily of aerobic chemoorganotrophs that metabolize sugars, amino acids, organic acids, or complex nutrient mixtures. It includes two major orders: *Deinococcales*, typically isolated from diverse environments such as arid soils, radioactive sites, nuclear waste-contaminated sediments, water, air, and even the human gut, and *Thermales*, whose members are generally thermophiles or hyperthermophiles inhabiting natural and artificial thermal environments (Goodfellow et al. 2012; Theodorakopoulos et al. 2013). The type genus *Thermus* is of particular importance, as *Thermus aquaticus* provided the Taq DNA polymerase enzyme, which revolutionized molecular biology by enabling the polymerase chain reaction (PCR) technique (Ishino and Ishino 2014). Despite their Gram-negative cell wall structure, members of *Deinococcales* often stain Gram-positive. Their cell envelopes are unusual, lacking lipid A in the outer membrane and containing a unique form of peptidoglycan. Many species also produce carotenoid pigments, giving them characteristic red or pink coloration (Goodfellow et al. 2012; Tian and Hua 2010). A hallmark feature of *Deinococcales* is their extraordinary resistance to radiation and desiccation. For instance, *Deinococcus radiodurans* survives extreme levels of ionizing radiation and UV exposure, allowing selective isolation from soil, meat, dust, and filtered air (Theodorakopoulos et al. 2013). Unlike most bacteria, *D. radiodurans* resists numerous mutagenic agents, with the exception of nitrosoguanidine. Its resilience is attributed to a combination of efficient DNA repair systems and unique nucleoid organization. In addition to classical RecA-dependent repair, *D. radiodurans* employs RecA-independent mechanisms, excision repair pathways, and an ability to reassemble fragmented chromosomes (Makarova et al. 2001). Another factor contributing to radiation resistance is the toroidal arrangement of DNA within paired or tetrad cell structures. This ordered nucleoid architecture enables merging and recombination across nucleoids of neighboring compartments, effectively reconstructing entire chromosomes after fragmentation. This remarkable capacity for large-scale homologous recombination ensures genome integrity, allowing for recovery, growth, and division even after extreme DNA damage (Makarova et al. 2001; Goodfellow et al. 2012).

2.7.2 Acidobacteriota and Nitrospirota

The phylum *Acidobacteriota* is widely distributed in the environment and has been detected in soils, freshwater, hot spring microbial mats, wastewater treatment reactors, and sewage sludge (Krieg et al. 2010; Kielak et al. 2016). Despite their abundance, they remain poorly understood due to difficulties in cultivation, and much of our knowledge comes from metagenomic surveys. Most cultured representatives are Gram-negative, acid-tolerant, aerobic chemoorganotrophs. For instance, *Acidobacterium capsulatum*, an encapsulated acidophile isolated from acid mine drainage, utilizes various sugars and organic acids; *Acidisarcina polymorpha*, from acidic boreal forest soil, grows in cuboidal packets of 6–8 cells and can degrade polymers such as cellulose and chitin; while *Pyrinomonas methylaliphatogenes*, a thermophilic strain, oxidizes trace levels of atmospheric hydrogen (H2) for energy, although this process does not support growth (Krieg et al. 2010). Such traits may help explain their ecological success under conditions of nutrient limitation. Beyond these acidophilic, aerobic lineages, *Acidobacteriota* exhibit remarkable metabolic diversity, encompassing at least 26 subgroups. Some members are strict anaerobes, such as *Geothrix fermentans*, capable of dissimilatory iron reduction and citrate fermentation, and *Holophaga foetida*, an obligate acetogen that degrades methylated aromatic compounds into acetate. Phototrophy has also been reported in *Candidatus chloracidobacterium* (Kielak et al. 2016). This wide range of metabolic strategies highlights their ecological importance across diverse environments.

The phylum *Nitrospirota* is named after the type genus *Nitrospira*, which consists of chemolithoautotrophs oxidizing nitrite to nitrate. These organisms are widespread in terrestrial environments similar to those inhabited by *Nitrobacter*, but *Nitrospira* species are generally more abundant, leading to the hypothesis that they perform most of the nitrite oxidation in nitrogen-rich habitats such as wastewater treatment plants and ammonia-rich soils (Boone et al. 2001). A notable feature of some *Nitrospira* is their ability to perform complete nitrification, directly oxidizing ammonia to nitrate in a single organism. This capacity expands their ecological role in nitrogen cycling and sets them apart from canonical nitrifiers. Other significant members of *Nitrospirota* include *Leptospirillum*, an aerobic, acidophilic, iron-oxidizing chemolithotroph frequently found in acid mine drainage associated with coal and iron mining (Boone et al. 2001; Bandopadhyay and Shade 2024). Collectively, *Nitrospirota* contribute crucially to global nitrogen and iron cycling, often thriving in extreme or nutrient-rich environments.

2.7.3 Fusobacteriota

The *Fusobacteriota* phylum comprises Gram-negative, non-sporulating, obligate anaerobic rods that primarily ferment carbohydrates, peptides, and amino acids. Members are commonly found in sediments, the gastrointestinal tracts of animals, and oral cavities (Krieg et al. 2010). Species of the genus *Fusobacterium* are integral components of the human microbiome, particularly in the oral cavity. Their adhesins facilitate biofilm formation, contributing to the development of dental plaque. While prevalent in healthy individuals, certain species are opportunistic pathogens associated with oral and systemic diseases. For example, they play a role in periodontitis and can cause systemic infections when oral bacteria access the bloodstream (Brennan and Garrett 2018). Fusobacteria have also been detected in arterial plaques, suggesting potential involvement in cardiovascular diseases and adverse pregnancy outcomes. A clinically significant species, *Fusobacterium nucleatum*, has been strongly linked to colorectal cancer. This organism adheres to tumor cells and promotes their proliferation. In murine models, treatment with metronidazole reduced *F. nucleatum* abundance and tumor growth, supporting its role in colorectal carcinogenesis (McIlvanna et al. 2021). Collectively, *Fusobacteriota* represent a phylum that bridges ecology and human health, acting as both commensals in natural microbiomes and opportunistic pathogens with implications in chronic disease and cancer.

2.7.4 *Fibrobacterota and Synergistota*

The phyla *Fibrobacterota* and *Synergistota* comprise relatively few described species, most of which are obligate fermentative anaerobes commonly inhabiting animal gastrointestinal tracts (Krieg et al. 2010). Despite their limited taxonomic diversity, both groups play important ecological and medical roles. *Fibrobacterota* are represented primarily by the genus *Fibrobacter*, which includes Gram-negative, strictly anaerobic bacteria specialized in carbohydrate fermentation. Unlike *Fusobacteriota* and *Synergistota*, *Fibrobacter* species cannot ferment proteins or amino acids, relying instead on the degradation of cellulose as their primary metabolic strategy. In the rumen of herbivorous animals, where cellulose constitutes the major energy source, *Fibrobacter* species convert cellulose into glucose, which in turn fuels both cellulolytic and non-cellulolytic anaerobes. Thus, they occupy a pivotal niche in the ruminal carbon cycle, supporting energy flow and symbiotic microbial interactions (Ransom-Jones et al. 2012). *Synergistota*, by contrast, are Gram-negative, non-sporulating rods that inhabit both animal-associated environments and anoxic terrestrial or marine habitats. Most characterized species are obligate anaerobes that degrade proteins and ferment amino acids. In ruminants, *Synergistes jonesii* is a notable representative, whereas in humans, *Synergistota* species are frequently associated with soft tissue infections, abscesses, dental plaque, and periodontal diseases (El Houari et al. 2025; Brenner et al. 2005). Their prevalence in both commensal and pathogenic contexts highlights their dual ecological and clinical significance. Together, *Fibrobacterota* and *Synergistota* exemplify anaerobic bacterial lineages that contribute to both nutrient turnover in animal digestive systems and, in the case of *Synergistota*, human health and disease.

2.7.5 Deferribacterota and Chrysiogenota

The phyla *Deferribacterota* and *Chrysiogenota* comprise anaerobic chemoorganotrophs with versatile metabolic strategies, particularly in their use of diverse electron acceptors for anaerobic respiration (Qi et al. 2023; Boone et al. 2001). While many species reduce nitrate to nitrite or ammonium, others utilize metal oxides, elemental sulfur, or arsenate, highlighting their ecological importance in biogeochemical cycling. *Deferribacterota* were isolated from extreme environments, such as high-temperature seawater-flooded oil reservoirs, deep-sea hydrothermal vents, hydrothermal chimneys, and even the intestinal microbiota of the hydrothermal shrimp *Rimicaris* (Qi et al. 2023). A representative species, *Deferribacter abyssi*, is capable of using molecular hydrogen, acetate, succinate, pyruvate, and proteinaceous substrates as electron donors while reducing elemental sulfur, nitrate, or Fe(III). Similarly, the related genus *Geovibrio* employs nitrate and elemental sulfur as electron acceptors to support growth (Qi et al. 2023; Boone et al. 2001). *Chrysiogenota* are notable for their ability to couple acetate oxidation to the reduction of toxic oxyanions. The type species, *Chrysiogenes arsenatis*, originally isolated from arsenic-contaminated mud in a wastewater treatment system, oxidizes acetate while reducing arsenate to arsenite. Related species further expand this metabolic versatility by reducing selenate, nitrite, thiosulfate, and elemental sulfur (Boone et al. 2001). Together, *Deferribacterota* and *Chrysiogenota* exemplify lineages adapted to anaerobic and metal-rich environments, contributing to elemental cycling through their unique metabolic pathways.

2.8 Conclusion

The survey of bacterial diversity presented in this chapter highlights the extraordinary evolutionary breadth, ecological importance, and physiological specialization of prokaryotes. Bacteria reveal remarkable versatility in form and function, ranging from well-studied phyla with extensive representation, such as *Pseudomonadota, Bacillota, Actinomycetota*, and *Bacteroidota*, to more recently defined or poorly cultivated lineages, like *Verrucomicrobiota*, *Chlamydiae*, *Acidobacteria*, and *Nitrospirae*. Specialized groups, including hyperthermophiles such as *Thermotogota, Aquificota*, and *Thermodesulfobacteriota*, demonstrate adaptations to extreme conditions. Meanwhile, host-associated taxa such as *Fusobacteriota*, *Fibrobacterota*, and *Synergistota* play crucial roles in symbiosis, digestion, and disease. Even less-represented phyla, such as *Deferribacterota* and *Chrysiogenota*, exemplify unique metabolic pathways that expand our understanding of microbial energy-generation strategies.

Together, these lineages demonstrate that bacterial diversity is vast and fundamental to shaping ecosystems, biogeochemical cycles, and host health. Despite advances in genomics and cultivation-independent approaches, a substantial proportion of bacterial diversity remains uncultured and functionally uncharacterized. Continued exploration of these lineages, including major groups such as *Pseudomonadota* dominating environmental and host-associated niches, as well as minor phyla with only a few representatives, promises to refine bacterial systematics and uncover novel biological processes with implications for ecology, biotechnology, and medicine.

References

Adeolu M, Alnajar S, Naushad S, Gupta RS (2016) Genome-based phylogeny and taxonomy of the 'Enterobacteriales': Proposal for Enterobacterales ord. nov. divided into the families Enterobacteriaceae, Erwiniaceae fam. nov., Pectobacteriaceae fam. nov., Yersiniaceae fam. nov., Hafniaceae fam. nov., Morganellaceae fam. nov., and Budviciaceae fam. nov. Int J Syst Evol Microbiol 66:5575–5599. https://doi.org/10.1099/IJSEM.0.001485

Adeolu M, Gupta RS (2013) Phylogenomics and molecular signatures for the order Neisseriales: proposal for division of the order Neisseriales into the emended family Neisseriaceae and Chromobacteriaceae fam. nov. Antonie van Leeuwenhoek 104:1–24. https://doi.org/10.1007/S10482-013-9920-6

Agbaje M, Begum RH, Oyekunle MA et al (2011) Evolution of Salmonella nomenclature: a critical note. Folia Microbiol 56:497–503. https://doi.org/10.1007/S12223-011-0075-4

Aguilar C, Alwali A, Mair M et al (2024) Actinomycetota bioprospecting from ore-forming environments. Microb Genom 10. https://doi.org/10.1099/MGEN.0.001253

Ahmed MA, Campbell BJ (2025) Genome-resolved adaptation strategies of Rhodobacterales to changing conditions in the Chesapeake and Delaware Bays. Appl Environ Microbiol 91. https://doi.org/10.1128/AEM.02357-24

Aivelo T, Norberg A (2018) Parasite–microbiota interactions potentially affect intestinal communities in wild mammals. J Animal Ecol 87:438–447. https://doi.org/10.1111/1365-2656.12708

Al-Awthan YS, Mir R, Alatawi FA et al (2024) Metagenome analysis identified novel microbial diversity of sandy soils surrounded by natural lakes and artificial water points in King Salman Bin Abdulaziz Royal Natural Reserve. Saudi Arabia 1692(14):1692. https://doi.org/10.3390/LIFE14121692

Alsukaibi AK, Mechi L, Alimi FR et al (2023) Bioremediation potential of *Pseudomonas fluorescens* strain isolated from the Ha'il region of Saudi Arabia. Bioinformation 19:901–907. https://doi.org/10.6026/97320630019901

Alton GG, Forsyth JRL (1996) Brucella. Medical microbiology 4th edn. In: Baron S (ed) University of Texas Medical Branch at Galveston, Galveston, TX. Available from: https://www.ncbi.nlm.nih.gov/books/NBK8572/. Accessed 15 Aug 2025

An X, Tian C, Xu J et al (2020) Characterization of hexaconazole-degrading strain *Sphingobacterium multivorum* and analysis of transcriptome for biodegradation mechanism. Sci Total Environ 722:137171. https://doi.org/10.1016/J.SCITOTENV.2020.137171

Argentino ICV, Godoy MG, Seldin L, Jurelevicius D (2025) Distribution of bacillota in water and sediments from aquatic environments. Microb Ecol 88:1–10. https://doi.org/10.1007/S00248-024-02482-0

Arthofer P, Delafont V, Willemsen A et al (2022) Defensive symbiosis against giant viruses in amoebae. Proc Natl Acad Sci USA 119:e2205856119. https://doi.org/10.1073/PNAS.2205856119

Baker BJ, Lazar CS, Teske AP, Dick GJ (2015) Genomic resolution of linkages in carbon, nitrogen, and sulfur cycling among widespread estuary sediment bacteria. Microbiome 2015 3(1):1–12. https://doi.org/10.1186/S40168-015-0077-6

Baker-Austin C, Oliver JD, Alam M et al (2018) Vibrio spp. infections. Nat Rev Dis Primers 4. https://doi.org/10.1038/S41572-018-0005-8

Bandopadhyay S, Shade A (2024) Soil bacteria and archaea. In: Soil microbiology, ecology and biochemistry, 5th edn. Elsevier, The Netherlands, pp 41–74

Baranowski K, Huang B (2023) Cat scratch disease. StatPearls [Internet]. StatPearls Publishing, Treasure Island (FL), Available from https://www.ncbi.nlm.nih.gov/books/NBK482139/. Accessed 15 Aug 2025

Barka EA, Vatsa P, Sanchez L et al (2016) Taxonomy, physiology, and natural products of actinobacteria. Microbiol Mol Biol Rev 80:1–43. https://doi.org/10.1128/MMBR.00019-15

Basile A, Riggio FP, Tescari M et al (2025) Metagenome-resolved functional traits of Rubrobacter species implicated in rosy discoloration of ancient frescoes in two Georgian Cathedrals. Sci Total Environ. 958. https://doi.org/10.1016/j.scitotenv.2024.178135

Batista JH, Neto JF, da S (2017) Chromobacterium violaceum pathogenicity: updates and insights from genome sequencing of novel Chromobacterium species. Front Microbiol 8:295050. https://doi.org/10.3389/FMICB.2017.02213/BIBTEX

Batut J, Andersson SGE, O'Callaghan D (2004) The evolution of chronic infection strategies in the α-proteobacteria. Nat Rev Microbiol 2:933–945. https://doi.org/10.1038/NRMICRO1044

Bertaccini A, Arocha-Rosete Y, Contaldo N et al (2022) Revision of the 'Candidatus Phytoplasma' species description guidelines. Int J Syst Evol Microbiol 72. https://doi.org/10.1099/IJSEM.0.005353

Bhandari V, Gupta RS (2014) Molecular signatures for the phylum (class) Thermotogae and a proposal for its division into three orders (Thermotogales, Kosmotogales ord. Nov. and Petrotogales ord. Nov.) containing four families (Thermotogaceae, Fervidobacteriaceae fam. Nov., Kosmotogaceae fam. Nov. and Petrotogaceae fam. Nov.) and a new genus Pseudothermotoga gen. Nov. with five new combinations. Antonie Van Leeuwenhoek 105:143–168. https://doi.org/10.1007/S10482-013-0062-7

Bhatti AA, Haq S, Bhat RA (2017) Actinomycetes benefaction role in soil and plant health. Microb Pathog 111:458–467. https://doi.org/10.1016/j.micpath.09.036

Billah MM, Rahman MS (2024) Salmonella in the environment: a review on ecology, antimicrobial resistance, seafood contaminations, and human health implications. J Hazard Mater Adv 13:100407. https://doi.org/10.1016/J.HAZADV.2024.100407

Boedeker C, Schüler M, Reintjes G et al (2017) Determining the bacterial cell biology of Planctomycetes. Nat Commun 8:1–14. https://doi.org/10.1038/NCOMMS14853

Boniface A, Parquet C, Arthur M et al (2009) The elucidation of the structure of Thermotoga maritima peptidoglycan reveals two novel types of cross-link. J Biol Chem 284:21856. https://doi.org/10.1074/JBC.M109.034363

Boone DR, Castenholz RW, Garrity GM (2001) Bergey's Manual of Systematic Bacteriology, Volume One: the Archaea and the Deeply Branching and Phototrophic . Springer, New York, https://doi.org/10.1007/978-0-387-21609-6

Brennan CA, Garrett WS (2018) Fusobacterium nucleatum—symbiont, opportunist and oncobacterium. Nat Rev Microbiol 17(3):156–166. https://doi.org/10.1038/s41579-018-0129-6

Brenner DJ, Krieg NR, Staley JT et al 2005. Bergey's Manual of Systematic Bacteriology. Springer, New York, NY

Brinkmann S, Kurz M, Patras MA et al (2022) Genomic and chemical decryption of the bacteroidetes phylum for its potential to biosynthesize natural products. Microbiol Spectr 10:e02479–21. https://doi.org/10.1128/SPECTRUM.02479-21

Brown EM, Temple ER, Jeanfavre S et al (2025) Bacteroides sphingolipids promote anti-inflammatory responses through the mevalonate pathway. Cell Host Microbe 33:901–914. https://doi.org/10.1016/j.chom.2025.05.007e6

Brown PJB, De Pedro MA, Kysela DT et al (2012) Polar growth in the Alphaproteobacterial order Rhizobiales. Proc Natl Acad Sci USA 109:1697–1701. https://doi.org/10.1073/PNAS.1114476109

Brown PJB, Kysela DT, Brun YV (2011) Polarity and the diversity of growth mechanisms in bacteria. Semin Cell Dev Biol 22:790–798. https://doi.org/10.1016/j.semcdb.2011.06.006

Cabello-Yeves PJ, Ghai R, Mehrshad M et al (2017) Reconstruction of diverse verrucomicrobial genomes from metagenome datasets of freshwater reservoirs. Front Microbiol 8:303740. https://doi.org/10.3389/FMICB.2017.02131/XML

Candeliere F, Musmeci E, Amaretti A et al (2023) Profiling of the intestinal community of Clostridia: taxonomy and evolutionary analysis. Microbiome Res Rep 2. https://doi.org/10.20517/MRR.2022.19,

Chen D, Di Tian Y, Jiao JY et al (2020) Comparative genomics analysis of Nitriliruptoria reveals the genomic differences and salt adaptation strategies. Extremophiles 24:249–264. https://doi.org/10.1007/S00792-019-01150-3

Chiang E, Schmidt ML, Berry MA et al (2018) Verrucomicrobia are prevalent in north-temperate freshwater lakes and display class-level preferences between lake habitats. PLoS One 13:e0195112. https://doi.org/10.1371/JOURNAL.PONE.0195112

Chilton MD, Drummond MH, Merlo DJ et al (1977) Stable incorporation of plasmid DNA into higher plant cells: the molecular basis of crown gall tumorigenesis. Cell 11:263–271. https://doi.org/10.1016/0092-8674(77)90043-5

Choi JK, Poudel S, Yee N, Goff JL (2024) Deeply branching Bacillota species exhibit atypical Gram-negative staining. Microbiol Spectr 12. https://doi.org/10.1128/SPECTRUM.00732-24

Cicala F, Moore JD, Cáaceres-Martínez J et al (2017) Multigenetic characterization of 'Candidatus Xenohaliotis californiensis. Int J Syst Evol Microbiol 67:42–49. https://doi.org/10.1099/IJSEM.0.001563

Clark A, Mach N (2023) The gut mucin-microbiota interactions: a missing key to optimizing endurance performance. Front Physiol 14. https://doi.org/10.3389/FPHYS.2023.1284423,

Collingro A, Köstlbacher S, Horn M (2020) Chlamydiae in the environment. Trends Microbiol 28:877–888. https://doi.org/10.1016/j.tim.2020.05.020

da Silva K, Florentino LA, da Silva KB et al (2012) Cupriavidus necator isolates are able to fix nitrogen in symbiosis with different legume species. Syst Appl Microbiol 35:175–182. https://doi.org/10.1016/j.syapm.2011.10.005

Darby AC, Armstrong SD, Bah GS et al (2012) Analysis of gene expression from the Wolbachia genome of a filarial nematode supports both metabolic and defensive roles within the symbiosis. Genome Res 22:2467. https://doi.org/10.1101/GR.138420.112

Davin-Regli A, Lavigne JP, Pagès JM (2019) Enterobacter spp.: update on taxonomy, clinical aspects, and emerging antimicrobial resistance. Clin Microbiol Rev 32. https://doi.org/10.1128/CMR.00002-19

De Angelis M, Gobbetti M (2016) Lactobacillus SPP.: general characteristics. Reference module in food science. https://doi.org/10.1016/B978-0-08-100596-5.00851-9

de Sousa T, Hébraud M, Enes Dapkevicius MLN et al (2021) Genomic and metabolic characteristics of the pathogenicity in *Pseudomonas aeruginosa*. Int J Mol Sci 22. https://doi.org/10.3390/IJMS222312892

Declercq AM, Haesebrouck F, Van Den Broeck W et al (2013) Columnaris disease in fish: a review with emphasis on bacterium-host interactions. Vet Res 44(1):1–17. https://doi.org/10.1186/1297-9716-44-27

Dharamshi JE, Köstlbacher S, Schön ME et al (2023) Gene gain facilitated endosymbiotic evolution of Chlamydiae. Nat Microbiol 8:40–54. https://doi.org/10.1038/S41564-022-01284-9

Doern CD (2023) Classification of medically important bacteria. In: Molecular medical microbiology, third edition. pp. 9–21. https://doi.org/10.1016/B978-0-12-818619-0.00029-0

Dong N, Yang X, Chan EWC et al (2022) Klebsiella species: taxonomy, hypervirulence and multidrug resistance. EBioMedicine 79. https://doi.org/10.1016/j.ebiom.2022.103998

Dourado MN, Camargo Neves AA, Santos DS, Araújo WL (2015) Biotechnological and agronomic potential of endophytic pink-pigmented methylotrophic Methylobacterium spp. Biomed Res Int. https://doi.org/10.1155/2015/909016

Drews G (2000) The roots of microbiology and the influence of Ferdinand Cohn on microbiology of the 19th century. FEMS Microbiol Rev 24:225–249. https://doi.org/10.1111/J.1574-6976.2000.TB00540.X

Duar RM, Lin XB, Zheng J et al (2017) Lifestyles in transition: evolution and natural history of the genus Lactobacillus. FEMS Microbiol Rev 41:S27–S48. https://doi.org/10.1093/FEMSRE/FUX030

Egas C, Barroso C, Froufe HJC et al (2015) Complete genome sequence of the radiation-resistant bacterium Rubrobacter radiotolerans RSPS-4. Stand Genomic Sci 9:1062–1075. https://doi.org/10.4056/SIGS.5661021

El Houari A, Carpenter M, Chaplin D, Golyshin P, McDonald JE (2025) Taxonomic description and genome sequence of Anaerorudis cellulosivorans gen. Nov. sp. nov., a novel cellulose-and Xylan-degrading bacterium of the Bacteroidota phylum isolated from a lab-scale methanogenic landfill bioreactor digesting municipal solid waste. Syst Appl Microbiol 48(2):126590

Embley TM, Stackebrandt E (1994) The molecular phylogeny and systematics of the actinomycetes. Annu Rev Microbiol 48:257–289. https://doi.org/10.1146/ANNUREV.MI.48.100194.001353

Erlacher A, Cernava T, Cardinale M et al (2015) Rhizobiales as functional and endosymbiontic members in the lichen symbiosis of Lobaria pulmonaria L. Front Microbiol 6:126536. https://doi.org/10.3389/FMICB.2015.00053

Estrada-De Los Santos P, Bustillos-Cristales R, Caballero-Mellado J (2001) Burkholderia, a genus rich in plant-associated nitrogen fixers with wide environmental and geographic distribution. Appl Environ Microbiol 67:2790–2798. https://doi.org/10.1128/AEM.67.6.2790-2798.2001

Faburay B, McVey DS (2022) Mollicutes. In: Veterinary microbiology: fourth edition. pp 364–376. https://doi.org/10.1002/9781119650836.CH38

Ferraz Helene LC, Klepa MS, Hungria M (2022) New insights into the taxonomy of bacteria in the genomic era and a case study with Rhizobia. Int J Microbiol. https://doi.org/10.1155/2022/4623713

Firrao G, Andersen M, Bertaccini A, Boudon E, Bové JM, Daire X, et al (2004) Candidatus Phytoplasma', a taxon for the wall-less, non-helical prokaryotes that colonize plant phloem and insects. Int J Syst Evol Microbiol 54(4):1243–1255

Flores-Félix JD, García-Fraile P, Saati-Santamaría Z (2025) Harmony in diversity: reorganizing the families within the order Pseudomonadales. Mol Phylogenet Evol 206. https://doi.org/10.1016/j.ympev.2025.108321

Fuerst JA (2019) Phylum Verrucomicrobia. In: Encyclopedia of Microbiology. pp 551–563. https://doi.org/10.1016/B978-0-12-809633-8.20772-3

Fuerst AJ (2017) Planctomycetes–new models for microbial cells and activities. In: Microbial resources: from functional existence in nature to applications. pp 1–27. https://doi.org/10.1016/B978-0-12-804765-1.00001-1

Galperin MY (2013) Genome diversity of spore-forming firmicutes. Microbiol Spectr 1:TBS-0015-2012. https://doi.org/10.1128/MICROBIOLSPECTRUM.TBS-0015-2012

Galperin MY, Mekhedov SL, Puigbo P et al (2012) Genomic determinants of sporulation in Bacilli and Clostridia: towards the minimal set of sporulation-specific genes. Environ Microbiol 14:2870–2890. https://doi.org/10.1111/J.1462-2920.2012.02841.X

Galperin MY, Yutin N, Wolf YI et al (2022) Conservation and evolution of the sporulation gene set in diverse members of the firmicutes. J Bacteriol 204. https://doi.org/10.1128/JB.00079-22

Gao L, Tan Y, Zhang W et al (2021) *Cytophaga hutchinsonii* SprA and SprT are essential components of the Type IX secretion system required for Ca2+ acquisition, cellulose degradation, and cell motility. Front Microbiol 12:628555. https://doi.org/10.3389/FMICB.2021.628555

García-López M, Meier-Kolthoff JP, Tindall BJ et al (2019) Analysis of 1,000 type-strain genomes improves taxonomic classification of bacteroidetes. Front Microbiol 10:2083. https://doi.org/10.3389/FMICB.2019.02083

García-Romero I, Dios R de, Reyes-Ramírez F (2024) An improved genome editing system for Sphingomonadaceae. Access Microbiol 6. https://doi.org/10.1099/ACMI.0.000755.V3

Gatsios A, Kim CS, Crawford JM (2021) *Escherichia coli* small molecule metabolism at the host–microorganism interface. Nat Chem Biol 17:1016–1026. https://doi.org/10.1038/S41589-021-00807-5

Glaeser SP, Kämpfer P (2014) The family Sphingomonadaceae. In: The prokaryotes: alphaproteobacteria and betaproteobacteria. pp. 641–707. https://doi.org/10.1007/978-3-642-30197-1_302

Goodfellow M, Kämpfer P, Busse HJ (2012) Bergey's Manual of Systematic Bacteriology, Volume 5: the Actinobacteria. Springer, New York, NY

Guillotte ML, Gillespie JJ, Chandler CE et al (eds) (2018) Rickettsia lipid A biosynthesis utilizes the late acyltransferase LpxJ for secondary fatty acid addition. J Bacteriol 200. https://doi.org/10.1128/JB.00334-18

Gupta RS (2014) The phylum Aquificae. In: The prokaryotes: other major lineages of bacteria and the archaea. pp 417–445. https://doi.org/10.1007/978-3-642-38954-2_119

Gupta RS, Chen WJ, Adeolu M, Chai Y (2013) Molecular signatures for the class Coriobacteriia and its different clades; proposal for division of the class Coriobacteriia into the emended order Coriobacteriales, containing the emended family Coriobacteriaceae and Atopobiaceae fam. nov., and Eggerthellales ord. nov., containing the family Eggerthellaceae fam. nov. Int J Syst Evol Microbiol 63:3379–3397. https://doi.org/10.1099/IJS.0.048371-0

Gupta RS, Lo B, Son J (2018) Phylogenomics and comparative genomic studies robustly support division of the genus Mycobacterium into an emended genus Mycobacterium and four novel genera. Front Microbiol 9:306849. https://doi.org/10.3389/FMICB.2018.00067

Gutiérrez Santana JC, Coria Jiménez VR (2024) *Burkholderia cepacia* complex in cystic fibrosis: critical gaps in diagnosis and therapy. Ann Med 56:2307503. https://doi.org/10.1080/07853890.2024.2307503

Hakobyan A, Liesack W (2020) Unexpected metabolic versatility among type II methanotrophs in the Alphaproteobacteria. Biol Chem 401:1469–1477. https://doi.org/10.1515/HSZ-2020-0200

Hahnke RL, Meier-Kolthoff JP, García-López M et al (2016) Genome-based taxonomic classification of Bacteroidetes. Front Microbiol 7:208377. https://doi.org/10.3389/FMICB.2016.02003

Hale TL, Keusch GT (1996) Shigella. In: Baron S (ed). In: Medical microbiology. 4th edn. University of Texas Medical Branch at Galveston, Galveston (TX), Available from: https://www.ncbi.nlm.nih.gov/books/NBK8038/. Accessed 15 Aug 2025

Hamedi J, Poorinmohammad N (2017) The cellular structure of actinobacteria. I: Biology and Biotechnology of Actinobacteria. pp. 5–28. https://doi.org/10.1007/978-3-319-60339-1_2

Harirchi S, Sar T, Ramezani M et al (2022) Bacillales: from taxonomy to biotechnological and industrial perspectives. Microorganisms 10. https://doi.org/10.3390/MICROORGANISMS10122355

Harne S, Gayathri P, Béven L (2020) Exploring spiroplasma biology: opportunities and challenges. Front Microbiol 11. https://doi.org/10.3389/FMICB.2020.589279

Hashimoto Y, Shimamura S, Tame A et al (2022) Physiological and comparative proteomic characterization of Desulfolithobacter dissulfuricans gen. nov., sp. nov., a novel mesophilic, sulfur-disproportionating chemolithoautotroph from a deep-sea hydrothermal vent. Front Microbiol 13:1042116. https://doi.org/10.3389/FMICB.2022.1042116

Hashmi I, Bindschedler S, Junier P (2020) Firmicutes. In: Beneficial microbes in agro-ecology: bacteria and fungi. pp 363–396. https://doi.org/10.1016/B978-0-12-823414-3.00018-6

Haverkamp THA, Lossouarn J, Zhaxybayeva O et al (2021) Newly identified proviruses in Thermotogota suggest that viruses are the vehicles on the highways of interphylum gene sharing. Environ Microbiol 23:7105–7120. https://doi.org/10.1111/1462-2920.15723

Helmi NR (2025) Exploring the diversity and antimicrobial potential of actinomycetes isolated from different environments in Saudi Arabia: a systematic review. Front Microbiol 16:1568899. https://doi.org/10.3389/FMICB.2025.1568899

Herlemann DPR, Lundin D, Labrenz M et al (2013) Metagenomic de novo assembly of an aquatic representative of the verrucomicrobial class Spartobacteria. mBio 4:e00569–12. https://doi.org/10.1128/MBIO.00569-12

Hesselmann RPX, Werlen C, Hahn D et al (1999) Enrichment, phylogenetic analysis and detection of a bacterium that performs enhanced biological phosphate removal in activated sludge. Syst Appl Microbiol 22:454–465. https://doi.org/10.1016/S0723-2020(99)80055-1

Hilgenboecker K, Hammerstein P, Schlattmann P et al (2008) How many species are infected with Wolbachia?-A statistical analysis of current data. FEMS Microbiol Lett 281:215–220. https://doi.org/10.1111/J.1574-6968.2008.01110.X

Holt KE, Wertheim H, Zadoks RN et al (2015) Genomic analysis of diversity, population structure, virulence, and antimicrobial resistance in Klebsiella pneumoniae, an urgent threat to public health. Proc Natl Acad Sci USA 112:E3574–E3581. https://doi.org/10.1073/PNAS.1501049112

Hoover RL, Keffer JL, Polson SW, Chan CS (2023) Gallionellaceae pangenomic analysis reveals insight into phylogeny, metabolic flexibility, and iron oxidation mechanisms.https://doi.org/10.1128/MSYSTEMS.00038-23

Hoyles L (2019) Diversity of the class Coriobacteriia within different ecosystems. Access Microbiol 1. https://doi.org/10.1099/ACMI.AFM2019.PO0004

Hu D, Cha G, Gao B (2018) A phylogenomic and molecular markers based analysis of the class Acidimicrobiia. Front Microbiol 9:361535. https://doi.org/10.3389/FMICB.2018.00987

Hu D, Zang Y, Mao Y, Gao B (2019) Identification of molecular markers that are specific to the class Thermoleophilia. Front Microbiol 10:450275. https://doi.org/10.3389/FMICB.2019.01185

Huang J, Wiese J, Steiner LX et al (2025) Genomic characterisation of Actinospongicola halichondriae gen. nov., sp. nov., the first sponge-derived cultivated representative of a new genus within the class Acidimicrobiia. Antonie van Leeuwenhoek 118. https://doi.org/10.1007/S10482-025-02126-4

Ishino S, Ishino Y (2014) DNA polymerases as useful reagents for biotechnology—the history of developmental research in the field. Front Microbiol 5:465. https://doi.org/10.3389/FMICB.2014.00465

Islam MS, Yamamoto K, Morita N et al (2025) Characterization of *Congregicoccus parvus* gen. nov., sp. nov., a novel slow-growing ultramicrobacterium of the phylum Verrucomicrobiota. PLoS One 20:e0326734. https://doi.org/10.1371/JOURNAL.PONE.0326734

Jenal U, Stephens C (2002) The Caulobacter cell cycle: timing, spatial organization and checkpoints. Curr Opin Microbiol 5:558–563. https://doi.org/10.1016/S1369-5274(02)00378-8

Johnson EA (2009) Clostridia. In: Encyclopedia of microbiology, 3rd edn. pp. 87–93. https://doi.org/10.1016/B978-012373944-5.00139-5

Kallscheuer N, Jogler C (2021) The bacterial phylum Planctomycetes as novel source for bioactive small molecules. Biotechnol Adv 53. https://doi.org/10.1016/j.biotechadv.2021.107818

Kaper JB, Nataro JP, Mobley HLT (2004) Pathogenic *Escherichia coli*. Nat Rev Microbiol 2(2):123–140. https://doi.org/10.1038/nrmicro818

Kartal B, Van Niftrik L, Rattray J et al (2008) Candidatus 'Brocadia fulgida': an autofluorescent anaerobic ammonium oxidizing bacterium. FEMS Microbiol Ecol 63:46–55. https://doi.org/10.1111/J.1574-6941.2007.00408.X

Keller LM, Colman DR, Boyd ES (2025) Simultaneous aerobic and anaerobic respiration in hot spring chemolithotrophic bacteria. Nat Commun 16:1–14. https://doi.org/10.1038/S41467-025-56418-4

Kersters K, De Vos P, Gillis M et al (2006) Introduction to the Proteobacteria. In: The Prokaryotes. pp 3–37. https://doi.org/10.1007/0-387-30745-1_1

Khanna A, Khanna M, Aggarwal A (2013) Serratia marcescens-a rare opportunistic nosocomial pathogen and measures to limit its spread in hospitalized patients. J Clin Diagn Res 7:243–246. https://doi.org/10.7860/JCDR/2013/5010.2737

Khomyakova MA, Zavarzina DG, Merkel AY et al (2022) The first cultivated representatives of the actinobacterial lineage OPB41 isolated from subsurface environments constitute a novel order Anaerosomatales. Front Microbiol 13:1047580. https://doi.org/10.3389/FMICB.2022.1047580

Kielak AM, Barreto CC, Kowalchuk GA et al (2016) The ecology of Acidobacteria: moving beyond genes and genomes. Front Microbiol 7:744. https://doi.org/10.3389/FMICB.2016.00744

Klimek D, Herold M, Calusinska M (2024) Comparative genomic analysis of Planctomycetota potential for polysaccharide degradation identifies biotechnologically relevant microbes. BMC Genomics 25:1–16. https://doi.org/10.1186/S12864-024-10413-Z

König L, Wentrup C, Schulz F et al (2019) Symbiont-mediated defense against Legionella pneumophila in amoebae. mBio 10. https://doi.org/10.1128/MBIO.00333-19

Kopejtka K, Tomasch J, Zeng Y et al (2017) Genomic analysis of the evolution of phototrophy among haloalkaliphilic rhodobacterales. Genome Biol Evol 9:1950. https://doi.org/10.1093/GBE/EVX141

Kotloff KL, Riddle MS, Platts-Mills JA et al (2018) Shigellosis. Lancet 391:801–812. https://doi.org/10.1016/S0140-6736(17)33296-8

Kouřilová X, Schwarzerová J, Pernicová I et al (2021) The first insight into polyhydroxyalkanoates accumulation in multi-extremophilic *Rubrobacter xylanophilus* and *Rubrobacter spartanus*. Microorganisms 9. https://doi.org/10.3390/MICROORGANISMS9050909

Krieg, NR, Stanley, JT, Brown, DR et al (2010) Bergey's Manual of Systematic Bacteriology, Volume 4: The Bacteroidetes, Spirochaetes, Tenericutes (Mollicutes), Acidobacteria, Fibrobacteres, Fusobacteria, Dictyoglomi, Gemmatimonadetes, Lentisphaerae, Verrucomicrobia, Chlamydiae, and Planctomycetes. Springer, New York, NY

Ladewig L, Tufail MA, Kinfu BM et al (2025) Exploring new Bacteroidota strains: functional diversity and probiotic characteristics. Preprint at https://doi.org/10.1101/2024.12.18.628894

Lamichhane B, Mawad AMM, Saleh M et al (2024) Salmonellosis: an overview of epidemiology, pathogenesis, and innovative approaches to mitigate the antimicrobial resistant infections. Antibiotics 76(13):76. https://doi.org/10.3390/ANTIBIOTICS130100762024

Lamichhane JR, Varvaro L, Parisi L et al (2014) Disease and frost damage of woody plants caused by *Pseudomonas syringae*: seeing the forest for the trees. Adv Agron 126:235–295. https://doi.org/10.1016/B978-0-12-800132-5.00004-3

Langwig MV, De Anda V, Dombrowski N et al (2022) Large-scale protein level comparison of Deltaproteobacteria reveals cohesive metabolic groups. ISME Journal 16:307–320. https://doi.org/10.1038/S41396-021-01057-Y

Lazarev VN, Levitskii SA, Basovskii YI et al (2011) Complete genome and proteome of *Acholeplasma laidlawii*. J Bacteriol 193:4943–4953. https://doi.org/10.1128/JB.05059-11

Lee KC, Webb RI, Janssen PH et al (2009) Phylum Verrucomicrobia representatives share a compartmentalized cell plan with members of bacterial phylum Planctomycetes. BMC Microbiol 9. https://doi.org/10.1186/1471-2180-9-5

Lenferink WB, van Alen TA, Jetten MSM et al (2024) Genomic analysis of the class Phycisphaerae reveals a versatile group of complex carbon-degrading bacteria. Antonie Van Leeuwenhoek. 117. https://doi.org/10.1007/S10482-024-02002-7

Lertsethtakarn P, Ottemann KM, Hendrixson DR (2011) Motility and chemotaxis in Campylobacter and Helicobacter. Annu Rev Microbiol 65:389–410. https://doi.org/10.1146/ANNUREV-MICRO-090110-102908

Letizia M, Diggle SP, Whiteley M (2025) *Pseudomonas aeruginosa*: ecology, evolution, pathogenesis and antimicrobial susceptibility. Nat Rev Microbiol. https://doi.org/10.1038/S41579-025-01193-8

Lewin GR, Carlos C, Chevrette MG et al (2016) Evolution and ecology of Actinobacteria and their bioenergy applications. Annu Rev Microbiol 70:235–254. https://doi.org/10.1146/ANNUREV-MICRO-102215-095748

Li C, He YQ, Cui LQ et al (2021) Miltoncostaea marina gen. nov. sp. nov., and Miltoncostaea oceani sp. nov., a novel deep branching phylogenetic lineage within the class Thermoleophilia isolated from marine environments, and proposal of Miltoncostaeaceae fam. nov. and Miltoncostaeales ord. nov. Syst Appl Microbiol 44. https://doi.org/10.1016/j.syapm.2021.126216

Li Y, Zhang D, Bo D et al (2024) A taxonomic note on the order Caryophanales: description of 12 novel families and emended description of 21 families. Int J Syst Evol Microbiol 74:006539. https://doi.org/10.1099/IJSEM.0.006539

Liechti GW (2021) Localized peptidoglycan biosynthesis in chlamydia trachomatis conforms to the polarized division and cell size reduction developmental models. Front Microbiol 12:733850. https://doi.org/10.3389/FMICB.2021.733850

Liu R, Wang X, Huang R et al (2023) Profundirhabdus halotolerans gen. nov., sp. nov., an haloalkaliphilic actinobacterium isolated from seawater of the Mariana Trench. Int J Syst Evol Microbiol 73. https://doi.org/10.1099/IJSEM.0.006016

Ludfiani DD, Asmara W, Wahyuni AETH, Astuti P (2021) Identification of Lactobacillus spp. on basis morphological, physiological, and biochemical characteristic from Jawa Super Chicken Excreta. BIO Web Conf 33:06012. https://doi.org/10.1051/BIOCONF/20213306012

Luu LDW, Kasimov V, Phillips S et al (2023) Genome organization and genomics in Chlamydia: whole genome sequencing increases understanding of chlamydial virulence, evolution, and phylogeny. Front Cell Infect Microbiol 13:1178736. https://doi.org/10.3389/FCIMB.2023.1178736
Madigan, MT, Bender, KS, Buckler, DH et al (2022) Brock biology of microorganisms, 16th edn. Pearson, Harlow, pp 555–591
Maita C, Matsushita M, Miyoshi M et al (2018) Amoebal endosymbiont Neochlamydia protects host amoebae against *Legionella pneumophila* infection by preventing Legionella entry. Microbes Infect 20:236–244. https://doi.org/10.1016/j.micinf.2017.12.012
Makarova KS, Aravind L, Wolf YI et al (2001) Genome of the extremely radiation-resistant bacterium *Deinococcus radiodurans* viewed from the perspective of comparative genomics. Microbiol Mol Biol Rev 65:44. https://doi.org/10.1128/MMBR.65.1.44-79.2001
Mardanov AV, Beletsky AV, Kadnikov VV et al (2016) Genome analysis of *Thermosulfurimonas dismutans*, the first thermophilic sulfur-disproportionating bacterium of the phylum Thermodesulfobacteria. Front Microbiol 7:208689. https://doi.org/10.3389/FMICB.2016.00950
Martijn J, Vosseberg J, Guy L et al (2018) Deep mitochondrial origin outside the sampled alphaproteobacteria. Nature 557:101–105. https://doi.org/10.1038/S41586-018-0059-5
May M, Balish MF, Blanchard A (2014) The order mycoplasmatales. In: The Prokaryotes: firmicutes and tenericutes. pp 515–550. https://doi.org/10.1007/978-3-642-30120-9_289
McIlvanna E, Linden GJ, Craig SG et al (2021) Fusobacterium nucleatum and oral cancer: a critical review. BMC Cancer 21:1212. https://doi.org/10.1186/S12885-021-08903-4
Mehmood N, Saeed M, Zafarullah S et al (2023) Multifaceted impacts of plant-beneficial Pseudomonas spp. in managing various plant diseases and crop yield improvement. ACS Omega 8:22296. https://doi.org/10.1021/ACSOMEGA.3C00870
Meier DV, Imminger S, Gillor O, Woebken D (2021) Distribution of mixotrophy and desiccation survival mechanisms across microbial genomes in an arid biological soil crust community. mSystems 6. https://doi.org/10.1128/MSYSTEMS.00786-20
Meumann EM, Limmathurotsakul D, Dunachie SJ et al (2024) *Burkholderia pseudomallei* and melioidosis. Nat Rev Microbiol 22:155–169. https://doi.org/10.1038/S41579-023-00972-5
Mohammadipanah F, Dehhaghi M (2017) Classification and taxonomy of actinobacteria. In: Biology and Biotechnology of Actinobacteria. pp 51–77. https://doi.org/10.1007/978-3-319-60339-1_4
Mondal N, Dutta S, Chatterjee S et al (2024) Aquificae overcomes competition by archaeal thermophiles, and crowding by bacterial mesophiles, to dominate the boiling vent-water of a Trans-Himalayan sulfur-borax spring. PLoS One 19:e0310595. https://doi.org/10.1371/JOURNAL.PONE.0310595
Mortazavi A, Hassanshahian M, Ali E et al (2023) Isolation and identification of phenol-degrading bacteria from Iranian soil and leaf samples. Front Biosci 15. https://doi.org/10.31083/J.FBE1504029
Morya R, Salvachúa D, Thakur IS (2020) Burkholderia: an untapped but promising bacterial genus for the conversion of aromatic compounds. Trends Biotechnol 38:963–975. https://doi.org/10.1016/j.tibtech.2020.02.008
Nataro JP, Kaper JB (1998) Diarrheagenic *Escherichia coli*. Clin Microbiol Rev 11:142–201. https://doi.org/10.1128/CMR.11.1.142
Nemoto D, Hitomi S, Moriyama Y et al (2019) Cellulitis complicated with bacteremia due to Sphingobacterium species: a report of two cases and a literature review. Intern Med 58:2573. https://doi.org/10.2169/INTERNALMEDICINE.2178-18
Nunoura T, Chikaraishi Y, Izaki R et al (2018) A primordial and reversible TCA cycle in a facultatively chemolithoautotrophic thermophile. Science 359:559–563. https://doi.org/10.1126/SCIENCE.AAO3407
O'Callaghan A, van Sinderen D (2016) Bifiidobacteria and their role as members of the human gut microbiota. Front Microbiol 7. https://doi.org/10.3389/FMICB.2016.00925

Oren A (2024) On validly published names, correct names, and changes in the nomenclature of phyla and genera of prokaryotes: a guide for the perplexed. NPJ Biofilms Microbiomes 10:1–4. https://doi.org/10.1038/S41522-024-00494-9

Oren A, Arahal DR, Christensen H et al (2025) Preparing the 2025 revision of the International Code of Nomenclature of Prokaryotes. Int J Syst Evol Microbiol 75:006666. https://doi.org/10.1099/IJSEM.0.006666

Oren A, Garrity GM (2021) Valid publication of the names of forty-two phyla of prokaryotes. Int J Syst Evol Microbiol 71. https://doi.org/10.1099/IJSEM.0.005056

Pace NR, Sapp J, Goldenfeld N (2012) Phylogeny and beyond: scientific, historical, and conceptual significance of the first tree of life. Proc Natl Acad Sci USA 109:1011–1018. https://doi.org/10.1073/PNAS.1109716109

Park J, Roh HJ, Lee Y et al (2023) Characterization and pathogenicity of *Flavobacterium psychrophilum* isolated from Rainbow Trout (Oncorhynchus mykiss) in Korea. Microorganisms 11:2546. https://doi.org/10.3390/MICROORGANISMS11102546

Parks DH, Chuvochina M, Rinke C et al (2022) GTDB: an ongoing census of bacterial and archaeal diversity through a phylogenetically consistent, rank normalized and complete genome-based taxonomy. Nucleic Acids Res 50:D785–D794. https://doi.org/10.1093/NAR/GKAB776

Parola P, Paddock CD, Socolovschi C et al (2013) Update on tick-borne rickettsioses around the world: a geographic approach. Clin Microbiol Rev 26:657–702. https://doi.org/10.1128/CMR.00032-13

Parte AC, Carbasse JS, Meier-Kolthoff JP et al (2020) List of prokaryotic names with standing in nomenclature (LPSN) moves to the DSMZ. Int J Syst Evol Microbiol 70:5607–5612. https://doi.org/10.1099/IJSEM.0.004332

Pechter KB, Gallagher L, Pyles H et al (2016) Essential genome of the metabolically versatile alphaproteobacterium *Rhodopseudomonas palustris*. J Bacteriol 198:867. https://doi.org/10.1128/JB.00771-15

Peña-Salinas ME, Speth DR, Utter DR et al (2024) Thermotogota diversity and distribution patterns revealed in Auka and JaichMaa 'ja 'ag hydrothermal vent fields in the Pescadero Basin, Gulf of California. PeerJ 12. https://doi.org/10.7717/PEERJ.17724

Poole P, Ramachandran V, Terpolilli J (2018) Rhizobia: from saprophytes to endosymbionts. Nat Rev Microbiol 16:291–303. https://doi.org/10.1038/NRMICRO.2017.171

Power JF, Carere CR, Welford HE et al (2024) A genus in the bacterial phylum Aquificota appears to be endemic to Aotearoa-New Zealand. Nat Commun 15. https://doi.org/10.1038/S41467-023-43960-2

Prosser JI, Head IM, Stein LY (2014) The family Nitrosomonadaceae. In: The Prokaryotes: Alphaproteobacteria and Betaproteobacteria. pp. 901–918. https://doi.org/10.1007/978-3-642-30197-1_372

Qi L, Shi M, Zhu FC et al (2023) Genomic evidence for the first symbiotic Deferribacterota, a novel gut symbiont from the deep-sea hydrothermal vent shrimp *Rimicaris kairei*. Front Microbiol 14:1179935. https://doi.org/10.3389/FMICB.2023.1179935/

Qian F, Morse DE (2011) Miniaturizing microbial fuel cells. Trends Biotechnol 29:62–69. https://doi.org/10.1016/j.tibtech.2010.10.003

Ramos-Silva P, Serrano M, Henriques AO (2019) From root to tips: sporulation evolution and specialization in *Bacillus subtilis* and the intestinal pathogen *Clostridioides difficile*. Mol Biol Evol 36:2714–2736. https://doi.org/10.1093/MOLBEV/MSZ175

Ransom-Jones E, Jones DL, McCarthy AJ, McDonald JE (2012) The fibrobacteres: an important phylum of cellulose-degrading bacteria. Microb Ecol 2011 63(2):267–281. https://doi.org/10.1007/S00248-011-9998-163

Ratner H (1984) Flavobacterium meningosepticum. Infect Contr 5:237–239

Rogers TJ, Buongiorno J, Jessen GL et al (2023) Chemolithoautotroph distributions across the subsurface of a convergent margin. ISME J 17:140–150. https://doi.org/10.1038/S41396-022-01331-7

Saito T, Terato H, Yamamoto O (1994) Pigments of *Rubrobacter radiotolerans*. Arch Microbiol 162:414–421. https://doi.org/10.1007/BF00282106
Saksena S, Forbes K, Rajan N, Giles D (2023) Phylogenetic investigation of Gammaproteobacteria proteins involved in exogenous long-chain fatty acid acquisition and assimilation. Biochem Biophys Rep 35. https://doi.org/10.1016/j.bbrep.2023.101504
Salcher MM, Schaefle D, Kaspar M et al (2019) Evolution in action: habitat transition from sediment to the pelagial leads to slophilaceae. The ISME J 13(11):2764–2777. https://doi.org/10.1038/s41396-019-0471-3
Salje J (2021) Cells within cells: Rickettsiales and the obligate intracellular bacterial lifestyle. Nat Rev Microbiol 19:375–390. https://doi.org/10.1038/S41579-020-00507-2
Santinello D, Zampieri G, Agostini S et al (2024) Process stability in anaerobicdigestion: unveiling microbial signatures of full-scale reactor performance. Chem Eng J 497:154962. https://doi.org/10.1016/J.CEJ.2024.154962
Schön ME, Martijn J, Vosseberg J et al (2022) The evolutionary origin of host association in the Rickettsiales. Nat Microbiol 2022 7(8):1189–1199. https://doi.org/10.1038/s41564-022-01169-x
Schulze-Makuch D, Wagner D, Kounaves SP et al (2018) Transitory microbial habitat in the hyperarid Atacama Desert. Proc Natl Acad Sci USA 115:2670–2675. https://doi.org/10.1073/PNAS.1714341115
Sexton DL, Hashimi A, Beskrovnaya P et al (2023) The cell envelope of Thermotogae suggests a mechanism for outer membrane biogenesis. Proc Natl Acad Sci USA 120:e2303275120. https://doi.org/10.1073/PNAS.2303275120
Shao Y, Chung BS, Lee SS et al (2009) Zoogloea caeni sp. nov., a floc-forming bacterium isolated from activated sludge. Int J Syst Evol Microbiol 59:526–530. https://doi.org/10.1099/IJS.0.65670-0
Simon M, Scheuner C, Meier-Kolthoff JP et al (2017) Phylogenomics of Rhodobacteraceae reveals evolutionary adaptation to marine and non-marine habitats. The ISME J 11(6):1483–1499. https://doi.org/10.1038/ismej.2016.198
Sokolova T, Hanel J, Onyenwoke RU et al (2007) Novel chemolithotrophic, thermophilic, anaerobic bacteria *Thermolithobacter ferrireducens* gen. nov., sp. nov. and Thermolithobacter carboxydivorans sp. nov. Extremophiles 11:145–157. https://doi.org/10.1007/S00792-006-0022-5
Stackebrandt E, Otten LG (2014) The class Nitriliruptoria. In: The prokaryotes: actinobacteria. pp. 587–594. https://doi.org/10.1007/978-3-642-30138-4_197
Starkenburg SR, Chain PSG, Sayavedra-Soto LA et al (2006) Genome sequence of the chemolithoautotrophic nitrite-oxidizing bacterium Nitrobacter winogradskyi Nb-255. Appl Environ Microbiol 72:2050–2063. https://doi.org/10.1128/AEM.72.3.2050-2063.2006
Stewart V (1994) Regulation of nitrate and nitrite reductase synthesis in enterobacteria. Antonie Van Leeuwenhoek 66:37–45. https://doi.org/10.1007/BF00871631
Sutton GG, Brinkac LM, Clarke TH, Fouts DE (2018) *Enterobacter hormaechei* subsp. Hoffmannii subsp. nov., *Enterobacter hormaechei* subsp. Xiangfangensis comb. nov., *Enterobacter roggenkampii* sp. nov., and *Enterobacter muelleri* is a later heterotypic synonym of Enterobacter asburiae based on computational analysis of sequenced Enterobacter genomes. F1000Res 7. https://doi.org/10.12688/F1000RESEARCH.14566.2
Tan XY, Liu XJ, Lu DC et al (2025) Insights into the physiological and metabolic features of Thalassobacterium, a novel genus of Verrucomicrobiota with the potential to drive the carbon cycle. mBio 16. https://doi.org/10.1128/MBIO.00305-25
Theodorakopoulos N, Bachar D, Christen R et al (2013) Exploration of Deinococcus-Thermus molecular diversity by novel group-specific PCR primers. Microbiologyopen 2:862. https://doi.org/10.1002/MBO3.119
Tian B, Hua Y (2010) Carotenoid biosynthesis in extremophilic Deinococcus-Thermus bacteria. Trends Microbiol 18:512–520. https://doi.org/10.1016/j.tim.2010.07.007

Tiffany CR, Lee JY, Rogers AWL et al (2021) The metabolic footprint of Clostridia and Erysipelotrichia reveals their role in depleting sugar alcohols in the cecum. Microbiome 9. https://doi.org/10.1186/S40168-021-01123-9

Tuon FF, Dantas LR, Suss PH, Tasca Ribeiro VS (2022) Pathogenesis of the *Pseudomonas aeruginosa* biofilm: a review. Pathogens 11. https://doi.org/10.3390/PATHOGENS11030300

Uesugi JHE, dos Santos Caldas D, Coelho BBF et al (2024) Morphological diversity of actinobacteria isolated from oil palm compost (*Elaeis guineensis*). Braz J Microbiol 55:455–469. https://doi.org/10.1007/S42770-023-01178-W

Van Zandt KE, Greer MT, Gelhaus HC (2013) Glanders:an overview of infection in humans. Orphanet J Rare Dis 8:1–7. https://doi.org/10.1186/1750-1172-8-131

Vitorino IR, Gambardella N, Semedo M et al (2025) Diversity and vertical distribution of Planctomycetota in the water column of the remote North Pacific. Environ Microbiol Rep 17:e70063. https://doi.org/10.1111/1758-2229.70063

Vivijs B, Moons P, Geeraerd AH et al (2014) 2,3-Butanediol fermentation promotes growth of Serratia plymuthica at low pH but not survival of extreme acid challenge. Int J Food Microbiol 175:36–44. https://doi.org/10.1016/j.ijfoodmicro.2014.01.017

Vorobev A, Beck DAC, Kalyuzhnaya MG et al (2013) Comparative transcriptomics in three Methylophilaceae species uncover different strategies for environmental adaptation. PeerJ 2013. https://doi.org/10.7717/PEERJ.115

Vos, P, Garrity, GM, Jones, D et al (2009) Bergey's Manual of Systematic Bacteriology Volume 3: The Firmicutes. Springer, New York, NY

Waite DW, Chuvochina M, Pelikan C et al (2020) Proposal to reclassify the proteobacterial classes deltaproteobacteria and oligoflexia, and the phylum thermodesulfobacteria into four phyla reflecting major functional capabilities. Int J Syst Evol Microbiol 70:5972–6016. https://doi.org/10.1099/IJSEM.0.004213

Waite DW, Vanwonterghem I, Rinke C et al (2017) Comparative genomic analysis of the class Epsilonproteobacteria and proposed reclassification to epsilonbacteraeota (phyl. nov.). Front Microbiol 8:264454. https://doi.org/10.3389/FMICB.2017.00682

Wang S, Meade A, Lam H-M, Luo H (2020a) Evolutionary timeline and genomic plasticity underlying the lifestyle diversity in Rhizobiales. mSystems 5. https://doi.org/10.1128/MSYSTEMS.00438-20

Wang Z, Li W, Li H et al (2020b) Phylogenomics of Rhodocyclales and its distribution in wastewater treatment systems. Sci Rep 10:1–12. https://doi.org/10.1038/s41598-020-60723-x

Whitman WB, Bergey's Manual Trust, contributor (2015) Bergey's Manual of Systematics of Archaea and Bacteria. In: Whitman WB (ed). Wiley, USA

Williams KP, Gillespie JJ, Sobral BWS et al (2010) Phylogeny of gammaproteobacteria. J Bacteriol 192:2305–2314. https://doi.org/10.1128/JB.01480-09

Williams KP, Sobral BW, Dickerman AW (2007) A robust species tree forthe Alphaproteobacteria. J Bacteriol 189:4578. https://doi.org/10.1128/JB.00269-07

Williamson NR, Fineran PC, Leeper FJ, Salmond GPC (2006) The biosynthesis and regulation of bacterial prodiginines. Nat Rev Microbiol 4:887–899. https://doi.org/10.1038/NRMICRO1531

Woese CR (1987) Bacterial evolution. Microbiol Rev 51:221–271. https://doi.org/10.1128/MR.51.2.221-271.1987

Yan XH, Pei SC, Yen HC et al (2024) Delineating bacterial genera based on gene content analysis: a case study of the Mycoplasmatales–Entomoplasmatales clade within the class Mollicutes. Microb Genom 10. https://doi.org/10.1099/MGEN.0.001321

Ye YF, Ming H, Yu-feng D (2004) Characterization of a strain of Sphingobacterium sp. and its degradation to herbicide mefenacet. J Environ Sci 2:343–347. https://pubmed.ncbi.nlm.nih.gov/15137667/Available from Accessed 15 Aug 2025

Young JM, Park DC (2007) Probable synonymy of the nitrogen-fixing genus Azotobacter and the genus Pseudomonas. Int J Syst Evol Microbiol 57:2894–2901. https://doi.org/10.1099/IJS.0.64969-0

Zeikus JG, Dawson MA, Thompson TE (1983) Microbial ecology of volcanic sulphidogenesis: isolation and characterization of Thermodesulfobacterium commune gen. nov. and sp. nov. Microbiology 129:1159–1169. https://doi.org/10.1099/00221287-129-4-1159

Zheng R, Liu R, Shan Y et al (2021) Characterization of the first cultured free-living representative of *Candidatus izemoplasma* uncovers its unique biology. ISME J 15(9):2676–2691. https://doi.org/10.1038/s41396-021-00961-7

Chapter 3
Diversity of Archaea

Ayten Kimiran, Merve Mücahide Çağ and Nihal Doğruöz Güngör

Abstract Archaea are prokaryotes that represent a unique evolutionary lineage distinct from bacteria and eukaryotes. Once thought to exist only in extreme environments, archaea inhabit a wide range of habitats, including soil, marine environments, and the digestive systems of living organisms. This section highlights the systematic diversity of archaea, particularly the superphyla Euryarchaeota, TACK, DPANN, and Asgard, along with their respective phyla. Additionally, the ecological distribution of archaea, their morphological and structural differences, their extensive metabolic properties, and their adaptation mechanisms in extreme conditions are discussed to highlight their biological importance and contribute to our understanding of life on our planet.

Keywords Archaea · Diversity · Extreme environments · Metabolism · Adaptation

A. Kimiran (✉) · N. Doğruöz Güngör
Department of Biology, Section of Fundamental and Industrial Microbiology, Istanbul University, Faculty of Science, Istanbul, Türkiye

A. Kimiran
e-mail: kimiran@istanbul.edu.tr

M. M. Çağ
Department of Biology, Istanbul University, Institute of Graduate Studies in Science, Istanbul, Türkiye

M. Çalışkan and M. Arıkan (eds.), *Microbial World: Bacteria and Archaea*, Progress in Molecular and Subcellular Biology 62, https://doi.org/10.1007/978-3-032-11438-9_3

3.1 Introduction

Archaea, one of the three major domains of life, was discovered in 1977 by Carl Woese and George Fox. Molecular analyses in 1990 revealed that it is evolutionarily distinct and unique from Bacteria and Eukarya (Woese and Fox 1977; Woese et al. 1990). Initially, it was thought that archaea only lived in extreme environments because they were isolated in such conditions. However, with the development of molecular methods, it was determined that they are also widespread in non-extreme environments. Archaea are not only found in hot springs, volcanic mud, hydrothermal vents, Antarctic glaciers, cold deep seas, acidic/alkaline, and high-salt environments, but also in agricultural soils, forest soils, ocean surfaces/depths, lakes, rivers, plant rhizosphere, and animal/ human bodies, particularly in the digestive system (Zou et al. 2025; Moissl-Eichinger et al. 2018).

Archaea are morphologically and metabolically similar to the Bacteria domain; however, in terms of genetic information processing mechanisms, such as DNA replication, transcription, and translation, they are more closely related to the Eukarya domain (Albers and Meyer 2011; Valentine 2007; Bell and Jackson 2001). After it was determined that many genes of the Asgard superphylum members discovered in recent years are specific to the Eukarya domain, these data supported the hypothesis that eukaryotes may have evolved from an archaeal species (Eme et al. 2017). With the advancement of molecular biology techniques, particularly next-generation sequencing techniques, the genomes of uncultivable microorganisms have been sequenced, leading to the discovery of numerous previously unknown archaea phyla (Spang et al. 2017; Baker et al. 2020). The archaea domain, initially identified as comprising two phyla, now encompasses 30 phyla and over 200,000 species. As a result, their phylogenies have been frequently revised and continue to be updated due to discoveries (Tahon et al. 2021). Based on recent phylogenetic analyses, the archaea domain is currently divided into four major superphyla: Euryarchaeota, TACK, DPANN, and Asgard. These superphyla also contain phyla with various physiological and metabolic characteristics (Adam et al. 2017; Castelle and Banfield 2018).

Archaea play a crucial role in the carbon and nutrient cycles through their unique metabolic pathways. For example, methanogenic archaea produce methane through a unique metabolism that is exclusive to them. Archaea sustain their existence in a wide variety of ecosystems through the production of methane, oxidation of ammonia, carbon fixation, and sulfur metabolism (Ma et al. 2025; Jiang et al. 2025; Karishma et al. 2024; Zou et al. 2025). They can adapt to extremely high/low temperatures, high pressure, extremely high salt concentrations, highly acidic/alkaline environments, and radiation, which the Bacteria domain cannot. These adaptations include molecular modifications that ensure the structural stability of DNA, RNA, and proteins (Baker et al. 2024; Amangeldina et al. 2024; Díaz-Rullo and González-Pastor 2025; Sun et al. 2023).

To provide a better understanding of the roles of archaea in microbial ecosystems and biogeochemical cycles, this section will examine the taxonomic, ecological, metabolic, and morphological diversity of archaea, as well as their adaptations to extreme environments. Finally, the future prospects of archaea research will be discussed.

3.2 Archaea Superphylum and Basic Features

Archaea constitute a significant portion of Earth's microbial diversity, serving as dominant members of both marine and terrestrial environments. Archaea are classified into superphyla such as Euryarchaeota, TACK (Thaumarchaeota, Aigarchaeota, Crenarchaeota, Korarchaeota, Bathyarchaeota, Hydrothermarchaeota, Geothermarchaeota, and Verstraetearchaeota), DPANN (Diapherotrites, Parvarchaeota, Aenigmarchaeota, Nanoarchaeota, Nanohaloarchaeota, Woesearchaeota, Pacearchaeota, and potentially Altiarchaea) superphyla, and the recently defined Asgard superphylum, including Lokiarchaeota, Thorarchaeota, Odinarchaeota, and Heimdallarchaeota (MacLeod et al. 2019).

3.2.1 Euryarchaeota Superphylum

Euryarchaeota is one of the best characterized superphyla of archaea. It is a physiologically and morphologically heterogeneous superphylum consisting of organisms with versatile metabolic properties that thrive in extreme conditions. While Euryarchaeota is one of two superphyla with cultured representatives, culture-independent methods have revealed that many Euryarchaeota species, which have not yet been cultured, are present in marine and terrestrial ecosystems (Oren 2019).

Euryarchaeota employ various metabolic pathways, including methanogenesis, halophily, and thermophily, which demonstrate their ability to adapt to extreme environments (Amils 2022). Methanogenesis is a crucial metabolic pathway, particularly for organisms living under anaerobic conditions, where they convert CO_2 and H_2 into methane (Liu and Whitman 2008; Amils 2022). Overall, Euryarchaeota play a crucial role in global methane emissions and biogeochemical cycles across various environments, including geothermal areas (Ou et al. 2022). Euryarchaeota also utilize different energy sources through various metabolic pathways (Dombrowski et al. 2021). Some are chemotrophs that utilize compounds such as sulfur and ammonia, while others are phototrophs that produce ATP without oxygen-producing photosynthesis. Halophilic Euryarchaeota can tolerate extreme salt levels due to their unique metabolism and ability to utilize light. The extremophiles that Woese used to discover the existence of archaea also belong to this superphylum (Amils 2022).

There are six phyla associated with the Euryarchaeota: Halobacteriota, Hydrothermarchaeota, Methanobacteriota, Methanobacteriota_B, Thermoplasmatota, and Hadarchaeota (Baker et al. 2025).

3.2.1.1 Halobacteriota

Halobacteriota is a phylum characterized by extreme halophilic and alkaline lifestyles (Sorokin et al. 2022a). Members of this phylum are commonly found in hypersaline ecosystems such as lakes and saline waters. Halobacteriota members employ strategies such as the synthesis of trehalose and proline, as well as the uptake of compatible solutes like glycine betaine, to cope with osmotic stress. Additionally, they utilize various electron transport systems for energy conservation. The presence of genes encoding microbial rhodopsins suggests that rhodopsin-based phototrophy may contribute to auxiliary energy acquisition in the photic and osmotically dynamic water column (Xamxidin et al. 2025). They exhibit methanogenesis activity, which is crucial for global methane cycles (Ou et al. 2022). Methanogenic archaea are divided into two subgroups based on whether they use type B cytochromes for energy conservation. Since methanogens with type B cytochromes are found only in the Halobacteriota phylum, it is proposed that metabolism originated within this phylum. Members of this phylum with type B cytochromes are reported to have a broader substrate range and higher reproductive efficiency, and are able to conserve energy through membrane-bound electron transport chains (Ou et al. 2022; Xamxidin et al. 2025).

3.2.1.2 Hydrothermarchaeota

Previously known as the Marine Benthic Group E, Hydrothermarchaeota is a phylum of archaea that is abundant in extreme marine environments such as hydrothermal vents and cold seeps. Hydrothermal vents release reduced compounds and small organic carbon compounds into the surrounding seawater, providing the substrates necessary for microbial growth and bioenergy conversion. Genomic analyses have revealed that they can synergistically participate in the sulfur and nitrogen cycles through processes such as carbon monoxide oxidation (carboxyotrophy), sulfate and nitrate reduction, and various carbon metabolism pathways (Zhou et al. 2020). Despite their ecological importance, cultured representatives of Hydrothermarchaeota are still lacking, so current knowledge relies on phylogenomic studies (Carr et al. 2017; Zhou et al. 2020). It has been suggested that this potential metabolic diversity could be advantageous for hydrogeologically and geochemically dynamic underground aquifers, which are considered to be energy- and nutrient-limited environments (Carr et al. 2017).

3.2.1.3 Methanobacteriota

Methanobacteriota is an archaeal phylum that produces methane, a key component of the global carbon cycle. Depending on substrate utilization, its members use metabolic pathways such as hydrogenotrophic, acetoclastic, and methylotrophic methanogenesis, which affect microbial community dynamics and ecosystem functioning. Recent studies have identified Methanobacteriota as a crucial component of microbial communities involved in methanogenesis, demonstrating their resilience to substrate changes and operational stresses (Khesali Aghtaei et al. 2022). These microorganisms primarily thrive anaerobically in environmental and biological settings such as wetlands, rice fields, and the gastrointestinal tracts of ruminants. While they are recognized as members of the intestinal and mucosal microbiota of mammals, they have also been reported in human milk and skin (Umbach et al. 2021). In general, Methanobacteriota are important not only in biogas production but also in the field of health through their interactions within microbial communities (Khesali Aghtaei et al. 2022).

3.2.1.4 Thermoplasmatota

Thermoplasmatota is a globally distributed and ecologically important phylum comprising various classes (Zinke et al. 2021; Zheng et al. 2022; Kadnikov et al. 2024). The presence of photolyase and rhodopsin genes, along with genes for peptide and lipid degradation, in the metagenome of *Candidatus Pontarchaea* suggests the possibility of photoheterotrophic and heterotrophic metabolism in archaea (Zheng et al. 2022; Kadnikov et al. 2024). The traditional view is that methanogenesis is the primary metabolism of the Thermoplasmatota phylum. However, genomic analyses have revealed that there is no evidence of mcrA homologs outside of Methanomassiliicoccales within Thermoplasmatota, and that only Methanomassiliicoccales are methanogens utilizing a truncated methylotrophic methanogenesis pathway. Other Thermoplasmatota members have been found to employ various metabolic strategies for energy conservation, including heterotrophy, sulfur and hydrogen metabolism, denitrification, and fermentation. The presence of members of this phylum in ecosystems such as geothermal sources, alpine caves, hypersaline and subarctic lake waters, and sediments underscores their ecological importance in carbon fixation processes and their metabolic diversity (Zinke et al. 2021; Zheng et al. 2022; Kadnikov et al. 2024). A genomic analysis revealed that Thermoplasmatota genomes are the third most abundant in unknown genes (Zheng et al. 2022).

3.2.1.5 Hadarchaeota

Hadarchaeota inhabit various anoxic underground habitats, including hot springs, hydrothermal sediments, deep-sea sediments, aquifers, and cold seeps. Genomic analyses indicate that most Hadarchaeota are heterotrophs that grow on sugars and amino acids, or oxidize carbon monoxide, and possess genes encoding alkyl-coenzyme M reductases (Acrs), β-oxidation, and the Wood-Ljungdahl pathway. Both unique metabolic capabilities, along with their ability to break down hydrocarbons and their symbiotic relationships with sulfate-reducing bacteria, suggest that they play an important role in petroleum degradation. The active participation of Hadarchaeota in the methane cycle and hydrocarbon degradation generally indicates that they are vital components in biogeochemical processes in extreme environments (Benito Merino et al. 2024; Yu et al. 2024).

3.2.2 *TACK Superphylum*

The TACK superphylum is a significant superphylum that encompasses the Thaumarchaeota, Aigarchaeota, Crenarchaeota, and Korarchaeota phyla, exhibiting a diverse range of physiological and ecological adaptations (Guy and Ettema 2011; Ren et al. 2018; Zhou et al. 2022; Amils 2023). Subsequently, the phyla Bathyarchaeota, Hydrothermarchaeota, Geothermarchaeota, and Verstraetearchaeota were added to TACK (Baker et al. 2020). This superphylum is particularly notable for its distinct metabolic capabilities and ecological roles in carbon and nitrogen biogeochemical cycles. Phylogenetic studies have revealed numerous shared features between the TACK superphylum and eukaryotes, including proteins involved in cytokinesis, membrane remodeling, cell shape determination, and protein recycling (Guy and Ettema 2011; MacLeod et al. 2019). Studies have shown that eukaryotic signature proteins (ESP) typically include the true archaeal orthologs of actin, tubulin, ESCRT (endosomal sorting complex required for transport) proteins, and several proteins involved in transcription and translation (Bulzu et al. 2019). Metagenomic studies have also revealed that several lineages within the TACK superphylum encode methyl-coenzyme M reductase (MCR), an enzyme specific to the alkane cycle that was previously thought to be unique to Euryarchaeota. This suggests that the potential for methanogenesis and alkane oxidation is more widespread than previously recognized (Kohtz et al. 2022). The TACK superphylum has been identified in various extremophilic environments, including high-temperature, low-temperature, and low-pH conditions, with most members of the group being anaerobic (Amils 2023). Many thermophilic archaea, including members of the TACK superphylum, are known to be heterotrophs or chemolithotrophs with sulfur-dependent metabolisms and a free-living lifestyle (Qu et al. 2023).

3.2.2.1 Thaumarchaeota

In the early 1990s, they were discovered as uncultured archaeal lineages distantly related to the Crenarchaeota in both marine and terrestrial ecosystems (Dombrowski et al. 2021). In 2008, they were proposed and accepted as a new archaeal phylum based on the analysis of the genome of "*Ca. Cenarchaeum symbiosum*." The position of Thaumarchaeota within the tree of life, its genetic markers, and the presence of specific membrane lipids support its status as a unique organism (Hatzenpichler 2012). Prior to the discovery of Thaumarchaeota, it was assumed that all archaeal lineages were obligate anaerobes and lived in extreme environments. The first isolated strain of Thaumarchaeota, *Nitrosopumilus maritimus*, represents an aerobic lifestyle dependent on ammonia and nitrite oxidation (Zou et al. 2025).

Most Thaumarchaeota members cultured to date are chemolithotrophic ammonia-oxidizing archaea (AOA) (Dombrowski et al. 2021). AOA are one of the most common yet least explored archaeal groups, having adapted remarkably well to the harsh conditions of the subsurface biosphere (Schleper and Nicol 2010; Kerou et al.

2021). The discovery of archaeal ammonia oxidizers has led to a fundamental shift in the bacterial-centric view of nitrification. The first evidence of archaeal involvement in ammonia oxidation emerged with the discovery of ammonia monooxygenase (AMO)-like genes associated with archaeal skeletons in surface seawater and soil. Thaumarchaeota play a crucial role in the global nitrogen cycle, as they can aerobically oxidize ammonia under diverse conditions. Studies indicate that they contribute to nitrification by expressing ammonia-oxidizing genes in soil ecosystems and also influence global nitrogen transformations (Hatzenpichler 2012). They are hypothesized to have played a key role in the global nitrogen cycle during Earth's early history (Kerou et al. 2021).

It is believed that AOA originated in terrestrial habitats, then spread to soil, hot springs, and freshwater environments across different biomes, before extending to shallow waters and deeper oceans. AOA are well-equipped to cope with the challenges of the oxygenated deep-sea environment, encoding the most energy-efficient aerobic carbon fixation pathway and exhibiting a high affinity for ammonia. Some populations have also been reported to be mixotrophic (Kerou et al. 2021). Molecular studies indicate that these archaea are also present in environments considered extreme, such as hot springs. AOA constitutes one of the archaeal phyla found in nearly every environment on Earth, having colonized almost every oxygenated environment on the planet. They are known to be many times more abundant than bacterial ammonia oxidizers in environments such as marine plankton, soil, and estuaries (Schleper and Nicol 2010; Kerou et al. 2021). The fact that ammonia-oxidizing archaea have adapted to thrive at ammonia concentrations 100 times lower than those required by bacteria explains their widespread distribution in the open ocean (Zou et al. 2022).

3.2.2.2 Aigarchaeota

Aigarchaeota is a phylum found in various environments, including terrestrial and underground geothermal systems, as well as marine hydrothermal environments (Dombrowski et al. 2021). However, its ecology, physiology, and evolutionary history remain largely unknown. Genomic analyses have revealed that Aigarchaeota are both facultative and obligate anaerobes, with some species being anaerobic chemolithotrophs capable of utilizing other electron acceptors, such as oxidized sulfur or nitrogen compounds. Comparative genomic analyses suggest that they share a common evolutionary history with Thaumarchaeota, and that horizontal gene transfer has contributed to their genetic diversity. Metabolic profiling has demonstrated that Aigarchaeota possess versatile metabolic capabilities, including the potential for autotrophy and the utilization of various organic substrates (Beam et al. 2016; Hua et al. 2018).

3.2.2.3 Crenarchaeota

Crenarchaeota is an archaeal phylum characterized by members that reproduce in high-temperature habitats, such as hot springs and hydrothermal vents, which were discovered through culture-based approaches (Baker et al. 2020; Dombrowski et al. 2021). Crenarchaeota was initially classified as a superphylum based on 16S rRNA gene phylogeny and the absence of histones (Dombrowski et al. 2021). However, archaeal histone-coding genes have been found in marine Crenarchaea. The detection of histones in some Crenarchaeotes also supports the view that histones evolved before the divergence of archaea and Eukaryota (Čubonová et al. 2005).

Crenarchaeota are known for their thermophilic and acidophilic properties, but some species are also known as hyperthermophiles, which can live at temperatures exceeding 113°C. Nearly all cultured species are obligate hyperthermophiles, with optimal growth temperatures ranging from 70 to 113°C. In addition to adapting to high temperatures, hyperthermophiles have developed various metabolic strategies, such as lithotrophy, which derives energy from inorganic compounds (Baker et al. 2020; Dombrowski et al. 2021). Crenarchaeota play a vital role in various biogeochemical cycles, particularly in extreme environments such as hot springs and deep-sea hydrothermal vents. They are crucial for the global nutrient cycle, particularly in nitrogen and sulfur metabolism, which are essential for ecosystems (Dombrowski et al. 2021).

3.2.2.4 Korarchaeota

Korarchaeota is an uncultured archaea group that has existed since the late 1990s. They have been found in both terrestrial and marine thermal environments (Dombrowski et al. 2021; Amils 2023). Genomic analysis of Ca. *Korarchaeum cryptofilum*, isolated from Obsidian Pool in Yellowstone, is a unique peptidolytic fermenter with a distinct gene set (Dombrowski et al. 2021). Data from deep-sea hydrothermal vent sediments and hot spring environments have revealed new metabolic features of this group. Key genes involved in methane metabolism have been identified in the genomes of certain members (Dombrowski et al. 2021; Amils 2023). The detection of methanogenesis in one of the most ancient archaeal phyla supports the hypothesis that this metabolism represents an early energy conservation system (Amils 2023).

3.2.2.5 Bathyarchaeota

Bathyarchaeota is a newly proposed phylum mostly abundant in anoxic sediments worldwide and plays a crucial role in biogeochemical cycles (Zhou et al. 2018; Dombrowski et al. 2021; Xue et al. 2023). Bathyarchaeota were first identified in hot springs and designated as the Miscellaneous Crenarchaeota Group (MCG) (Xue et al. 2023). It is proposed that acetyl-coenzyme A-centered heterotrophic energy conservation pathways function in Bathyarchaeota (Zhou et al. 2018). Many Bathyarchaeota can utilize cellulose, chitin, acetate, proteins, and aromatic compounds as nutrients (Zhou et al. 2018; Xue et al. 2023). Members of this phylum possess capabilities such as acetogenesis, methane metabolism, and interactions with other microbial groups. As the dominant archaeal lineage in sedimentary environments, Bathyarchaeota play a vital role in the carbon cycle. These archaea are often found in association with methanogens and ammonia-oxidizing archaea (Xue et al. 2023).

3.2.2.6 Geoarchaeota

Geoarchaeota, also known as New Archaeal Group 1 (NAG1), is typically found in sources containing varying levels of oxygen, heat, acidity, and iron. Genomic studies have shown that Geoarchaeota possess pathways necessary for the catabolism of peptides and carbohydrates, as well as a Form I carbon monoxide dehydrogenase complex that is thought to enable energy conservation (Kozubal et al. 2013). NAG1 also possesses unique bacterial-like biosynthesis and transport genes, as well as a Type 3 CRISPR system. Discovering NAG1 was crucial for understanding the microbial communities in the current thermophilic iron oxide mats at Yellowstone (Kozubal et al. 2013; Dombrowski et al. 2021).

3.2.2.7 Verstraetearchaeota

Verstraetearchaeota, discovered in microbial communities in South African gold mines, is also known as the Terrestrial Miscellaneous Crenarchaeota Group (TMCG). Members of this group are found in hydrocarbon-rich soils and wetlands (Dombrowski et al. 2021). Verstraetearchaeota is a recently discovered archaeal lineage with unique methane metabolism capabilities that emphasize hydrogenotrophic and methylotrophic pathways. This phylum possesses methyl-coenzyme M reductase (Mcr) genes, which are involved in the production of methane. Their metabolic diversity suggests that Verstraetearchaeota are adapted to utilize different substrates, supporting the view that archaea play a role in the global carbon cycle (Vanwonterghem et al. 2016).

3.2.2.8 Nezhaarchaeota

Nezhaarchaeota represents a lineage of non-traditional methanogenic archaea that contribute significantly to methane production in geothermal sources. Genomic analyses indicate that Nezhaarchaeota possesses the methyl-coenzyme M reductase (Mcr) gene and generally performs hydrogenotrophic methanogenesis under specific anaerobic conditions (Wang et al. 2019). Since their methanogenic activities are affected by temperature, methylotrophic pathways are preferred at higher temperatures (Welte et al. 2021).

3.2.2.9 Marsarchaeota

Marsarchaeota, or New Archaeal Group 2 (NAG2), is found in iron oxide-rich mats. The first genomes of Marsarchaeota were obtained from thermal and acidic microbial mats in Yellowstone National Park. These are aerobic chemoorganotrophs capable of degrading lipids, peptides, and carbohydrates, as well as reducing ferric oxide (Dombrowski et al. 2021). Phylogenetically, Marsarchaeota is the sister group of Geoarchaeota. Metagenomic and metatranscriptomic analyses indicate that Marsarchaeota are aerobic chemoheterotrophs that take up sugars and peptides/amino acids, oxidizing and assimilating them via glycolysis, the pentose phosphate pathway, and the oxidative tricarboxylic acid cycle (Teske 2018).

3.2.2.10 Geothermarchaeota

Geothermarchaeota is one of the most recently added phyla of archaea. Their genomes have been reconstructed from metagenomes of the seafloor at the Juan de Fuca Ridge in the Guaymas Basin (Jungbluth et al. 2017) and from hydrothermal vent sediments (Dombrowski et al. 2020). Knowledge about the lifestyle and ecological roles of Geothermarchaeota is still limited (Dombrowski et al. 2021).

3.2.3 DPANN Superphylum

They are characterized by their tiny size ($<0.1\ \mu m^3$) and small genomes, which limit their metabolic capacity (Sakai et al. 2022; Amils 2023). The DPANN superphylum comprises archaeal phyla that are believed to rely on other microorganisms for reproduction and survival (Dombrowski et al. 2021; Baker et al. 2025). Although initially included phyla Diapherotrites, Parvarchaeota, Aenigmarchaeota, Nanoarchaeota, and Nanohaloarchaeota (DPANN) (Zhou et al. 2022; Amils 2023; Baker et al. 2025), phyla such as Altiarchaeota, Huberarchaea, Micrarchaeota, Pacearchaeota, Undinarchaeota, and Woesearchaeota have been added (Zhou et al. 2022; Baker et al. 2025).

DPANN genomes lack genes necessary for amino acid, nucleotide, and phospholipid biosynthesis, as well as for free-living lifestyles. This indicates dependence on other organisms for basic metabolic requirements and the potential for (epi)symbiotic lifestyles (Baker et al. 2025). While many DPANN members are associated with symbiotic/parasitic interactions with other organisms, some maintain a free-living lifestyle, primarily through fermentation for cellular energy production (Sakai et al. 2022; Zhou et al. 2022; Amils 2023). Although DPANNs are metabolically constrained, some possess the genomic capacity to utilize organic compounds, such as carbon and lipids, via glycolysis, beta oxidation, and other pathways (Baker et al. 2020). DPANN archaea have been found to possess genes for sulfur cycling and the degradation of organic substrates (Zhang et al. 2024). Many of them possess genes for fermentation products such as lactate, acetate, formate, and ethanol, indicating that they have evolved to utilize substrate-level phosphorylation for energy (Cai et al. 2021b; Cloarec et al. 2024). However, additional catabolic pathways, such as the Embden-Meyerhof-Parnas pathway, an incomplete Entner-Doudoroff pathway, a beta-oxidation pathway, and a nucleoside degradation pathway linked to RubisCO, have also been reported, highlighting the genomic diversity within this taxonomically large superphylum (Cloarec et al. 2024). In anoxic environments, certain DPANN archaea can form consortia with methanogens, contributing to the anaerobic carbon cycle (Zhang et al. 2024).

This superphylum accounts for approximately half of the archaea diversity in the biosphere. Since their discovery, they have been found in both aerobic and anaerobic biomes, including acidic, alkaline, and hypersaline ecosystems, such as underground water layers, fresh, marine, and hypersaline waters and sediments, hydrothermal vents, hot springs, acid mine drainage, and the human microbiome (Sakai et al. 2022; Amils 2023; Cloarec et al. 2024).

3.2.3.1 Nanoarchaeota

Nanoarchaeota are a unique, largely uncultured phylum of archaea that live as obligate symbionts on the surfaces of diverse archaeal hosts (Jarett et al. 2018). The first DPANN archaeum is associated with the species *Ignicoccus hospitalis*, a hyperthermophilic Crenarchaeota. This is a hyperthermophilic, anaerobic, small-celled organism called *Nanoarchaeum equitans* (Huber et al. 2003; Baker et al. 2020). The genome of *N. equitans* is one of the smallest genomes known (490 kb). Studies have shown that *N. equitans* requires cell–cell contact with its host to survive; the host does not need any nanoarchaeota. Whole-genome sequencing reveals that *N. equitans* lacks many essential genes, including those involved in lipid, cofactor, amino acid, and nucleotide biosynthesis, consistent with its host-dependent lifestyle. This indicates that *N. equitans* is an obligate symbiont (St. John 2019; Baker et al. 2020).

It is hypothesized that members of the Nanoarchaeota, which are present in various thermal environments, infect different crenarchaeal hosts. For instance, Nanoarchaeota species, including *Candidatus* Nanobsidianus stetteri, *Candidatus* Nanopusillus acidilobi, *Candidatus* Nanoclepta minutus, *Acidolobus* sp., *Acidilobus* sp. 7A, and *Zestosphaera tikiterensis*, have recently been found to co-produce with their crenarchaeal hosts (St. John 2019). Genomic studies demonstrate that Nanoarchaeota interact with viral populations in geothermal environments. CRISPR-Cas cassettes have been identified in *N. equitans* and *C. minutus*, and Cas genes have been discovered in deep-sea nanoarchaeotal metagenome-assembled genomes (MAGs). While viruses can pose a threat to these cells, they can also serve as a source of genetic material to replace functions lost through HGT (St. John 2019).

3.2.3.2 Diapherotrites

The phylum Diapherotrites has been proposed based on genome analysis of a groundwater seep from an abandoned gold mine in the USA. Phylogenetic analyses indicate that the Diapherotrites 16S rRNA genes are incompatible with most archaeal 16S rRNA primers, whereas most of its anabolic genes originated from bacteria through horizontal gene transfer. This suggests that Diapherotrites evolved from an obligate symbiotic ancestor, acquiring anabolic genes from bacteria and biosynthesizing molecules from their hosts (Youssef et al. 2015). Members of Diapherotrites can conserve energy through fermentation. However, the lack of specific biosynthetic pathways suggests that they still depend on the external acquisition of amino acids, vitamins, and/or cofactors (Dombrowski et al. 2021).

3.2.3.3 Nanohaloarchaeota

Nanohaloarchaeota are highly halophilic archaea characterized by obligate symbiotic relationships with other halophilic archaea, especially in extreme saline environments. Nanohaloarchaeota possess "SPEARE" proteins that can function in attaching to and invading hosts (Youssef et al. 2015; Xie et al. 2022; Reva et al. 2023). As a result of genomic analyses, Nanohaloarchaeota are considered to be aerotolerant anaerobes, whereas their hosts generally perform aerobic respiration (Youssef et al. 2015). While cell volumes approach the theoretical limit for a unicellular life form, nanohaloarchaea exhibit multiple autotrophies, suggesting a dependence on haloarchaeal hosts (Reva et al. 2023). The discovery of a new group of Nanohaloarchaeota in deep-underground marine sediment ecosystems suggests that Nanohaloarchaeota may inhabit more habitats than previously thought and may have new functional niches that allow them to reproduce in extreme environments (Xie et al. 2022).

3.2.3.4 Parvarchaeota

The phylum Parvarchaeota is known to physically interact with some members of Thermoplasmatales in nature (Chen et al. 2018). Despite their small size (0.64–1.08 Mb), they can play a crucial role in the carbon and nitrogen cycles, break down sugars and proteins, and produce ATP. However, they lack pathways to produce amino acids and nucleotides, so they must obtain these molecules from the environment (Chen et al. 2018; Luo et al. 2020). Therefore, these ultra-small archaea may adopt a lifestyle dependent on other community members. Parvarchaeota of minimal size have an incomplete tricarboxylic acid cycle. However, NADH dehydrogenase and electron transport chains are absent. Environmental adaptation capabilities, such as carbohydrate and protein utilization, as well as resistance to metal and oxidative stress, may enable these microbes to survive in harsh environmental conditions (Luo et al. 2020).

3.2.3.5 Aenigmarchaeota

The Aenigmarchaeota phylum comprises uncultivable archaea primarily found in various thermal and anoxic environments, including brackish and freshwater, as well as hydrothermal environments (Li et al. 2021a). Aenigmarchaeota exhibits distinct characteristics indicative of an anaerobic, syntrophic lifestyle that relies heavily on symbiotic relationships for energy production and nutrient acquisition. Aenigmarchaeota can form symbiotic relationships with microorganisms, such as *Pyrobaculum*, but host specificity can be broad and variable (Cai et al. 2020; Li et al. 2021b). Metagenomic analyses have shown that, although Aenigmarchaeota genomes generally lack many genes associated with basic metabolic pathways, they possess important features in carbohydrate metabolism, enabling them to adapt to various environmental stress conditions (Cai et al. 2021b; Li et al. 2021b). Evolutionary analyses indicate that horizontally acquired genes from the TACK superphylum and Euryarchaeota members significantly contribute to the niche expansion of Aigarchaeota by conferring the ability to survive various environmental stresses, including high temperatures and oxidative stress (Li et al. 2021b).

3.2.3.6 Altiarchaeota

Altiarchaeota have been proposed as free-living organisms inhabiting environments such as sulfur swamps, sulfur-rich water sources, marine sediments, hot springs, and aquifers (Li et al. 2021a). However, some Altiarchaeota species serve as hosts to Huberarchaeota, another DPANN phylum (Baker et al. 2025). Altiarchaeota possess hook-like attachment extensions called "hammocks" that play a role in binding for biofilm formation. These extensions are believed to be related to Altiarchaeota's adaptation to riverine environments (Baker et al. 2020; Li et al. 2021a). Some Altiarchaeota members form "pearl string"-like microbial consortia composed of small white pearls (0.5–3 mm in diameter) connected by thin filaments. The outer part of the bead consists of bacteria such as Gammaproteobacterium *Thiotrix uunzi* or *Epsilonproteobacterium* IMB1, while Altiarchaeota dominate the inner part. This provides an excellent environment for horizontal transfer of genetic material between different areas (Rudolph et al. 2001).

Altiarchaeota, which has the largest genome among the DPANN phyla (Dombrowski et al. 2021; Li et al. 2021a), also possesses many pathways not found in other phyla, such as formaldehyde assimilation and the Wood-Ljungdahl pathway. Additionally, lateral gene transfer enables the genetic flexibility of DPANN archaea, allowing them to thrive in diverse ecological niches (Li et al. 2021a).

3.2.3.7 Woesearchaeota

Woesearchaeota has not yet been cultured, meaning its lifestyle and metabolism remain a mystery. Analysis of some genomes suggests that Woesearchaeota may utilize anaerobic processes and fermentative metabolism via lactate dehydrogenase, [FeFe] hydrogenase, and flavin-based mechanisms. In contrast, some genomes from marine and hypersaline environments contain genes involved in the sulfur cycle (Cloarec et al. 2024). Woesearchaeota are abundant in environments such as deep oil reservoirs, sulfurous sources, and anoxic aquifers. This highlights the diversity of their roles in global biogeochemical cycles (Liu et al. 2021). In Woesearchaeota, most key metabolic pathways, such as glycolysis, the TCA cycle, amino acid synthesis, and lipid metabolism, are either absent or incomplete. This suggests a strong dependence on the host or environment for their requirements. However, analyses suggest that genomic and cellular characteristics may stem from ancestral traits (Cloarec et al. 2024). Both the abundance distribution and simultaneous network analysis studies in various biotopes have revealed a potential symbiotic relationship between Woesearchaeota and methanogens (Liu et al. 2021).

3.2.3.8 Micrarchaeota

Micrarchaeota are poorly characterized archaea with reduced genomes whose survival depends on host interactions (Krause et al. 2022). It is known that small acidophilic archaea belonging to the Micrarchaeota physically interact with certain members of the Thermoplasmatales in nature (Chen et al. 2018; Golyshina et al. 2019). Recent studies have demonstrated that Micrarchaeota can be cultured alongside Thermoplasmatales hosts, underscoring the role of biofilm formation in their interactions (Golyshina et al. 2019; Krause et al. 2022). Genomic analyses have revealed that Micrarchaeota acquire metabolites from their hosts, indicating a process of nutrient exchange (Krause et al. 2022). Micrarchaeota have been identified in various habitats, including acidic environments, where they play ecological roles such as carbon and nitrogen cycling (Golyshina et al. 2019; Krause et al. 2022). Genomic analyses indicate that Micrarchaeota possess both microaerophilic and anaerobic lifestyles, as evidenced by the presence of aerobic respiratory chains, fermentation potential, and genes encoding cytochrome bd-II (Golyshina et al. 2019).

3.2.3.9 Huberarchaeota

The proposed symbiotic archaeon of Altiarchaeum, Candidatus Huberarchaea, has recently been added to the DPANN superphylum (Li et al. 2021a). Among the genome-reduced DPANN archaea, the ectoparasitic Nanoarchaeota and Huberarchaea have the smallest known genomes (Kellner et al. 2018). Based on varying cell abundance profiles and microscopic imaging, Huberarchaea is thought to live in symbiosis with Altiarchaeota (Li et al. 2021a).

3.2.3.10 Pacearchaeota

Metagenomic analysis has revealed the absence of biosynthetic genes for nucleotides, amino acids, cofactors, and lipids in Pacearchaeota genomes. Pacearchaeota, which have small genomes, are often found alongside other archaeal groups such as Woesearchaeota. They are a stable component of microbial communities in oxygen-depleted regions of the ocean, playing important roles in the carbon and nitrogen cycles (Takamiya et al. 2024).

3.2.3.11 Undinarchaeota

Undinarchaeota have small genomes and may rely on symbiotic organisms to meet their basic nutritional needs. Phylogenomic studies place Undinarchaeota as an independent DPANN lineage among other groups. These archaea likely possess limited metabolic capabilities, including energy conservation through fermentation. Horizontal gene transfer complicates the phylogenetic placement of DPANN within the archaeal tree of life, including Undinarchaeota (Dombrowski et al. 2020).

3.2.4 Asgard Superphylum

Asgard archaea, also known as Asgardarchaeota, is a crucial superphylum for understanding the origins of eukaryotes (Cai et al. 2021a; Amils 2023). Except for the strain "*Candidatus* Prometheoarchaeum syntrophicum" found in the Lokiarchaeota phylum, no Asgard members have been cultured in the laboratory to date, so most of the information about this group is based on metagenomic studies (Cai et al. 2021a; Banciu et al. 2022). Within Asgardarchaeota, numerous metagenome-assembled genomes (MAGs) have been taxonomically classified in publicly available databases (MacLeod et al. 2019). Phylogenetic analyses indicate that Asgard archaea genomes contain eukaryotic signature proteins (ESPs), suggesting that they are the closest relatives of archaea to eukaryotes. Most of these ESPs are proteins associated with highly organized cellular features, such as transport mechanism proteins, cytoskeletal proteins, proteins involved in the ubiquitination modification system, and eukaryotic-like ribosomal proteins (MacLeod et al. 2019; Zhou et al. 2022; Amils 2023). Metagenomic and 16S rRNA gene analyses have shown that this archaeal group is present in various microbiomes, including marine sediments, hydrothermal vents, rivers, freshwater lakes, and terrestrial soils (MacLeod et al. 2019; Amils 2023), and factors such as water depth and salinity have been identified as driving forces influencing the distribution of Asgardarchaeota (Cai et al. 2021a). Asgard archaeal genomes possess metabolic pathways such as the Wood-Ljungdahl pathway and nucleotide salvage pathways (MacLeod et al. 2019). Asgard archaea are generally obligate anaerobes, though some members can grow in the presence of O2. Based on carbon sources, there are two main groups: autotrophic and heterotrophic.

Additionally, the identification of a new rhodopsin class in Asgard archaeal genomes suggests a phototrophic lifestyle for Asgard (MacLeod et al. 2019; Amils 2023). Some Asgard archaea play important roles in the nitrogen and sulfur cycles (MacLeod et al. 2019) and are considered key participants in the degradation of organic carbon in sediments. Additionally, metagenomic studies have shown that Asgard archaea possess various physiological characteristics, including hydrogen-dependent anaerobic autotrophy, peptide-dependent organotrophy, and short-chain hydrocarbon-dependent organotrophy (Cai et al. 2021a).

Newly defined phyla, including Asgard archaea, Lokiarchaeota, Thorarchaeota, Odinarchaeota, Heimdallarchaeota, Helarchaeota, and Gerdarchaeota, have been identified (Dombrowski et al. 2021; Amils 2023).

3.2.4.1 Lokiarchaeota

Lokiarchaeota represents an archaeal lineage that is abundant in many coastal and various marine sediments (Yin et al. 2021). It is known that Lokiarchaeota possesses ESP involved in various eukaryotic processes, such as membrane-related processes and cell shape dynamics. Additionally, similarities have been identified between eukaryotic and Asgard archaeal RNA structures in Asgard genomes. The mechanisms for selenocysteine (SEC) insertion during translation vary significantly among bacteria, archaea, and eukaryotes. The selenocysteine insertion sequence (SECIS) of Lokiarchaeota resembles both the RNA structures of eukaryotic SECIS and the SECIS of archaeal selenoproteins. Therefore, it has been proposed that Lokiarchaeota encodes an intermediate archaeal-eukaryotic selenocysteine insertion system and that the eukaryotic SECIS has archaeal evolutionary origins (MacLeod et al. 2019).

They use the tetrahydromethanopterin-dependent Wood-Ljungdahl (H4MPT-WL) pathway for inorganic carbon fixation. This is a well-documented feature of Lokiarchaeota (Dombrowski et al. 2021; Yin et al. 2021). The Wood–Ljungdahl pathway (WLP) enables the reduction of carbon dioxide to acetyl-coenzyme A (CoA) and can be used to support both autotrophic carbon fixation and energy conservation when coupled with chemiosmotic processes. However, genomic analyses have also revealed the potential for utilizing various organic carbon substrates. Lokiarchaeota primarily reproduce as fermentative organoheterotrophs (Spang et al. 2017). Additionally, some studies have noted that certain Lokiarchaeota members are homoacetogenic and that a Lokiarchaeum reproduces syntrophically with methanogens or sulfate-reducing bacteria while breaking down amino acids. These findings indicate significant diversity in the physiology of Lokiarchaeota (Dombrowski et al. 2021; Yin et al. 2021).

3.2.4.2 Thorarchaeota

Although they have not yet been produced in a laboratory setting, recent genomic analyses have revealed overlapping genomic content between Thorarchaeota and Lokiarchaeota (Manoharan et al. 2019). Additionally, they utilize the Wood-Ljungdahl pathway for carbon fixation or during growth on organics (Dombrowski et al. 2021). These archaea can utilize a variety of inorganic and organic carbon sources. They are thought to possess genes encoding ribulose bisphosphate carboxylase-like proteins and are estimated to have a nearly complete Calvin-Benson-Bassham cycle (Liu et al. 2018). Additionally, a methyl-CoM reductase-like enzyme has been identified, indicating the potential for anaerobic oxidation of short-chain hydrocarbons (Manoharan et al. 2019).

The presence of eukaryotic selenocysteine insertion sequences and numerous genes encoding proteins previously thought to be unique to eukaryotes in Thorarchaeota genomes provides new insights into their evolutionary roles. Among Asgard archaea, Thorarchaeota genomes encode unique proteins for eukaryotic membrane trafficking machinery and vesicle biogenesis, suggesting a potential role for Thorarchaeota in eukaryotic evolution (Liu et al. 2018). Thorarchaeota are found in various microbiomes, particularly in marine, lake, and estuarine sediments (MacLeod et al. 2019).

3.2.4.3 Odinarchaeota

Odinarchaeota has been identified in hot spring metagenomes (Yellowstone National Park and Radiata Pool) (Zaremba-Niedzwiedzka et al. 2017). The only available MAG indicates a thermophilic fermentation lifestyle (Zaremba-Niedzwiedzka et al. 2017; Tahon et al. 2021). A true tubulin ortholog related to eukaryotic tubulins, rather than the artubulins previously reported in Thaumarchaeota, has been identified in Odinarchaeota members (Zaremba-Niedzwiedzka et al. 2017). Odinarchaeum, like other Asgard archaea, has the ability to use organic compounds as substrates. It lacks the key enzyme of the Wood-Ljungdahl pathway and the nitrite reductase (NADH) enzyme. Members of the Odinarchaeota are thought to primarily inhabit thermal environments, such as hot spring sediments and hydrothermal vents (Dombrowski et al. 2021).

3.2.4.4 Heimdallarchaeota

Heimdallarchaeota members are considered the most likely candidates for the archaeal proto-eukaryotic ancestor identified to date (Dombrowski et al. 2021; Liu et al. 2024). The Heimdallarchaeota members identified to date are metabolically distinct from other Asgard lineages. Genomic analyses indicate that Heimdallarchaeota members, like other Asgard members, can utilize a variety of organic substrates. However, the existing Heimdallarchaeota members are not fermentative organisms, as indicated by the absence of the Wood-Ljungdahl (WL) pathway in their metagenome-assembled genomes (MAGs). Wood-Ljungdahl (WL) pathway (Dombrowski et al. 2021), sulfur metabolism, and the reductive citrate cycle (rTCA) (Liu et al. 2024). Heimdallarchaeota members have been shown to possess the ability to reduce sulfate (assimilator) and nitrate (disassimilator) to sulfur and ammonia, respectively, thereby playing a crucial role in the biogeochemical cycle. Additionally, it has been noted that they can synthesize bacteriochlorophyll and carotenoids and utilize light through these pigments, and that the effect of a new multifunctional MetH enzyme enables Heimdallarchaeota phyla to fix CO_2 via an atypical Wood-Ljungdahl pathway. Considering its phototrophic lifestyle alongside other anaerobic and aerobic metabolic pathways, it is likely that Heimdallarchaeota is photomixotrophic, which provides flexibility in adapting to harsh deep-sea conditions (Liu et al. 2020; Liu et al. 2024).

3.2.4.5 Helarchaeota

Although Helarchaeota and Loki, and Thorarchaeota share similar gene sets, the known Helarchaeota representatives identified to date contain mcr gene clusters (Dombrowski et al. 2021; Zhao and Biddle 2021). The presence of these mcr genes is limited to Helarchaeota among the known Asgard archaea. Therefore, these genes were likely transferred to Helarchaeota and do not constitute an ancestral feature within the Asgard superphylum. Based on the current phylogenetic analysis, the Helarchaeota mcr gene cluster may have been horizontally acquired from Bathyarchaeota (Seitz et al. 2019). Genomic analysis of Helarchaeota reveals the presence of three variants of a methyl-CoM reductase-like complex that catalyzes the anaerobic oxidation of short-chain alkanes. Most of these characterized alkane-oxidizing archaea form syntrophic associations with sulfate-reducing bacteria (SRB), which utilize the reducing equivalents released during alkane oxidation for sulfate reduction (Zhao and Biddle 2021). Some methanogenic archaea utilize formate for energy transfer to a symbiotic partner; therefore, it has been hypothesized that the reverse reaction could be harnessed as an energy source for methane oxidation (Seitz et al. 2019).

3.2.4.6 Gerdarchaeota

Gerdarchaeota is a newly reconstructed phylum from metagenomic analyses of coastal sediments. This group has the capacity to utilize both organic and inorganic carbon and exhibits facultative anaerobic properties. While sharing some similarities with Heimdallarchaeota, Gerdarchaeota possesses distinct genes for cellulase and enzymes for the Wood-Ljungdahl pathway. This expanded understanding of Gerdarchaeota contributes to the overall diversity of Asgard archaea and highlights their ecological roles in the degradation of organic matter in sediment layers (Cai et al. 2020).

3.3 Metabolic and Biochemical Diversity

The domain Archaea occupies a central position in the evolutionary diversity of life. In terms of molecular biology, cell architecture, and metabolism, archaea differ substantially from both bacteria and eukaryotes. Their metabolic and biochemical diversity not only enables them to thrive under extreme conditions but also offers exceptional potential for biotechnological applications. This section surveys archaeal energy metabolisms, pathways of carbon and nitrogen fixation, coenzyme systems, membrane architectures, and biochemical adaptations to extreme environments.

3.3.1 Energy Metabolism

3.3.1.1 Methanogenic Energy System

Methanogenic metabolism is unique to archaea and yields methane (CH_4) by reducing substrates, including carbon dioxide (CO_2), acetate, formate, or methanol. Although the energy yield is modest, ATP synthesis suffices for survival in anoxic habitats with low redox potential (e.g., sediments and animal guts). Key coenzymes include coenzyme M (CoM), coenzyme B (CoB), methanofuran, tetrahydromethanopterin, and the low-potential redox carrier coenzyme F_{420}; methanophenazine serves as a lipid-soluble electron carrier within the membrane. Methane formation is catalyzed by methyl-coenzyme M reductase (MCR), and the free-energy change is conserved by generating a proton or sodium ion gradient that drives ATP synthase. Moreover, several anaerobes increase energetic efficiency by electron bifurcation, allowing endergonic reductions to proceed by coupling them to exergonic ones (Buckel and Thauer 2013; Deppenmeier 2002; Deppenmeier 2004; Greening et al. 2016; Ney et al. 2017).

3.3.1.2 Phototrophic Energy System

Halophilic archaea lack chlorophyll-based photosystems yet harness light to make ATP using bacteriorhodopsin, a retinal-binding transmembrane protein. Upon illumination, retinal isomerization triggers outward proton pumping, generating a proton-motive force that powers ATP synthesis. These organisms may also deploy halorhodopsin, a light-driven chloride pump that contributes to ion homeostasis (Oesterhelt and Stoeckenius 1971; Kolbe et al. 2000).

3.3.1.3 Chemolithotrophic Energy System

Many archaea—especially Crenarchaeota and some Euryarchaeota—oxidize inorganic substrates such as elemental sulfur (S^0), thiosulfate ($S_2O_3^{2-}$), hydrogen (H_2), or ferrous iron (Fe^{2+}). Sulfur-metabolizing thermoacidophiles (e.g., members of Sulfolobales) typically use flexible respiratory chains and can perform aerobic sulfur oxidation and, in some taxa, anaerobic sulfur respiration with hydrogen as an electron donor. Their electron transport features alternatives to the "classical" bacterial modules, relying on ferredoxin, menaquinone, cytochrome c, and other carriers adapted to high-temperature, low-redox-potential niches (Kletzin et al. 2015).

3.3.1.4 Alternative Energy Systems and Redox Flexibility

Extremophilic archaea often tune ATP synthase ion specificity: in several haloarchaea and hyperthermophiles, A_1A_0-type ATP synthases can be Na^+-coupled, H^+-coupled, or even promiscuous (able to use both ions), enhancing robustness under high salinity or pH stress (Pisa et al. 2007; McMillan et al. 2011; Schlegel et al. 2012; Mayer et al. 2015).

3.3.2 Carbon Metabolism

Archaea display both autotrophic and heterotrophic strategies for carbon acquisition. Autotrophs (e.g., hydrogenotrophic methanogens; ammonia-oxidizing archaea, AOA) reduce CO_2 to organic carbon. Heterotrophs (e.g., many haloarchaea) utilize organic molecules, such as glucose, peptides, or acetate (Tourna et al. 2011).

3.3.2.1 Autotrophic Carbon Fixation Pathways

Wood–Ljungdahl (reductive acetyl-CoA) pathway
In methanogens, this pathway reduces two CO_2 molecules via distinct branches (one to a methyl group on tetrahydromethanopterin/F_{420} cofactor sets, and the other to CO_2) that combine to form acetyl-CoA. It is ATP-efficient and intimately coupled to energy conservation. Comparative analyses suggest non-homologous solutions in archaea and bacteria, consistent with convergent evolution and/or horizontal transfer (Sousa and Martin 2014).
3-Hydroxypropionate/4-Hydroxybutyrate (3HP/4HB) cycle
Characteristic of many thermophilic Crenarchaeota (e.g., *Metallosphaera*, *Sulfolobus*), this cycle fixes CO_2 via acetyl-CoA and thermally stable enzymes, supporting growth under high temperatures and often in low-oxygen environments (Berg et al. 2010).
Dicarboxylate/4-Hydroxybutyrate (DC/4HB) cycle
Found in anaerobic hyperthermophiles (e.g., *Ignicoccus*), this route incorporates CO_2through intermediates such as phosphoenolpyruvate and succinyl-CoA, regenerating acetyl-CoA via 4-hydroxybutyrate (Huber et al. 2008; Ramos-Vera et al. 2009).

3.3.2.2 Glycolysis and the Pentose Phosphate Pathway: Archaeal Variants

In heterotrophic archaea, glucose catabolism often diverges from the canonical Embden–Meyerhof–Parnas (EMP) route:

- Modified EMP: Some enzymes differ in cofactor specificity; in *Pyrococcus* species, ferredoxin-dependent steps replace NAD(H) reactions, altering redox balance (Verhees et al. 2003; Kengen 2017).
- Entner–Doudoroff (ED) variants: Thermoacidophiles such as *Sulfolobus solfataricus* run branched semi-phosphorylative and non-phosphorylative ED pathways (Ahmed et al. 2005; Lamble et al. 2005).

- Pentose phosphate pathway (PPP): Many archaea harbor modified oxidative/non-oxidative modules to supply NADPH and ribose-5-phosphate, with genomic support from *Sulfolobus* models (She et al. 2001).

3.3.2.3 Fermentation and Anaerobic Carbon Metabolism

Some archaea are fermentative. *Thermococcus* spp. ferment carbohydrates and peptides to acetate, CO_2, and H_2, frequently coupling growth to elemental sulfur reduction; in the absence of sulfur, several strains evolve H_2 as a significant product. Haloarchaea can generate ATP by the arginine deiminase (ADI) pathway, and many perform anaerobic respiration using nitrate, sulfate, or ferric iron as terminal acceptors (Hensley et al. 2016; Monstadt et al. 1991; Falb et al. 2008).

3.3.2.4 Methanogenesis and Carbon Flow

Archaeal methanogenesis converts CO_2 to CH_4 via three principal substrate classes: (i) CO_2+H_2 (hydrogenotrophic), (ii) acetate (aceticlastic), and (iii) methylated compounds (methylotrophic). Distinct coenzymes (e.g., methanofuran, CoM) and archaeal enzymes (e.g., MCR) orchestrate these routes (Deppenmeier 2002).

3.3.2.5 Carbon Storage Compounds

Many archaea store carbon as polyhydroxybutyrate (PHB), glycogen, or polyphosphate, conferring resilience during stress. The halophile *Haloferax mediterranei* is a noted PHB producer with biotechnological relevance (Lillo and Rodríguez-Valera 1990).

3.3.3 Nitrogen Metabolism

3.3.3.1 Ammonia Oxidation

Since 2005, AOA have been recognized as chemolithoautotrophs that oxidize NH_3 to NO_2 via ammonia monooxygenase (AMO). The first cultured AOA, *Nitrosopumilus maritimus* SCM1, established archaeal nitrification; soil AOA, such as *Nitrososphaera viennensis*, broadened the ecological scope. AOA typically possess high-affinity AMO systems, conferring advantages in oligotrophic settings (Könneke et al. 2005; Martens-Habbena et al. 2009; Tourna et al. 2011).

3.3.3.2 Nitrogen Fixation

Nitrogen fixation ($N_2 \rightarrow NH_3$) occurs in a subset of methanogens (e.g., *Methanococcus maripaludis*), typically via the oxygen-sensitive MoFe nitrogenase and under anoxic conditions. Transcriptional regulation—e.g., repression of nif by the nitrogen regulator NrpR and "ammonia switch-off"—is well characterized in *M. maripaludis* (Leigh 2000; Lie and Leigh 2003).

3.3.3.3 Denitrification

While denitrification is widespread among bacteria, multiple haloarchaea encode nitrate/nitrite reductases and can reduce NO_3^- to NO_2^- to N_2O to N_2 under oxygen limitation. *Haloferax mediterranei* has emerged as a model for archaeal denitrification physiology (Torregrosa-Crespo et al. 2020).

3.3.4 Electron Transport and Coenzyme Systems

Archaea deploy distinctive electron-transport components and cofactors. In methanogens, the lipid-like methanophenazine functions as a membrane electron carrier, while coenzyme F_{420}—a low-potential deazaflavin—serves broadly in redox catalysis (including several steps of C1 metabolism and H_2 turnover). Compared to canonical NAD(P) systems, archaeal chains frequently rely on ferredoxin-based modules and alternative ATPases, reflecting adaptation to low-potential and high-temperature environments (Deppenmeier 2004; Hendrickson et al. 2008; Greening et al. 2016).

3.3.5 Membrane Architecture and Lipid Chemistry

Archaea membranes are distinguished by ether-linked isoprenoid chains bound to sn-glycerol-1-phosphate, which enhances stability under high temperature and low pH. Many archaeal lipids are tetraethers (GDGTs) that span the membrane to form monolayers, conferring exceptional thermal and acid tolerance—standard in hyperthermophiles and several Thaumarchaeota. Environmental and experimental studies demonstrate that GDGT composition (e.g., cyclization) shifts in response to temperature and pH, supporting energy metabolism under stress (Schouten et al. 2007; Boyd et al. 2013; Bale et al. 2019).

3.4 Ecological Distribution of Archaea

Terrestrial hot springs and deep-sea hydrothermal vents, glaciers and permafrost, hypersaline habitats, acid mine drainage (AMD), and underground environments are widely distributed throughout the world. Although these habitats are challenging in terms of their physicochemical conditions, they provide a variety of ecological niches for diverse microorganisms from all three domains of life. Archaea are highly adapted to thrive in extreme environments, such as those with high salinity, acidity, and temperature, due to their unique biological and structural characteristics (Shu and Huang 2022).

3.4.1 Extreme Environments

Archaea are categorized according to the environmental conditions in which they live. Thermophiles/hyperthermophiles live in habitats with high or very high temperatures, such as hydrothermal vents, volcanic areas, and hot springs; psychrophiles live in cold habitats, such as polar regions and the depths of the ocean; acidophiles in habitats with a pH below 5.0 or even close to 0.0, such as acid mine drainage areas and acidic lakes; alkaliphiles in habitats with high alkaline pH, such as sodium-rich lakes; and halophiles in habitats with very high salt concentrations, such as the sea and saline alkaline lakes. Sometimes, certain groups can thrive in habitats where multiple stress factors are present simultaneously. These extremophiles are referred to as polyextremophiles (thermoacidophiles, haloalkaliphiles, etc.) (Mehta and Satyanarayana 2013).

3.4.1.1 Thermophilic Archaea (Hot Springs, Hydrothermal Vents)

Thermophilic microorganisms are widely found on the Earth's surface. The microbial communities in hot springs are exciting because they are analogs of the primitive Earth. The wide range of physicochemical properties of hot springs, with pH values ranging from −0.8 to 10.5 and temperatures ranging from <10 to >100°C, significantly influences the composition of microbial communities (Podar et al. 2020; Zou et al. 2025). Sulfur-rich hot water sources with pH values below 4 and temperatures exceeding 37°C, commonly found in volcanic areas worldwide, are suitable habitats for extremophiles (Mehta and Satyanarayana 2013). As demonstrated by 16S rRNA gene and metagenomic studies, common archaeal lineages in terrestrial hot springs include Korarchaeia, Thermoplasmata, Thermoprotei, Archaeoglobi, and Thermococci. The diversity of archaea in hot springs has been confirmed by the discovery of phyla such as Aigarchaeota, Geothermarchaeota, and Methanomethylicia, thereby enhancing our understanding of thermophilic archaeal communities (Zou et al. 2025).

Marine hydrothermal systems (shallow and deep-sea vents) are areas of the seafloor that host different groups of microorganisms, where hot, anoxic, and chemically rich water is released into the cold, oxygenated, and deep ocean (Dick 2019; Zou et al. 2025). Deep-sea hydrothermal vent fluids exhibit a wide temperature range (from just a few degrees above ambient temperature to below 100°C and up to 400°C), depending on the type of vent. When these thermally charged fluids mix with oxygenated seawater, they provide a plentiful source of various reduced oxidants and reducing agents for microbial metabolism (Dick 2019; Shu and Huang 2022; Zou et al. 2025). These chemosynthetic ecosystems host a rich diversity of archaea. The steep thermal and chemical gradients in active hydrothermal vents create diverse niches. Ultramafic-hosted vents support anaerobic methanotrophic archaea (Dick 2019). In basalt-based systems, low pH, metal-rich, and sulfur-rich thermal fluids are dominated by methanogenic and sulfate-reducing Euryarchaeota, while deep-sea hydrothermal vents harbor Euryarchaeota.

In contrast, vent fluids from hyperthermophilic habitats in this area host a community containing taxa belonging to Thermococcales and uncultured Crenarchaeota (Shu and Huang 2022). Other important groups found in hydrothermal vents include members of Hydrothermarchaeia, Bathyarchaeia, and Nanoarchaeia, which are reported to be dominant in deep-sea hydrothermal sediments and are also believed to exist in non-extreme environments (Zou et al. 2025). From these environments, the important genera *Sulfolobus* and *Acidianus* of the Sulfolobales order have been isolated (Dick 2019).

3.4.1.2 Halophilic Archaea (Salt Lakes, Solar Salterns)

Extremely saline environments (environments with salt concentrations exceeding 35 g/L) are found all over the world. These include saline-alkaline soils, salt mines, and salt lakes characterized by high salinity, high osmotic pressure, low water activity, temperature, and pH fluctuations (Ding et al. 2025). Archaea are of key importance for saline lake communities, particularly in extremely saline lakes (salinity >50.0 mg/L) (Ding et al. 2025; Zou et al. 2025). Recent studies have revealed significant diversity, including various genera such as *Natronomonas, Halorubrum*, and *Haloarcula*, with unique dominant species found in different habitats (Shu and Huang 2022; Sorokin et al. 2022b; Ding et al. 2025). Specifically, the square archaeon *Haloquadratum walsbyi* and the bacteroidete *Salinibacter ruber* are dominant in saturated saline waters, while more diverse bacterial and archaeal taxa are observed in moderately saline conditions. Metagenomic studies of the hypersaline surface waters of Lake Tyrrell have led to the discovery of Nanohaloarchaeota, an ultra-small but metabolically versatile halophile within the DPANN superphylum (Shu and Huang 2022). Halophilic methanogens, such as *Methanosarcina* and *Methanomicrobia*, which thrive in nearly saturated salt concentrations, are a crucial component of the archaeal community in salt lakes (Zou et al. 2025). Salty-alkaline soils have been shown to have a higher diversity of halophilic archaea compared to salt lakes, where specific genera are more abundant (Ding et al. 2025; Sorokin et al. 2022b).

3.4.1.3 Acidophilic/Alkaliphilic Archaea (Acid Mine Drainage, Soda Lakes)

Acidophilic and alkalophilic archaea are specialized microorganisms that thrive in extreme pH environments. Acidophilic archaea are a group of microorganisms that thrive in highly acidic environments such as acidic hot springs, geothermal fields, and acid mine drainage (Johnson et al. 2012).

Acidic mine drainage (AMD) environments are particularly stressful due to low pH, high levels of toxic metals, and a limited range of electron donors and acceptors, resulting in limited microbial species richness. AMD communities are composed of bacteria and members of the Euryarchaeota. Among these, the most extensively studied group comprises iron- and/or sulfur-oxidizing bacteria (Shu and Huang 2022). Acidophilic microorganisms exhibit significant metabolic diversity. They possess the ability to utilize both organic and inorganic energy sources, various electron acceptors, and both inorganic and organic carbon sources, in some cases.

Additionally, some species contribute to the carbon cycle by converting organic carbon into methane through a process known as methanogenesis. Taxonomically, acidophilic archaea belong to several distinct groups, including Crenarchaeota, Euryarchaeota, and the recently described Asgard archaea (Johnson et al. 2012; Kadnikov et al. 2020). *Sulfolobus* spp., which utilize reduced forms of sulfur, and two genera of iron-oxidizing Euryarchaeota, *Ferroplasma* and *Acidiplasma*, are included in this group (Johnson et al. 2012).

Salty soda lakes offer a unique "double extreme" habitat characterized by high salt content (saturation level) and high pH values (9.0–12.0) (Sorokin et al. 2015; Shu and Huang 2022). Naturally occurring soda lakes are extremely alkaline environments containing high concentrations of sodium carbonate. The chemical composition of common salts creates ideal conditions for the development of haloalkali-loving organisms (Sorokin et al. 2014). Metagenomic and metatranscriptomic analyses of soda lakes across different continents have revealed that salt concentrations are a key factor influencing microbial diversity and distribution (Sorokin et al. 2014; Banda et al. 2021). Taxonomically, alkaliphile archaea are predominantly members of the Euryarchaeota phylum, and important genera such as *Natronobacterium* and *Natronococcus* are well-characterized (Sorokin et al. 2014; Banda et al. 2021). Studies conducted in various saline alkaline lakes have also revealed the presence of two distinct subgroups of obligate haloalkaliphiles, lithotrophs, and methanogens belonging to the genus *Methanocalculus* (Sorokin et al. 2015).

3.4.1.4 Psychrophilic Archaea (Polar Regions, Deep Sea Sediments)

Recent microbial ecology studies have focused on cold ecosystems with temperatures below 5°C, such as glacial habitats and permafrost. Glacial and subglacial ecosystems provide diverse habitats that support a significant number of microbial cells (Shu and Huang 2022). The dominant prokaryotes in polar environments are psychrophilic and psychrotolerant bacteria and archaea (Zou et al. 2025). Recent studies have identified various archaeal groups in subsurface sediments, indicating significant diversity within these communities (Shu and Huang 2022; Nozhevnikova et al. 2003; Morozova et al. 2006; Karr et al. 2005; Zou et al. 2025). A study by Nozhevnikova et al. (2003) demonstrated the presence of psychrophilic methanogenic communities in anoxic sediments of deep lakes with low temperatures. Methanogenic research for polar environments also holds great interest for astrobiology. Morozova et al. (2006) found that methanogenic archaea from Siberian permafrost survived under simulated Martian thermal conditions, suggesting they could be suitable candidates for microbial life on Mars. In a study on the biodiversity of methanogenic and other archaea in the frozen Fryxell Lake in Antarctica, it was found that archaea belonging to the Euryarchaeota and Crenarchaeota groups live in Fryxell Lake. At least two methanogenic clusters have been identified within the Euryarchaeota (Karr et al. 2005). Sun et al. (2025) demonstrated that archaeal communities in Arctic marine sediments exhibit distinct geographical distribution differences. That temperature has a positive effect on microbial diversity and the community composition of bacteria, archaea, and fungi.

3.4.2 Moderate Environments

Although Archaea are generally considered extremophiles inhabiting extreme environments, mesophilic archaea inhabiting moderate environments, such as soil and oceans, have also been identified. Additionally, the presence of archaea has been confirmed in human skin, the mouth, and the intestines. They are a crucial component of the human microbiome (Kim et al. 2020).

3.4.2.1 Soil Archaea

Terrestrial areas are an important source of the Earth's prokaryotic biomass. Archaea constitute a significant portion of the soil microbial community and biomass (Jiao et al. 2019; Shu and Huang 2022). Studies have shown that soil archaeal communities exhibit strong habitat specificity, with distinct community structures in arid and wet areas influenced by factors such as soil pH and moisture content. Archaea communities in soil environments exhibit more complex networks and higher metabolic activity compared to freshwater biofilms and estuarine waters (Jiao et al. 2019). Important archaeal groups, such as Thaumarchaeota and Euryarchaeota, are widely present in these environments (Essel et al. 2017; Jiao et al. 2019; Shu and Huang 2022). In general, archaea play a vital role in the cycling of nutrients such as nitrogen, carbon, and sulfur in soil and freshwater ecosystems, and they also have the potential to contribute to atmospheric greenhouse gas emissions (Essel et al. 2017; Gubry-Rangin et al. 2018).

Autotrophic AOA in the Thaumarchaeota phylum are ecologically important due to their role in nitrification, a key biogeochemical process (Gubry-Rangin et al. 2018; Jiao et al. 2019). In the nitrogen cycle, Crenarchaeota, including Thaumarchaeota and *Candidatus* Nitrososphaera, have been reported to play a role in dinitrogen gas fixation, ammonia oxidation to nitrite, and denitrification (Essel et al. 2017).

3.4.2.2 Marine Archaea (Pelagic and Benthic Communities)

Marine archaea are crucial components of both pelagic and benthic microbial communities, playing distinct roles in their respective environments. Planktonic archaea are among the fundamental life forms in the ocean, playing crucial roles in ecological functions and biogeochemical cycles (Lauer et al. 2016; Chen et al. 2022). Studies have identified various uncultivable archaeal lineages in deep-sea sediments, including Marine Group I and Marine Benthic Groups A and B, which exhibit a depth-dependent distribution and community structure distinct from their pelagic counterparts (Lauer et al. 2016). Low cell numbers and insufficient DNA recovery limit the analysis of organically poor open-ocean sediments (Chen et al. 2022). AOA, particularly Ca. *Nitrosopelagicus* and *Nitrosopumilus* are widely distributed in the oceans and play a crucial role in the marine nitrogen cycle through the process of nitrification.

Marine sediments represent a unique ecosystem distinct from the water column. These sediments harbor greater microbial diversity and abundance than the overlying waters and play important roles in nutrient cycling and biogeochemical processes. Recent studies on the diversity and distribution of microorganisms in marine sediments have revealed that the geographical distribution of microorganisms in these marine sediments exhibits firm spatial heterogeneity (Shu and Huang 2022).

3.4.2.3 Host-Associated Archaea

Host-associated archaea are commonly referred to as archaeoms. They play crucial roles in shaping the health of various organisms, including humans, animals, and plants (Borrel et al. 2020; Cai and Tang 2022; Feehan et al. 2023). These archaea are primarily composed of methanogens, such as those belonging to the Methanobacteriales and Methanomassiliicoccales (Euryarchaeota), as well as representatives of Thaumarchaeota (Borrel et al. 2020; Cai and Tang 2022). Research has highlighted that archaea may play a crucial role in maintaining host health through functions such as hydrogen utilization, which supports bacterial growth, and may also contribute to various diseases (Cai and Tang 2022; Feehan et al. 2023). Among these archaea, methanogens play important roles in processes such as methane production in the gastrointestinal tract and can influence host health and disease (Feehan et al. 2023). Archaea protists, like bacteria, can harbor archaea (Methanobacteriales, Methanomicrobiales, Methanosarcinales, and Methanomassiliicoccales) in their cytoplasm. These methane-producing archaea can also serve as endosymbionts and ectosymbionts in other organisms, such as termites, cockroaches, amphibians, and ruminants (Borrel et al. 2020).

Archaea are among the least studied members of the indigenous microbiota residing in the intestine (Kim et al. 2020). Approximately 1.2% of the microbial community in the human gastrointestinal tract consists of archaea, with methanogenic archaea playing a significant role. Among methanogens, the most common and abundant archaea in the digestive tract are *Methanobrevibacter smithii* and *Methanosphaera*, representatives of the Methanobacteriales and Methanomassiliicoccales orders (Borrel et al. 2020;

Kim et al. 2020; Khalil et al. 2024). In a study conducted with East Asian subjects, it was found that 42.47% of fecal samples were colonized by archaea, with an average relative abundance of 10.24%. Among the archaea, Haloarchaea were found to be more abundant than methanogens in specific samples (Kim et al. 2020).

3.5 Adaptations to Extreme Environments

3.5.1 Molecular Adaptations to High Temperatures

The proteins possessed by thermophilic and hyperthermophilic archaea are highly resistant to denaturation compared to those of mesophilic microorganisms. Thermophilic and hyperthermophilic archaea have developed special strategies to maintain protein stability. This ensures that the structural integrity of proteins is preserved, allowing them to remain functional. Thermophilic and hyperthermophilic archaeal proteins are rich in strong hydrophobic amino acids such as leucine, isoleucine, and valine, as well as charged amino acids such as arginine and glutamic acid. This allows the side chains, which are brought together by hydrophobic interactions in the protein core, to be stabilized by Van der Waals forces (Amangeldina et al. 2024). The presence of more hydrophobic amino acids in the protein core and a higher number of ionic bonds on the surface prevents the denaturation of the protein's folded structure at high temperatures. This ensures thermal stability. Asparagine is a thermolabile amino acid that is sensitive to high temperatures and easily denatures. The low asparagine content in thermophilic and hyperthermophilic archaeal proteins reduces the likelihood of protein degradation at high temperatures, thereby contributing to the protein's heat resistance (Rahimzadeh et al. 2012; Reed et al. 2013).

Additionally, thermophilic and hyperthermophilic archaeal proteins contain a higher number of disulfide bonds, which stabilize their three-dimensional structure. This helps the protein maintain its structural integrity and reduce structural flexibility even at high temperatures. Shaperonin molecules, known as heat shock proteins, also facilitate the proper refolding of proteins whose structure has been disrupted at high temperatures (Marzban and Tesei 2025).

3.5.1.1 DNA Stability

The DNA of thermophilic and hyperthermophilic archaea contains a high proportion of guanine-cytosine base pairs. Additionally, the DNA structure can be protected at high temperatures by forming positive supercoils with the help of reverse DNA gyrase enzymes. Furthermore, archaea possess special histone-like DNA-binding proteins that prevent denaturation and damage to their DNA at high temperatures (Danson et al. 1998; Hocher et al. 2022).

3.5.1.2 Cell Membrane Structures

The lipids found in archaeal cell membranes do not contain ester-linked fatty acids, as in bacteria and eukaryotes, but instead contain ether-linked isoprenoid chains. This structure ensures that the cell membrane's stability is maintained even at very high temperatures (Siliakus et al. 2017). Isoprenoid chains have a branched structure, unlike the straight-chain structure of fatty acids. This feature allows the membranes to form a tighter and more compact structure, reducing membrane fluidity and maintaining cell membrane integrity at high temperatures. Many extremophilic archaea synthesize C40 tetraether lipids that connect two glycerol structures and completely cover the cell membrane. These lipids form a monolayer cell membrane, rather than the traditional bilayer membrane. This single-layer membrane is less permeable, thermally more stable, and more compact than bilayer membranes. Some thermophilic/hyperthermophilic archaea have cyclopentane rings attached to isoprenoid chains in their membrane lipids. These rings, which are added to the cell membrane, increase in number as the temperature rises, thereby reducing the fluidity of the membrane and supporting the thermal stability of the cell membrane (Madigan et al. 2021).

3.5.2 Adaptation to Cold

Psychrophilic archaea, which live in cold environments, have optimal growth temperatures below 15°C and are typically found at temperatures of 0°C or lower. Psychrotolerant archaea can grow at low temperatures such as 0°C, but their optimal growth temperatures are above 15°C (Pathania et al. 2021). Psychrophilic archaea accumulate compatible solutes such as glycerol, trehalose, betaine, and ectoine at high concentrations within the cell to maintain intracellular osmotic balance and provide protection against cold stress. These compounds prevent cell freezing and increase the stability of proteins and other cellular components, ensuring their continued function even at low temperatures (Bowman 2017).

3.5.2.1 Protein and Enzyme Adaptations (Cold-Active Enzymes)

Archaea living in the polar regions, the deep seas, and extremely high-altitude cold regions must maintain the flexibility and activity of their enzymes even at extremely low temperatures. Therefore, unlike thermophiles and hyperthermophiles, they possess proteins with a less hydrophobic core and a less charged surface. The proportion of proline amino acids, which provide structural rigidity, is also low. These features enable them to maintain their flexibility at low temperatures while also allowing for the conformational changes necessary for enzymatic activity (Collins and Feller 2023). The antifreeze proteins produced by psychrophilic archaea prevent the freezing of water inside the cell, thereby preventing ice crystals from forming and damaging the cell. Antifreeze proteins are key to their survival at temperatures below the freezing point (Yusof et al. 2021).

The adaptation of psychrophiles to high temperatures increases their competitiveness and ensures their survival during seasonal temperature changes. The glycosylation of S-layer proteins is a mechanism that enables psychrophiles to survive at increasing temperatures. Increased glycosylation supports this adaptation. It has been observed that the psychrophilic methanogen named *Methanolobus psychrophilus* R15 can grow at 30°C when a high amount of nutrients is present. Li et al. (2020) determined in their study that the glycosylation of S-layer proteins is crucial for survival and adaptation at 30°C. It was found that N94 and 879N glycosylation changes depending on high temperature and nutrient-rich conditions. Inhibiting glycosylation also prevents the microorganism from growing at higher temperatures. This finding demonstrates a direct link between glycosylation and temperature adaptation.

3.5.2.2 Cell Membrane Adaptations

Psychrophilic archaea contain unsaturated isoprenoid chains with a higher number of double bonds in their cell membrane lipids, branched and short-chain isoprenoid lipids, isoprenoid chains containing fewer cyclopentane rings, or diether lipids (Oger and Cario 2013).

3.5.2.3 DNA and RNA Adaptations

Low temperatures can increase the stability of the DNA double helix structure of psychrophilic archaea and make transcription more difficult. To counteract this, psychrophilic archaea have a lower G-C content in their DNA, allowing the DNA chain to open more easily. Additionally, they can produce special DNA-binding proteins and RNA helicases that protect DNA from cold and facilitate transcription. This ensures that genetic material is read and processed correctly, even at low temperatures (Collins and Feller 2023; Xu et al. 2024). Furthermore, cold shock proteins produced during sudden temperature drops bind to RNA molecules, preventing them from forming secondary structures and ensuring the continuity of protein synthesis (Weidenbach et al. 2022).

3.5.3 Halophilic Adaptation Mechanisms

Halophilic archaea grow best at 2.5–5.5 M NaCl concentrations and can even survive at salt concentrations above 30%. Halophilic archaea living in extremely salty environments have developed some mechanisms to maintain osmotic balance and protein stability (Baker et al. 2024). Halophilic archaea balance the salt concentration between the inside and outside of the cell by accumulating high concentrations of potassium chloride (KCl) inside the cell. With this adaptation, intracellular proteins and enzymes have evolved to be compatible with this highly ionic environment. Intracellular potassium concentration can vary between 1.2 and 4.5 M, depending on the species. At the same time, intracellular sodium (Na^+) concentration is kept low, which is achieved through Na^+/K^+ antiport systems. Halophilic proteins contain abundant acidic amino acids such as aspartate and glutamate on their surfaces. This increases protein solubility, prevents aggregation, and allows them to remain stable and functional under high ionic strength conditions by forming ionic bridges. The best example of this is the Halaoarchaea class, which are the most extensively studied and therefore the best-known extreme halophilic archaea (Baker et al. 2024). Most of this group are aerobic or facultative aerobic heterotrophs. While they typically prefer amino acids as an energy and carbon source, some can also metabolize carbohydrates, alcohols, organic acids, and even hydrocarbons (Saini et al. 2023).

Nanohaloarchaeota are considered small, symbiotic archaea commonly found in highly saline environments (Shu and Huang 2022). Their lack of essential biosynthetic pathways suggests that they are dependent on a host for their survival (Gutiérrez-Preciado et al. 2024; Baker et al. 2024). Like Halaoarchaea, they adopt the "Salt-in" osmoregulation strategy and possess acidic proteomes. *Candidatus Methanonatronarchaeia*, which shares a similar strategy, is a member of the extreme halophilic methanogen class. It is generally believed to have evolved through gene loss (Shu and Huang 2022).

Some halophilic archaea synthesize organic molecules, such as glycerol, sucrose, and glycine betaine, within the cell to lower the water potential and maintain osmotic balance. The accumulation of these molecules within the cell ensures the preservation of protein three-dimensional structures and biological functions even at high salt concentrations. Ding et al. (2022) found that the primary osmoprotective strategy of the halophilic archaeon *H. kocurii* 2020YC7 is K^+ uptake and glycine betaine transport, they found that glycine betaine is the primary osmotic balance-maintaining substance in an environment containing 200–250 g/L NaCl, and that under conditions of 50–100 g/L NaCl, intracellular polysaccharides are broken down to increase trehalose production, which in turn provides osmoprotection.

3.5.3.1 Cell Membrane Adaptations

Halophilic archaea have cell membrane lipids that contain a high proportion of negatively charged groups. These negatively charged groups increase the surface charge of the cell membrane, thereby ensuring membrane stability (Baker et al. 2024).

3.5.4 Adaptation to Extreme pH (Acidophilia and Alkaliphilia)

Acidophilic archaea grow at pH values between 0 and 4 while maintaining their intracellular pH between 4.5 and 6.5, while alkalophilic archaea grow optimally at pH values between 9 and 11 while maintaining their intracellular pH between 7.5 and 8.5. Acidophiles exhibit low permeability to protons in order to maintain intracellular pH at a level close to neutral. Alkaliphiles, on the other hand, maintain an intracellular pH lower than the external environment by taking up H^+ from the outside, reducing H^+ leakage, and producing organic acids to maintain pH homeostasis (Madigan et al. 2021). Intracellular pH regulation is achieved through the action of ion pumps and ion transporters. Acidophilic archaea raise intracellular pH by pumping protons out of the cell via proton pumps. The energy required for this process is provided by ATP hydrolysis or the electron transport chain. Alkalophilic archaea, on the other hand, lower intracellular pH by taking protons into the cell. The ion transporters contained in acidophilic and alkalophilic archaea regulate intracellular ion concentrations and maintain pH balance. Na^+/H^+ antiport systems play a crucial role in regulating intracellular pH (Amangeldina et al. 2024).

Archaea living in highly acidic or basic environments modify their amino acid sequences by reducing basic amino acids to adapt to low pH levels and reducing acidic amino acids to adapt to high pH levels. For example, it causes amino acids with acidic properties, such as glutamate (Glu), to be replaced with more neutral and stable derivatives, such as glutamine (Gln). As a result of this change, the protein becomes more resistant to pH changes (Blachier 2025).

3.5.4.1 Membrane Adaptations

The membranes of acidophilic archaea contain lipids with low proton permeability. These lipids limit proton entry and help maintain intracellular pH. The cell membranes of alkalophilic archaea, on the other hand, have a more permeable structure, facilitating proton uptake in basic environments. The lipids found in the cell membranes of acidophilic archaea contain more cyclopentane rings and longer isoprenoid chains, which reduce proton permeability. In contrast, the membrane lipids of alkalophilic archaea contain fewer cyclopentane rings, making their membranes more permeable (Sharma et al. 2021).

A unique five-membered ring structure with a cyclopentyl head group, which is ether-bonded to glycerol dibutyltriester glycerol tetraether, is found in thermoacidophilic archaea within the Sulfolobales order and is named calditol. The calditol synthase mutant of *S. acidocaldarius* exhibited similar growth to the wild type at the optimal pH of 3.5, but showed no growth at pH 1.6. This result directly demonstrates that calditol-linked membrane lipids play a vital protective role under very low pH conditions. Additionally, bioinformatics analyses have revealed that calditol synthase homologs are predominantly found in metagenomes derived from acidic ecosystems (pH 1–4). Calditol synthase homologs have also been identified in different archaea phyla. This suggests that these archaea have spread through horizontal gene transfer due to their close spatial proximity in extreme environments (Zeng et al. 2018).

3.5.4.2 Protein Adaptations

Acidophilic proteins contain more acidic amino acids and fewer basic amino acids. As a result, their net charge is more negative, which increases their stability in acidic environments. Alkalophilic proteins, on the other hand, contain more basic amino acids and fewer acidic amino acids. Their net charge is more positive, increasing their stability in basic environments (Blachier 2025).

3.5.5 Adaptation to High Pressure

Piezophilic archaea are found in environments with high hydrostatic pressure, such as the deep sea or underground, and are usually also adapted to high temperatures. Piezophilic proteins fold into a more compact three-dimensional structure. They contain fewer voids, which prevents water from seeping in and makes them more resistant to pressure. They contain fewer amino acids, such as proline and glycine, which could destabilize the protein at high pressure by increasing its flexibility (Brininger et al. 2018).

High pressure reduces the fluidity of the membrane lipids in piezophilic archaea, thereby increasing the membrane's stiffness. This negatively impacts the cell membrane's functions. Piezophilic archaea have adapted to maintain membrane fluidity even under high pressure by altering the composition of membrane lipids. The branching of isoprenoid chains and the decrease in the ratio of shorter isoprenoid chains to cyclic structures increase membrane fluidity (Ando et al. 2021).

3.6 Conclusions and Future Perspectives

Since the discovery of archaea in 1977, significant information has been gathered about the ecological, metabolic, and genetic differences and diversity of these microorganisms, as well as their potential for biotechnological applications. New insights are continually being added to this knowledge. Since the regions where archaea were first discovered were extreme environments, it was initially believed that they only existed in such conditions. However, over time, it has been determined that they inhabit every environment where bacteria are present, including soil, oceans, lakes, and the intestines of highly organized organisms. With the development of new technologies, such as metagenomics and single-cell DNA genome sequencing, it has been determined that archaea exhibit a high level of diversity. It would not be incorrect to say that there are still many undiscovered archaea phyla and species in understudied habitats, such as the deep seas and polar regions. Research in this area will contribute to a better understanding of their diversity.

A considerable portion of archaea cannot be cultured. Although the physiological and metabolic characteristics of uncultivable archaea can be determined using methods such as metagenomics, metatranscriptomics, metaproteomics, and metabolomics, they must be cultured to confirm their metabolic capacities and functions. To this end, more isolation studies should be conducted by preparing environments that mimic the conditions in which archaea are found. For symbiotic archaea, existing methods used for symbiotic bacteria should be modified, or new, different methods should be developed. The more archaea that can be cultured, the better their physiological and metabolic characteristics will be understood. As a result, they will provide and continue to provide important resources in fields such as biotechnology, bioremediation, energy production, and drug discovery.

No disease-causing species of archaea has yet been identified. Archaea are found in the microbiomes of animals and humans and are critical components of the gut microbiota. Although they have not been identified as pathogens, some studies suggest that they may play a role in inflammatory bowel diseases and periodontal diseases; however, indirect evidence is available, and convincing evidence has yet to be obtained. The difficulties in isolating and culturing them hinder the attainment of definitive evidence, but research in this area continues intensively.

In conclusion, all future studies focusing on the archaea domain will contribute to a better understanding and evaluation of the diversity and ecological roles of these unique microorganisms. All these developments will facilitate the full utilization of the biotechnological potential of archaea (Karr et al. 2005; Nozhevnikova et al. 2003).

References

Adam PS, Borrel G, Brochier-Armanet C, Gribaldo S (2017) The growing tree of Archaea: new perspectives on their diversity, evolution and ecology. ISME J 11(11):2407–2424

Ahmed H, Ettema TJG, Tjaden B, Geerling ACM, van der Oost J, Siebers B (2005) The semi-phosphorylative Entner–Doudoroff pathway in *Thermoproteus tenax* and *Sulfolobus solfataricus*. J Bacteriol 187(21):7545–7552

Albers SV, Meyer BH (2011) The archaeal cell envelope. Nat Rev Microbiol 9(6):414–426

Amangeldina A, Tan ZW, Berezovsky IN (2024) Living in trinity of extremes: Genomic and proteomic signatures of halophilic, thermophilic, and pH adaptation. Curr Res Struct Biol 7:100129

Amils R (2022) Euryarchaeota. In: Gargaud M et al (eds) Encyclopedia of Astrobiology. Springer, Berlin, Heidelberg, pp 1-1

Amils R (2023) DPANN, Archaea. In: Gargaud M et al (eds) Encyclopedia of Astrobiology. Springer, Berlin, Heidelberg, pp 3136-3136

Ando N, Barquera B, Bartlett DH, Boyd E, Burnim AA, Byer AS, Watkins MB (2021) Molecular basis for life in extreme environments: genomic and structural perspectives. Annu Rev Biophys 50(1):343–372

Baker BA, Gutiérrez-Preciado A, Rodríguez del Río Á, McCarthy CG, López-García P, Huerta-Cepas J, Roger AJ, Eme L, Moreira D (2024) Expanded phylogeny of extremely halophilic archaea shows multiple independent adaptations to hypersaline environments. Nat Microbiol 9:964–975

Baker BA, McCarthy CG, López-García P, Leroy RB, Susko E, Roger AJ, Eme L, Moreira D (2025) Phylogenomic analyses indicate the archaeal superphylum DPANN originated from free-living euryarchaeal-like ancestors. Nat Microbiol 10:1593–1604

Baker BJ, De Anda V, Seitz KW, Dombrowski N, Santoro AE, Lloyd KG (2020) Diversity, ecology and evolution of Archaea. Nat Microbiol 5(7):887–900

Bale NJ, Palatinszky M, Rijpstra WIC, Herbold CW, Wagner M, Sinninghe Damsté JS (2019) Membrane lipid composition of the moderately thermophilic ammonia-oxidizing Archaeon "*Candidatus* Nitrosotenuis uzonensis" at different growth temperatures. Appl Environ Microbiol 85(20):e01332–19

Banciu HL, Gridan IM, Zety AV, Baricz A (2022) Asgard archaea in saline environments. Extremophiles 26(2):21

Banda JF, Zhang Q, Ma L, Pei L, Du Z, Hao C, Dong H (2021) Both pH and salinity shape the microbial communities of the lakes in Badain Jaran Desert, NW China. Sci Total Environ 791:148108

Beam JP, Jay ZJ, Schmid MC, Rusch DB, Romine MF, Jennings RDM, Kozubal MA, Tringe SG, Wagner M, Inskeep WP (2016) Ecophysiology of an uncultivated lineage of Aigarchaeota from an oxic, hot spring filamentous 'streamer'community. ISME J 10(1):210–224

Bell SD, Jackson SP (2001) Mechanism and regulation of transcription in archaea. Curr Opin Microbiol 4(2):208–213

Benito Merino D, Lipp JS, Borrel G, Boetius A, Wegener G (2024) Anaerobic hexadecane degradation by a thermophilic Hadarchaeon from Guaymas Basin. ISME J 18(1):wrad004

Berg IA, Kockelkorn D, Ramos-Vera WH, Say RF, Zarzycki J, Hügler M, Alber BE, Fuchs G (2010) Autotrophic carbon fixation in archaea. Nat Rev Microbiol 8(6):447–460

Blachier F (2025) Amino acid metabolism for archaeal physiology. In: The evolutionary journey of amino acids: from the origin of life to human metabolism. Springer Nature, Cham, Switzerland. pp. 39–62

Borrel G, Brugère JF, Gribaldo S, Schmitz RA, Moissl-Eichinger C (2020) The host-associated archaeome. Nat Rev Microbiol 18(11):622–636

Bowman JP (2017) Genomics of psychrophilic bacteria and archaea. In: Margesin R (ed) Psychrophiles: from biodiversity to biotechnology. Springer, Cham, pp. 345–387e

Boyd ES, Hamilton TL, Wang J, He L, Zhang CL (2013) The role of tetraether lipid composition in the adaptation of thermophilic archaea to acidity. Front Microbiol 4:62

Brininger C, Spradlin S, Cobani L, Evilia C (2018) The more adaptive to change, the more likely you are to survive: Protein adaptation in extremophiles. Semin Cell Dev Biol 84:158–169

Buckel W, Thauer RK (2013) Energy conservation via electron bifurcating ferredoxin reduction and proton/Na(+) translocating ferredoxin oxidation. Biochim Biophys Acta 1827(2):94–113

Bulzu PA, Cristea A, Buda DM, Banciu HL (2019) Asgardarchaeota: A novel prokaryotic group discovered in aquatic sediments that might shed light on the origin and early evolution of eukaryotes. In: Air and Water Components of the Environment 2019 Conference. Conference Proceedings, Cluj-Napoca, Romania, pp 435–446

Cai M, Liu Y, Yin X, Zhou Z, Friedrich MW, Richter-Heitmann T, Nimzyk R, Kulkarni A, Wang X, Li W, Pan J, Yang Y, Gu D, Li M (2020) Diverse Asgard archaea including the novel phylum Gerdarchaeota participate in organic matter degradation. Sci China Life Sci 63:886–897

Cai M, Richter-Heitmann T, Yin X, Huang WC, Yang Y, Zhang C, Duan C, Pan J, Liu Y, Liu Y, Michae IW, Li M (2021) Ecological features and global distribution of Asgard archaea. Sci Total Environ 58:143581

Cai M, Tang X (2022) Human archaea and associated metabolites in health and disease. Biochem 61(24):2835–2840

Cai R, Zhang J, Liu R, Sun C (2021) Metagenomic insights into the metabolic and ecological functions of abundant deep-sea hydrothermal vent DPANN archaea. Appl Environ Microbiol 87(9):e03009–20

Carr SA, Jungbluth S, Rappe MS, Orcutt B (2017) Carboxydotrophy potential of uncultivated Hydrothermarchaeota from the oceanic crust deep biosphere. In: AGU Fall Meeting Abstracts 2017. pp H31F–1574

Castelle CJ, Banfield JF (2018) Major new microbial groups expand diversity and alter our understanding of the tree of life. Cell 172(6):1181–1197

Chen LX, Méndez-García C, Dombrowski N, Servín-Garcidueñas LE, Eloe-Fadrosh EA, Fang BZ, Luo BZ, Luo ZH, Tan S, Zhi XY, Hua ZS, Martinez-Romero E, Woyke T, Huang LN, Sánchez J, Peláez AI, Ferrer M, Shu WS (2018) Metabolic versatility of small archaea Micrarchaeota and Parvarchaeota. ISME J 12(3):756–775

Chen S, Tao J, Chen Y, Wang W, Fan L, Zhang C (2022) Interactions between marine group II Archaea and phytoplankton revealed by population correlations in the northern coast of South China Sea. Front Microbiol 12:785532

Cloarec LA, Bacchetta T, Bruto M, Leboulanger C, Grossi V, Brochier-Armanet C, Flandrois JP, Zurmely A, Bernard C, Troussellier M, Agogué H, Ader M, Oger-Desfeux C, Oger PM, Hugoni M (2024) Lineage-dependent partitioning of activities in chemoclines defines Woesearchaeota ecotypes in an extreme aquatic ecosystem. Microbiome 12(1):249

Collins T, Feller G (2023) Psychrophilic enzymes: strategies for cold-adaptation. Essays Biochem 67(4):701–713

Čubonová LU, Sandman K, Hallam SJ, DeLong EF, & Reeve JN (2005) Histones in crenarchaea. J Bacteriol 187(15):5482–5485

Danson MJ, Hough DW (1998) Structure, function and stability of enzymes from the Archaea. Trends Microbiol 6(8):307–314

Deppenmeier U (2002) The unique biochemistry of methanogenesis. Prog Nucleic Acid Res Mol Biol 71:223–283

Deppenmeier U (2004) The membrane-bound electron transport system of *Methanosarcina* species. J Bioenerg Biomembr 36(1):55–64

Díaz-Rullo J, González-Pastor JE (2025) Molecular adaptations specific to extreme halophilic archaea could promote high perchlorate tolerance. Appl Environ Microbiol 91(6):e00512–25

Dick GJ (2019) The microbiomes of deep-sea hydrothermal vents: distributed globally, shaped locally. Nat Rev Microbiol 17(5):271–283

Ding R, Yang N, Liu J (2022) The osmoprotectant switch of potassium to compatible solutes in an extremely halophilic archaea *Halorubrum kocurii* 2020YC7. Genes 13(6):939

Ding Y, Ke J, Hong T, Zhang A, Wu X, Jiang X, Chen S (2025) Microbial diversity and ecological roles of halophilic microorganisms in Dingbian (Shaanxi, China) saline-alkali soils and salt lakes. BMC Microbiol 25(1):287

Dombrowski N, Williams TA, Sun J, Woodcroft BJ, Lee JH, Minh BQ, Rinke C, Spang A (2020) Undinarchaeota illuminate DPANN phylogeny and the impact of gene transfer on archaeal evolution. Nat Commun 11(1):3939

Dombrowski N, Mahendrarajah T, Gross ST, Eme L, Spang A (2021) Archaea. In: Green LH, Goldman E (eds.) Practical handbook of microbiology. CRC Press, London/New York, pp. 229–248

Eme L, Spang A, Lombard J, Stairs CW, Ettema TJ (2017) Archaea and the origin of eukaryotes. Nat Rev Microbiol 15(12):711–723

Essel E, Li L, Deng C, Xie J, Zhang R, Luo Z, Cai L (2017) Effect of conservational and conventional tillage systems on functional soil archaea diversity in wheat-pea rotation field. J Agricul Scien 9(11):34–44

Falb M, Müller K, Königsmaier L, Oberwinkler T, Horn P, von Gronau S, Gonzalez O, Pfeiffer F, Bornberg-Bauer E, Oesterhelt D (2008) Metabolism of halophilic archaea. Extremophiles 12(2):177–196

Feehan B, Ran Q, Dorman V, Rumback K, Pogranichniy S, Ward K, Lee ST (2023) Novel complete methanogenic pathways in longitudinal genomic study of monogastric age-associated archaea. Anim Microbiome 5(1):35

Greening C, Ahmed FH, Mohamed AE, Lee BM, Pandey G, Warden AC, Scott, Oakeshott JG, Taylor MC, Jackson CJ (2016) Physiology, biochemistry, and applications of F420- and Fo-dependent redox reactions. Microbiol Mol Biol Rev 80(2):451–493

Golyshina OV, Bargiela R, Toshchakov SV, Chernyh NA, Ramayah S, Korzhenkov AA, Kunlanov IV, Golyshin PN (2019) Diversity of "Ca. Micrarchaeota" in two distinct types of acidic environments and their associations with Thermoplasmatales. Genes 10(6):461

Gubry-Rangin C, Williams W, Prosser JI (2018) Approaches to understanding the ecology and evolution of understudied terrestrial archaeal ammonia-oxidisers. Emerg Top Life Sci 2(4):619–628

Gutiérrez-Preciado A, Dede B, Baker BA, Eme L, Moreira D, López-García P (2024) Extremely acidic proteomes and metabolic flexibility in bacteria and highly diversified archaea thriving in geothermal chaotropic brines. Nat Ecol Evol 8(10):1856–1869

Guy L, Ettema TJ (2011) The archaeal 'TACK'superphylum and the origin of eukaryotes. Trends Microbiol 19(12):580–587

Hatzenpichler R (2012) Diversity, physiology, and niche differentiation of ammonia-oxidizing archaea. Appl Environ Microbiol 78(21):7501–7510

Hensley SA, Moreira E, Holden JF (2016) Hydrogen production and enzyme activities in the Hyperthermophile *Thermococcus paralvinellae* grown on maltose, tryptone, and agricultural waste. Front Microbiol 7:167

Hendrickson EL, Leigh JA (2008) Roles of coenzyme F420-reducing hydrogenases and hydrogen- and F420-dependent methylenetetrahydromethanopterin dehydrogenases in reduction of F420 and production of hydrogen during methanogenesis. J Bacteriol 190(14):4818–4821

Hocher A, Borrel G, Fadhlaoui K, Brugère JF, Gribaldo S, Warnecke T (2022) Growth temperature and chromatinization in archaea. Nat Microbiol 7(11):1932–1942

Hua ZS, Qu YN, Zhu Q, Zhou EM, Qi YL, Yin YR, Rao YZ, Tian Y, Li YX, Liu L, Castella CJ, Hedlund BP, Shu WS, Knight R, Li WJ (2018) Genomic inference of the metabolism and evolution of the archaeal phylum Aigarchaeota. Nat Commun 9(1):2832

Huber H, Hohn MJ, Stetter KO, Rachel R (2003) The phylum Nanoarchaeota: present knowledge and future perspectives of a unique form of life. Res Microbiol 154(3):165–171

Huber H, Gallenberger M, Jahn U, Eylert E, Berg IA, Kockelkorn D, Eisenreich W, Fuchs G (2008) A dicarboxylate/4-hydroxybutyrate autotrophic carbon assimilation cycle in the hyperthermophilic Archaeum *Ignicoccus hospitalis*. Proc Natl Acad Sci USA 105(22):7851–7856

Jarett JK, Nayfach S, Podar M, Inskeep W, Ivanova NN, Munson-McGee J, Schulz F, Young M, Jay ZJ, Beam JP, Kyrpides NC, Malmstrom RR, Stepanauskas R, Woyke T (2018) Single-cell genomics of co-sorted Nanoarchaeota suggests novel putative host associations and diversification of proteins involved in symbiosis. Microbiome 6:1–14

Jiang M, Xu P, Su R, Xiao H, Khan I, Hu R (2025) Ammonia-oxidizing archaea rather than ammonia-oxidizing bacteria drives methane emissions in a wheat-paddy rice rotation system. Appl Soil Ecol 209:106015

Jiao S, Xu Y, Zhang J, Lu Y (2019) Environmental filtering drives distinct continental atlases of soil archaea between dryland and wetland agricultural ecosystems. Microbiome 7(1):15

Johnson DB, Kanao T, Hedrich S (2012) Redox transformations of iron at extremely low pH: fundamental and applied aspects. Front Microbiol 3:96

Jungbluth SP, Amend JP, Rappé MS (2017) Metagenome sequencing and 98 microbial genomes from Juan de Fuca Ridge flank subsurface fluids. Sci Data 4(1):1–11

Kadnikov VV, Savvichev AS, Mardanov AV, Beletsky AV, Chupakov AV, Kokryatskaya NM et al. (2020) Metabolic diversity and evolutionary history of the archaeal phylum "Candidatus Micrarchaeota" uncovered from a freshwater lake metagenome. Appl Environ Microbiol 86(23):e02199–20

Kadnikov VV, Savvichev AS, Rusanov II, Beletskii AV, Pimenov NV (2024) Metagenomic analysis of bottom sediments of the karst meromictic lake Black Kichier revealed abundant unculturable Thermoplasmatota. Microbiol 93(2):128–133

Karishma S, Kamalesh R, Saravanan A, Deivayanai VC, Yaashikaa PR, Vickram AS (2024) A review on recent advancements in biochemical fixation and transformation of CO_2 into constructive products. Biochem Eng J 208:109366

Karr EA, Sattley WM, Rice MR, Jung DO, Madigan MT, Achenbach L (2005) Diversity and distribution of sulphate-reducing bacteria in permanently frozen Lake Fryxell, McMurdo Dry Valleys, Antarctica. Appl Environ Microbiol 71:6353–6359

Kellner S, Spang A, Offre P, Szöllosi GJ, Petitjean C, Williams TA (2018) Genome size evolution in the Archaea. Emerg Top Life Sci 2(4):595–605

Kerou M, Ponce-Toledo RI, Zhao R, Abby SS, Hirai M, Nomaki H, Takaki Y, Nunoura T, Jørgensen SL, Schleper C (2021) Genomes of Thaumarchaeota from deep sea sediments reveal specific adaptations of three independently evolved lineages. ISME J 15(9):2792–2808

Khalil M, Di Ciaula A, Mahdi L, Jaber N, Di Palo DM, Graziani A, Baffy G, Portincasa P (2024) Unraveling the role of the human gut microbiome in health and diseases. Microorganisms 12(11):2333

Khesali Aghtaei H, Püttker S, Maus I, Heyer R, Huang L, Sczyrba A, Reichl U, Benndorf D (2022) Adaptation of a microbial community to demand-oriented biological methanation. Biotechnol Biof Biop 15(1):125

Kim JY, Whon TW, Lim MY, Kim YB, Kim N, Kwon MS, Nam YD (2020) The human gut archaeome: identification of diverse haloarchaea in Korean subjects. Microbiome 8(1):114

Kletzin A, Heimerl T, Flechsler J, van Niftrik L, Rachel R, Klingl A (2015) Cytochromes c in Archaea: distribution, maturation, cell architecture, and the special case of *Ignicoccus hospitalis*. Front Microbiol 6:439

Kohtz AJ, Jay ZJ, Lynes MM, Krukenberg V, Hatzenpichler R (2022) Culexarchaeia, a novel archaeal class of anaerobic generalists inhabiting geothermal environments. ISME commun 2(1):86

Kolbe M, Besir H, Essen LO, Oesterhelt D (2000) Structure of the light-driven chloride pump halorhodopsin at 1.8 A resolution. Science 288(5470):1390–1396

Kozubal MA, Romine M, Jennings RD, Jay ZJ, Tringe SG, Rusch DB, Beam JP, McCue LA, Inskeep WP (2013) Geoarchaeota: a new candidate phylum in the Archaea from high-temperature acidic iron mats in Yellowstone National Park. ISME J 7(3):622–634

Könneke M, Bernhard AE, de la Torre JR, Walker CB, Waterbury JB, Stahl DA (2005) Isolation of an autotrophic ammonia-oxidizing marine archaeon. Nature 437(7058):543–546

Krause S, Gfrerer S, von Kügelgen A, Reuse C, Dombrowski N, Villanueva L, Gescher J (2022) The importance of biofilm formation for cultivation of a micrarchaeon and its interactions with its thermoplasmatales host. Nat Commun 13: 1735

Lamble HJ, Theodossis A, Milburn CC, Taylor GL, Bull SD, Hough DW, Danson MJ (2005) Promiscuity in the part-phosphorylative Entner-Doudoroff pathway of the archaeon *Sulfolobus solfataricus*. FEBS Lett 579(30):6865–6869

Lauer A, Sørensen KB, Teske A (2016) Phylogenetic characterization of marine benthic archaea in organic-poor sediments of the eastern equatorial Pacific Ocean (ODP site 1225). Microorganisms 4(3):32

Leigh JA (2000) Nitrogen fixation in methanogens: the archaeal perspective. Curr Issues Mol Biol 2(4):125–131

Lie TJ, Leigh JA (2003) A novel repressor of nif and glnA expression in *Methanococcus maripaludis*. Mol Microbiol 47(1):235–246

Li L, Liu Z, Zhou Z, Zhang M, Meng D, Liu X, Yin H (2021a) Comparative genomics provides insights into the genetic diversity and evolution of the DPANN superphylum. mSystems 6(4):10–1128

Li L, Ren M, Xu Y, Jin C, Zhang W, Dong X (2020) Enhanced glycosylation of an S-layer protein enables a psychrophilic methanogenic archaeon to adapt to elevated temperatures in abundant substrates. FEBS Lett 594(4):665–677

Li YX, Rao YZ, Qi YL, Qu YN, Chen YT, Jiao JY, Li WJ (2021b) Deciphering symbiotic interactions of "*Candidatus Aenigmarchaeota*" with inferred horizontal gene transfers and co-occurrence networks. mSystems 6(4):10–1128

Lillo JG, Rodríguez-Valera F (1990) Effects of culture conditions on poly-β-hydroxybutyrate production by Haloferax mediterranei. Appl Environ Microbiol 56(8):2517–2521

Liu R, Cai R, Wang M, Zhang J, Zhang H, Li C, Sun C (2024) Metagenomic insights into Heimdallarchaeia clades from the deep-sea cold seep and hydrothermal vent. Environ. Microbiome 19(1):43

Liu R, Cai R, Zhang J, Sun C (2020) Heimdallarchaeota harness light energy through photosynthesis. Preprint at https://doi.org/10.1101/2020.02.20.957134

Liu X, Wang Y, Gu JD (2021) Ecological distribution and potential roles of Woesearchaeota in anaerobic biogeochemical cycling unveiled by genomic analysis. Comput Struct Biotechnol J 19:794–800

Liu Y, Whitman WB (2008) Metabolic, phylogenetic, and ecological diversity of the methanogenic archaea. Ann N Y Acad Sci 1:171–189

Liu Y, Zhou Z, Pan J, Baker BJ, Gu JD, Li M (2018) Comparative genomic inference suggests mixotrophic lifestyle for Thorarchaeota. ISME J 12(4):1021–1031

Luo ZH, Li Q, Lai Y, Chen H, Liao B, Huang LN (2020) Diversity and genomic characterization of a novel Parvarchaeota family in acid mine drainage sediments. Front Microbiol 11:612257

Ma W, Lin M, Shen P, Chi H, Zhang W, Zhu J, Liu P (2025) Exploring methanogenic archaea and their thermal responses in the glacier-fed stream sediments of Rongbuk River basin, Mt. Everest. FEMS Microbiol Ecol 101(5):fiaf044

MacLeod F, Kindler GS, Wong HL, Chen R, Burns BP (2019) Asgard archaea: diversity, function, and evolutionary implications in a range of microbiomes. AIMS Microbiol 5(1):48

Madigan MT, Bender KS, Buckley DH, Sattley WM, Stahl DA (2021) Brock biology of microorganisms, 16th edn. Pearson, Hoboken (NJ)

Manoharan L, Kozlowski JA, Murdoch RW, Löffler FE, Sousa FL, Schleper C (2019) Metagenomes from coastal marine sediments give insights into the ecological role and cellular features of Loki-and Thorarchaeota. mBio 10(5):10–1128

Martens-Habbena W, Berube PM, Urakawa H, de la Torre JR, Stahl DA (2009) Ammonia oxidation kinetics determine niche separation of nitrifying Archaea and Bacteria. Nature 461(7266):976–979

Marzban G, Tesei D (2025) The extremophiles: adaptation mechanisms and biotechnological applications. Biology 14(4):412

Mayer F, Lim JK, Langer JD, Kang SG, Müller V (2015) Na+ transport by the A1A0-ATP synthase purified from Thermococcus onnurineus and reconstituted into liposomes. J Biol Chem 290(11):6994–7002

McMillan DG, Ferguson SA, Dey D, Schröder K, Aung HL, Carbone V, Attwood GT, Ronimus RS, Meier T, Janssen PH, Cook GM (2011) A1A0-ATP synthase of *Methanobrevibacter ruminantium* couples sodium ions for ATP synthesis under physiological conditions. J Biol Chem 286(46):39882–39892

Mehta D, Satyanarayana T (2013) Diversity of hot environments and thermophilic microbes. In: Satyanarayana T, Littlechild J, Kawarabayasi Y (eds) Thermophilic microbes in environmental and industrial biotechnology: biotechnology of thermophiles. Springer, Dordrecht, pp 3–60

Moissl-Eichinger C, Pausan M, Taffner J, Berg G, Bang C, Schmitz RA (2018) Archaea are interactive components of complex microbiomes. Trends Microbiol 26(1):70–85

Monstadt GM, Holldorf AW (1991) Arginine deiminase from *Halobacterium salinarium*. Purification and properties. Biochem J 273:739–745

Morozova D, Möhlmann D, Wagner D (2006) Survival of methanogenic archaea from Siberian permafrost under simulated martian thermal conditions. Orig Life Evol Biosph 37:189–200

Ney B, Ahmed FH, Carere CR, Biswas A, Warden AC, Morales SE, Pandey G, Watt SJ, Oakeshott JG, Taylor MC, Stott MB, Jackson CJ, Greening C (2017) The methanogenic redox cofactor F420 is widely synthesized by aerobic soil bacteria. ISME J 11(1):125–137

Nozhevnikova AN, Zepp K, Vazquez F, Zehnder AJ, Holliger C (2003) Evidence for the existence of psychrophilic methanogenic communities in anoxic sediments of deep lakes. Appl Microbiol 69(3):1832–1835

Oesterhelt D, Stoeckenius W (1971) Rhodopsin-like protein from the purple membrane of *Halobacterium halobium*. Nat New Biol 233(39):149–152

Oger PM, Cario A (2013) Adaptation of the membrane in Archaea. Biophys Chem 183:42–56

Oren A (2019) Euryarchaeota. Encyclopedia of Life Sciences, 1-17. https://doi.org/10.1002/9780470015902.a0004243.pub3.

Ou YF, Dong HP, McIlroy SJ, Crowe SA, Hallam SJ, Han P, Hou LJ (2022) Expanding the phylogenetic distribution of cytochrome b-containing methanogenic archaea sheds light on the evolution of methanogenesis. ISME J 16(10):2373–2387

Pathania S, Solanki P, Putatunda C, Bhatia RK, Walia A (2021) Adaptation to cold environment: the survival strategy of psychrophiles. In: Goel R, Soni R, Suyal DC, Khan M (eds) Survival strategies in cold-adapted microorganisms. Springer, Singapore, pp 87–111

Pisa KY, Weidner C, Maischak H, Kavermann H, Müller V (2007) The coupling ion in the methanoarchaeal ATP synthases: H+ vs. Na+ in the A1A0-ATP synthase from the archaeon *Methanosarcina mazei* Gö1. FEMS Microbiol Lett 277(1):56–63

Podar PT, Yang Z, Björnsdóttir SH, Podar M (2020) Comparative analysis of microbial diversity across temperature gradients in hot springs from Yellowstone and Iceland. Front Microbiol 11:1625

Qu YN, Rao YZ, Qi YL, Li YX, Li A, Palmer M, Li WJ (2023) *Panguiarchaeum* symbiosum, a potential hyperthermophilic symbiont in the TACK superphylum. Cell Rep 42(3):112158

Rahimzadeh M, Khajeh K, Mirshahi M, Khayatian M, Schwarzenbacher R (2012) Probing the role of asparagine mutation in thermostability of *Bacillus* KR-8104 α-amylase. Int J Biol Macromol 50(4):1175–1182

Ramos-Vera WH, Berg IA, Fuchs G (2009) Autotrophic carbon dioxide assimilation in Thermoproteales revisited. J Bacteriol 191(13):4286–4297

Reed CJ, Lewis H, Trejo E, Winston V, Evilia C (2013) Protein adaptations in archaeal extremophiles. Archaea 2013(1):373275

Ren M, Zhang Z, Wang X, Zhou Z, Chen D, Zeng H, Peng N (2018) Diversity and contributions to nitrogen cycling and carbon fixation of soil salinity shaped microbial communities in Tarim Basin. Front Microbiol 9:431

Reva O, Messina E, La Cono V, Crisafi F, Smedile F, La Spada G, Yakimov MM (2023) Functional diversity of nanohaloarchaea within xylan-degrading consortia. Front Microbiol 14:1182464

Rudolph C, Wanner G, Huber R (2001) Natural communities of novel archaea and bacteria growing in cold sulfurous springs with a string-of-pearls-like morphology. Appl Environ Microbiol 67(5):2336–2344

Saini A, Kumar A, Singh G, Giri SK (2023) Survival strategies and stress adaptations in halophilic Archaebacteria. In: Microbial stress response: mechanisms and data science. ACS Symposium Series, 1434:1–21

Sakai HD, Nur N, Kato S, Yuki M, Shimizu M, Itoh T, Ohkuma M, Suwanto A, Kurosawa N (2022) Insight into the symbiotic lifestyle of DPANN archaea revealed by cultivation and genome analyses. Proc Natl Acad Sci USA 119(3):e2115449119

Schlegel K, Leone V, Faraldo-Gómez JD, Müller V (2012) Promiscuous archaeal ATP synthase concurrently coupled to Na+ and H+ translocation. Proc Natl Acad Sci USA 109(3):947–952

Schleper C, Nicol GW (2010) Ammonia-oxidising archaea–physiology, ecology and evolution. Adv Microb Physiol 57:1–41

Schouten S, van der Meer MT, Hopmans EC, Rijpstra WI, Reysenbach AL, Ward DM, Sinninghe Damsté JS (2007) Archaeal and bacterial glycerol dialkyl glycerol tetraether lipids in hot springs of Yellowstone National Park. Appl Environ Microbiol 73(19):6181–6191

Seitz KW, Dombrowski N, Eme L, Spang A, Lombard J, Sieber JR, Baker BJ (2019) Asgard archaea capable of anaerobic hydrocarbon cycling. Nat Commun 10(1):1822

Sharma S, Chaturvedi U, Sharma K, Vaishnav A, Singh HB (2021) An overview of survival strategies of psychrophiles and their applications. In: Goel R, Soni R, Suyal DC, Khan M (eds) Survival strategies in cold-adapted microorganisms. Springer, Singapore, pp 133–151

She Q, Singh RK, Confalonieri F, Zivanovic Y, Allard G, Awayez MJ, Chan-Weiher CC, Clausen IG, Curtis BA, De Moors A, Erauso G, Fletcher C, Gordon PM, Heikamp-de Jong I, Jeffries AC, Kozera CJ, Medina N, Peng X, Thi-Ngoc HP, Redder P, Van der Oost J (2001) The complete genome of the crenarchaeon Sulfolobus solfataricus P2. Proc Natl Acad Sci USA 98(14):7835–7840

Shu WS, Huang LN (2022) Microbial diversity in extreme environments. Nat Rev Microbiol 20(4):219–235

Siliakus MF, van der Oost J, Kengen SW (2017) Adaptations of archaeal and bacterial membranes to variations in temperature, pH and pressure. Extremophiles 21(4):651–670

Sorokin DY, Abbas B, Merkel AY, Rijpstra WIC, Damsté JSS, Sukhacheva MV, van Loosdrecht MC (2015) *Methanosalsum natronophilum* sp. nov., and *Methanocalculus alkaliphilus* sp. nov., haloalkaliphilic methanogens from hypersaline soda lakes. Int J Syst Evol Microbiol 65:3739–3745

Sorokin DY, Berben T, Melton ED, Overmars L, Vavourakis CD, Muyzer G (2014) Microbial diversity and biogeochemical cycling in soda lakes. Extremophiles 18(5):791–809

Sorokin DY, Elcheninov AG, Khijniak TV, Kolganova TV, Kublanov IV (2022) Selective enrichment on a wide polysaccharide spectrum allowed isolation of novel metabolic and taxonomic groups of haloarchaea from hypersaline lakes. Front Microbiol 13:1059347

Sorokin DY, Merkel AY, Abbas B (2022) Ecology of Methanonatronarchaeia. Environ Microbiol 24(11):5217–5229

Sousa FL, Martin WF (2014) Biochemical fossils of the ancient transition from geoenergetics to bioenergetics in prokaryotic one carbon compound metabolism. BBA-Bioenergetics 1837(7):964–981

Spang A, Caceres EF, Ettema TJ (2017) Genomic exploration of the diversity, ecology, and evolution of the archaeal domain of life. Science 357(6351):eaaf3883

St John EJ (2019) Symbiosis in Archaea: functional and phylogenetic diversity of marine and terrestrial nanoarchaeota and their hosts. Dissertations and Theses. Paper no. 4939, Portland State University

Sun J, Zhang A, Zhang Z, Liu Y, Zhou H, Cheng H, Wang Y (2023) Distinct assembly processes and environmental adaptation of abundant and rare archaea in Arctic marine sediments. Mar Environ Res 190:106082

Sun J, Zhou H, Cheng H, Chen Z, Wang Y (2025) Archaea show different geographical distribution patterns compared to bacteria and fungi in Arctic marine sediments. mLife 4(2):205–218

Tahon G, Geesink P, Ettema TJ (2021) Expanding archaeal diversity and phylogeny: past, present, and future. Annu Rev Microbiol 75(1):359–381

Takamiya H, Kouduka M, Kato S, Suga H, Oura M, Yokoyama T, Suzuki Y (2024) Genome-resolved metaproteogenomic and nanosolid characterization of an inactive vent chimney densely colonized by enigmatic DPANN archaea. ISME J 18(1):wrae207

Teske A (2018) Aerobic Archaea in iron-rich springs. Nat Microbiol 3(6):646–647

Torregrosa-Crespo J, Pire C, Bergaust L, Martínez-Espinosa RM (2020) Haloferax mediterranei, an archaeal model for denitrification in saline systems, characterized through integrated physiological and transcriptional analyses. Front Microbiol 11:768

Tourna M, Stieglmeier M, Spang A, Könneke M, Schintlmeister A, Urich T, Engel M, Schloter M, Wagner M, Richter A, Schleper C (2011) Nitrososphaera viennensis, an ammonia oxidizing archaeon from soil. Proc Natl Acad Sci USA 108(20):8420–8425

Umbach AK, Stegelmeier AA, Neufeld JD (2021) Archaea are rare and uncommon members of the mammalian skin microbiome. mSystems 6(4):10–1128

Valentine DL (2007) Adaptations to energy stress dictate the ecology and evolution of the Archaea. Nat Rev Microbiol 5(4):316–323

Vanwonterghem I, Evans PN, Parks DH, Jensen PD, Woodcroft BJ, Hugenholtz P, Tyson GW (2016) Methylotrophic methanogenesis discovered in the archaeal phylum Verstraetearchaeota. Nat Microbiol 1(12):1–9

Verhees CH, Kengen SW, Tuininga JE, Schut GJ, Adams MW, De Vos WM, Van Der Oost J (2003) The unique features of glycolytic pathways in Archaea. Biochem J 375(Pt 2):231–246

Wang Y, Wegener G, Hou J, Wang F, Xiao X (2019) Expanding anaerobic alkane metabolism in the domain of Archaea. Nat Microbiol 4(4):595–602

Weidenbach K, Gutt M, Cassidy L, Chibani C, Schmitz RA (2022) Small proteins in archaea, a mainly unexplored world. J Bacteriol 204(1):e00313–21

Welte CU, de Graaf R, Dalcin Martins P, Jansen RS, Jetten MS, Kurth JM (2021) A novel methoxydotrophic metabolism discovered in the hyperthermophilic archaeon *Archaeoglobus fulgidus*. Environ Microbiol 23(7):4017–4033

Woese CR, Fox GE (1977) Phylogenetic structure of the prokaryotic domain: the primary kingdoms. Proc Natl Acad Sci USA 11:5088–5090
Woese CR, Kandler O, Wheelis ML (1990) Towards a natural system of organisms: proposal for the domains Archaea, Bacteria, and Eucarya. Proc Natl Acad Sci USA 87(12):4576–4579
Xamxidin M, Zhang X, Zheng G, Chen C, Wu M (2025) Metagenomics-assembled genomes reveal microbial metabolic adaptation to athalassohaline environment, the case Lake Barkol, China. Front Microbiol 16:1550346
Xie YG, Luo ZH, Fang BZ, Jiao JY, Xie QJ, Cao XR, Hua ZS (2022) Functional differentiation determines the molecular basis of the symbiotic lifestyle of *Ca.* Nanohaloarchaeota. Microbiome 10(1):172
Xu H, Xu D, Liu Y (2024) Molecular biology applications of psychrophilic enzymes: adaptations, advantages, expression, and prospective. Appl Biochem Biotechnol 196(9):5765–5789
Xue SD, Yi XY, Cui HL, Li M, Peng JJ, Zhu YG, Duan GL (2023) Global biogeographic distribution of Bathyarchaeota in paddy soils. mSystems 8(3):e00143–23
Yin X, Cai M, Liu Y, Zhou G, Richter-Heitmann T, Aromokeye DA, Friedrich MW (2021) Subgroup level differences of physiological activities in marine Lokiarchaeota. ISME J 15(3):848–861
Youssef NH, Rinke C, Stepanauskas R, Farag I, Woyke T, Elshahed MS (2015) Insights into the metabolism, lifestyle and putative evolutionary history of the novel archaeal phylum 'Diapherotrites'. ISME J 9(2):447–460
Yu T, Fu L, Wang Y, Dong Y, Chen Y, Wegener G, Wang F (2024) Thermophilic Hadarchaeota grow on long-chain alkanes in syntrophy with methanogens. Nat Commun 15(1):6560
Yusof NA, Hashim NHF, Bharudin I (2021) Cold adaptation strategies and the potential of psychrophilic enzymes from the antarctic yeast, Glaciozyma antarctica PI12. J Fungi 7(7):528
Zaremba-Niedzwiedzka K, Caceres EF, Saw JH, Bäckström D, Juzokaite L, Vancaester E, Ettema TJ (2017) Asgard archaea illuminate the origin of eukaryotic cellular complexity. Nature 541(7637):353–358
Zeng Z, Liu XL, Wei JH, Summons RE, Welander PV (2018) Calditol-linked membrane lipids are required for acid tolerance in *Sulfolobus acidocaldarius*. Proc Natl Acad Sci USA 115(51):12932–12937
Zhang IH, Borer B, Zhao R, Wilbert S, Newman DK, Babbin AR (2024) Uncultivated DPANN archaea are ubiquitous inhabitants of global oxygen-deficient zones with diverse metabolic potential. mBio 15(3):e02918–23
Zhao R, Biddle JF (2021) Helarchaeota and co-occurring sulfate-reducing bacteria in subseafloor sediments from the Costa Rica Margin. ISME Commun 1(1):25
Zheng PF, Wei Z, Zhou Y, Li Q, Qi Z, Diao X, Wang Y (2022) Genomic evidence for the recycling of complex organic carbon by novel Thermoplasmatota clades in deep-sea sediments. mSystems 7(3):e00077–22
Zhou Z, Liu Y, Anantharaman K, Li M (2022) The expanding Asgard archaea invoke novel insights into Tree of Life and eukaryogenesis. Mlife 1(4):374–381
Zhou Z, Liu Y, Xu W, Pan J, Luo ZH, Li M (2020) Genome-and community-level interaction insights into carbon utilization and element cycling functions of Hydrothermarchaeota in hydrothermal sediment. mSystems 5(1):10–1128
Zhou Z, Pan J, Wang F, Gu JD, Li M (2018) Bathyarchaeota: globally distributed metabolic generalists in anoxic environments. FEMS Microbiol Rev 42(5):639–655
Zinke LA, Evans PN, Santos-Medellín C, Schroeder AL, Parks DH, Varner RK, Emerson JB (2021) Evidence for non-methanogenic metabolisms in globally distributed archaeal clades basal to the Methanomassiliicoccales. Environ Microbiol 23(1):340–357
Zou D, Qi Y, Zhou J, Liu Y, Li M (2025) Unveiling the life of archaea in sediments: diversity, metabolic potentials, and ecological roles. iMetaOmics 2(1):e56

Chapter 4
Microbial Metabolism: Primary and Secondary Metabolites

Nazmiye Ozlem Sanli and Elif Özlem Arslan Aydoğdu

Abstract Microbial metabolism encompasses the biochemical networks that enable cells to convert nutrients into the energy and biosynthetic precursors necessary for cellular function. Understanding microbial metabolism is crucial for ecology, microbiology, and applications in biotechnology, agriculture, and medicine, as it underpins the biological processes that sustain life in various environments. Microbial metabolic products can be broadly classified as primary or secondary metabolites. Primary metabolites, such as amino acids, carbohydrates, organic acids, and alcohols, are essential for energy generation, biosynthesis, and cellular maintenance, typically accumulating during the exponential growth phase. In contrast, secondary metabolites, including compounds such as antibiotics, toxins, and alkaloids, are predominantly synthesised during the stationary growth phase and are not essential for basic cellular functions. However, they are often involved in ecological interactions and may provide adaptive advantages to the organisms. This distinction emphasises the dual function of microbial metabolism in supporting life and influencing ecological processes. The regulation of microbial metabolite biosynthesis is influenced by nutrient availability, environmental stresses, and intricate genetic networks. Advances in metabolic engineering, synthetic biology, and CRISPR-based technologies have enabled the redirection of central carbon fluxes to enhance the yield of both primary and secondary metabolites. Multi-omics approaches and computational modeling now provide deeper insights into metabolic regulation, pathway optimization, and

N. O. Sanli (✉)
Istanbul University, Faculty of Science, Department of Biology, Section of Biotechnology, Vezneciler, Istanbul, Türkiye
e-mail: nosanli@istanbul.edu.tr

E. Ö. Arslan Aydoğdu
Istanbul University, Faculty of Science, Department of Biology, Section of Fundamental and Industrial Microbiology, Vezneciler, Istanbul, Türkiye

M. Çalışkan and M. Arıkan (eds.), *Microbial World: Bacteria and Archaea*, Progress in Molecular and Subcellular Biology 62, https://doi.org/10.1007/978-3-032-11438-9_4

the discovery of novel bioactive molecules. In conclusion, microbial metabolites are an invaluable resource for biotechnological innovation. While primary metabolites are essential for metabolic physiology, secondary metabolites offer unique bioactivities that have transformed pharmaceutical development, sustainable agriculture and industrial microbiology. Research focusing on uncovering new microbial diversity and optimising biosynthetic pathways is expected to continue providing insights that could deliver sustainable strategies and innovative solutions to urgent global challenges such as food security, healthcare, and environmental restoration.

Keywords Primary metabolites · Secondary metabolites · Bioactive compound sources · Biotechnology · Industrial microbiology

4.1 Introduction

4.1.1 Basic Principles of Microbial Metabolism

Microbial metabolism is a fundamental biological process that involves the chemical reactions occurring within microorganisms, enabling them to grow, reproduce, and survive (Pierre 2024). This essential biological process can be categorized into two main pathways: catabolism (the breakdown of substances to release energy) and anabolism (the use of this energy to synthesize complex molecules for cellular structures and functions) (Ebenhöh et al. 2024). Understanding microbial metabolism is crucial for ecology, microbiology, and applications in biotechnology, agriculture and medicine, as it underpins biological processes that sustain life in various environments (Holbrook-Smith et al. 2023).

Microbial metabolism consists of various interconnected biochemical pathways that enable cells to convert nutrients into the energy and biosynthetic precursors necessary for cellular function. Primary metabolic pathways, however, differ between prokaryotic and eukaryotic microorganisms in distinct and measurable ways. Prokaryotes optimize metabolic pathways for resource efficiency, whereas eukaryotes use compartmentalization and regulatory complexity to better integrate energy production with biomass generation (Flamholz et al. 2013). Despite these differences, key metabolic pathways in microorganisms include glycolysis, the citric acid cycle, and the pentose phosphate pathway (Fuhrmann 2021).

Studies of prokaryotic organisms indicate that they choose between the Embden-Meyerhof-Parnas (EMP) and Entner-Doudoroff (ED) glycolytic pathways depending on the availability of oxygen and the cost of proteins. It has also been observed that most bacteria exhibit complete tricarboxylic acid (TCA) cycles, with exceptions such as cyanobacteria that exhibit incomplete cycles. The diversity of electron transport chains enabling flexible aerobic and anaerobic energy production is emphasized.

A universal theme that emerges from these systems is the trade-off between enzyme investment and ATP yield (Ng et al. 2019; Kopp and Sunna 2020; Kopp et al. 2020; Veaudor et al. 2020; Peretó 2023). In contrast, studies of eukaryotes describe a strategy centered on regulatory complexity and compartmentalization. In yeasts and protists, for instance, glycolytic processes may be segregated within organelles such as glycosomes or mitochondria, and the TCA cycle may be bifurcated or rewired (Ginger et al. 2010; Müller et al. 2012; Martin et al. 2020). These adaptations facilitate the integration of energy production with biomass generation, including the use of unique organelles such as hydrogenosomes, as well as the influence of lateral gene transfer in parasitic protists. In summary, prokaryotes tend to optimize metabolic pathways for rapid resource and protein cost efficiency, whereas eukaryotes favor spatial and regulatory innovation to balance high-efficiency respiration with the demands of biosynthesis (Rodríguez-Gijón et al. 2022; Taha et al. 2023).

The production of primary and secondary metabolites during microbial metabolism is of particular significance. In bacteria and fungi, primary metabolites such as amino acids, carbohydrates, and nucleotides are synthesized by conserved central pathways such as glycolysis, the tricarboxylic acid cycle, and fatty acid biosynthesis. These metabolites are crucial for energy production, cellular maintenance, and the overall metabolic physiology of microorganisms (Mccann and Rawls 2023; Al-Ishaq 2024). In contrast, secondary metabolites arise from specialized biosynthetic routes (for example, via polyketide synthases, non-ribosomal peptide synthetases, and terpene synthases) that branch from these central pathways by redirecting simple precursors like acetyl coenzyme A, amino acids, and NADPH into more complex, species-specific compounds.

4.1.2 Overview of Central Metabolic Pathways in Microorganisms

Central carbon metabolic (CCM) pathways in microorganisms, including glycolysis, gluconeogenesis, the pentose phosphate pathway, and the tricarboxylic acid (TCA) cycle, form an interconnected network of reactions that convert substrates into energy and building blocks (Papagianni 2012; Wu et al. 2023). These interconnected pathways can be manipulated to increase the production of metabolites, including proteins, amino acids, and organic acids (Papagianni 2012). While central carbon metabolism is largely conserved across species, it manifests differently depending on the organism and its ecological niche. Studies have demonstrated that transitions between glycolysis and gluconeogenesis involve dynamic shifts in intracellular metabolite concentrations that influence the regulatory capabilities of microbial cells. Furthermore, managing these metabolic processes often results in trade-offs that affect growth rates and enzyme cycling, which explains why certain microorganisms tend to specialize in glycolytic or gluconeogenic pathways (Suarez-Mendez 2015; Toya and Shimizu 2020; Schink et al. 2022). Understanding these dynamics is crucial for improving biotechnological applications, particularly when microorganisms are exposed to fluctuating environmental conditions in industrial settings (Papagianni 2012; Liu et al. 2021). Advances in synthetic biology facilitate the redesign and optimization of these pathways by enabling the precise modulation of genetic control elements, thereby improving metabolic process efficiency (Alper 2011; Papagianni 2012). Emerging approaches focus on ensuring optimal carbon and energy flow within these pathways to maximize yield and productivity. The integration of genetic and biochemical engineering tools is an ongoing process, providing new strategies for understanding and applying central metabolism in industrial microbiology (Alper 2011; Papagianni 2012).

Glycolysis is the process by which glucose is broken down to produce pyruvate, which generates energy in the form of ATP and NADH. This process plays a key role in the primary metabolism of microorganisms, acting as an important energy production pathway while also providing the necessary precursors for secondary metabolite biosynthesis (Borodina et al. 2008). For example, in Streptomycetes, modifications to glycolytic pathways—such as the deletion of phosphofructokinase (pfkA2)—have been shown to enhance the production of secondary metabolites, such as antibiotics, due to increased carbon flux through the pentose phosphate pathway. Similarly, microorganisms can adapt their glycolytic flux in response to environmental stressors, which in turn influences the levels of primary and secondary metabolites, such as amino acids and phenolic compounds, that are crucial for survival and growth. Metabolomic studies in different microbial species indicate that primary metabolites play a crucial role in supporting the biosynthesis of secondary metabolites, particularly during critical growth phases. Glycolysis is identified as a key metabolic pathway that orchestrates this interplay between primary and secondary metabolism. As primary metabolites are generated through glycolytic pathways, they serve as precursors or energy sources necessary for the production of secondary metabolites. This relationship emphasizes the significance of metabolic flux and regulation in microbial growth and the synthesis of secondary metabolites (Zhang and Davies 2016; Castro-Moretti et al. 2020). Therefore, engineering glycolytic pathways in microorganisms has significant potential for optimizing secondary metabolite production in industrial applications (Borodina et al. 2008; Liu et al. 2021).

The TCA cycle, which is conserved in organisms ranging from bacteria to humans, begins with the combination of acetyl-CoA and oxaloacetate to form citric acid (MacLean et al. 2023; Börekçi et al. 2021). This cycle is widely regarded as vital for energy production. It oxidizes acetyl-CoA, releasing the energy stored in food molecules and generating ATP, NADH, and $FADH_2$ through a process known as oxidative phosphorylation (Kumar and Dubey 2019). Furthermore, the cycle plays a crucial role in determining the production of various metabolites among different microbial species by providing essential building blocks for biosynthesis. The TCA cycle is also a key pathway for the biosynthesis of organic acids, such as citric acid, α-ketoglutaric acid, succinic acid, fumaric acid and malic acid, in microorganisms (Yin et al. 2015).

Pentose phosphate pathway (PPP) is a metabolic route that is essential for nucleotide synthesis, as it generates ribose-5-phosphate and NADPH (Fuentes-Lemus et al. 2023). Operating in parallel with glycolysis, the pathway provides critical components for biosynthetic processes, including fatty acid synthesis and antioxidant defenses (Gupta and Gupta 2021). The PPP's ability to supply nucleotides and maintain redox balance is crucial for efficient growth and metabolite synthesis under various metabolic conditions. Overall, the PPP plays a pivotal role in metabolic reprogramming, supporting the anabolic needs of microbial cells while responding to oxidative stress.

Microbial metabolic pathways evolve in response to environmental changes, such as exposure to xenobiotics and antibiotic pressure. These selective forces also drive the evolution of metabolic plasticity at the species level and through cooperative interactions within microbial consortia, where enzymatic capabilities are often shared to enhance survival and functional efficiency. Trade-offs in microbial life history traits and ecological niche compatibility contribute to the stability of these communities (Kavvas et al. 2022; Kehila et al. 2025). Furthermore, horizontal gene transfer accelerates the acquisition of novel metabolic traits, facilitating rapid adaptation and influencing ecological interactions (Khan 2025). Distinct differences emerge between prokaryotic and eukaryotic microorganisms in the regulation of primary and secondary metabolism (Shibl et al. 2020). Studies on prokaryotes suggest that primary metabolic pathways, such as central carbon and amino acid metabolism, are modulated by direct changes in gene expression, operon structure, and gene duplication or horizontal transfer (Schink et al. 2022; Sharma and Curtis 2022). In prokaryotes, the production of secondary metabolites is characterized by the diversification of extensive gene clusters (e.g., non-ribosomal peptide and polyketide synthases), which is influenced by environmental pressures (Schniete et al. 2018). By contrast, studies of eukaryotic microbes—from protists and diatoms to fungi—show that primary metabolism is altered through subcellular compartmentalization (Ginger et al. 2010; Gabaldón and Pittis 2015). Glycolytic pathways localized in glycosomes and plastidial isoprenoid synthesis illustrate how eukaryotes integrate metabolism with cell cycle and redox balance (Ginger et al. 2010). Secondary metabolism in eukaryotes appears to be regulated by ecological and developmental cues (Shibl et al. 2020). In summary, prokaryotes alter their metabolism through swift genetic adjustments and the expansion of metabolic clusters, whereas eukaryotic microorganisms regulate both primary and secondary metabolism through complex regulatory networks that are linked to cellular organization and compartmentalization.

Primary and secondary metabolites produced by microorganisms play a significant role in various biotechnological applications. These metabolites have a wide range of applications in the food, pharmaceutical, and chemical industries. They address nutritional and health needs by acting as vital sources of bioactive compounds. Primary metabolites, such as amino acids and organic acids, are vital for essential metabolic processes. Meanwhile, secondary metabolites, including antibiotics and biopesticides, are being investigated for their therapeutic potential and effectiveness in pest management. Furthermore, these metabolites can improve crop health and yield, enhance nutritional quality, and promote the development of environmentally friendly agricultural practices. Overall, microbial metabolites play a key role in bioprocess innovation, helping to effectively address health and environmental challenges (Chen et al. 2024; Ofoedum et al. 2025; Prabhu et al. 2025).

4.2 Microbial Primary Metabolites

Microbial primary metabolites, like those in all other living organisms, serve as the building blocks and energy sources that form the basis of microorganisms' growth, development, and fundamental life processes (Chubukov et al. 2014). These metabolites are both inputs and outputs of catabolic and anabolic processes. These compounds are typically structural metabolites characterized by their low molecular weight, such as amino acids, proteins, nucleic acids, vitamins, alcohols, purines, and pyrimidines (Çetiz and Memon 2021). These metabolites, synthesized in varying amounts and types throughout the microorganism's life cycle, are particularly known for determining the rate of reproduction. The accumulation of these metabolites and the subsequent readiness of the cell mass for proliferation is a fundamental principle. Therefore, unlike secondary metabolites, the production of primary metabolites is vital for the survival and proliferation of microorganisms.

These compounds, generally classified as anabolic and catabolic metabolites, are regulators responsible not only for reproduction but also for adaptation to environmental conditions (Madigan et al. 2012). They regulate nutrient uptake, coordinate monomer synthesis and macromolecular polymerization, and oversee central metabolism. Their production mostly occurs during the exponential phase or reaches a maximum level (Sanchez and Demain 2008). Synthesized by almost all microorganisms and through conserved metabolic pathways across species, these compounds play a key role in energy production and storage, as well as in fundamental metabolic pathways such as cell growth and division, glycolysis, the TCA cycle, and protein synthesis, play a key role in energy production and storage, and serve as building blocks for macromolecules such as nucleic acids and proteins. Additionally, their polar or amphipathic structures, high solubility in water, stability, and generally small molecular weights (typically less than 1000 Da) are fundamental chemical properties (Demain and Adrio 2008; Sanchez and Demain 2008).

4.2.1 Functions of Primary Metabolites

The primary function of primary metabolites is to regulate the growth and basic life activities of microorganisms. These molecules coordinate central metabolic functions such as monomer synthesis, macromolecular polymerization, and nutrient storage, and also coordinate the cell's needs, including energy metabolism. For example, limiting glucose, the primary carbon source for microbial life, affects fermentation and mitochondrial respiration in microorganisms. Amino acids, which play a key role in reproduction, also regulate many vital activities in the cell, so failure to produce even one of them can cause the cell to be unable to reproduce and die (Sanchez and Demain 2008; Madigan et al. 2012; Chubukov et al. 2014).

Primary metabolites associated with the exponential phase are involved in all life activities and therefore play a role in every phase of reproduction. Limited primary production is observed in the lag phase, where environmental adaptation and mass increase occur. During this period, enzyme regulation and the activation of fundamental metabolic pathways are among the functions of primary metabolites (Nielsen and Keasling 2016). The exponential phase, of course, is the period during which primary metabolites, such as amino acids, nucleotides, vitamins, and organic acids, are produced at maximum levels. To rapidly increase the number of cells exponentially, building blocks, cofactors, and energy reserves are required. For this purpose, microorganisms rapidly consume the nutrients they have stored during the stationary phase to synthesize primary metabolites. For example, organic compounds are catabolized through pathways such as glycolysis and the Krebs cycle to obtain ATP, NADPH, and various precursor molecules. Since these precursor molecules will be utilized in reproduction and subsequently in the production of essential metabolites and building blocks, the efficiency of these catabolic pathways directly impacts the growth rate and quantity. In other words, under optimal conditions with access to necessary nutrients, a cell achieves its maximum reproduction rate and primary metabolite production efficiency (Dale et al. 2008; Chai et al. 2022; Wang and Kuzyakov 2023).

The next stage is characterized by a significant decrease in reproduction, resulting from both food scarcity and an increase in toxic byproducts. The continued production of storage compounds particularly marks this stage. The death phase, where primary metabolite production reaches a minimum, is the stage at which the number of cells decreases significantly. Once cells enter the stationary phase, primary metabolite production is limited, while the production of secondary metabolites, which are responsible for resistance to adverse conditions, begins to increase rapidly (Sanchez and Demain 2008; Davies 2013).

4.2.2 Synthesis and Regulation of Primary Metabolites

Understanding the mechanisms of primary metabolite synthesis and utilization, which regulate the growth kinetics and physiology of microorganisms, and thereby regulating microbial mass, is a fundamental goal in all disciplines that utilize microorganisms, including biotechnology. Understanding the dynamics of survival and reproduction is crucial for obtaining maximum yield from a cell. Several regulatory mechanisms are involved in the synthesis of primary metabolites. These include feedback inhibition, where the synthesis of a metabolite is regulated based on its concentration within the cell; catabolic repression, where alternative metabolic pathways are suppressed in the presence of the target carbon source; and induction and repression mechanisms, where the production of enzymes targeting specific substrates is regulated based on their presence or absence.

Feedback inhibition can be defined as the regulation of production by a product reaching a particular concentration, thereby inhibiting the activity or synthesis of the key enzymes responsible for its biosynthesis. This mechanism is effective in maintaining homeostasis by controlling resource efficiency and metabolic flows. In this mechanism, the regulatory end product either binds to the allosteric site of the first enzyme in the pathway, thereby suppressing its activity, or suppresses enzyme synthesis. In enzyme inactivation, production continues until the intermediate products are depleted (Kulis-Horn et al. 2014; Galili et al. 2016; Reznik et al. 2017). Enzyme synthesis regulation occurs in two ways: feedback repression and attenuation. In the first mechanism, the compound whose production is to be stopped binds to the apo-receptor protein as a co-receptor. This complex binds to the operator sequence of the gene region responsible for producing the target enzyme, preventing RNA polymerase from initiating transcription. In attenuation, the product binds to the regulatory operon, causing mRNA to be terminated prematurely during the process of amino acid biosynthesis. This results in the release of a tRNA that lacks sufficient information for the enzyme and cannot initiate protein synthesis. In enzyme suppression, only one of these mechanisms may be functional, while for some operons, such as the tryptophan operon, regulation by both mechanisms is involved (Glazer and Nikaido 2007; Sanchez and Demain, 2008; Naz et al. 2023).

These regulatory mechanisms are very common in amino acid synthesis. For example, when arginine levels increase in bacteria such as Corynebacteria, it inhibits the enzymes N-acetyl-L-glutamate kinase and N-acetyl-L-glutamate synthase, which play a key role in biosynthesis (Xu et al. 2012). In *E. coli*, the ArgR protein suppresses the biosynthesis of genes, thereby halting the pathway (Glazer and Nikaido 2007). In lysine biosynthesis, inhibition of enzymes such as dihydrodipicolinate synthase and aspartokinase is observed. Numerous studies have been conducted on C*orynebacterium glutamicum*, a Gram-positive organism, for lysine production. Inhibition of these regulatory mechanisms is crucial for the high-yield production of L-lysine, an important amino acid in the food and pharmaceutical industries (Liu et al. 2023). The main deregulation strategies developed to prevent feedback inhibition include creating mutations, steric hindrances, or charge changes in key amino acids in the regulatory domain of the enzyme to prevent the inhibitor from binding, blocking the inhibitor's access to the binding site, interface mutations in allosteric enzymes with oligomeric structures, and production of enzymes with reduced sensitivity to the end product, such as the use of an aspartokinase III mutant with reduced lysine sensitivity in *E. coli* (Chubukov et al. 2014; Huccetogullari et al. 2019; Naz et al. 2023).

Catabolic repression is another regulatory mechanism for maintaining homeostasis. The purpose of this mechanism is to prioritize the utilization of the carbon source that is most abundant in the environment. This ensures that the enzymes required for the primary substrate are produced and that resources are utilized with maximum efficiency, minimizing waste (Görke and Stülke 2008; Sanchez and Demain 2008). Catabolic repression occurs through pathways that vary depending on the organism. For example, in *E. coli*, glucose transported into the cell via the phosphotransferase system (PTS) is a priority carbohydrate. The EIIA protein, a component of PTS, becomes phosphorylated in the presence of glucose and cannot transport other carbon sources. Meanwhile, the catabolic pathways of other carbohydrates not transported via the PTS are suppressed by the cAMP receptor protein, which is induced by carbon limitation and repressed by nitrogen or sulfur limitation (You et al. 2013; Pal et al. 2022).

In Gram-positive bacteria, catabolic repression occurs through a complex interaction between two proteins: Hpr (a PTS transporter protein) and CcpA (catabolite control protein A). Rapid sugar metabolism results in excessive accumulation of glycolytic intermediates. These products cause phosphorylation at the Ser-46 position in the Hpr protein in bacteria such as *Bacillus subtilis*. Phosphorylated Hpr then binds to CcpA and the *cis*-acting operator sequence, thereby inhibiting gene expression (Singh et al. 2008).

Catabolic suppression mechanisms are also present in yeasts. For example, in *Saccharomyces* species, glucose plays a role in metabolic inhibition. Unlike bacteria, this mechanism occurs via hexokinase rather than glucokinase. Studies have shown that α-glucosidase, maltase, and invertase are not inhibited by glucose in hexokinase mutants (Entian and Schuller 2024). An important metabolic regulator in yeasts is the Crabtree effect. Some yeasts, such as *Saccharomyces cerevisiae*, *S. eubayanus*, *Candida glabrata*, *Lachancea fermentati*, and *L. thermotolerans*, break down glucose through fermentation instead of respiration in the presence of oxygen. In these species, glucose suppresses mitochondrial genes, and ethanol and acetate fermentation occur instead of oxidative phosphorylation of pyruvate. This characteristic means that these species are preferred in the ethanol production industry and in biomedical research because they are similar to glucose metabolism in cancer cells. Furthermore, Crabtree-positive species are valuable in industry for heterologous protein production, isopropanol, biomass-based products, lactate, and glutathione production due to their high ATP and biomass synthesis yields. In these species, the *GCR1* gene regulates the expression of genes encoding glycolytic enzymes while also increasing the expression of fermentation genes, such as alcohol dehydrogenase (Genç et al. 2021; Guo et al. 2025).

Another mechanism involved in regulating primary metabolism in microorganisms is the induction and repression of enzyme synthesis. Inducible enzymes are produced in the presence of their substrates, thereby avoiding unnecessary energy expenditure. For example, fatty acids induce lipase, thiamine pyruvate decarboxylase, galacturonic acid polygalacturonase, and maltodextrins induce amylase. In contrast to this mechanism, repression is the process of suppressing enzyme synthesis or activity due to substrate scarcity or the accumulation of end products. These processes occur in the form of catabolic repression or feedback inhibition. The best-known examples are the cessation of production of the synthesis enzymes for these amino acids following binding to the Trp operon in the case of excessive tryptophan production, or repression of the His operon following histidine accumulation (Chubukov et al. 2014; Dev et al. 2022).

4.2.3 Important Primary Metabolites and Their Industrial Significance

4.2.3.1 Amino Acids

Amino acids, nucleotides, vitamins, solvents, alcohols, and organic acids are the most important primary metabolites. Amino acids, which are the building blocks of proteins and therefore play a regulatory role in protein homeostasis, are also involved in cellular signaling and microbial adaptation. Although microorganisms synthesize these amino acids for their own use, it is known that when grown in culture media, central metabolic intermediates and amino acids are released into the environment at varying concentrations. This phenomenon, known as extended transport metabolism, arises from the organism's combination of passive and active transport mechanisms (Sanchez and Demain 2008; Dai et al. 2011; Chai et al. 2022). This phenomenon is a crucial component of the mutualistic relationship between the host animal and its microbiota (Wu et al. 2021).

They are typically produced through a complex biosynthesis process that starts from precursor molecules derived from central metabolic pathways, such as glycolysis, the pentose phosphate pathway, and the Krebs cycle. Examples include pyruvate serving as a precursor metabolite in the synthesis of amino acids such as alanine and leucine, and oxaloacetate playing a role in the synthesis of aspartate and lysine. Various regulatory, suppressive, and inducible mechanisms come into play in these processes. For example, the aspartate family amino acids, such as methionine, lysine, and threonine, are synthesized in *Corynebacterium glutamicum* via a complex pathway regulated by feedback inhibition. This intense control of amino acid synthesis allows microorganisms to survive in different environments. Various regulatory, suppression, and induction mechanisms come into play in these mechanisms. For example, amino acids of the aspartate family, such as methionine, lysine, and threonine, are synthesized in *C. glutamicum* through a complex pathway regulated by feedback inhibition. This intense control of amino acid synthesis is a crucial aspect of the survival ability of microorganisms in various environments. This enables microorganisms to maintain homeostasis while utilizing resources at maximum efficiency (Paczia et al. 2012; Chandel 2021; Chai et al. 2022).

Amino acids with quite diverse structures and functions have been identified as microbial metabolites. One group consists of aromatic amino acids such as phenylalanine, tyrosine, or tryptophan, which are industrially produced by *E. coli* and *C. glutamicum* (Braga and Faria 2022). Another group consists of branched-chain amino acids such as valine and leucine. These amino acids, which are crucial for spore formation, also serve as signaling molecules involved in nutrition and chemotaxis. Sulfur-containing amino acids such as methionine and cysteine, which are involved in the formation of disulfide bridges in protein structure, and amino acids such as lysine and arginine, which are derived from various carbon skeletons, are also of industrial importance. Another important amino acid is glutamic acid. Known as the most industrially produced amino acid worldwide and used as a flavor enhancer in the food industry, notably as monosodium glutamate, this amino acid is obtained from *C. glutamicum* (Bajić et al. 2022; Korytko 2024; Nie et al. 2025).

4.2.3.2 Organic acids

Organic acids are another important family of primary metabolites produced by microorganisms. These are intermediate products of microbial processes. Due to their functional groups, they have a wide range of applications, primarily in the chemical industry. They maintain pH balance in the cell by facilitating the active transport of protons, thereby preventing excessive ATP consumption. Organic acids also play an important role in the food industry as they inhibit the growth of pathogens (Nair et al. 2017; Sallam et al. 2020).

Citric acid and lactic acid are organic acids with the most common applications and are widely used in research materials. Citric acid, used in countless industries such as food and beverages, pharmaceuticals, and cosmetics, is the organic acid with the highest production volume. This metabolite, which can be synthesized in different ways, is estimated to reach 2.91 million tons by 2026. 2-hydroxy-propane-1,2,3-tricarboxylic acid, or citric acid, obtained by fermentation from fungi such as *Aspergillus niger*, is produced through a complex metabolic flow of glycolysis and TCA cycle intermediates. First, glucose is catabolically converted to pyruvate. Part of this pyruvate is converted to acetyl-CoA in the mitochondria, while another part is converted to oxaloacetate (OAA) by cytosolic pyruvate carboxylase. This OAA produced in the cytosol is reduced to malate and transported to the mitochondria, where it is converted back to OAA and used in citric acid synthesis. The synthesized citric acid is transported out of the cell by a specific carrier molecule (Lambros et al. 2022; Latif et al. 2025).

Lactic acid is a metabolite associated with many microorganisms, primarily heterofermentative lactic acid bacteria (LAB), as well as *E. coli*, *B. licheniformis*, *S. cerevisiae*, and *C. glutamicum*. LAB, especially *Lactobacillus*, *Lactococcus*, *Leuconostoc*, *Streptococcus*, and *Pediococcus*, are important producers of organic acids. These bacteria are commonly used as starter cultures for the fermentation of milk, meat, and vegetable products. LAB are non-spore-forming, Gram-positive bacteria that produce organic acids, including lactic acid, acetic acid, succinic acid, citric acid, and propionic acid, through carbohydrate fermentation (Bangar et al. 2022; Zhou et al. 2025). Lactic acid, obtained by the glycolytic breakdown of hexoses such as glucose, mannose, galactose, and fructose under anaerobic conditions, is used (i) in the chemical industry as a green solvent, neutralizer, and cleaning agent, (ii) in the cosmetics industry as an anti-acne, anti-aging, moisturizing, and brightening agent, (iii) in agriculture for protection against pathogens, (iv) in coating materials in various industries, (v) in various biomedical fields such as dentistry, orthopedics, and radiology, and (vi) in the textile industry for the production of UV-resistant fabrics, among many other uses (Nair et al. 2017; Abedi and Hashemi 2020; Rana and Thakur 2025).

Another important organic acid is acetic acid. In addition to its antimicrobial properties, it also plays a role in the microorganism's resistance to toxic byproducts. This metabolite, which is a strong acid, can be produced by various bacteria such as *Lactobacillus acidophilus, Lactobacillus delbrueckii, Limosilactobacillus fermentum, Lactiplantibacillus plantarum, Lactobacillus acidophilus, Weissella cibaria, Pediococcus pentosaceus, Lactobacillus delbrueckii*, and *Lacticaseibacillus casei* (Ozcelik et al. 2016; Yang et al. 2018). Acetic acid, a key component of many fermented products, especially vinegar, is also utilized in the food industry. Additionally, it is used in the pharmaceutical industry for the production of vitamins, hormones, and antibiotics. Microorganisms produce acetic acid through a two-step process, involving both alcohol and acetic acid fermentation. Therefore, acetic acid bacteria and ethanol can be added to the mixture to accelerate the production process (Selvanathan et al. 2025).

4.2.3.3 Alcohols

Alcohols are the other important group of primary metabolites. These are important compounds used as fuel and solvents in many industries. The best-known microbial alcohol product is ethanol. Various bacteria and yeast species, primarily *S. cerevisiae, Zymomonas mobilis, E. coli*, and *Clostridium* species, convert sugars such as glucose and fructose into bioethanol through fermentation. These organisms also play a role in the food industry in the production of alcoholic beverages such as wine and beer (Glazer and Nikaido 2007; Toogood and Scrutton 2018; Ruslan et al. 2025). Butanol, another fermentative alcohol product, is a metabolite synthesized anaerobically by bacteria such as *E. coli*. Butanol, produced as a byproduct in the petrochemical industry, is a valuable biofuel that can also be produced biologically. Biobutanol is considered a sustainable energy source due to its high energy content, its ability to mix easily with gasoline, and its compatibility with traditional combustion engines, similar to bioethanol. Butanol fermentation typically begins during the exponential growth phase of microorganisms, where organic acids such as acetic acid and butyric acid are produced via the Embden-Meyerhof-Parnas pathway. At this stage, the pH drops and alcohol production begins. The process results in the production of intermediate products, such as acetone, butanol, ethanol, and butyrate (Ganeshan and Tülbek 2025).

4.3 Microbial Secondary Metabolites

Microbial secondary metabolites are a diverse group of low-molecular-weight, bioactive organic compounds produced by microorganisms during the late growth phase (idiophase). These compounds play crucial roles in ecological interactions and the survival of various species (Ullah et al. 2020; Reddy et al. 2021). Unlike primary metabolites, which are essential for growth and reproduction, secondary metabolites are synthesized in response to environmental factors, such as stress and competition, and contribute to microbial competition, communication, and defense mechanisms (Mitra et al. 2021). These compounds are generally categorized based on their chemical structure and biological functions, including antibiotics, food additives, pigments, toxins, enzyme inhibitors, biosurfactants, plant growth stimulators, herbicides, insecticides, antitumour agents, and immunomodulators (Demisie et al. 2024).

The synthesis of secondary metabolites in microbial cultures is influenced by various environmental factors throughout different developmental phases. The production of secondary metabolites is regulated by various factors, including nutrients, growth rate and feedback control. Carbon source type and concentration play a crucial role in regulating secondary metabolite production in microorganisms, particularly during the late growth phase (Ruiz-Villafán et al. 2022). Due to their unique structures and diverse functions, microbial secondary metabolites present promising opportunities for research and industrial applications in the fields of medicine and agriculture (Mitra et al. 2021).

4.3.1 Types of Secondary Microbial Metabolites and Their Biosynthesis Pathways

4.3.1.1 Antibiotics

Antibiotics, a crucial class of secondary metabolites produced by microorganisms, play vital roles in nature and medicine (Demisie et al. 2024; Assou Koffi et al. 2024). These compounds serve various functions, including competitive weapons against other organisms, symbiotic agents, and differentiation effectors (Demain and Fang 2001; Aladeboyeje and Sanli 2021; Aladeboyeje et al. 2022). Many antibiotics, such as penicillin and streptomycin, were originally derived from microbial sources. Streptomycetes and related actinomycetes are renowned for their ability to produce a wide array of novel antibiotics, making them a prolific source of these bioactive compounds (Demisie et al. 2024). The biosynthesis of antibiotics involves multistep pathways, complex regulatory mechanisms and gene clusters, with production typically occurring during the stationary growth phase (Niu and Tan 2013). One key aspect of antibiotic production is that horizontal gene transfer (HGT) mechanisms are influenced by several factors, including the host range of plasmids, restriction-modification systems, and the fitness costs associated with acquiring new genes (Peterson and Kaur 2018). Another prominent mechanism involves the enzymatic modification or degradation of antibiotics. This is particularly prevalent with aminoglycoside antibiotics, where specific enzymes, known as aminoglycoside modification enzymes (AMEs), can acetylate, phosphorylate, or adenylylate these compounds, rendering them ineffective. The biosynthesis of antibiotics in *Streptomyces* is largely regulated by biosynthetic gene clusters (BGCs), which encode the necessary enzymes for the production of secondary metabolites (Ueda and Beppu 2017). The biosynthesis of antibiotics in *Streptomyces* is largely regulated by cryptic biosynthetic gene clusters (BGCs) (Alam et al. 2022). Polyketide synthases (PKS) and non-ribosomal peptide synthetases (NRPS) are key molecular machines in antibiotic biosynthesis (Giordano et al. 2015). PKSs synthesize polyketides, while NRPSs produce peptides independent of ribosomal machinery. Both types of synthases can form hybrid systems that combine the features of PKSs and NRPS, thereby increasing the diversity of antimicrobial compounds produced. These enzymes operate within different structural frameworks, with Type I PKS being the most common in bacteria, followed by Type II and Type III PKS found primarily in fungi and plants, respectively. A better understanding of these biosynthetic pathways has led to the development of strategies for creating new antibiotic derivatives, including selective gene inactivation, mutasynthesis and combinatorial biosynthesis (Niu and Tan 2013). These approaches offer a promising way to develop new antibiotics to combat the growing threat of antibiotic resistance.

4.3.1.2 Pigments

Microbial pigments are sustainable alternatives to synthetic colorants thanks to their eco-friendly nature and bioactive properties (Grewal et al. 2022). Bacteria, in particular, are prominent sources of pigments owing to their genetic simplicity and ease of manipulation. Microbial pigments, such as flexirubin, carotenoid, violacein, melanins, astaxanthin, anthocyanin, phycocyanin, and prodigiosin, serve important roles in nature, from photoprotection and coloration to potential health benefits, and have garnered increasing interest across industries, such as food, medicine, and cosmetics. The biosynthesis of these pigments involves complex metabolic pathways that are intricately regulated by environmental factors, genetic modifications, and physiological conditions. Carotenoids are produced via the methylerythritol-4-phosphate (MEP) pathway in *Escherichia coli* and *Eubacteria*, while violacein pigments are synthesized through the condensation of two tryptophan molecules in a reaction catalyzed by molecular oxygen. Prodigiosin biosynthesis occurs when the compound 4-methoxy-2,2′-bipyridine-5-carbaldehyde (MBC) is concentrated with monopyrrole or bipyridine precursors, or with 2-methyl-3-pentylpyridine (MPP). Melanin pigments are formed through the oxidative polymerization of indolic or phenolic compounds, catalyzed by tyrosinase and laccase enzymes (Sundararajan and Ramasamy 2023).

4.3.1.3 Toxins

Microbial toxin production encompasses the biosynthesis of harmful substances by various microorganisms, including bacteria, cyanbobacteria, dinoflagellates and fungi, which can have significant implications for human health, agriculture, and industrial applications. Bacterial toxins (exotoxins and endotoxins), mycotoxins, cyanobacterial toxins and dinoflagellate toxins exhibit diverse mechanisms of action and effects, leading to notable health risks such as foodborne illnesses and environmental contamination.

Bacterial exotoxins damage host cells by either interfering with specific cellular functions, as with the diphtheria toxin produced by *Corynebacterium diphtheriae*, or by eliciting a strong immune response, as with the enterotoxins produced by *Staphylococcus aureus*. By contrast, endotoxins—components of the membrane of Gram-negative bacteria—are not actively secreted, but are released upon bacterial cell lysis, often triggering severe inflammatory responses in the host (Edae and Wabalo 2019).

Unlike bacterial toxins, mycotoxins, which are secondary metabolites produced by fungi, are not species-specific. Mycotoxins like aflatoxins, ochratoxins, and citrinin are carcinogenic and present in agricultural products, raising concerns over food safety and public health (Pandey et al. 2023).

Cyanobacteria produce a variety of toxic secondary metabolites, including neurotoxins and hepatotoxins, such as microcystins, anatoxins, and saxitoxins. Exposure to these toxins can have serious health consequences for humans and animals (Codd et al. 2020).

Marine dinoflagellates are a significant source of potent toxins, including saxitoxin, ciguatoxin and okadaic acid. They play a key ecological role in marine ecosystems, but also pose a substantial risk to human health through shellfish poisoning syndromes such as paralytic shellfish poisoning and ciguatera fish poisoning (Verma et al. 2019).

Carbohydrate-binding toxins, in particular, offer promising biotechnological applications in areas such as glycobiomarker detection, drug delivery systems, and anti-tumor biopharmaceuticals (Danielewicz et al. 2022).

The comprehension of toxin biosynthesis and production pathways has resulted in the development of enhanced detection methodologies and the potential for the formulation of future management strategies to address the challenges posed by harmful blooms and plant diseases.

4.3.1.4 Enzyme Inhibitors

Microbial enzyme inhibitors have attracted attention due to their potential therapeutic applications, particularly in combating antibiotic resistance, analyzing biological functions, and managing metabolic disorders, thus highlighting their importance in public health and ecological sustainability (Demain 2019). Enzyme inhibitors can be classified as reversible or irreversible, which function through distinct mechanisms. Reversible inhibitors can be subdivided further into competitive, non-competitive, uncompetitive, and mixed-type inhibitors, while irreversible inhibitors permanently inactivate enzymes through covalent bonding. Fungi, such as those from the *Aspergillus* and *Penicillium* genera, are well-known for their ability to produce a wide range of enzyme inhibitors. For instance, certain strains of *Aspergillus* produce β-galactosidase, which can inhibit the activity of glycosidic enzymes (Kumar et al. 2024). The *Bacillus* genus is another notable source of enzyme inhibitors, particularly proteolytic enzymes that exhibit high stability and activity under various conditions (Al-Madboly et al. 2024). Additionally, endophytic bacteria from marine environments, such as *Streptomyces* species, have been identified as promising producers of enzyme inhibitors with significant inhibitory effects on α-glucosidase and α-amylase, which are crucial targets in the treatment of diabetes (Begani et al. 2018).

4.3.1.5 Biosurfactants

Biosurfactants are surface-active secondary metabolites produced by various microorganisms, including bacteria, fungi, and yeasts, with diverse applications in industries, bioremediation, and medicine. These amphiphilic compounds offer advantages over synthetic surfactants, including biodegradability, low toxicity, and stability under extreme conditions. Biosurfactants are classified according to their microbial origin, molecular weight, and structure. Like synthetic surfactants, they comprise a hydrophilic and a hydrophobic moiety. Biosurfactant biosynthesis involves complex metabolic pathways, often mediated by non-ribosomal peptide synthetases for lipopeptides. The primary classes of biosurfactants include glycolipids, lipopeptides, lipoproteins, polymeric surfactants, and particulate surfactants (Patel et al. 2023).

Glycolipids are formed from one or more carbohydrates that are linked to fatty acids via ether or ester bonds. Notable examples include rhamnolipids, produced by *Pseudomonas* species; trehalose lipids, synthesized by *Rhodococcus* and

Mycobacterium species; and sophorolipids, generated by non-pathogenic yeasts such as *Candida bombicola* (Sharma et al. 2021).

Lipopeptides are characterized by amino acid chains that are attached to variable lipid tails. Common lipopeptides include surfactin, fengycin and polymyxins, which are all known for their antimicrobial properties (Simões et al. 2024). The efficacy of lipopeptides is influenced by the length of their lipid tails: shorter tails tend to exhibit antibacterial effects, whereas longer tails tend to exhibit antifungal activities. These biosurfactants interact with cell membranes, causing disruption and increasing permeability, which leads to cell lysis and death (Soberón-Chávez et al. 2021).

Polymeric biosurfactants are large molecules consisting of a combination of polysaccharides, proteins and peptides. Examples include emulsan, produced by *Acinetobacter calcoaceticus*, and liposan, synthesized by *Candida lipolytica*. These surfactants have a variety of applications, including emulsification and stabilization, as well as being used as additives in paints and coatings, thanks to their unique structural properties and functions (Sharma et al. 2021).

Particulate surfactants are composed of complex aggregates of biosurfactants and may contain structures such as exopolysaccharides. Although less common, these biosurfactants are important for specific applications in fields such as bioremediation and environmental management, due to their ability to interact with pollutants and enhance biodegradation processes (Gürkök and Özdal 2021).

4.3.1.6 Plant Growth Stimulators

Plant growth stimulators are essential compounds that enhance plant growth and resilience against environmental stresses. These metabolites are increasingly recognized for their significant roles in sustainable agriculture, as they can improve nutrient availability, stimulate root development, and induce plant defense mechanisms. The ability of these stimulators to promote crop productivity while minimizing reliance on chemical fertilizers has positioned them as pivotal in addressing the challenges posed by a growing global population and climate change (Marks et al. 2025; Singh et al. 2025). Microbial secondary metabolites can be classified into several categories, including peptides, polyketides, volatile compounds, terpenoids/steroids, and growth regulators, each playing distinct roles in plant physiology. Antimicrobial peptides produced by soil bacteria such as *Bacillus* and *Pseudomonas* not only inhibit phytopathogens but also facilitate important plant growth processes, such as nitrogen fixation and phosphate solubility (Marks et al. 2025). Polyketides are typically produced by fungi and certain bacterial species, and are renowned for their roles in protecting and promoting plant growth. Fungal polyketide synthase gene clusters can produce metabolites that enhance plant resistance to herbivory and environmental stresses (Lv et al. 2024). Volatile compounds produced through microbial activity can influence plant physiology, either enhancing growth or altering stress responses. These compounds often act as signaling molecules, facilitating communication between plants and microorganisms in the rhizosphere. Their presence can promote root development and improve nutrient absorption, thereby enhancing overall plant health and vigor

(Lv et al. 2024). Terpenoids and steroids have been shown to play a role in plant defense mechanisms, with some terpenoids exhibiting antifungal and antibacterial activities and helping plants to ward off pathogens (Marks et al. 2025). Bacteria belonging to the *Pseudomonas* and *Azospirillum* genera, as well as Rhizobia, produce growth regulators that enhance plant stress resistance, optimize nutrient absorption, and promote overall growth. These growth regulators, which include auxins, gibberellins, and cytokinins, are produced by microbial species and are very important for regulating plant growth and development (Egamberdieva et al. 2017).

4.3.1.7 Biocontrol Agents (Herbicides, Insecticides)

Herbicides produced by various microorganisms, including bacteria and fungi, are not essential for their growth, yet they play a vital role in ecological interactions, pest control and plant protection (Fouillaud and Dufossé 2022). Microbial herbicides primarily act through specific mechanisms that influence plant growth and development, particularly under environmental stress conditions. By modulating plant biochemical pathways, microbial metabolites can induce stress responses, inhibit cellular functions, and alter nutrient uptake, ultimately leading to the growth inhibition of target weeds (Wend et al. 2024). Noteworthy examples include herbicidal compounds derived from marine fungi and *Pseudomonas fluorescens* strains, which have been shown to be highly effective in controlling various weed species (Cheng et al. 2022). Despite their potential, microbial herbicides face challenges to their widespread adoption, including inconsistent performance in field conditions, regulatory obstacles, and competition from well-established chemical herbicides.

Microbial insecticides, which are derived from bacteria, fungi, viruses, and protozoa, play a crucial role in the sustainable management of insect populations through biological pest control. This innovative approach to pest management is gaining prominence as agricultural practices increasingly prioritize environmental safety and reducing chemical pesticide use (Beltrán Pineda and Castellanos-Rozo 2025). One of the most well-known bacterial insecticides is *Bacillus thuringiensis* (Bt), which produces crystalline proteins known as delta-endotoxins, which are toxic to certain insect larvae when ingested (Ayilara et al 2023). Other notable bacteria used as insecticides include *Lysinibacillus sphaericus*, which produces toxins that damage the epithelial cells of the insect midgut, leading to cessation of feeding and eventual death (Beltrán Pineda and Castellanos-Rozo 2025). Fungal insecticides include entomopathogenic fungi, which infect and kill insect hosts. These fungi employ various strategies, such as producing spores that attach to the insect cuticle and then germinate and penetrate the host. One notable example is *Beauveria bassiana*, which disrupts the insect's immune system and causes mortality through tissue colonization. Despite their potential, only a few fungal strains have been commercialized, indicating that this area of biopesticide development is still in its infancy (Ayilara et al. 2023). Viral insecticides can specifically target insect populations, effectively reducing their numbers without harming beneficial organisms. An example includes the Nuclear Polyhedrosis Virus (NPV), which infects caterpillars and can lead to widespread mortality in target pest populations when applied in appropriate formulations (Beltrán Pineda and Castellanos-Rozo 2025).

4.3.1.8 Antitumour Agents

Antitumour agents are promising candidates for developing novel chemotherapeutic agents because they exhibit significant cytotoxicity against cancer cells while often sparing normal cells (Liang et al. 2022). The historical use of certain natural products in traditional medicine has paved the way for modern research into their therapeutic applications, particularly following the discovery of antibiotics such as penicillin, which prompted a wider search for microbial derivatives with anti-cancer properties. Research shows that soil and marine fungi produce secondary metabolites with diverse anti-tumor mechanisms, including the inhibition of protein synthesis, the breaking of DNA, and the blocking of angiogenesis. A total of 53 compounds have been documented in the genera *Neosartorya* and *Aspergillus*. Two of these compounds, a cadinene sesquiterpene (terpenoid) and eurochevalierine (alkaloid), have been reported to exhibit significant cytostatic effects against glioblastoma (U373) and non-small cell lung cancer (A549) cell lines (War May Zin et al. 2016).

Streptomyces-derived compounds have been reported to exhibit antiproliferative, proapoptotic, antimetastatic, and antiangiogenic properties (Mukherjee and Sarkar 2025). Clinically proven microbial metabolites, such as actinomycin D, bleomycin, and doxorubicin, have been found to be effective against various cancers. Compounds such as alpha sarcin, gliotoxins, trichotillosinerol, L-lysine alpha-oxidase, fumagillin, decloxin, and chrysanthones have been reported to exhibit potent antitumour activity, daunorubicin, and doxorubicin, which are obtained from *Trichoderma viride*, exhibit potent anti-tumor activity by inhibiting protein synthesis and DNA breakage, inducing apoptosis and blocking angiogenesis. A new hepatoma growth inhibition factor has also been identified (Ahmed et al. 2010).

Gut microbial metabolites, including bacteriocins and short-chain fatty acids, have shown anticancer potential (Mustafayeva and Sanli 2022). However, some, such as secondary bile acids, may be pro-carcinogenic (Jaye et al. 2022). Researchers are exploring novel approaches, such as metagenomics, to discover new anti-cancer compounds from microbial sources (Mohan et al. 2022). Despite the challenges associated with drug development, microbial metabolites exhibit diverse mechanisms of action and hold great promise as potential cancer therapeutics (Giurini et al. 2024). Continued research in this field is warranted to uncover new molecules and translate findings into clinical applications for cancer patients (Mukherjee and Sarkar 2025).

4.3.1.9 Immunomodulators

Microbial secondary metabolites have emerged as promising immunomodulators with diverse therapeutic potential. *Actinobacteria*, particularly *Streptomyces* sp., produce a wide range of bioactive compounds that can regulate immune responses

(Gouveia et al. 2024). These metabolites can act as immunostimulants or immunosuppressants, offering advantages in the treatment of immune disorders, cancer and viral diseases (Mahmoudi et al. 2024). Microbes produce molecules that can suppress or activate immune responses; bacterial enterotoxins, parasite-derived products, and viral nucleic acids are potent activators of the immune system (Lavelle et al. 2004). The therapeutic exploitation of these microbial immunomodulators holds promise in enhancing vaccine efficacy and in the treatment of immune-mediated diseases. This offers the potential for the development of alternatives to conventional immunosuppressants and immunostimulants with fewer side effects (Mahmoudi et al. 2024).

4.4 Future Perspectives and Conclusion

Microbial metabolites are bioactive compounds produced by microorganisms, such as bacteria, fungi, and actinomycetes that play essential roles in various biological processes. These metabolites fall into two categories: primary metabolites, which are essential for microbial growth and development; and secondary metabolites, which frequently mediate ecological interactions, exhibit a variety of biological activities and are often utilized in various industrial applications. The study of microbial metabolites is significant due to their vast potential in pharmaceuticals, agriculture, and environmental sustainability. These metabolites play a key role in driving innovation in biotechnology and health-related fields as natural products. The industrial relevance of microbial metabolites is underscored by their application in drug development, particularly in the creation of antibiotics, antifungals, and anticancer agents. Secondary metabolites such as penicillin and erythromycin have revolutionized medical treatment, while primary metabolites serve crucial functions in nutrition and food production. Additionally, microbial metabolites promote soil health and enhance crop resilience, thereby contributing to sustainable agricultural practices and further emphasizing their ecological importance in nutrient cycling and ecosystem stability (Hashmi et al. 2023; Mccann and Rawls 2023; Al-Ishaq 2024). Despite their promising applications, research on microbial metabolites faces challenges, including funding constraints, regulatory hurdles, and the need for a better understanding of their safety profiles and ecological impacts. Furthermore, there is a growing interest in exploring the "microbial dark matter" in order to identify novel compounds with therapeutic potential that remain underexplored (Hashmi et al. 2023). This ongoing research aims to unlock the full capabilities of microbial metabolites, driving advances in health, agriculture, and environmental management, while addressing contemporary challenges in sustainability and public health challenges (Fouillaud and Dufossé 2022).

The concept of biomimicry in drug discovery is gaining traction, with microbial metabolites being studied for their potential to mimic host-endogenous molecules. This approach could lead to the development of new pharmaceuticals that target specific receptors, yielding powerful and non-toxic treatments. Future research may focus on optimizing these interactions to broaden the therapeutic arsenal available for treating various diseases (Suman et al. 2022).

Theoretical biology is expected to play a critical role in advancing microbial research. The integration of kinetic models, stochastic models, and network-based models will enhance our ability to predict microbial behavior and interactions. This interdisciplinary approach will ensure that experimental and theoretical methods complement each other, leading to improved experimental designs and more effective research outcomes (Marks et al. 2025).

Future research in microbial biotechnology is set to greatly improve our understanding of, and ability to manipulate, microbial metabolites (Kimiran et al. 2023). Key areas of focus include genetically engineering microbial strains to optimize the production of primary and secondary metabolites, particularly those with health-promoting or industrial properties. These metabolites hold significant promise for use in drug discovery and biotechnological applications due to their diverse bioactive properties. Future research in this field is set to greatly enhance our understanding of and ability to manipulate microbial metabolites (Demisie et al. 2024).

Recent advances in microbiome research emphasize the importance of multi-omics approaches, combining metagenomics, metatranscriptomics, metaproteomics, and metabolomics. This integration enables researchers to gain a comprehensive understanding of microbial functions and interactions, offering deeper insights into the influence of microbial communities on host health and ecological dynamics. It is expected that this approach will unveil the complex interplay between microbial metabolites and host physiological processes, potentially leading to innovative therapeutic strategies, particularly in areas such as cancer prevention and treatment (Martínez-Espinosa 2020; Tavşanlı et al. 2024).

Metabolic engineering in microbial metabolite production is a rapidly evolving field focused on the genetic optimization of microorganisms to enhance the production of valuable compounds. This discipline integrates genetic manipulations, bioinformatics, and systems biology to manipulate microbial pathways effectively, enabling the sustainable generation of biofuels, pharmaceuticals, and specialty chemicals. The advent of powerful techniques, particularly CRISPR/Cas systems, has transformed the precision and efficiency of genetic modifications, opening up new possibilities in industrial biotechnology and synthetic biology (Dhokane et al. 2023). The significance of metabolic engineering lies in its potential to address global challenges, such as resource scarcity and environmental sustainability. By leveraging microorganisms for the biosynthesis of complex metabolites, researchers can create sustainable alternatives to fossil fuels and reduce reliance on traditional chemical manufacturing processes. However, the field faces challenges, particularly in the production of metabolites, due to the intricate metabolic pathways and post-translational modifications (Kursheed et al. 2025).

The integration of bioinformatics and computational modeling has further enhanced the field, allowing for the simulation of metabolic networks and the optimization of production processes. Techniques such as metabolic flux analysis (MFA) and systems metabolic engineering enable researchers to quantify metabolic flows and refine genetic manipulations, leading to improved yields and efficiencies (Zhang et al. 2020; Van Beirs et al. 2025; Yadav et al. 2025). Nevertheless, ethical and regulatory considerations surrounding genetic engineering technologies pose significant challenges, necessitating responsible practices as the field continues to expand (Gao et al. 2022). In summary, metabolic engineering is a promising area of biotechnology that combines innovative genetic techniques with computational tools to improve the production of microbial metabolites. Its potential to provide sustainable solutions for various industries highlights its importance in addressing contemporary challenges related to resource utilization and environmental impact.

References

Abedi E, Hashemi SMB (2020) Lactic acid production–producing microorganisms and substrates sources-state of art. Heliyon 6(10):e04974

Ahmed HY, El Shikh HH, Ghoneimy EA, Ragab AM, Saad AA, Hashem AM, El Sayed RM, El Shikh MEM (2010) Fungal secondary metabolites: a promising source of antineoplastic drugs. Curr Cancer Ther Rev 6(3):222–228

Aladeboyeje O, Sanli NO (2021) Fermented traditional probiotic beverages of Turkish origin: a concise review. Int J Life Sci Biotechnol 4(3):546–564

Aladeboyeje OT, Sanli NO, Buyuk U (2022) Evaluation of the antimicrobial efficacy of some fermented traditional turkish beverages with probiotic potentials: the first study to evaluate antimicrobial properties of isolates from five beverages against a broad spectrum of human microbial pathogens. Johns Matthey Technol Rev 66(3):337–350

Alam K, Mazumder A, Sikdar S, Zhao YM, Hao J, Song C et al (2022) Streptomyces: the biofactory of secondary metabolites. Front Microbiol 13:968053

Al-Ishaq RK, Ferrara CR, Stephan N, Krumsiek J, Suhre K, Montrose DC (2024) A comprehensive metabolomic and microbial analysis following dietary amino acid reduction in mice. Metabolites 14(12):706

Al-Madboly LA, Aboulmagd A, El-Salam MA, Kushkevych I, El-Morsi RM (2024) Microbial enzymes as powerful natural anti-biofilm candidates. Microb Cell Fact 23(1):343

Alper H, Cirino P, Nevoigt E, Sriram G (2011) Applications of synthetic biology in microbial biotechnology. J Biomed Biotechnol 2010:918391

Assou Koffi AA, Sanli NO, Ciftcioglu G, Arda N (2024) A new *Enterococcus hirae* strain exhibits antibacterial activity against vancomycin-resistant enterococci. Biol Bull 51(Suppl 1):S175–S182

Ayilara MS, Adeleke BS, Akinola SA, Fayose CA, Adeyemi UT, Gbadegesin LA et al (2023) Biopesticides as a promising alternative to synthetic pesticides: a case for microbial pesticides, phytopesticides, and nanobiopesticides. Front Microbiol 14:1040901

Bajić B, Vučurović D, Vasić Đ, Jevtić-Mučibabić R, Dodić S (2022) Biotechnological production of sustainable microbial proteins from agro-industrial residues and by-products. Foods 12(1):107

Bangar SP, Suri S, Trif M, Ozogul F (2022) Organic acids production from lactic acid bacteria: a preservation approach. Food Biosci 46:101615

Begani J, Lakhani J, Harwani D (2018) Current strategies to induce secondary metabolites from microbial biosynthetic cryptic gene clusters. Ann Microbiology 68(7):419–432

Beltrán Pineda ME, Castellanos-Rozo J (2025) Bacterial insecticides beyond *Bacillus thuringiensis*. Phytopathology Res 7(1):19

Borodina I, Siebring J, Zhang J, Smith CP, van Keulen G, Dijkhuizen L, Nielsen J (2008) Antibiotic overproduction in *Streptomyces coelicolor* A3 (2) mediated by phosphofructokinase deletion. J Biol Chem 283(37):25186–25199

Börekçi BS, Kaban G, Kaya M (2021) Citric acid production of yeasts: an overview. EuroBiotech J 5:79–91

Braga A, Faria N (2022) Biotechnological production of specialty aromatic and aromatic-derivative compounds. World J Microbiol Biotechnol 38(5):80
Chai Y, Chen H, Tang X, Zhao J, Zhang H, Chen YQ, Chen W (2022) The relationship between amino acid and lipid metabolism in oleaginous eukaryotic microorganism. Appl Microbiol Biotechnol 106(9):3405–3417
Castro-Moretti FR, Gentzel IN, Mackey D, Alonso AP (2020) Metabolomics as an emerging tool for the study of plant–pathogen ınteractions. Metabolites 10(2):52
Chandel NS (2021) Amino acid metabolism. Cold Spring Harb Perspect Biol 13(4):a040584
Chen Q, Song Y, An Y, Lu Y, Zhong G (2024) Mechanisms and impact of rhizosphere microbial metabolites on crop health, traits, functional components: a comprehensive review. Molecules 29(24):5922
Cheng L, DiTommaso A, Kao-Kniffin J (2022) Opportunities for microbiome suppression of weeds using regenerative agricultural technologies. Front Soil Sci 2:838595
Chubukov V, Gerosa L, Kochanowski K, Sauer U (2014) Coordination of microbial metabolism. Nat Rev Microbiol 12(5):327–340
Codd GA, Testai E, Funari E, Svirčev Z (2020) Cyanobacteria, cyanotoxins, and human health. In: Water treatment for purification from Cyanobacteria and Cyanotoxins. John Wiley & Sons, USA, pp 37–68
Çetiz MV, Memon A (2021) Effect of the plant-microbe interaction on secondary metabolites. Turkish J Agriculture Food Sci Technol 9(2):281–287
Dai ZL, Wu G, Zhu WY (2011) Amino acid metabolism in intestinal bacteria: links between gut ecology and host health. Front Biosci 16(1):1768–1786
Dale AW, Van Cappellen P, Aguilera DR, Regnier P (2008) Methane efflux from marine sediments in passive and active margins: estimations from bioenergetic reaction-transport simulations. Earth Planet Sci Lett 265:329–344
Danielewicz N, Rosato F, Dai W, Römer W, Turnbull WB, Mairhofer J (2022) Microbial carbohydrate-binding toxins–from etiology to biotechnological application. Biotechnol Adv 59:107951
Davies J (2013) Specialized microbial metabolites: functions and origins. J Antibiot 66(7):361–364
Demain AL (2019) New applications of microbial products. In Biotechnology And Biological Frontiers, Routledge, New York, pp. 201–213
Demain AL, Adrio JL (2008) Contributions of microorganisms to industrial biology. Mol Biotechnol 38(1):41–55
Demain AL, Fang A (2001) The natural functions of secondary metabolites. Adv Biochem Eng Biotechnol 69:1–39
Demisie S, Oh DC, Abera A, Tasew G, Satessa GD, Fufa F et al (2024) Bioprospecting secondary metabolites with antimicrobial properties from soil bacteria in high-temperature ecosystems. Microb Cell Fact 23(1):332
Dev C, Jilani SB, Yazdani SS (2022) Adaptation on xylose improves glucose–xylose co-utilization and ethanol production in a carbon catabolite repression (CCR) compromised ethanologenic strain. Microb Cell Fact 21(1):154
Dhokane D, Shaikh A, Yadav A, Giri N, Bandyopadhyay A, Dasgupta S, Bhadra B (2023) CRISPR-based bioengineering in microalgae for production of industrially important biomolecules. Front Bioeng Biotechnol 11:1267826
Ebenhöh O, Ebeling J, Meyer R, Pohlkotte F, Nies T (2024) Microbial pathway thermodynamics: stoichiometric models unveil anabolic and catabolic processes. Life 14(2):247
Edae MCK, Wabalo EK (2019) Bacterial toxins and their modes of action: a review article. J Med Physiol Biophys 55:11-16
Egamberdieva D, Wirth S, Behrendt U, Ahmad P, Berg G (2017) Antimicrobial activity of medicinal plants correlates with the proportion of antagonistic endophytes. Front Microbiol 8:199.
Entian KD, Schuller HJ (2024) Genetics of di-and trisaccharide utilization. In: Yeast sugar metabolism. CRC Press, Boca Raton, pp 225–233
Flamholz A, Noor E, Bar-Even A, Liebermeister W, Milo R (2013) Glycolytic strategy as a tradeoff between energy yield and protein cost. Proc Natl Acad Sci USA 110(24):10039–10044

Fouillaud M, Dufossé L (2022) Microbial secondary metabolism and biotechnology. Microorganisms 10(1):123
Fuentes-Lemus E, Reyes JS, Figueroa JD, Davies MJ, López-Alarcón C (2023) The enzymes of the oxidative phase of the pentose phosphate pathway as targets of reactive species: consequences for NADPH production. Biochem Soc Trans 51(6):2173–2187
Fuhrmann JJ (2021) Microbial metabolism. In: Principles and applications of soil microbiology. Elsevier, Amsterdam, pp 57–87
Gabaldón T, Pittis AA (2015) Origin and evolution of metabolic sub-cellular compartmentalization in eukaryotes. Biochimie 119:262–268
Galili G, Amir R, Fernie AR (2016) The regulation of essential amino acid synthesis and accumulation in plants. Annu Rev Plant Biol 67(1):153–178
Ganeshan S, Tülbek MÇ (2025) Fermentative butanol production—perspectives and scale-up challenges. Encyclopedia 5(2):50
Gao HY, Zhao H, Hu TY, Jiang ZQ, Xia M, Zhang YF et al (2022) Metabolic engineering of *Saccharomyces cerevisiae* for high-level friedelin via genetic manipulation. Front Bioeng Biotechnol 10:805429
Genç TT, Çakan A, Günay M (2021) Crabtree pozitif ve crabtree negatif maya türlerinde gcr1 geninin in sılıco analizi. Euroasia J Mathematics, Engineering, Nat Med Sci 8(17):207–216
Ginger ML, Fritz-Laylin LK, Fulton C, Cande WZ, Dawson SC (2010) Intermediary metabolism in protists: a sequence-based view of facultative anaerobic metabolism in evolutionarily diverse eukaryotes. Protist 161(5):642–671
Ginger ML, McFadden GI, Michels PA (2010) Rewiring and regulation of cross-compartmentalized metabolism in protists. Philos Trans R Soc Lond B Biol Sci 365(1541):831–845
Giordano D, Coppola D, Russo R, Denaro R, Giuliano L, Lauro FM et al (2015) Marine microbial secondary metabolites: pathways, evolution and physiological roles. Adv Microb Physiol 66:357–428
Giurini EF, Godla A, Gupta KH (2024) Redefining bioactive small molecules from microbial metabolites as revolutionary anticancer agents. Cancer Gene Ther 31(2):187–206
Glazer AN, Nikaido H (2007) Microbial biotechnology: fundamentals of applied microbiology. Cambridge University Press, Cambridge
Gouveia SDSA, Santos DCP, de Almeida, IA, de Miranda RDCM (2024) Analysis of the immunomodulator profile of secondary metabolites derived from Streptomyces sp. Rev de Gestão Soc e Ambiental 18(10):1–22
Görke B, Stülke J (2008) Carbon catabolite repression in bacteria: many ways to make the most out of nutrients. Nat Rev Microbiol 6(8):613–624
Grewal J, Wołacewicz M, Pyter W, Joshi N, Drewniak L, Pranaw K (2022) Colorful treasure from agro-industrial wastes: A sustainable chassis for microbial pigment production. Front Microbiol 13:832918
Guo Y, Xiong Z, Zhai H, Wang Y, Qi Q, Hou J (2025) The advances in creating Crabtree-negative *Saccharomyces cerevisiae* and the application for chemicals biosynthesis. FEMS Yeast Res 25:foaf014
Gupta R, Gupta N (2021) Pentose phosphate pathway. In: Fundamentals of bacterial physiology and metabolism. Springer Singapore, Singapore, pp 289–305
Gürkök S, Özdal M (2021) Microbial biosurfactants: properties, types, and production. Anatol J Biol 2(2):7–12
Hashmi MATS, Khawar MB, Naseem R, Afzal A, Shahid N, Abbasi MH et al (2023) Therapeutic potential of microbial metabolites: new insights and perspectives. Biomed Res Ther 10(4):5638–5653
Holbrook-Smith D, Trouillon J, Sauer U (2023) Metabolomics and microbial metabolism: toward a systematic understanding. Ann Rev Biophys 53:41-64
Huccetogullari D, Luo ZW, Lee SY (2019) Metabolic engineering of microorganisms for production of aromatic compounds. Microb Cell Fact 18(1):41

Jaye K, Li CG, Chang D, Bhuyan DJ (2022) The role of key gut microbial metabolites in the development and treatment of cancer. Gut Microbes 14(1):2038865

Kavvas ES, Long CP, Sastry AV, Poudel S, Antoniewicz MR, Ding Y, Mohamed ET, Szubin R, Monk JM, Feist AM, Palsson BO (2022) Experimental evolution reveals unifying systems-level adaptations but diversity in driving genotypes. Msystems 7:e00165-22

Kehila D, Tafreshi AG, Tokuriki N (2025) Stability of multi-species consortia during microbial metabolic evolution. Evolution 79(10):2156-2166

Khan N (2025) A narrative review on microbial response to antibiotic stress: mechanisms of resistance and adaptation. J Clin Diagn Res 19(1):DE01-DE05

Kimiran A, Battal G, Selim G, Ahamada Rachid N (2023) Biotechnological uses of marine microbial enzymes. In: Mega alliances with the micro-worlds. The biotechnological importance of aquatic microorganisms. Orient Publications, Ankara, pp 123–159

Kopp D, Bergquist PL, Sunna A (2020) Enzymology of alternative carbohydrate catabolic pathways. Catalysts 10(11):1231

Kopp D, Sunna A (2020) Alternative carbohydrate pathways–enzymes, functions and engineering. Crit Rev Biotechnol 40(7):895–912

Korytko OO (2024) Biological role and use of L-glutamic acid and its influence on the body. In: Scientific messenger of LNU of veterinary medicine and biotechnologies. Series: Agricultural sciences 26(101):242–250 Stepan Gzhytskyi National University of Veterinary Medicine and Biotechnologies Lviv. pp 242–250

Kulis-Horn RK, Persicke M, Kalinowski J (2014) Histidine biosynthesis, its regulation and biotechnological application in *Corynebacterium glutamicum*. Microb Biotechnol 7(1):5–25

Kumar A, Dhiman S, Krishan B, Samtiya M, Kumari A, Pathak N et al (2024) Microbial enzymes and major applications in the food industry: a concise review. Food Prod Process Nutr 6(1):85

Kumar P, Dubey KK (2019) Citric acid cycle regulation: back bone for secondary metabolite production. In: New and future developments in microbial biotechnology and bioengineering. Elsevier, pp 165–181

Kursheed F, Naz E, Mateen S, Kulsoom U (2025) CRISPR applications in microbial world: assessing the opportunities and challenges. Gene 935:149075

Lambros M, Tran T, Fei Q, Nicolaou M (2022) Citric acid: a multifunctional pharmaceutical excipient. Pharmaceutics 14(5):972

Latif A, Hassan N, Ali H, Niazi MBK, Jahan Z, Ghuman IL et al (2025) An overview of key industrial product citric acid production by *Aspergillus niger* and its application. J Ind Microbiol Biotechnol 52:kuaf007

Lavelle EC, McGuirk P, Mills KH (2004) Molecules of infectious agents as immunomodulatory drugs. Curr Top Med Chem 4(5):499–508

Liang S, Wang C, Shao Y, Wang Y, Xing D, Geng Z (2022) Recent advances in bacteria-mediated cancer therapy. Front Bioeng Biotechnol 10:1026248

Liu P, Wang S, Li C, Zhuang Y, Xia J, Noorman H (2021) Dynamic response of *Aspergillus niger* to periodical glucose pulse stimuli in chemostat cultures. Biotechnol Bioeng 118(6):2265–2282

Liu J, Ou Y, Xu JZ, Rao ZM, Zhang WG (2023) L-lysine production by systems metabolic engineering of an NADPH auto-regulated *Corynebacterium glutamicum*. Bioresour Technol 387:129701

Lv J, Yang S, Zhou W, Liu Z, Tan J, Wei M (2024) Microbial regulation of plant secondary metabolites: Impact, mechanisms and prospects. Microbiol Res 283:127688

MacLean A, Legendre F, Appanna V (2023) The tricarboxylic acid (TCA) cycle: a malleable metabolic network to counter cellular stress. Crit Rev Biochem Mol Biol 58:81–97

Madigan MT, Martinko JM, Stahl D, Clark DP (2012) Nutrition, culture, metabolism of microorganisms. In: Brock biology of microorganisms, 13th edn. Benjamin Cummings, San Francisco, pp 113–144

Mahmoudi F, Jalayeri MHT, Montaseri A, MohamedKhosroshahi L, Baradaran B (2024) Microbial natural compounds and secondary metabolites as immunomodulators: a review. Int J Biol Macromol 278:134778

Marks BB, Nogueira MA, Hungria M (2025) Microbial secondary metabolites and their use in achieving sustainable agriculture: present achievements and future challenges. Agronomy 15(6):1350

Martin WF, Tielens AG, Mentel M (2020) Mitochondria and anaerobic energy metabolism in eukaryotes: biochemistry and evolution. Walter de Gruyter GmbH, Berlin

Martínez-Espinosa RM (2020) Microorganisms and their metabolic capabilities in the context of the biogeochemical nitrogen cycle at extreme environments. Int J Mol Sci 21(12):4228

Mccann JR, Rawls JF (2023) Essential amino acid metabolites as chemical mediators of host-microbe interaction in the gut. Annu Rev Microbiol, 77(1): 479-497

Mitra R, Han J, Xiang H, Bera S (2021) Importance of microbial secondary metabolites in health care applications. In: Volatiles and metabolites of microbes. Academic Press. pp. 349–383

Mohan CD, Rangappa S, Nayak SC, Jadimurthy R, Wang L, Sethi G et al (2022) Bacteria as a treasure house of secondary metabolites with anticancer potential. In: Seminars in cancer biology. vol. 86. Academic Press, pp 998–1013

Mukherjee A, Sarkar R (2025) Unlocking the microbial treasure trove: advances in Streptomyces derived secondary metabolites in the battle against cancer. Naunyn Schmiedebergs Arch Pharmacol, 398: 9971–10005

Mustafayeva S, Şanlı NÖ (2022) Probiyotikler: kanserin önlenmesi ve tedavisinde yeni bir alternatif. Fonksiyonel Tıp. Orient Yayınları, Ankara. pp. 129–145

Müller M, Mentel M, van Hellemond JJ, Henze K, Woehle C, Gould SB et al (2012) Biochemistry and evolution of anaerobic energy metabolism in eukaryotes. Microbiol Mol Biol Rev 76(2):444–495

Nair MS, Amalaradjou MA, Venkitanarayanan K (2017) Antivirulence properties of probiotics in combating microbial pathogenesis. Adv Appl Microbiol 98:1–29

Naz S, Liu P, Farooq U, Ma H (2023) Insight into de-regulation of amino acid feedback inhibition: a focus on structure analysis method. Microb Cell Fact 22(1):161

Ng CY, Wang L, Chowdhury A, Maranas CD (2019) Pareto optimality explanation of the glycolytic alternatives in nature. Sci Rep 9(1):2633

Nie Y, Tian Y, Ren X, Liang J, Li B, Xiong ZQ (2025) Microbial production of l-threonine using metabolically engineered *Escherichia coli*. J Agri Food Chem, 73 (22): 13125-13141

Nielsen J, Keasling JD (2016) Engineering cellular metabolism. Cell 164(6):1185–1197

Niu G, Tan H (2013) Biosynthesis and regulation of secondary metabolites in microorganisms. Sci China Life Sci 56(7):581–583

Ofoedum AF, Uyanwa NC, Chikelu EC, Iroagba LN, Ugwoezuonu JN, Anaeke EJ, Odeyemi T, Okezie F, Nwuka MU (2024) Primary and secondary metabolites as products of microbial metabolism: uses and application in foods, pharmaceutical and allied industries. A review. Eur J Appl Sci Eng Technol 2(2):4–16

Ozcelik S, Kuley E, Özogul F (2016) Formation of lactic, acetic, succinic, propionic, formic and butyric acid by lactic acid bacteria. LWT 73:536–542

Paczia N, Nilgen A, Lehmann T, Gätgens J, Wiechert W, Noack S (2012) Extensive exometabolome analysis reveals extended overflow metabolism in various microorganisms. Microb Cell Fact 11(1):122

Pal A, Iyer MS, Srinivasan S, Narain Seshasayee AS, Venkatesh KV (2022) Global pleiotropic effects in adaptively evolved *Escherichia coli* lacking CRP reveal molecular mechanisms that define the growth physiology. Open Biol 12(2):210206

Pandey AK, Samota MK, Kumar A, Silva AS, Dubey NK (2023) Fungal mycotoxins in food commodities: present status and future concerns. Front Sustain Food Syst 7:1162595

Papagianni M (2012) Recent advances in engineering the central carbon metabolism of industrially important bacteria. Microb Cell Fact 11:1–13

Patel P, Patel H, Sharma J, Shrimali S, Sharma S, Saraf M (2023) Microbial biosurfactants: an overview of their uses, classification, types, properties, and biosynthesis. Acta Sci Microbiol 6(6): 61–71
Peretó J (2023) Embden-meyerhof-parnas pathway. In: Encyclopedia of astrobiology. Springer Berlin Heidelberg, Berlin, Heidelberg, pp 888
Peterson E, Kaur P (2018) Antibiotic resistance mechanisms in bacteria: relationships between resistance determinants of antibiotic producers, environmental bacteria, and clinical pathogens. Front Microbiol 9:2928
Pierre L (2024) Microbial metabolism: its role in biochemical cycles, industrial applications, and environmental impact. Appl Microbiol Open Access 10:323
Prabhu S, Poorniammal R, Dufossé L (2025) Microbial metabolites: a sustainable approach to combat plant pests. Metabolites 15(6):418
Rana AK, Thakur VK (2025) Advances and new horizons in metabolic engineering of heterotrophic bacteria and cyanobacteria for enhanced lactic acid production. Bioresour Technol 419:131951
Reddy S, Sinha A, Osborne WJ (2021) Microbial secondary metabolites: recent developments and technological challenges. In: Volatiles and metabolites of microbes. Academic Press, London, pp 1–22
Reznik E, Christodoulou D, Goldford JE, Briars E, Sauer U, Segrè D, Noor E (2017) Genome-scale architecture of small molecule regulatory networks and the fundamental trade-off between regulation and enzymatic activity. Cell Rep 20(11):2666–2677
Rodríguez-Gijón A, Nuy JK, Mehrshad M, Buck M, Schulz F, Woyke T, Garcia SL (2022) A genomic perspective across Earth's microbiomes reveals that genome size in Archaea and Bacteria is linked to ecosystem type and trophic strategy. Front Microbiol 12:761869
Ruiz-Villafán B, Cruz-Bautista R, Manzo-Ruiz M, Passari AK, Villarreal-Gómez K, Rodríguez-Sanoja R, Sánchez S (2022) Carbon catabolite regulation of secondary metabolite formation, an old but not well-established regulatory system. Microb Biotechnol 15(4):1058–1072
Ruslan NF, Ahmad N, Abas A, Sanfilippo A, Mahmoud K, Munaim MSA, Nour AH (2025) Sustainable bioethanol production by solid-state fermentation: a systematic review. Environ Sci Pollut Res 32(22):13140–13158
Sallam KI, Abd-Elghany SM, Hussein MA, Imre K, Morar A, Morshdy AE, Sayed-Ahmed MZ (2020) Microbial decontamination of beef carcass surfaces by lactic acid, acetic acid, trisodium phosphate sprays. Biomed Res Int 2020(1):2324358
Sanchez S, Demain AL (2008) Metabolic regulation and overproduction of primary metabolites. Microb Biotechnol 1(4):283–319
Schink SJ, Christodoulou D, Mukherjee A, Athaide E, Brunner V, Fuhrer T et al (2022) Glycolysis/gluconeogenesis specialization in microbes is driven by biochemical constraints of flux sensing. Mol Syst Biol 18(1):e10704
Schniete JK, Cruz-Morales P, Selem-Mojica N, Fernández-Martínez LT, Hunter IS, Barona-Gómez F, Hoskisson PA (2018) Expanding primary metabolism helps generate the metabolic robustness to facilitate antibiotic biosynthesis in Streptomyces. mBio 9(1):10–1128
Selvanathan Y, Shamsudin S, Zainol N, Masngut N (2025) Nutritional requirements and fermentation condition for acetic acid production from agricultural fruit waste. Chem Eng Technol 48(2):e202300581
Sharma G, Curtis PD (2022) The impacts of microgravity on bacterial metabolism. Life 12(6):774
Sharma J, Sundar D, Srivastava P (2021) Biosurfactants: potential agents for controlling cellular communication, motility, and antagonism. Front Mol Biosci 8:727070
Shibl AA, Isaac A, Ochsenkuhn MA, Cárdenas A, Fei C, Behringer G et al (2020) Diatom modulation of select bacteria through use of two unique secondary metabolites. Proc Natl Acad Sci USA 117(44):27445–27455
Simões CR, da Silva MWP, de Souza RFM, Hacha RR, Merma AG, Torem ML, Silvas FPC (2024) Biosurfactants: an overview of their properties, production, and application in mineral flotation. Resources 13(6):81

Singh KD, Schmalisch MH, Stülke J, Görke B (2008) Carbon catabolite repression in *Bacillus subtilis*: quantitative analysis of repression exerted by different carbon sources. J Bacteriol 190(21):7275–7284

Singh R, Priyanka S, Rajul J, Chauhan A, Kumar R, Gunsola D et al (2025) Microbial metabolites: a sustainable solution for agricultural challenges. Pedosphere, doi:10.1016/j.pedsph.2025.06.003, Epub ahead of print

Soberón-Chávez G, Hausmann R, Déziel E (2021) Biosurfactants: new insights in their biosynthesis, production and applications. Front Bioeng Biotechnol 9:769899

Suarez-Mendez CA (2015) Dynamics of storage carbohydrates metabolism in *Saccharomyces cerevisiae*: a quantitative. TU Delft Repos, Delft

Suman J, Rakshit A, Ogireddy SD, Singh S, Gupta C, Chandrakala J (2022) Microbiome as a key player in sustainable agriculture and human health. Front Soil Sci 2:821589

Sundararajan P, Ramasamy SP (2023) Bacteria as promising biofactory for pigment production: a prospective insights into production strategies and industrial applications. AsPac J Mol Biol Biotechnol 31:53–61

Taha A, Patón M, Penas DR, Banga JR, Rodríguez J (2023) Optimal evaluation of energy yield and driving force in microbial metabolic pathway variants. PLoS Comput Biol 19(7):e1011264

Tavşanlı N, Yıldız S, Çalışkan M, Aydin S (2024) Evaluating the potential effect of microalgae on soymilk vegan kefir in terms of physical, chemical, microbiological properties. Algal Res-Biomass Biofuels And Bioproducts 82:103630

Toogood HS, Scrutton NS (2018) Retooling microorganisms for the fermentative production of alcohols. Curr Opin Biotechnol 50:1–10

Toya Y, Shimizu H (2020) Flux controlling technology for central carbon metabolism for efficient microbial bio-production. Curr Opin Biotechnol 64:169–174

Ueda K, Beppu T (2017) Antibiotics in microbial coculture. J Antibiot 70(4):361–365

Ullah A, Bano A, Janjua HT (2020) Microbial secondary metabolites and defense of plant stress. In: Microbial services in restoration ecology. Elsevier, Amsterdam, pp 37–46

Van Beirs C, El Houari I, Vanholme B (2025) Nature's laboratory: plant metabolic engineering methods using phenylpropanoids as a case study. Biotechnol Biofuels Bioprod 18(1):81

Veaudor T, Blanc-Garin V, Chenebault C, Diaz-Santos E, Sassi JF, Cassier-Chauvat C, Chauvat F (2020) Recent advances in the photoautotrophic metabolism of cyanobacteria: biotechnological implications. Life 10(5):71

Verma A, Barua A, Ruvindy R, Savela H, Ajani PA, Murray SA (2019) The genetic basis of toxin biosynthesis in dinoflagellates. Microorganisms 7(8):222

Wang C, Kuzyakov Y (2023) Energy use efficiency of soil microorganisms: Driven by carbon recycling and reduction. Glob Change Biol 29(22):6170–6187

War May Zin W, Prompanya C, Buttachon S, Kijjoa A (2016) Bioactive secondary metabolites from a Thai collection of soil and marine-derived fungi of the genera *Neosartorya* and *Aspergillus*. Curr Drug Deliv 13(3):378–388

Wend K, Zorrilla L, Freimoser FM, Gallet A (2024) Microbial pesticides–challenges and future perspectives for testing and safety assessment with respect to human health. Env Health 23(1):49

Wu Z, Liang X, Li M, Ma M, Zheng Q, Li D et al (2023) Advances in the optimization of central carbon metabolism in metabolic engineering. Microb Cell Fact 22(1):76

Wu L, Tang Z, Chen H, Ren Z, Ding Q, Liang K, Sun Z (2021) Mutual interaction between gut microbiota and protein/amino acid metabolism for host mucosal immunity and health. Anim Nutr 7(1):11–16

Xu M, Rao Z, Dou W, Yang J, Jin J, Xu Z (2012) Site-directed mutagenesis and feedback-resistant N-acetyl-L-glutamate kinase (NAGK) increase *Corynebacterium crenatum* L-arginine production. Amino Acids 43(1):255–266

Yadav J, Marwah H, Kumar C (2025) Synthetic biology and metabolic engineering paving the way for sustainable next-gen biofuels: a comprehensive review. Energy Adv 4(10):1209-1228

Yang E, Fan L, Yan J, Jiang Y, Doucette C, Fillmore S, Walker B (2018) Influence of culture media, pH and temperature on growth and bacteriocin production of bacteriocinogenic lactic acid bacteria. AMB Express 8(1):10

Yin X, Li J, Shin HD, Du G, Liu L, Chen J (2015) Metabolic engineering in the biotechnological production of organic acids in the tricarboxylic acid cycle of microorganisms: advances and prospects. Biotechnol Adv 33(6):830–841

You C, Okano H, Hui S, Zhang Z, Kim M, Gunderson CW et al (2013) Coordination of bacterial proteome with metabolism by cyclic AMP signalling. Nature 500(7462):301–306

Zhang LS, Davies SS (2016) Microbial metabolism of dietary components to bioactive metabolites: opportunities for new therapeutic interventions. Genome Med 8:1–18

Zhang S, Guo F, Yan W, Dai Z, Dong W, Zhou J et al (2020) Recent advances of CRISPR/Cas9-based genetic engineering and transcriptional regulation in industrial biology. Front Bioeng Biotechnol 7:459

Zhou X, Zhu X, Zhou L, Zhang P, Xue B, Ren F, Liu H (2025) The potential contribution of lactic acid bacteria in the fermentation of mixed grain foods. Food Rev Int 41(9): 3165-3189

Chapter 5
Biotechnological Applications of Bacteria

İbrahim Halil Kılıç

Abstract Microbial biotechnology has historically relied heavily on bacteria, both in fundamental science and applied research. Their rapid growth capacity, ease of genetic manipulation, and extraordinary metabolic diversity enable them to offer a wide range of applications across all areas of modern biotechnology. Today, model bacteria such as *Escherichia coli*, *Bacillus subtilis*, and *Pseudomonas* species are at the center of multidimensional research ranging from recombinant protein production to biofuel synthesis, bioremediation, and probiotic therapies. The biotechnological potential of bacteria extends beyond industrial product development to enable the creation of sustainable solutions for global health, food safety, and environmental issues. In the medical field, bacteria serve as cellular factories for the large-scale production of recombinant proteins, including insulin, growth hormones, and monoclonal antibodies. Additionally, bacterial vectors are utilized as platforms in vaccine development and gene therapy applications. Probiotic bacteria are used to regulate the gut microbiota, strengthen the immune system, and prevent infections. In the agricultural sector, bacteria are utilized as biofertilizers, nitrogen-fixing agents (Rhizobium, Azotobacter), and biocontrol agents. They also contribute to reducing chemical fertilizer use by increasing productivity through the production of phytohormones that promote plant growth and by enhancing the availability of nutrients to plants. Bacteria have gained an important place in industrial applications by being used in the production of valuable metabolites such as antibiotics, vitamins, amino acids, and organic acids. With advances in metabolic engineering and synthetic biology, they are also involved in the production of biofuels, biodegradable plastics, and enzymes used in the food and textile industries. In environmental biotechnology, bacteria have a wide

İ. H. Kılıç (✉)
Biology Department, Faculty of Arts and Science, Gaziantep University, Gaziantep, Turkey
e-mail: kilic@gantep.edu.tr

M. Çalışkan and M. Arıkan (eds.), *Microbial World: Bacteria and Archaea*, Progress in Molecular and Subcellular Biology 62, https://doi.org/10.1007/978-3-032-11438-9_5

range of applications in bioremediation processes. Natural or genetically modified bacteria make important contributions to the degradation or detoxification of petroleum derivatives, heavy metals, and industrial waste. They also play critical roles in wastewater treatment and sustainable environmental management. In conclusion, bacteria serve as biological tools that offer innovative solutions in various areas of biotechnology. Advances in genomics, CRISPR-based gene editing, and systems biology are expanding the biotechnological applications of bacteria, increasing their importance by offering sustainable solutions to global issues such as health, food safety, and environmental protection.

Keywords Bacteria · Agricultural biotechnology · Environmental biotechnology · Medical biotechnology

5.1 The General Role of Bacteria in Biotechnology

5.1.1 Introduction

The metabolic capabilities and adaptability of bacteria have made them indispensable tools in various fields, including biofuel production, enzyme synthesis, environmental cleanup, wastewater treatment, heavy metal removal, and health applications. Due to their versatile properties, bacteria play a key role in biotechnology (Ifiok Mfon 2024; Yang et al. 2024).

In the food industry, bacteria are utilized to produce fermented foods and beverages; in agriculture, bacteria are used as biofertilizers and biopesticides to enhance soil fertility and protect plants from pests and diseases (Chaudhary et al. 2024; Yang et al. 2024).

Bacteria function as mini bioreactors, synthesizing enzymes, organic acids, vitamins, antibiotics, antibodies, hormones, carotenoids, steroids, alkaloids, alcoholic beverages, interferons, vaccines, and economically important products. In addition to synthesizing these products, bacteria also play a role in the production of numerous industrial and pharmaceutical products, such as wastewater treatment, bioremediation, biomineralization, bio-recovery of valuable metals and metalloids, biocontrol agents, prebiotic and probiotic production, biogas, and environmentally friendly nanomaterials (Akoijam et al. 2022; Yang et al. 2024).

The biodiversity of each bacterium is different (Chaudhary et al. 2024). For example, Streptomyces bacteria are used for the production of life-saving antibiotics, while model bacteria such as *Escherichia coli* are widely used in recombinant protein production (Santos-Beneit 2024; Yang et al. 2024).

In conclusion, bacteria emerge as indispensable biotechnological tools in various sectors, including food, agriculture, health, energy, and the environment (Santos-Beneit

2024; Yang et al. 2024). The increasing use of bacteria in industrial applications has driven the expansion of microbial industries toward more diverse and valuable horizons (Akoijam et al. 2022).

5.1.2 Historical Development

Bacteria play a crucial role in sustaining life. In addition to their roles in geochemical cycles and ecological restoration of the environment, bacteria are widely used in industries to synthesize products beneficial to humans (Akoijam et al. 2022).

The biotechnological benefits of bacteria have been known for thousands of years. Since ancient times, when the phenomenon of fermentation was first recognized, bacteria such as *Lactobacillus* have played a significant role in the production of fermented foods like bread, cheese, and yogurt, while *Saccharomyces* yeast has been crucial in the production of beer and wine (Chaudhary et al. 2024). Long before their discovery, microorganisms were utilized to meet human needs and desires, despite their previously unknown existence. These applications can be considered the first examples of biotechnology (Verma et al. 2011; Demain et al. 2016).

The term "biotechnology" in its modern sense was first used in 1919 by Hungarian engineer Karl Ereky (Verma et al. 2011). The scientific and industrial importance of bacteria was understood in the late nineteenth and early twentieth centuries by pioneering scientists such as Louis Pasteur and Robert Koch, who demonstrated the role of microorganisms in fermentation and disease. Throughout the twentieth century, the biotechnological use of bacteria advanced rapidly (Santos-Beneit 2024).

Due to the necessities brought about by World War I, Chaim Weizmann succeeded in producing acetone and butanol from corn starch using the bacterium *Clostridium acetobutylicum*, demonstrating that pure bacterial cultures could be used in industrial chemical production (Adams 2017).

In the 1940s, fermenters were used on an industrial scale primarily for the mass production of antibiotics, particularly penicillin. In these fermenters, microorganisms were developed, and their metabolic products were obtained, making production possible (Landecker 2016). Additionally, large-scale fermentation processes developed using bacteria and fungi during this period shaped the modern pharmaceutical industry and revolutionized the treatment of infectious diseases.

In the 1970s, the emergence of recombinant DNA technology and genetic engineering took bacterial biotechnological applications to the next level. By the end of the 1970s, efforts to transfer different genes into bacterial cells were successfully achieved, leading to the production of the first recombinant human insulin in 1978 using *E. coli* bacteria. In 1982, the Food and Drug Administration (FDA) approved the general use of insulin produced from bacteria using recombinant DNA procedures for the treatment of people with diabetes. With this production, the modern biopharmaceutical era began (Alyas et al. 2021).

Today, the use of bacteria has expanded significantly. While they were previously used to obtain natural products, they are now employed both for this purpose and to shape biotechnology through tools developed from their genetic systems. For example, heat-resistant enzymes such as bacterial restriction enzymes and Taq DNA polymerase have made fundamental techniques in molecular biology, including PCR, possible.

Finally, in the early 2020s, advancements in gene editing, such as CRISPR-Cas 9 (clustered regularly interspaced short palindromic repeats) technology, have enabled scientists to correct genetic errors causing conditions like sickle cell anemia and many other disorders. This discovery is one of the most significant breakthroughs of the twenty-first century and has revolutionized biotechnology today. These historical developments demonstrate that bacteria have been at the center of biotechnology for centuries, gaining increasing importance alongside scientific progress (Yang et al. 2024; Li et al. 2023).

5.1.3 Genetic and Metabolic Advantages

The genetic structure and metabolic properties of bacteria provide important advantages for biotechnological applications. Their ability to multiply rapidly enables them to be produced on a large scale both in laboratories and in industry (Santos-Beneit 2024).

Bacterial genomes are relatively small and have a regular structure compared to other organisms. Bacterial genomes allow for easy gene transfer and cloning through the use of extrachromosomal plasmids. The discovery of gene transfer via plasmids has paved the way for research in areas such as gene cloning, recombinant protein production (e.g., human insulin), and gene therapy (Al Doghaither and Gull 2019).

Bacteria can acquire new genetic traits through three primary mechanisms: conjugation, transformation, and transduction. In the laboratory, these genetic material exchange techniques are often used to transfer a desired gene into a bacterial cell, thereby conferring new traits on the isolate (Rafeeq et al. 2023). In this way, bacteria can be engineered to produce a variety of enzymes. The CRISPR-Cas defense system discovered in bacteria has been reengineered by scientists into precise genome editing techniques, enabling the creation of desired genetic changes in plant, animal, or human cells (Santos-Beneit 2024). All this genetic flexibility and intervention potential make bacteria indispensable for the development of new biotechnological products and processes (Yang et al. 2024). Another advantage that makes bacteria unique is their metabolic diversity. Prokaryotes encompass various species that can utilize nearly all organic (and even inorganic) substances in nature as energy sources. For example, while some bacterial species break down chemicals found in soil that are harmful to the soil, other bacterial species can produce biogas by utilizing organic waste materials. Some bacteria possess the ability to synthesize various compounds such as ethyl alcohol, lactic acid, amino acids, and antibiotics through their diverse biochemical pathways. Additionally, enzymes derived from bacteria capable of surviving under extreme conditions (e.g., *Thermus aquaticus*, which thrives at high temperatures, or

Halomonas species that can survive in high salt concentrations) are used as biocatalysts in industry under challenging conditions (Yang et al. 2024).

Thanks to all these metabolic advantages, bacteria offer both efficient and sustainable solutions in many fields, such as biofuel production, waste conversion, drug raw material synthesis, and environmental cleanup (Yang et al. 2024).

In conclusion, with their susceptibility to genetic manipulation and metabolic flexibility, bacteria form the foundation of many current biotechnological applications and will likely remain at the forefront of innovative technologies in the future.

5.2 Industrial Applications

5.2.1 Enzyme Production

Microbial enzymes play a significant role in the development of industrial bioprocesses (Adrio and Demain 2014). Over 500 industrial products are produced using various microbial enzymes. Industrial enzymes are used in many sectors, such as detergent, food processing, textile, paper, and pharmaceutical industries and are produced on a large scale through microbial systems (Liu and Kokare 2017).

One of the most commonly preferred microorganisms for the production of these enzymes is the genus *Bacillus*. These bacteria are preferred due to their rapid growth, high enzyme secretion levels, and ease of genetic manipulation (Danilova and Sharipova 2020).

5.2.1.1 Proteases

Enzymes that catalyze the hydrolysis of peptide bonds in proteins and polypeptides, breaking them down into amino acids. They are widely used in the detergent and pharmaceutical industries (Raveendran et al. 2018). Among the numerous enzymes produced on a large scale by *Bacillus* species, proteases are important for various industrial applications (Degering et al. 2010). Among proteases, alkaline proteases account for two-thirds of the protease market due to their unique properties, such as high stability and relatively low substrate specificity (Matkawala et al. 2021). Currently, *B. subtilis*, *B. licheniformis*, *Bacillus pumilus*, and *B. amyloliquefaciens* have the capacity to produce alkaline proteases on an industrial scale (Chen et al. 2022).

5.2.1.2 Amylases

They hydrolyze starch molecules into polymers composed of glucose units. They find applications in industries such as food, paper, and biofuel production. α-Amylase can be produced by various microbial species, but for commercial applications, α-amylase is primarily derived from the *Bacillus* genus (de Souza and de Oliveira Magalhães 2010). α-Amylases produced by *Bacillus licheniformis, Bacillus stearothermophilus*, and *Bacillus amyloliquefaciens* have potential applications in various industrial processes, such as food, fermentation, textile, and paper industries (Kunamneni et al. 2005). *Bacillus subtilis, Bacillus stearothermophilus, Bacillus licheniformis*, and *Bacillus amyloliquefaciens* are known to be good producers of thermostable α-amylase and are widely used in the commercial production of the enzyme for various applications (Prakash and Jaiswal 2009).

5.2.1.3 Cellulases

Microorganism-derived cellulases have a wide range of applications, including biofuel production, food and wine biotechnology, pulp and paper production, biological ink removal, textile and laundry industries, and the conversion of cellulosic biomass. Cellulases are enzymes that convert cellulose into glucose. These enzymes are particularly important in the biofuel industry, where they are used to ferment lignocellulosic waste into sugar (Menendez et al. 2015).

5.2.2 Organic Acid and Fermented Product Production

Organic acids are traditional products of food technology and biotechnology. The production of organic acids through microbial fermentation using economical raw materials is one successful area of industrial microbiology (Yadav et al. 2022).

The production of organic acids through fermentation is a significant economic factor. Organic acids have various applications in industries such as food, pharmaceuticals, cosmetics, and textiles. In addition to their traditional applications, they are also used in bioplastic production (Singh et al. 2022).

The reasons for preferring microbial processes for organic acid production are their high efficiency and environmentally friendly nature (Singh et al. 2022).

Lactobacillus species, including *Lactobacillus plantarum, L. delbrueckii*, and *L. casei*, are effective strains for lactic acid production using both homolactic and heterolactic fermentation pathways (Wu et al. 2021). These species are widely used in industry due to their low pH tolerance, high production yield, and GRAS (Generally Recognized As Safe) status (Yadav et al. 2022).

Lactic acid is widely used in the food, pharmaceutical, cosmetic, and textile industries. Its applications in the synthesis of biodegradable, biocompatible plastics and coatings have increasingly heightened interest in lactic acid. L-lactic acid is a compound that can be polymerized to form polylactic acid. This polymer can be used in the production of biodegradable plastics. Biodegradable plastics offer an important alternative to non-biodegradable plastics in addressing environmental pollution issues (Yadav et al. 2022).

5.2.3 *Bioplastic Production*

Bioplastics are a family of new-generation bio-based, biodegradable, or both plastics (Gill 2014). They serve as an environmentally friendly alternative to plastics and can be biodegraded by microbial enzymes into CO_2, methane, H_2O, and biomass. Although bioplastic production reduces fossil fuel consumption, CO_2 emissions, and plastic waste generation, the high cost of producing bioplastics from bacteria remains the most significant limiting factor when compared to the production costs of petroleum-derived polymers (Gill 2014; Wala'A et al. 2017).

Bioplastics, developed as alternatives to petroleum-based plastics, are of great importance, particularly in terms of reducing environmental pollution and lowering the carbon footprint. Polymers containing polyhydroxybutyrate (PHB), a type of bioplastic derived from bacterial metabolic processes, are the most typical examples of polyhydroxyalkanoates (PHAs). These are polyesters produced by microbial cells during the storage of carbon sources. In nature, they are completely biodegradable molecules (de Souza et al. 2024).

One of the most studied bacteria for PHA production is *Ralstonia eutropha* (now known as *Cupriavidus necator*). This bacterium can synthesize PHA at high yields from various carbon sources and accumulate it in granules within the cell (Kourmentza et al. 2017). Industrially, PHA can be produced from glucose, sucrose, fatty acids, and industrial waste.

Bioplastics are widely used in the packaging industry, agricultural films, medical applications (such as suture threads and skeletal systems), and the production of single-use products. Additionally, increasing environmental regulations and sustainability goals are continuously enhancing the importance of microbial bioplastics (Kourmentza et al. 2017).

5.3 Agricultural Biotechnology

5.3.1 *Biofertilizers*

Biofertilizers are formulations of beneficial microorganisms that, when applied, can enhance the availability of nutrients through their biological activities and contribute to improving soil health (Mosa et al. 2016).

Chemical fertilizers have a wide range of harmful effects on the environment. These include soil and food contamination due to heavy metals and radioactive substances, air pollution caused by gases such as NO, N_2O, NO_2, and others, groundwater contamination caused by nitrates, and changes in soil pH and structure (Zhao et al. 2024). Environmentally friendly biofertilizers can serve as an alternative to chemical fertilizers and support sustainable agriculture (Olanrewaju et al. 2017). Biofertilizers enhance soil microbial activity, promote plant growth, increase yield, and help maintain ecological balance (Bahuguna et al. 2025).

Rhizobium, a member of *the Rhizobiaceae* bacterial family, is the best example of symbiotic nitrogen fixation. They can fix N2 in legumes and non-legume plants (Pindi 2012). The bacteria infect legume roots and form nodules; within these nodules, they reduce molecular nitrogen to ammonia, which the plant uses to produce proteins, vitamins, and other nitrogen-containing compounds. Thus, these root nodules act as ammonia production factories. *Rhizobium* species improve the growth of non-legume plants by causing changes in root morphology and growth physiology (Nosheen et al. 2021). The use of these nitrogen-fixing bacterial products has the potential to reduce environmental risks and costs by decreasing dependence on nitrogen-based chemical fertilizers in agriculture (Bahuguna et al. 2025). Indeed, it has been reported that inoculating legume plants with *Rhizobium* can introduce approximately 50–100 kg/ha of biological nitrogen into the root zone annually, resulting in a 10–35% increase in crop yield (Bahuguna et al. 2025).

Rhizobia contribute to plant growth in legumes not only by increasing nitrogen fixation but also through various other mechanisms such as phytohormone production, phosphate solubilization, siderophore production, and reduction of ethylene levels. *Rhizobium* species not only increase nutrient levels and promote plant growth but also have the advantage of improving soil structure (Pereira et al. 2024). As a result, the use of *Rhizobium* biofertilizers supports plant development while contributing to sustainable agriculture by replacing synthetic fertilizers (Bahuguna et al. 2025).

Bacteria such as *Azotobacter*, which have the potential for free-living nitrogen fixation, are widely used as biofertilizers. These bacteria can fix nitrogen without establishing a symbiotic relationship with plants. Unlike *Rhizobium*, which forms a symbiotic relationship with legume roots, *Azotobacter* can live in the rhizosphere of a wide variety of plants, including cereals and vegetables. *Azotobacter chroococcum, Azotobacter vinelandii,* and *Azotobacter beijerinckii* are *Azotobacter* species that have nitrogen-fixing properties under aerobic conditions. *Azotobacter's* ability to tolerate oxygen, which typically inhibits nitrogen fixation, is an important characteristic (Nida et al. 2024; Wani et al. 2013; Sokra et al. 2025).

Azotobacter not only enriches the soil with nitrogen but also produces plant growth hormones such as indole-3-acetic acid (IAA) and siderophores that promote plant growth. Additionally, the exopolysaccharides they secrete into the root zone improve soil structure (Gauri et al. 2012).

For example, the application of *Azotobacter chroococcum* has been reported to have positive effects on root and shoot growth in various plants, increase plant nutrient uptake, and support soil fertility (Wani et al. 2013). The use of bacteria such as *Rhizobium* and *Azotobacter* as biofertilizers reduces the need for chemical fertilizers and has positive effects on sustainability and ecological balance in agriculture (Bahuguna et al. 2025; Wani et al. 2013).

5.3.2 Biopesticides

Biopesticides are microbial, plant-based, or naturally derived pesticides used to control plant pests. The most commonly used bacterial agent in biopesticide production is

Bacillus thuringiensis. These bacteria produce δ-endotoxins (Cry proteins) during the spore formation stage, infecting various insect species and exerting their effect (Kumar et al. 2021). The insecticidal proteins produced by *Bacillus thuringiensis* (*Bt*) have been successfully used in agricultural fields for many years (Kumar et al. 2021). The active ingredient in many commercial preparations developed against harmful insects in vegetables and field crops is the Cry proteins produced by *Bt*. Cry proteins, after being ingested by target insect larvae, are converted into their active form by proteolytic enzymes present in the insect's midgut and cause the formation of pores in the epithelial cells of the insect, thereby exerting their effect. As a result, paralysis develops in the intestinal cells, and the insect larva dies rapidly (Vachon et al. 2012). *Bt*-derived Cry toxins are effective against a wide range of insect species, primarily *Lepidoptera* (butterflies and moths), *Coleoptera* (beetles), and *Diptera* (flies and mosquitoes) (Vachon et al. 2012).

The long-term and intensive use of chemical pesticides can result in serious adverse effects such as persistent environmental pollution, accumulation of high toxicity levels harmful to human health, and the development of resistance to pesticides in target pest species (Mishra et al. 2020). Due to these issues, biopesticides, particularly formulations based on *B. thuringiensis*, are gaining increasing importance as sustainable alternatives to synthetic chemicals in agriculture (Kumar et al. 2021). Biopesticides offer significant advantages over chemical pesticides due to their specificity in targeting pests, their efficacy, their selectivity toward beneficial insects, and their lack of persistence in the environment (Daraban et al. 2023). However, biopesticides based on *Bt* have become an indispensable part of Integrated Pest Management programs due to their high levels of effectiveness. While as effective as chemical insecticides, their greater selectivity and environmental friendliness make them a preferred choice (Matyjaszczyk 2018).

Indeed, most biopesticides are easily biodegradable in nature and affect only target organisms, causing minimal harm to beneficial insects and other living organisms. Additionally, the use of biopesticides helps slow the development of pesticide resistance in harmful organisms, thereby increasing the likelihood of long-term effectiveness and success in pest control (Mishra et al. 2020). In conclusion, biopesticides, which replace chemical pesticides, are considered an important biotechnological approach that supports both environmental sustainability and food safety.

5.4 Environmental Biotechnology

Environmental biotechnology relies on microorganisms and enzymes to protect, improve, or restore environmental quality. Its applications range from traditional municipal wastewater treatment processes to the latest bioremediation technologies aimed at cleaning up areas contaminated with hazardous industrial chemicals (Agathos 2011).

These approaches offer more sustainable and environmentally friendly solutions compared to traditional chemical or physical methods. In particular, by-products such as biogas generated during waste treatment can be utilized as renewable energy sources, thereby contributing to sustainable development through environmental biotechnology (Song et al. 2025).

5.4.1 Bioremediation

Bioremediation is a method that uses microorganisms to convert toxic pollutants into harmless compounds, thereby purifying contaminated water and soil (Hlihor et al. 2017). Various microorganisms, such as *Pseudomonas*, *Deinococcus*, yeast, and algae, have the ability to convert complex organic compounds, petroleum hydrocarbons, pesticides, and even some heavy metals into less harmful forms (Bala et al. 2022). This enables the biological remediation of contaminated soil and water environments (Bala et al. 2022; Hlihor et al. 2017).

In particular, their metabolic flexibility and ability to adapt to harsh conditions have brought *Pseudomonas* bacteria to the forefront in the field of bioremediation (Song et al. 2025). In particular, *Pseudomonas aeruginosa* possesses a wide range of enzymes. This property enables it to convert various organic pollutants, including petroleum hydrocarbons, chlorinated compounds, and pesticides (Song et al. 2025). This bacterium also produces biosurfactants such as rhamnolipids, which increase the solubility of hydrophobic pollutants, thereby enhancing the biodegradability of pollutants like petroleum derivatives or dyes in soil and water and facilitating microbial degradation (Hu et al. 2023).

Extremophilic microorganisms play important roles in the remediation of pollution under extreme conditions. *Deinococcus radiodurans* is highly resistant to ionizing radiation and can survive in environments containing radioactive waste. It can actively remove pollutants from these environments (Lakshmi et al. 2022). The strong antioxidants and DNA repair mechanisms of *Deinococcus radiodurans* are indicative of its metal reduction ability.

In high-radiation environments, uranium and radioactive iodine, which are released into the environment, are absorbed by the organism like a sponge, reducing their bioavailability and helping to clean the environment of these radioactive contaminants (Lakshmi et al. 2022).

However, genetic engineering studies on *D. radiodurans* have aimed to create a recombinant strain capable of both breaking down toxic organic solvents and detoxifying heavy metals. For example, by transferring a mercuric reductase gene from *E. coli to D. radiodurans*, the ability to reduce mercury(II) to elemental mercury even under high radiation and, additionally, to metabolize organic solvents such as toluene has been conferred (Brim et al. 2000).

5.4.2 Wastewater Treatment

Biological treatment involves the removal of pollutants from wastewater using microorganisms or processes. Microorganisms play a critical role in wastewater treatment and recovery, representing a promising green technology tool. In biological treatment, bacteria, fungi, microalgae, yeast, and other microbial groups are utilized. Biological treatments are more cost-effective than physical and chemical treatments (Sharma et al. 2023).

While the activated sludge process using aerobic bacteria is widely used in conventional wastewater treatment plants, anaerobic biological treatment systems also play a significant role, particularly in the treatment of high-organic-load wastewater and energy recovery (Vítězová et al. 2020). Anaerobic digestion has two unique advantages over aerobic biological treatment systems: no energy is required for aeration, and organic pollutants are primarily converted into biogas, which can be used as fuel (Sham 1984). This method is increasingly preferred in industrial wastewater treatment due to its advantages, such as low energy consumption and operating costs (Vítězová et al. 2020).

5.5 Antibiotic Production

Antibiotics are an indispensable and important pharmacological group in modern medicine, used to combat bacterial infections and have been in use to this day. Antibiotics are compounds obtained from natural substances produced by microorganisms to inhibit or eliminate the growth of other microorganisms, or developed through chemical modification of these structures (Katz and Baltz 2016).

Since the beginning of the twentieth century, antibiotics have revolutionized human health. They play a critical role not only in the treatment of infectious diseases but also in surgical interventions, organ transplants, and immunosuppressive treatments such as cancer chemotherapy (Ventola 2015). However, antibiotics are not only used in human health but also widely in sectors such as veterinary medicine, agriculture, and fisheries. Therefore, this situation has made both the production of antibiotics and the emerging resistance problems a central issue in global health policies (Michael et al. 2014).

Microorganism-derived antibiotics have emerged from compounds obtained from actinomycetes such as *Streptomyces* species, firmicutes such as *Bacillus*, and certain fungi. These compounds have been used against a wide range of bacterial targets both in their natural form and as semi-synthetic derivatives (Demain and Sanchez 2009).

Although the modern use of antibiotics began in the early twentieth century, the use of microorganisms against diseases dates back to earlier periods. For example, records from ancient Egyptian, Chinese, and Greek medicine mention the use of molds for wound healing, indicating an awareness of their natural antibacterial effects (Forrest 1982). However, the first scientifically defined antibiotic was discovered in 1928 when Alexander Fleming isolated penicillin from a culture of *Penicillium notatum*. Fleming observed that this mold inhibited the growth of *Staphylococcus aureus* colonies, marking the beginning of the antibiotic era (Fleming 1929).

Penicillin was purified and made available for large-scale production in the 1940s through the efforts of Howard Florey, Ernst Boris Chain, and Norman Heatley (Lax 2004). The use of penicillin in the military during World War II clearly demonstrated the remarkable impact of antibiotics on public health. These three scientists were awarded the Nobel Prize in Medicine in 1945. The success of penicillin increased interest in natural product research and played a significant role in the isolation of new antibiotics from the soil microbiome. In 1943, Selman Waksman and Albert Schatz achieved a major breakthrough in the treatment of tuberculosis (tuberculosis) with streptomycin, obtained from *Streptomyces griseus* (Waksman 1953). During the same period, Waksman coined the term "antibiotic," defining it as bactericidal substances obtained from natural sources and produced by microorganisms. The 1940s and 1950s, considered a golden age, saw the discovery of many new antibiotics, including erythromycin, chloramphenicol, tetracycline, and vancomycin, which were isolated and began to be produced industrially (Demain and Sanchez 2009). A significant portion of these drugs was obtained from actinomycetes such as *Streptomyces* or fungus-like organisms. These developments not only proved effective in treating infections but also facilitated scientific advancements in fields such as microbial ecology, biosynthetic processes, and fermentation technologies. In particular, structural analyses of natural antibiotics laid the groundwork for the development of semi-synthetic antibiotics in subsequent periods (Katz and Baltz 2016).

5.6 Bacteria in Energy Production

One of the most pressing issues of our time, the growing global energy demand and the climate crisis, is accelerating the shift toward environmentally friendly and renewable energy sources as alternatives to fossil fuels (Serrano-Echeverry et al. 2025). In this context, biofuels emerge as one of the most viable solutions for providing sustainable energy and play an indispensable role in clean energy production (Serrano-Echeverry et al. 2025). In particular, hydrogen is considered the "energy carrier of the future" due to its high energy content of 122 kJ per unit mass (approximately 2.75 times higher than that of fossil fuels) and the fact that it produces only water instead of harmful emissions during combustion (Albuquerque et al. 2024). This view is widely accepted in the scientific community, and it is emphasized that biological hydrogen production represents a carbon-neutral and scalable energy alternative (Sterner et al. 2024). Similarly, biobutanol has gained significance as a promising biofuel with the potential to replace fossil fuels due to its high energy density and physical–chemical properties similar to gasoline. Bacteria serve as the primary organisms in the production of such biofuels (Serrano-Echeverry et al. 2025). Additionally, bacteria play a crucial role in integrated processes such as the conversion of organic waste into energy and wastewater treatment, thereby constituting an indispensable component of the renewable energy portfolio (Dakal et al. 2025). This study focuses on biohydrogen and biobutanol production through bacteria, while also examining the general role and potential of bacteria in renewable energy systems.

5.6.1 Biohydrogen and Biobutanol Production

5.6.1.1 Biohydrogen Production

The production of hydrogen gas through biological methods, i.e., biohydrogen production, is being intensively studied as a sustainable alternative to fossil fuels. Biological hydrogen production processes are carried out through bacterial activities and can be broadly categorized into two main categories: light-dependent and light-independent (Albuquerque et al. 2024). Among light-dependent approaches, photofermentation using photosynthetic bacteria is particularly noteworthy. For example, species such as *Rhodopseudomonas palustris*, which belong to the group of non-purple sulfur bacteria, can produce hydrogen by oxidizing organic acids in the presence of light with the help of nitrogenase enzymes (Li et al. 2022). In this photobiological process, bacteria fix nitrogen while releasing hydrogen gas as a byproduct; however, it is known that the process is negatively affected by the presence of ammonia. Another light-dependent pathway, biophotolysis, involves the direct decomposition of water by cyanobacteria and algae to produce hydrogen (Albuquerque et al. 2024).

Additionally, among light-independent methods, dark fermentation is the most widely studied method. In dark fermentation, heterotrophic anaerobic bacteria, particularly *Clostridium* species, ferment organic matter in the absence of light, transferring electrons generated during metabolic processes to protons, ultimately resulting in the release of hydrogen gas (Albuquerque et al. 2024). In this process, substrates such as glucose are broken down to pyruvate; In *Clostridium* species, pyruvate is converted to acetyl-CoA by enzymes such as pyruvate: ferredoxin oxidoreductase, and the reduced electron carriers (e.g., ferredoxin) released during this process are reduced by hydrogenase enzymes, converting protons into H_2 gas (Albuquerque et al. 2024). The attractive aspect of dark fermentation is its potential to produce hydrogen from waste organic materials under simpler bioreactor conditions compared to complex and energy-intensive industrial processes (Albuquerque et al. 2024).

Dark fermentation is one of the most widely studied biogas production methods due to its advantages such as the absence of light requirements and the ability to achieve higher H_2 efficiency rates compared to other biological methods. However, this process can also produce byproducts such as organic acids, alcohol, and CO_2; if these byproducts are not removed from the system, they can lower the pH of the environment and slow down the process (Albuquerque et al. 2024).

As an advantage, the output products of the dark fermentation process can be utilized in a second stage to integrate and increase hydrogen production. Indeed, two-stage systems combining dark fermentation and photofermentation can significantly increase total hydrogen yield by reprocessing the organic acids formed in the first stage with photosynthetic bacteria in the second stage (Chen et al. 2008). In the study, when dark fermentation was performed alone with *Clostridium pasteurianum*, a maximum of ~3.8 mol $H_{(2)}$/mol glucose was obtained, while subsequent application of photofermentation with *R. palustris* increased the total yield to 10.0 mol H_2/mol glucose. Such integrated bioprocesses not only enhance hydrogen production by fully utilizing the metabolic capacity of bacteria but also contribute to waste utilization (Chen et al. 2008).

Finally, hydrogen production can also be achieved using bacteria through a bio-electrochemical approach known as microbial electrolysis cells (MEC). In MEC systems, hydrogen gas can be produced from organic waste instead of water using an external low-voltage electric current, thanks to electrode-supported bacteria that break down organic matter (Albuquerque et al. 2024). In studies conducted in 2025, it was reported that when real wastewater was used in dual-chamber MEC systems, a daily production of approximately 0.7 L H_2/L could be sustainably achieved, with a mass-based current density of ~60 A/m^3 (Baía et al. 2025).

In summary, hydrogen-producing bacteria such as *Clostridium butyricum*, *C. pasteurianum*, and *Rhodopseudomonas palustris* offer significant potential for producing clean hydrogen as a sustainable alternative to fossil fuels. However, reducing feed substrate costs, optimizing synergy in mixed cultures, and enhancing the efficiency of hydrogenase enzymes through genetic engineering remain active research areas to improve biogas production efficiency and make the processes economically viable at industrial scales (Li et al. 2022; Albuquerque et al. 2024).

5.6.1.2 Biobutanol Production

Biobutanol is another promising bacterial biofuel product, particularly as a transportation fuel. This alcohol, with the chemical formula n-butanol ($C_4H_{10}O$), this alcohol is considered a superior alternative to first-generation biofuels like ethanol due to its benzine-like properties, compatibility with gasoline, high energy content, and minimal modifications required for existing internal combustion engines (Serrano-Echeverry et al. 2025). Additionally, biobutanol's energy density and low water affinity enhance fuel performance by providing approximately 30% higher energy content than ethanol (Kazmi et al. 2025).

The production of biobutanol is associated with a process traditionally referred to as acetone-butanol-ethanol (ABE) fermentation. During this fermentation, solventogenic *Clostridium* bacteria ferment various carbon sources to produce acidic products first, followed by solvent products (Serrano-Echeverry et al. 2025).

The most well-known example in this field is the *Clostridium acetobutylicum* ATCC 824 strain, isolated by Chaim Weizmann in the 1910s (Serrano-Echeverry et al. 2025). This bacterium has historically been used in ABE fermentation for industrial production and played a strategic role in acetone production during World War I (Serrano-Echeverry et al. 2025).

ABE fermentation is a characteristic process comprising two distinct stages: acidogenesis and solventogenesis. During the first stage, acidogenesis, bacteria proliferate rapidly and synthesize organic acids such as acetic and butyric acids from substrates like glucose. While these acids are essential for ATP production, their accumulation in the environment lowers the pH level (Serrano-Echeverry et al. 2025). Upon reaching a certain acid concentration (typically triggered by a decrease in environmental pH), cells transition to the second phase, the solventogenesis phase. During solventogenesis, bacterial growth slows down or stops, and the accumulated acidic products are re-metabolized to neutral solvents such as butanol, acetone, and ethanol. These solvents are typically produced in a molar ratio of approximately 6:3:1 (butanol:acetone:ethanol) (Serrano-Echeverry et al. 2025). Additionally, gases such as H2 and CO2 are released as natural byproducts of this fermentation process (Serrano-Echeverry et al. 2025). However, the produced butanol and other solvents begin to accumulate on the cell membrane, causing damage to the bacteria. Generally, when the butanol concentration in the culture exceeds ~13–20 g/L, bacterial growth and solvent production are significantly inhibited (Serrano-Echeverry et al. 2025).

In classical ABE fermentation, the butanol concentration in the product mixture is only approximately 2% by volume, with the remaining majority consisting of water. This makes the separation of butanol from the fermentation product mixture a highly energy-intensive and costly process. The boiling point of the butanol-water azeotrope at atmospheric pressure is 93°C, and the low butanol concentration in the mixture at the end of fermentation renders traditional separation techniques such as distillation economically unfeasible.

In this context, both biological and engineering approaches are needed to improve the efficiency of biobutanol production. Current research has focused on various methods to increase butanol yield:

- Development of new *Clostridium* strains with high solvent tolerance.
- Conducting co-culture fermentation with genetically engineered yeast or other bacteria.
- Use of continuous or fed-batch fermentation modes.
- In-situ product recovery (Serrano-Echeverry et al. 2025).

For example, substrate inhibition can be reduced, and bacteria can be made to continue producing solvents for longer periods using fed-batch culture strategies. Additionally, the removal of butanol during fermentation using integrated membrane separation systems can mitigate the toxic effects. Furthermore, processes involving the use of co-cultures of different *Clostridium* species, where one species breaks down lignocellulosic raw materials to provide sugar and the other contributes to solvent production, are being investigated (Serrano-Echeverry et al. 2025).

With all these developments, scaling up, reducing raw material costs, and optimizing the process are necessary to enable biobutanol to compete with petroleum-based fuels on cost. Nevertheless, due to its superior fuel properties, biobutanol has the potential to make significant contributions as a carbon-neutral alternative fuel in the transportation sector under suitable conditions (Serrano-Echeverry et al. 2025).

5.6.2 The Role and Potential of Bacteria in Renewable Energy Sources

Bacteria play critical roles in renewable energy ecosystems and various bioenergy systems. These microorganisms have an important place in both energy production and environmental remediation functions thanks to their biochemical conversion capabilities (Dakal et al. 2025).

Especially in anaerobic digestion processes, bacteria break down organic waste, laying the foundation for the formation of methane-containing biogas. Biogas is a flammable gas mixture primarily composed of methane (CH_4) and carbon dioxide (CO_2), produced through a four-stage decomposition process by microorganisms in an oxygen-free environment from agricultural, domestic, and industrial waste (Mignogna et al. 2023). In this process:

- Various bacterial groups are involved in the hydrolysis, acidogenesis, and acetogenesis stages,
- In the final stage, methanogenesis, archaea convert the produced acetate, hydrogen, and carbon dioxide into methane gas (Mignogna et al. 2023).

Hydrolytic and acid-producing bacteria convert complex polymeric organic materials (polysaccharides, proteins, lipids, etc.) into simple organic acids and intermediate products such as H_2 and CO_2, thereby preparing a suitable environment for methanogenic archaea. In this way, bacteria perform a circular biorefinery function that combines waste management with energy production (Mignogna et al. 2023).

Hydrolytic and acid-producing bacteria convert complex polymeric organic materials (e.g., polysaccharides, proteins, and lipids) into simple organic acids and intermediate products such as H_2 and CO_2, thereby preparing a suitable environment for methanogenic archaea. In this way, bacteria perform a circular biorefinery function that integrates waste management with energy production. For example, it has been reported that a biogas plant with a capacity of 100 tons per day processing food waste can produce enough energy to supply over a thousand households annually (Mignogna et al. 2023). Biomethane purified from biogas is a renewable energy carrier that can be fed into the grid in a manner similar to natural gas or used as vehicle fuel (Mignogna et al. 2023). In this context, bacterial communities involved in anaerobic digestion processes contribute significantly to waste conversion by reducing greenhouse gas emissions and play a major role in renewable energy production.

5.6.3 *Innovative Applications of Bacteria in the Renewable Energy Sector*

Microbial fuel cells (MFCs) are an innovative application of bacteria in the field of renewable energy. These systems are biotechnological systems that enable the electrons produced by the oxidation of organic waste by bacteria to be transferred directly to an electrical circuit (Dakal et al. 2025). The basic working principle of an MFC is as follows:

In the anode compartment of the system, there are electroactive bacteria that break down organic substrates (such as wastewater, agricultural waste, or food waste). Here, the electrons produced as a result of bacterial metabolism are transferred directly to the anode electrode or to the anode via a mediator molecule. To complete the electrical circuit, a reaction that reduces oxygen typically occurs in the cathode compartment. Thus, the electrons stored by the bacteria in the form of chemical energy are converted into electrical energy. Throughout the process, protons are also transported from the anode to the cathode via a proton exchange membrane, enabling the system to operate as a closed electrical circuit (Dakal et al. 2025).

MFC technology is a dual-purpose method (both purification and energy production) as it removes organic contamination from wastewater while also generating electricity. One of the innovative energy applications of bacteria is microbial fuel cells (MFCs). These systems are biotechnological systems that directly generate electricity by utilizing electrons from organic waste. In MFCs, bacteria break down waste and produce electricity (Dakal et al. 2025). The main advantages of MFCs are as follows:

- Combines waste treatment with electricity production (dual benefit). For example, it can provide energy in wastewater treatment plants.
- It can utilize various and inexpensive wastes such as domestic wastewater and food scraps, thereby providing both energy production and waste management.
- It is considered an important bridge technology for transitioning from fossil fuels to clean energy, as it generates power from renewable biomass.
- It can serve as a low-cost, continuous energy source for off-grid areas and small devices like sensors (Dakal et al. 2025).

In addition to its advantages, MFCs have disadvantages, including low power density, scalability issues, and membrane fouling (biofouling). To overcome these issues, nanotechnological electrodes, genetically improved bacteria, and new design approaches are being researched.

Another system similar to MFCs is the microbial electrolysis cell (MEC) method. MEC systems produce hydrogen gas from organic waste using a small electric current. This is a renewable energy storage method that requires less energy compared to water electrolysis.

In general, the concept of "microbial energy" demonstrates that bacteria play a versatile and effective role in producing both direct energy carriers (electricity and hydrogen) and indirect energy carriers (biofuels such as biogas, biomethane, and biobutanol) (Dakal et al. 2025).

Microbial fuel cells (MFCs) have the potential to serve as low-cost power sources in off-grid and rural areas, while also providing continuous energy for sensor systems and small-scale devices.

As a derivative of the MFC concept, microbial electrolysis cells (MECs) also exist. MECs operate in a similar configuration to MFCs but require an external small voltage applied to the system, enabling the production of hydrogen gas from electrons and protons generated at the bacterial anode. Thanks to MEC, hydrogen production from organic waste is possible with lower energy input compared to classical water electrolysis, thereby offering a biological alternative to renewable energy storage methods (Albuquerque et al. 2024).

Recent studies have demonstrated that the concept of "microbial energy" can be applied in a versatile manner, with bacteria playing an active role in both direct energy production (electricity, hydrogen) and the production of indirect energy carriers (biofuels, biogas, biomethane) (Albuquerque et al. 2024; Dakal et al. 2025).

The potential of bacteria in the field of renewable energy is expanding further with promising new approaches, as mentioned. One such approach involves systems that aim to achieve both waste treatment and biofuel production through the symbiotic use of photosynthetic bacteria and microalgae. In these microalgae-bacteria co-cultures, the process proceeds as follows:

- Bacteria provide mineral nutrients to support algal growth.
- Microalgae, in turn, produce oxygen through photosynthesis and supply it to the bacteria (Mignogna et al. 2023).

Recent studies have demonstrated significant advancements in achieving carbon neutrality in integrated energy systems through microbial electrosynthesis and in scaling up the technology (Singh et al. 2022). In particular, CO_2 conversion efficiencies of up to 40% have been reported with advanced electrode materials and optimized process control (Horrillo-Quintero et al. 2025).

These technologies are in the research phase and have the potential to convert electricity from intermittent energy sources (such as wind and solar) into chemical fuels for storage. They could offer significant advantages for decarbonizing energy systems. The recognition of hydrogen as the "fuel of the future" and the importance of other biofuels, like biobutanol, highlight the significance of these microbial approaches in the renewable energy sector.

The role of bacteria in the renewable energy sector is significant. With their biochemical conversion capabilities, they contribute to both energy production and environmental improvement. Key areas where they play a major role: Bacteria have significant potential in the field of renewable energy. Their biochemical conversion capabilities contribute to both energy production and environmental improvement. Their primary roles can be summarized as biofuel production, biogas production, microbial fuel cells (MFC), electrolysis cells (MEC), and microbial electrosynthesis (Dakal et al. 2025). The efficiency and scalability challenges of these biological systems can be addressed through genetic and process engineering improvements supported by nanotechnology. In conclusion, bacterial processes hold groundbreaking potential for clean and renewable energy production. In the future, the role of bacteria in energy production will become strategic and play a key role in sustainable energy conversion.

5.7 Future Perspectives

In recent years, significant progress has been made in synthetic biology, a field that has found widespread application in metabolic engineering. These studies involve reprogramming microorganisms to produce desired substances. However, to efficiently reconfigure metabolic networks, it is necessary to ensure the dynamic distribution of cellular resources, create entirely new metabolic pathways, and control these pathways optimally (Lee et al. 2018). Standard component and circuit design principles developed in synthetic biology are being applied to create flexible and innovative metabolic pathways and achieve higher product yields by adjusting intracellular flows in real-time (Lee et al. 2018; García-Granados et al. 2019). The emergence of synthetic biology, which utilizes DNA synthesis and assembly techniques, has enabled the implementation of "extremely efficient" cycles known as "design-build-test-learn (DBTL)" cycles. A comprehensive, multi-step metabolic pathway design can be rapidly transferred into a production cell using synthetic gene sequences and traditional assembly methods (e.g., Gibson assembly) (Voigt 2020). As reported by Voigt, advanced strategies such as computer-aided enzyme engineering, the development of part libraries to balance gene expressions, and the integration of multi-gene constructs in a single step have made the functional transfer of enzymes from source organisms to producer organisms a standard practice. Approaches inspired by synthetic biology enable the synthesis of compounds that are naturally produced in low quantities or not at all in microorganisms. This development has led to an expansion of the product range in industrial biotechnology (Voigt 2020). In summary, the standardization, modularity, and modeling capabilities provided by synthetic biology have given metabolic engineering a more rational design infrastructure (García-Granados et al. 2019).

The future potential of bacterial platforms: Traditionally, model bacteria such as *Escherichia coli* and *Bacillus subtilis* have been used as chassis organisms in industrial biotechnology. However, the contemporary "design once, use many times" paradigm has led to the development of various bacterial chassis types tailored to specific production targets (Calero and Nikel 2019). For example, non-traditional microorganisms such as *Pseudomonas putida*, which exhibit high solvent tolerance, are emerging as promising platforms for the production or degradation of toxic chemicals (Calero and Nikel 2019). The concept of a bacterial chassis refers to the design of a microorganism as a standard biological framework into which desired genetic components can be "plugged" (Calero and Nikel 2019). In the future, the use of specialized chassis that are resistant to different environments, exhibit rapid growth, or provide dietary diversity will become more widespread for metabolic engineering applications. Additionally, removing unnecessary genetic elements from the chassis, such as those found in minimal genome bacteria, could result in more predictable platforms for engineering. Indeed, the creation of a synthetic minimal bacterial genome containing 473 genes (Hutchison et al. 2016) and the synthesis of a completely recoded *E. coli* genome (Fredens et al. 2019) demonstrate that fully customized microbial platforms may be feasible in the long term. The design of synthetic cell chassis as ideal bioreactors requires the elimination of unwanted side pathways and the redirection of all resources toward the desired product. The potential of bacterial platforms extends beyond chemical production, opening the door to revolutionary applications in agriculture and medicine. For example, innovative approaches such as reducing chemical fertilizer use through the application of engineered bacteria applied to plant roots or treating diseases with probiotic bacteria designed for the human microbiome are already being developed (Voigt 2020). In the coming years, "living bacterial systems" specifically programmed to address a particular problem will be at the heart of sustainable biotechnology applications.

The impact of genome editing technologies: Advances in genome editing over the past decade have acted as a catalyst, accelerating progress in synthetic biology and metabolic engineering. The emergence of the CRISPR/Cas system has brought about a paradigm shift in genome manipulation, enabling rapid and precise changes to be made with high accuracy in a wide range of organisms, including bacteria. CRISPR-based tools have become the primary genome engineering method in many organisms, including those of industrial importance (Nishida and Kondo 2021). This technology facilitates interventions at various scales, ranging from the inactivation of a single gene to the addition of specific point mutations and even the reprogramming of entire metabolic pathways (Nishida and Kondo 2021). In addition to classical CRISPR/Cas9 nuclease cleavage, derivative approaches developed in recent years offer improved regulation by reducing toxicity in microorganisms (Nishida and Kondo 2021). For example, base editing methodologies can achieve targeted changes in the base sequence without compromising the integrity of the DNA double helix, while CRISPRi/a systems have the capacity to temporarily suppress or enhance gene expression without causing changes in the genome sequence (Nishida and Kondo 2021). Additionally, the emergence of guide RNA libraries capable of targeting multiple genes has facilitated the simultaneous modification of multiple genes or pathways, thereby accelerating the screening of new metabolic designs (Nishida and Kondo 2021). As a result, the cell's metabolic network can be optimized in a short time using simultaneous and combinatorial regulation techniques. These advancements in genome editing technologies have enabled the genetic engineering of a wide range of bacterial species that were previously considered difficult to engineer. For example, some soil or marine bacteria that were previously difficult to genetically manipulate using traditional methods can now be converted into new production platforms using CRISPR intervention. The speed and flexibility that CRISPR has brought to the scientific world earned the researchers who pioneered this technology the Nobel Prize in 2020, highlighting the critical role of genome editing in the future of biotechnology (Voigt 2020).

In summary, advancements in synthetic biology and metabolic engineering are poised to create a new wave of innovation in biotechnology in the near future. The emergence of standardized bacterial chassis and advanced genome manipulation tools has enabled the programming of microorganisms as designer cells, allowing them to perform a range of desired functions. The developments mentioned above have led to numerous revolutionary applications across a wide range of fields, including sustainable chemical production, renewable energy, environmental sustainability, and personalized medicine. This suggests a future where biological systems are viewed as engineering tools, resulting in more predictable, controllable, and efficient biological production platforms.

The versatile use of bacteria in biotechnology is evident across a wide range of applications, from environmental to industrial and medical uses. Today, microorganisms (especially bacteria) play critical roles in nearly every subdiscipline of biotechnology. For example, while fermentation is the oldest known microbial technology, there are countless new application areas today, including microbiome-based ventures (Singh and Timmis 2017). Bacteria have become versatile and indispensable tools in biotechnology, offering unique contributions across various sectors, such as waste management, industrial production, and healthcare (Singh and Timmis 2017).

5.8 Conclusion

5.8.1 Environmental Contributions

Bacteria hold an important place in the field of environmental biotechnology. In bioremediation, the use of plants and microorganisms to remediate toxic wastes and pollutants has become a widespread approach. The rapid growth and ease of genetic modification exhibited by bacteria make them highly effective bioremediation agents for the removal of pollutants (Ayilara and Babalola 2023). As demonstrated in the research conducted by Ayilara and Babalola, various bacterial groups have been successfully utilized in the transformation of a range of environmental pollutants, including heavy metals and petroleum derivatives. Bacteria also provide important benefits in ensuring environmental sustainability in agriculture. Biofertilizers containing rhizobacteria (PGPR) facilitate nutrient uptake by plants when applied to the soil, promote root development and growth hormones, and encourage plant growth, thereby reducing the need for chemical fertilizers (Aloo et al. 2022).

5.8.2 Industrial Applications

Bacteria represent a fundamental component of industrial biotechnology. The use of bacterial biocatalysts and processes has garnered significant interest across various industrial sectors, from food technology to the energy sector. In the field of industrial microbiology, model bacteria such as *Escherichia coli* and *Bacillus* species are used in fermentation processes to facilitate the synthesis of valuable metabolites, including enzymes, organic acids, amino acids, antibiotics, and vitamins. A significant portion of enzyme sources is directly obtained from bacteria, and these enzymes play a critical role in various sectors, such as detergent formulations, pulp processing, leather tanning, and textile processing (Sharma et al. 2025). For example, bacterial lipase enzymes are widely used in food processing, detergent production, and biofuel production. Their high efficiency and ease of production facilitate large-scale applications (Sharma et al. 2025). Similarly, lactic acid bacteria have been used for thousands of years in the fermentation of food products (yogurt, cheese, pickles, etc.) and constitute an important component in the production of traditional foods. The use of industrial bacterial processes extends beyond product production; these processes also serve the purpose of converting environmental waste. The production of renewable bioplastics, such as polihydroxyalkanoates, which offer a sustainable alternative to fossil-based plastics, is a notable example (Santos-Beneit 2024). This diversity demonstrates the versatile potential of bacteria in industrial biotechnology and continuously expands the possibilities for developing new niche products and processes (Singh and Timmis 2017).

5.8.3 Medical Contributions

The impact of bacteria in the fields of medicine and pharmacy is evident in many areas of modern healthcare. Natural products and metabolites derived from bacteria form the basis of a significant number of pharmaceutical drugs. Antibiotics are the most important example of this phenomenon. Secondary metabolites produced by soil bacteria belonging to the *Streptomyces* genus are the source of numerous critical antibiotics currently used in clinical settings (e.g., streptomycin, tetracycline, erythromycin). This discovery led to a paradigm shift in the approach to combating infectious diseases throughout the twentieth century (Santos-Beneit 2024). Additionally, the emergence of recombinant DNA technologies has enabled bacteria to function as direct factories for the production of therapeutic proteins. Since the 1980s, the production of human insulin in *E. coli* bacteria through genetic engineering has accelerated a paradigm shift in diabetes management, replacing animal insulin and providing a pure and immunologically compatible therapeutic alternative (Wackett 2022). Additionally, a significant number of therapeutic proteins, including growth hormones, interferons, and vaccine antigens, are produced using bacterial fermentation in a cost-effective and efficient manner.

The ability of bacteria to trigger the immune system has made them a potential tool in the development of many new treatment methods in cancer therapy, such as tumor-specific drug delivery (known as bacteriophages) (Santos-Beneit 2024). Probiotic bacteria may offer therapeutic benefits against inflammatory bowel diseases due to their modulatory effects on the human microbiome. Bacteria contribute significantly to medical biotechnology through genetic engineering. The CRISPR-Cas system has revolutionized genetic disease treatment and cell engineering with its ability to edit genomes (Santos-Beneit 2024). The heat-resistant Taq DNA polymerase enzyme obtained from the Thermus aquaticus bacterium has enabled the development of polymerase chain reaction (PCR) technology, which forms the foundation of molecular biology techniques (Santos-Beneit 2024).

References

Adams MR (2017) The birth of modern industrial microbiology: the acetone–butanol fermentation. Int J History Eng Technol 87(1):81–95. https://doi.org/10.1080/17581206.2017.1329970

Adrio JL, Demain AL (2014) Microbial enzymes: tools for biotechnological processes. Biomolecules 4(1):117–139. https://doi.org/10.3390/biom4010117

Agathos SN (2011) 6.01—Introduction. Comprehensive biotechnology (second edition). Academic. pp. 1–2. https://doi.org/10.1016/B978-0-08-088504-9.00363-9

Akoijam N, Kalita D, Joshi SR (2022) Bacteria and their industrial importance. In: Verma P (eds) Industrial microbiology and biotechnology. Springer, Singapore. https://doi.org/10.1007/978-981-16-5214-1_2

Al Doghaither H, Gull M (2019) Plasmids as genetic tools and their applications in ecology and evolution. Plasmid. IntechOpen. Available at: https://doi.org/10.5772/intechopen.85705

Albuquerque MM, Sartor GdB, Martinez-Burgos WJ, Scapini T, Edwiges T, Soccol CR, Medeiros ABP (2024) Biohydrogen produced via dark fermentation: a review. Methane 3(3):500–532. https://doi.org/10.3390/methane3030029

Aloo BN, Tripathi V, Makumba BA, Mbega ER (2022) Plant growth-promoting rhizobacterial biofertilizers for crop production: the past, present, and future. Front Plant Sci 13:1002448. https://doi.org/10.3389/fpls.2022.1002448

Alyas J, Rafiq A, Amir H, Khan SU, Sultana T, Ali A, Hameed A, Ahmad I, Kazmi A, Sajid T, Ahmad A (2021) Human insulin: history, recent advances, and expression systems for mass production. Biomed Res Ther 8(9):4540–4561. https://doi.org/10.15419/bmrat.v8i9.692

Ayilara MS, Babalola OO (2023) Bioremediation of environmental wastes: the role of microorganisms. Front Agron 5:1183691. https://doi.org/10.3389/fagro.2023.1183691

Bahuguna V, Matura R, Farswan AS, Naqvi SS, Sharma N, Chaudhary M (2025) Rhizobium as a potential biofertilizer and its quality control analysis for sustainable agriculture. J Appl Biol Biotechnol 13(3):97–105. https://doi.org/10.7324/JABB.2025.197428

Baía A, Arroyo-Escoto AI, Ramos N, Abdelkarim B, Pereira M, Fernandes MC et al (2025) Hydrogen production from winery wastewater through a dual-chamber microbial electrolysis cell. Energies 18(12):3043. https://doi.org/10.3390/en18123043

Bala S, Garg D, Thirumalesh BV, Sharma M, Sridhar K, Inbaraj BS, Tripathi M (2022) Recent strategies for bioremediation of emerging pollutants: a review for a green and sustainable environment. Toxics 10(8):484. https://doi.org/10.3390/toxics10080484

Brim H, McFarlan SC, Fredrickson JK, Minton KW, Zhai M, Wackett LP, Daly MJ (2000) Engineering *Deinococcus radiodurans* for metal remediation in radioactive mixed waste environments. Nat Biotechnol 18(1):85–90. https://doi.org/10.1038/71986

CaleroP,NikelPI(2019)MicrobBiotechnol12(1):98–124.https://doi.org/10.1111/1751-7915.13292

Chaudhary R, Nawaz A, Fouillaud M, Dufossé L, ul Haq I, Mukhtar H (2024) Microbial cell factories: biodiversity, pathway construction, robustness, and industrial applicability. Microbiology Res 15(1):247–272. https://doi.org/10.3390/microbiolres15010018

Chen C-Y, Yang M-H, Yeh K-L, Liu C-H, Chang J-S (2008) Biohydrogen production using sequential two-stage dark and photo fermentation processes. Int J Hydrog Energy 33(18):4755–4762. https://doi.org/10.1016/j.ijhydene.2008.06.055

Chen W, Li L, Ye C, Zhao Z, Huang K, Zou D, Wei X (2022) Efficient production of extracellular alkaline protease in *Bacillus amyloliquefaciens* by host strain construction, LWT 163. https://doi.org/10.1016/j.lwt.2022.113620

Dakal TC, Singh N, Kaur A, Dhillon PK, Bhatankar J, Meena R, Kumari K (2025) New horizons in microbial fuel cell technology: applications, challenges, and prospects. Biotechnol Biofuels Bioprod 18, Article 79. https://doi.org/10.1186/s13068-025-02649-y
Danilova I, Sharipova M (2020) The practical potential of bacilli and their enzymes for industrial production. Front Microbiol 4(11):1782. https://doi.org/10.3389/fmicb.2020.01782
Daraban GM, Hlihor RM, Suteu D (2023) Pesticides vs. biopesticides: from pest management to toxicity and impacts on the environment and human health. Toxics 11(12):983. https://doi.org/10.3390/toxics11120983.
de Souza AG, Silva CM, Oliveira RS, Pereira LM (2024) Bioplastic production from agro-industrial waste: a sustainable alternative to petroleum-based plastics. J Polym Environ 32(1):45–60
de Souza PM, de Oliveira Magalhães P (2010) Application of microbial α-amylase in industry—a review. Braz J Microbiol 41(4):850–861. https://doi.org/10.1590/S1517-83822010000400004
Degering C, Eggert T, Puls M, Bongaerts J, Evers S, Maurer KH, Jaeger KE (2010) Optimization of protease secretion in *Bacillus subtilis* and *Bacillus licheniformis* by screening of homologous and heterologous signal peptides. Appl Env Microbiol 76(19):6370– 6376. https://doi.org/10.1128/AEM.01146-10
Demain AL, Sanchez S (2009) Microbial drug discovery: 80 years of progress. J Antibiot 62(1):5–16. https://doi.org/10.1038/ja.2008.16
Demain A, Vandamme E, Collins J, Buchholz K (2016) History of industrial biotechnology: microorganisms. Wiley-VCH, pp 1–84. https://doi.org/10.1002/9783527807796.ch1
Fleming A (1929) On the antibacterial action of cultures of a penicillium, with special reference to their use in the isolation of *B. influenzae*. Br J Exp Pathol 10(3):226
Fredens J, Wang K, de la Torre D, Funke LFH, Robertson WE, Christova Y, Chin JW (2019) Total synthesis of *Escherichia coli* with a recoded genome. Nature 569(7757):514–518. https://doi.org/10.1038/s41586-019-1192-5
Forrest RD (1982) Development of wound therapy from dark ages to the present. J R Soc Med 75:268–273
García-Granados R, Lerma-Escalera JA, Morones-Ramírez JR (2019) Metabolic engineering and synthetic biology: synergies, future, and challenges. Front Bioeng Biotechnol 7:36. https://doi.org/10.3389/fbioe.2019.00036
Gauri SS, Mandal SM, Pati BR (2012) Impact of Azotobacter exopolysaccharides on sustainable agriculture. Appl Microbiol Biotechnol 95:331–338. https://doi.org/10.1007/s00253-012-4159-0
Gill M (2014) Bioplastic: a better alternative to plastics. Int J Res Appl Nat Soc Sci 2(8):115–120
Hlihor RM, Gavrilescu M, Tavares T, Favier L, Olivieri G (2017) Bioremediation: an overview on current practices, advances, and new perspectives in environmental pollution treatment. Biomed Res Int 2017:6327610. https://doi.org/10.1155/2017/6327610
Horrillo-Quintero P, De la Cruz-Loredo I, Garcia-Trivino P, Ugalde-Loo CE, Fernández-Ramírez LM (2025) A real-time combined dynamic control framework for multi-energy microgrids coupling hydrogen, electricity, heating, and cooling systems. Int J Hydrog Energy 106:454–470. https://doi.org/10.1016/j.ijhydene.2025.02.005
Hu F, Wang P, Li Y, Ling J, Ruan Y, Yu J, Zhang L (2023) Bioremediation of environmental organic pollutants by *Pseudomonas aeruginosa*: mechanisms, methods, and challenges. Env Res 239:117211. https://doi.org/10.1016/j.envres.2023.117211
Hutchison CAIII, Chuang RY, Noskov VN, Venter JC (2016) Design and synthesis of a minimal bacterial genome. Science 351(6280):aad6253. https://doi.org/10.1126/science.aad6253
Ifiok Mfon E (2024) Microbial biotechnology: application of bacteria in various industrial processes and environment remediation. Int J Dev Sustainy Environ Manag 4(1):16–24. https://doi.org/10.5281/zenodo.10511013
Katz L, Baltz RH (2016) Natural product discovery:past, present, and future. J Ind Microbiology Biotechnol 43(2–3):155–176. https://doi.org/10.1007/s10295-015-1723-5

Kazmi A, Sultana T, Ali A, Nijabat A, Li G, Hou H (2025) Innovations in bioethanol production: a comprehensive review of feedstock generations and technology advances. Energy Strat Rev 57:101634. https://doi.org/10.1016/j.esr.2024.101634

Kourmentza C, Plácido J, Venetsaneas N, Burniol-Figols A, Varrone C, Gavala HN, Reis MAM (2017) Recent advances and challenges towards sustainable polyhydroxyalkanoate (PHA) production. Bioengineering 4(2):55. https://doi.org/10.3390/bioengineering4020055

Kumar P, Kamle M, Borah R, Mahato DK, Sharma B (2021) *Bacillus thuringiensis* as a microbial biopesticide: uses and application for sustainable agriculture. Egypt J Biol Pest Control 31:95. https://doi.org/10.1186/s41938-021-00440-3

Kunamneni A, Permaul K, Singh S (2005) Amylase production in solid state fermentation by the thermophilic fungus *Thermomyces lanuginosus*. J Biosci Bioeng 100:168–171. https://doi.org/10.1263/jbb.100.168

Lakshmi PK, Abirami S, Meenakshi S, Usha C, Sakthieaswari P, Aarthy K, Gayathri SS, Baby S (2022) Chapter 39—Application of *Deinococcus radiodurans* for bioremediation of radioactive wastes. In: Malik JA (ed) Microbes and sustainable environment. Elsevier. pp. 717–732. Elsevier. https://doi.org/10.1016/B978-0-323-90452-0.00037-2

Landecker H (2016) Antibiotic resistance and the biology of history. Body Soc 22(4):19–52. https://doi.org/10.1177/1357034X14561341

Lax E (2004) The mold in Dr. Florey's coat: the story of the penicillin miracle. Henry Holt and Company, New York

Lee H-M, Vo PNL, Na D (2018) Advancement of metabolic engineering assisted by synthetic biology. Catalysts 8(12):619. https://doi.org/10.3390/catal8120619

Li M, Ning P, Sun Y, Luo J, Yang J (2022) Characteristics and application of *Rhodopseudomonas palustris* as a microbial cell factory. Front Bioeng Biotechnol 10:897003. https://doi.org/10.3389/fbioe.2022.897003

Li T, Yang Y, Qi H (2023) CRISPR/Cas9 therapeutics: progress and prospects. Signal Transduct Target Ther 8:36. https://doi.org/10.1038/s41392-023-01309-7

Liu X, Kokare C (2017) Chapter 11—Microbial enzymes of use in industry. production, biocatalysis and industrial applications. In: Biotechnology of microbial enzymes. pp. 267–298. https://doi.org/10.1016/B978-0-12-803725-6.00011-X.

Matkawala F, Nighojkar S, Kumar A, Nighojkar A (2021) Microbial alkaline serine proteases: production, properties and applications. World J Microbiol Biotechnol 37(4):63. https://doi.org/10.1007/s11274-021-03036-z

Matyjaszczyk E (2018) "Biorationals" in integrated pest management strategies. J Plant Dis Prot 125:523–527. https://doi.org/10.1007/s41348-018-0180-6

Menéndez C, García-Fernández S, Comino N, Esteban R, González M (2015) Cellulase production by *Trichoderma reesei* using lignocellulosic wastes as substrates. Bioresour Technol 192:150–156. https://doi.org/10.1016/j.biortech.2015.05.062

Michael CA, Dominey-Howes D, Labbate M (2014) The antimicrobial resistance crisis: causes, consequences, and management. Front Public Health 2:145. https://doi.org/10.3389/fpubh.2014.00145

Mignogna D, Ceci P, Cafaro C, Corazzi G, Avino P (2023) Production of biogas and biomethane as renewable energy sources: a review. Appl Sci 13(18):10219. https://doi.org/10.3390/app131810219

Mishra J, Dutta V, Arora NK (2020) Biopesticides in India: technology and sustainability linkages. Biotech 10:1–12. https://doi.org/10.1007/s13205-020-02192-7

Mosa WFAE, Sas-Paszt L, Frąc M, Trzciński P (2016) Microbial products and biofertilizers in improving growth and productivity of apple—a review. Pol J Microbiology 26(3):243–251. https://doi.org/10.5604/17331331.1215599

Nida K, Siddiqui ZS, Salman ZA, Aftab A, Abid R, Abideen Z, Siddiqui MH (2024) Scrutinize the integrated role of *Azotobacter vinelandii* in nitrogen assimilation, photosystem II functionality and aerenchyma formation of Zea mays under moisture stress environment. Plant Stress 11:100378. https://doi.org/10.1016/j.stress.2024.100378

Nishida K, Kondo A (2021) CRISPR-derived genome editing technologies for metabolic engineering. Metab Eng 63:141–147. https://doi.org/10.1016/j.ymben.2020.12.002
Nosheen S, Ajmal I, Song Y (2021) Microbes as biofertilizers, a potential approach for sustainable crop production. Sustainability 13(4):1868. https://doi.org/10.3390/su13041868
Olanrewaju OS, Glick BR, Babalola OO (2017) Mechanisms of action of plant growth promoting bacteria. World J Microbiol Biotechnol 33(11):197. https://doi.org/10.1007/s11274-017-2364-9
Pereira F, Pereira A, Menezes LD, Sawaikar R (2024) Rhizobium as a biofertilizer for non-leguminous plants. Discov Food 4:85. https://doi.org/10.1007/s44187-024-00167-8
Pindi PK (2012) Liquid microbial consortium-a potential tool for sustainable soil health. J Biofertil Biopestici 3(4):124. https://doi.org/10.4172/2155-6202.1000124 pp 2–9
Prakash O, Jaiswal N (2009) alpha-Amylase: an ideal representative of thermostable enzymes. Appl Biochem Biotechnol. https://doi.org/10.1007/s12010-009-8735-4
Rafeeq H, Afsheen N, Rafique S, Arshad A, Intisar M, Hussain A, Bilal M, Iqbal HMN (2023) Genetically engineered microorganisms for environmental remediation. Chemosphere 310. https://doi.org/10.1016/j.chemosphere.2022.136751
Raveendran S, Parameswaran B, Ummalyma SB, Abraham A, Mathew AK, Madhavan A, Pandey A (2018) Applications of microbial enzymes in the food industry. Food Technol Biotechnol 56(1):16–30. https://doi.org/10.17113/ftb.56.01.18.5511
Sahm H (1984) Anaerobic treatment of industrial wastewater, chemistry for the future. pp 81–90. Pergamon. https://doi.org/10.1016/B978-0-08-029249-6.50013-1
Santos-Beneit F (2024) What is the role of microbial biotechnology and genetic engineering in medicine? Microbiol Open 13(2):e1406. https://doi.org/10.1002/mbo3.1406
Serrano-Echeverry VA, Guerrero-Fajardo CA, Castro-Tibabisco KT (2025) A review of biobutanol: eco-friendly fuel of the future—history, current advances, and trends. Fuels 6(3):55. https://doi.org/10.3390/fuels6030055
Sharma M, Agarwal S, Agarwal Malik R, Kumar G, Pal DB, Mandal M et al (2023) Recent advances in microbial engineering approaches for wastewater treatment: a review. Bioengineered 14(1):2184518
Sharma N, Ahlawat YK, Stalin N, Mehmood S, Morya S, Malik A, Malathi H, Nellore J, Bhanot D (2025) Microbial enzymes in industrial biotechnology: sources, production, and significant applications of lipases. J Ind Microbiol Biotechnol 52:kuaf010. https://doi.org/10.1093/jimb/kuaf010
Singh BK, Timmis JK (2017) The contribution of microbial biotechnology to economic growth and employment creation. Microb Biotechnol 10(5):1137–1144. https://doi.org/10.1111/1751-7915.12845
Singh S, Srivastava P, Deep S, Priya Shukla A, Rizvi H, Sinha R (2022) Fermentation strategies for organic acid production. In: Verma P (ed) Industrial microbiology and biotechnology. Springer, Singapore. https://doi.org/10.1007/978-981-16-5214-1_14
Sokra I, Meta H, Linan H, Sarun H (2025) Different microorganisms used in fermentation for biofertilizer: a review 1:41–65. https://doi.org/10.5281/zenodo.15655617
Song Y-J, Zhao N-L, Dai D-R, Bao R (2025) Prospects of Pseudomonas in microbial fuel, bioremediation, and sustainability. ChemSusChem 18(2):e202401324. https://doi.org/10.1002/cssc.202401324
Sterner M, Hofrichter A, Meisinger A, Bauer F, Pinkwart K, Maletzko A, Cremers C (2024) Import options for green hydrogen and derivatives: an overview of efficiencies and technology readiness levels. Int J Hydrog Energy 90:1112–1127. https://doi.org/10.1016/j.ijhydene.2024.10.045
Vachon V, Laprade R, Schwartz J-L (2012) Current models of the mode of action of *Bacillus thuringiensis* insecticidal crystal proteins: a critical review. J Invertebr Pathol 111(1):1–12. https://doi.org/10.1016/j.jip.2012.05.001
Ventola CL (2015) The antibiotic resistance crisis: part 1: causes and threats. P T 40(4):277–283. https://www.ncbi.nlm.nih.gov/pmc/articles/PMC4378521/
Verma AS, Agrahari S, Rastogi S, Singh A (2011) Biotechnology in the realm of history. J Pharm Bioallied Sci 3(3):321–323. https://doi.org/10.4103/0975-7406.84430

Vítězová M, Kohoutová A, Vítěz T, Hanišáková N, Kushkevych I (2020) Methanogenic microorganisms in industrial wastewater anaerobic treatment. Processes 8(12):1546. https://doi.org/10.3390/pr8121546

Voigt CA (2020) Synthetic biology 2020–2030: six commercially-available products that are changing our world. Nat Commun 11:6379. https://doi.org/10.1038/s41467-020-20122-2

Wackett LP (2022) The future of microbial biotechnology. Microb Biotechnol 15(1):79–80. https://doi.org/10.1111/1751-7915.13920

Waksman SA (1953) My life with the microbes. Simon and Schuster, New York

Wala'A A, Neihaya Z, Somaya O (2017) Production of bioplastic by bacteria isolated from local soil and organic wastes. Curr Res Microbio Biotechnol 5:1012–1017

Wani SA, Chand S, Ali T (2013) Potential use of *Azotobacter chroococcum* in crop production: an overview. Curr Agriculture Res J 1(1):35–38. https://doi.org/10.12944/CARJ.1.1.04

Wu C, Tian Y, Wang L, Yang M (2021) Advances in lactic acid production by lactic acid bacteria: prospects and challenges. Bioresour Technol 329:124865. https://doi.org/10.1016/j.biortech.2021.124865

Yadav P, Chauhan AK, Singh RB, Khan S, Halabi G (2022) Chapter 22—Organic acids: microbial sources, production, and applications. In: Functional foods and nutraceuticals in metabolic and non-communicable diseases. pp 325–337. https://doi.org/10.1016/B978-0-12-819815-5.00053-7

Yang P, Condrich A, Lu L, Scranton S, Hebner C, Sheykhhasan M, Ali MA (2024) Genetic engineering in bacteria, fungi, and oomycetes, taking advantage of CRISPR. DNA 4(4):427–454. https://doi.org/10.3390/dna4040030

Zhao G, Zhu X, Zheng G, Meng G, Dong Z, Baek JH, Jeon CO, Yao Y, Xuan YH, Zhang J, Jia B (2024) Development of biofertilizers for sustainable agriculture over four decades (1980–2022). Geogr Sustainability 5(1):19–28. https://doi.org/10.1016/j.geosus.2023.09.006

Chapter 6
Biotechnological Applications of Archaea

Nalan Tavşanlı, Ahmet Arıhan Erözden, Muzaffer Arıkan and Mahmut Çalışkan

Abstract Archaea are the "silent and unseen majority of life" with an extraordinary capacity to live in some of the most unfavorable habitats in nature, and continue to exist in regions with extreme features that are inhospitable to most other living organisms. They are prokaryotic and frequently found in environments with high pH, salinity, and temperatures. Asgard, Euryarchaeota, TACK, and DPANN are the four main superphyla that make up the Archaea, one of the three domains of life on Earth. Extremophiles, the first Archaea found, are classified into four major physiological types. These are halophiles, thermophiles, alkaliphiles, and acidophiles. The Archaea domain is the subject of extensive research, particularly for potential biotechnological applications. Archaea play a crucial role in environmental sustainability and human health, particularly through their enzymes, metabolites, and biochemical processes. These are some core characteristics of Archaea enabling biotechnological usage. This chapter aims to provide an overview of the various roles that Archaea have in biotechnological usage, both current and potential. Their classification is based on the industry to which the application is targeted, which includes industrial enzyme and catalysis, energy production, and medical and pharmaceutical applications.

Keywords Archaea · Extremophiles · Methanogens · Biotechnology · Applications

---

N. Tavşanlı · A. A. Erözden
Division of Biotechnology, Biology Department, Faculty of Science, Istanbul University, Vezneciler, Istanbul, Türkiye

Program of Biotechnology, Biology Department, Institute of Graduate Studies in Sciences, Istanbul University, Vezneciler, Istanbul, Türkiye

M. Arıkan · M. Çalışkan (✉)
Division of Biotechnology, Biology Department, Faculty of Science, Istanbul University, Vezneciler, Istanbul, Türkiye

M. Çalışkan
e-mail: mahmut.caliskan@istanbul.edu.tr

M. Çalışkan and M. Arıkan (eds.), *Microbial World: Bacteria and Archaea*, Progress in Molecular and Subcellular Biology 62, https://doi.org/10.1007/978-3-032-11438-9_6

6.1 Overview

Archaea are one of the three major domains of life, along with Bacteria and Eukarya. Although they have a similar appearance to bacteria, they have a closer relationship with eukaryotes than bacteria. They have distinctive traits, which are unique metabolic, biochemical, and genetic traits. In the late 1970s, Carl Woese and his colleagues employed the small subunit rRNA as a molecular marker, revealing Archaea's individuality (De Lise et al. 2023). Despite their morphological appearance, they found that these microorganisms were not bacteria but rather belonged to a third domain of life called Archaea, which is completely different from bacteria and eukaryotes (Grivard et al. 2022). This groundbreaking finding transformed our knowledge of the living world, opened the door for the investigation and appreciation of these extreme microorganisms, and precisely reconstructed the tree of life's evolutionary history (Woose et al. 1990) (Fig. 6.1).

Archaea were once believed to exclusively live in extreme environments, such as those with high temperatures, high salinity, extreme pH levels, or high pressure, which are normally unsuitable for most other living forms. However, they have now been discovered in a wide variety of biological niches, including soil, oceans, and the human gut. Since they make up a significant portion of the human microbiota found in the gastrointestinal system and on the skin, their interactions with the human body are still limitedly understood, but they may have physiological and pathophysiological significance (Grivard et al. 2022). On the other hand, their ability to grow under harsh conditions and success in a wide range of environments is due to change their metabolism and adapt to different stressors. They commonly show polyextremophilic features. Archaea's resistance comes from adaptations in their cell membranes,

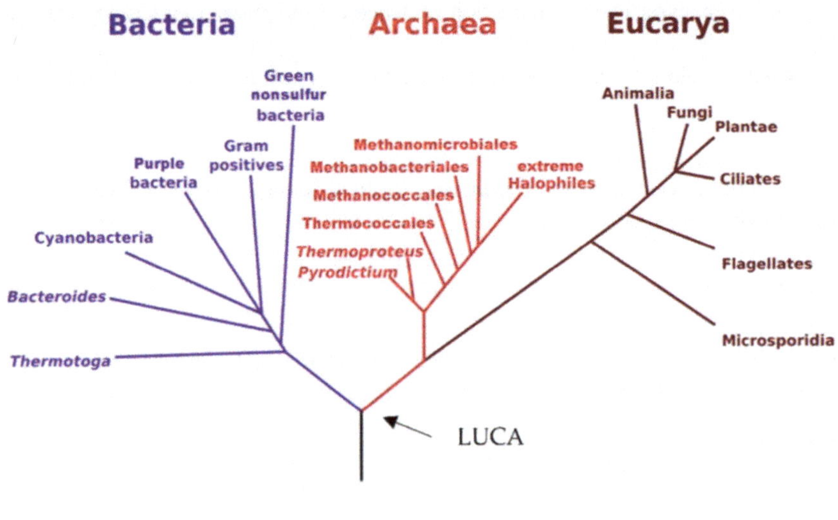

Fig 6.1 The phylogenetic tree of life. Retrieved from Woese et al. (1990)

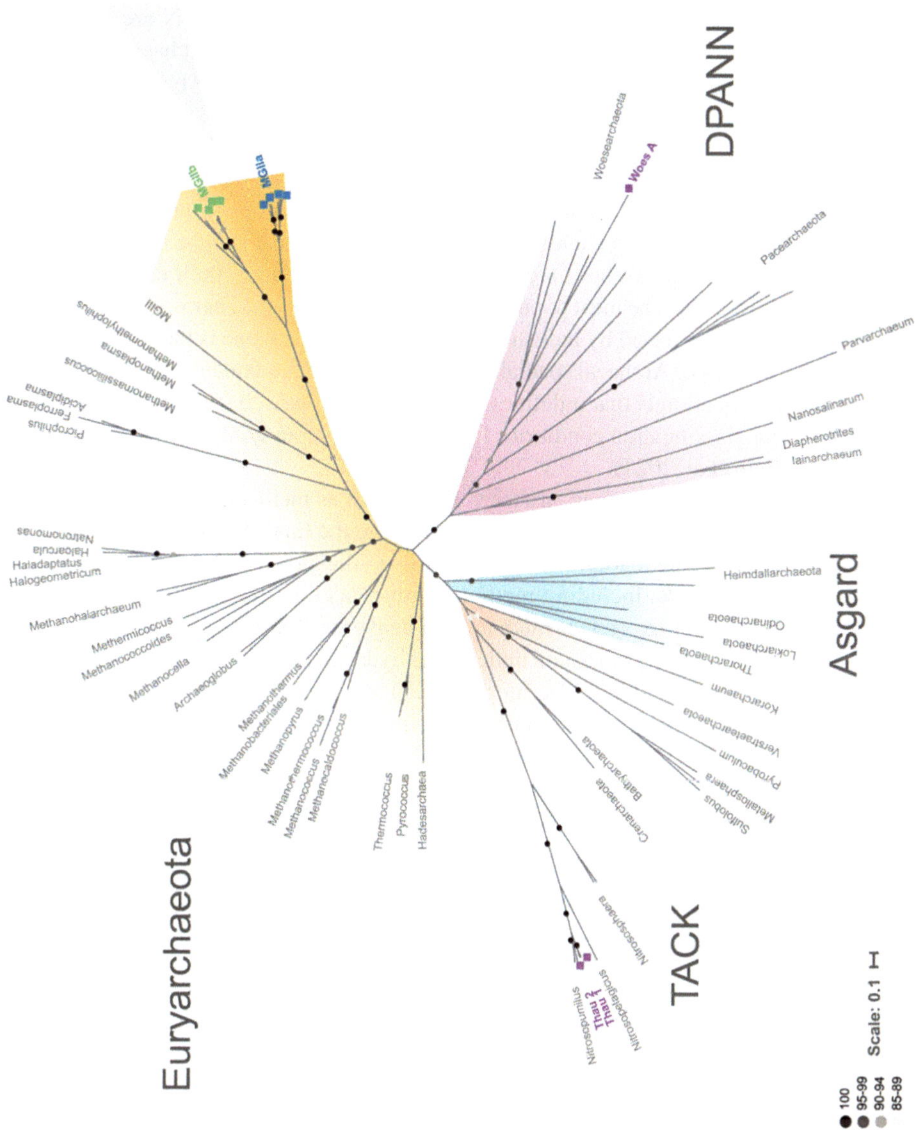

Fig 6.2 Archaea phylogenetic tree according to 16S rRNA. Retrieved from Orellana et al. (2019)

enzymes, and metabolic pathways. These characteristics make them desirable choices for biotechnological advancements in a variety of fields, such as industrial catalysis, environmental bioremediation, molecular biology, and pharmaceutical development (Cowan et al. 2024; Coker 2023).

Asgard, Euryarchaeota, TACK, and DPANN are the main superphyla of Archaea, and these four superphyla include a diverse range of species. DPANN are composed of Diapherotrites, Parvarchaeota, Aenigmarchaeota, Nanohaloarchaeota, and Nanoarchaeota, and TACK are composed of the Thaumarchaeota, Aigarchaeota, Crenarchaeota, and Korarchaeota (Baker et al. 2020) (Fig. 6.2).

Scientists have shown particular interest in the Asgard of Archaea due to its potential evolutionary relationship with eukaryotes (Xie et al. 2022). However, with the discovery of new members of Archaea with the help of new identification tools and a higher evolution rate, the official classification of Archaea is constantly being revised. In addition to taxonomy, Archaea are also classified according to their ecological function and physiology. The most popular one is the methanogens. Thermophiles and halophiles are most preferred by industrial demand, and the last one is the less well-known psychrophiles of Archaeal special classification. The fundamental obstacle of this limited classification is that it defines only extremophile Archaea. It is different from the normal classification conditions familiar to most commonly known microorganisms (Tahon et al. 2021).

Specifically, the methanogenic Archaea, also known as methanogens, are one of the most extensively researched groups. They catalyze the final stage of the anaerobic breakdown of organic matter, which is essential to the global carbon cycle. They employ various methods, including methylotrophic, aceticlastic, and hydrogenotrophic methanogenesis, to produce methane, a potent greenhouse gas and renewable energy source. The capacity to sustain underwater hydrothermal springs or geysers is a characteristic of thermophiles. Thermophiles can survive at temperatures ranging from 50 to 70°C, while hyperthermophiles may survive at temperatures higher than 70°C and up to 113°C. A minimum salinity of 1.5 M NaCl is required for Halophilic Archaea, which can thrive in hypersaline settings, including up to 5 M NaCl and KCl. Halophiles thrive in highly alkaline conditions, like potash mines and lakes, where they may endure up to 5 M NaCl or KCl. Some are referred to as alkalophiles because their pH is about 12. Psychrophilic Archaea, which have an optimal growth temperature between 0 and 10°C. They are expected to metabolize at −40°C and may survive at −45°C (Rekadwad et al. 2023).

6.2 Core Characteristics of Archaea Enabling Biotechnological Usage

There are some core characteristics of Archaea enabling biotechnological usage. Firstly, the ability of Archaea to produce extremozymes, enzymes that stay stable and active in harsh environments where other enzymes fail, is a key characteristic that makes them useful for biotechnology. These include DNA polymerases, ligases, xylanases, and proteases produced by thermophilic or halophilic Archaea. For example, DNA ligases from hyperthermophilic Archaea such as *Pyrococcus furiosus* have proved extraordinarily thermostable. Also, these enzymes are currently utilized in high-temperature molecular biology methods, including next-generation DNA assembly and ligation-based detection tests (Eichler 2001). Recent advancements in metagenomics and bioprospecting have identified novel Archaeal enzymes from previously unculturable strains. For instance, metagenomic investigations of hot spring and deep-sea vent communities have revealed Archaeal xylanases used in biomass conversion. These enzymes have strong thermostability and function at neutral to acidic pH levels, making them useful in industrial applications (Reichart et al. 2021).

Furthermore, unlike bacterial and eukaryotic lipids, Archaeal membranes are made up of ether-linked isoprenoid chains that frequently form monolayers. These properties provide greater stability in thermal and acidic environments. As a result, this adaptation extends the functional lifespan of Archaeal cells and enzymes under industrial conditions (Jain and Driessen 2014). Another valuable property comes from Archaeal metabolic variety. Hydrocarbons, methane, sulfur compounds, and cellulose derivatives are used in alternative energy production. Among extremophiles, methanogens serve important functions in anaerobic digestion and bioenergy production, while halophiles create pigments such as bacterioruberin and carotenoids with antioxidant and photoprotective qualities. These metabolites are used in medical applications. Also, Archaeal carotenoids and other secondary metabolites are being studied for pharmacological and nutraceutical purposes due to their stability and bioactivity (Pfeifer et al. 2021). Moreover, Archaea are now tractable for synthetic biology platforms because their genetic systems, particularly those of halophilic and methanogenic strains, have been modified for heterologous gene expression (De Lise et al. 2023).

Halophilic Archaea are also developing as powerful cell manufacturers. Their unique tolerance to high osmotic pressure and hazardous substances allows them to manufacture recombinant proteins, bioplastics, and biofuels in saline or waste-derived environments, lowering production costs and sterilization requirements. Archaea's unique structural and functional characteristics, together with their capacity to adapt to unfavorable conditions and compatibility with modern biotechnological instruments, present them as key players in the next generation of industrial and scientific advancements (Kasirajan and Maupin-Furlow 2021).

6.3 Key Biotechnological Application Areas of Archaea

The fundamental aim of the multidisciplinary science of biotechnology is to create or enhance processes that are beneficial to the environment and living things. Because biotechnological applications are so varied, they have been categorized using a variety of criteria. Archaea did not play a significant part in biotechnology at first, but they are now gaining popularity. It has become clear over the last few decades that Archaeal communities may survive under a variety of conditions, such as those with extremes in pH, salinity, temperature, and pressure. Especially, extremophiles are important in biotechnology because they provide unique biological resources, according to a large body of literature. They produce a wide range of specialized enzymes, metabolites, and macromolecules with remarkable qualities, providing a huge reservoir of genetic variation (Rawat et al. 2024).

This part of the chapter summarizes the primary uses of Archaea in biotechnology, both present and potential. The application is categorized based on the industry to which it is targeted, and its advantages as well as disadvantages are examined.

6.3.1 Industrial Enzymes and Biocatalysis

As biotechnology continues to advance, enzymes are gaining popularity as a biobased economy strategy. Enzymes are a more environmentally friendly and sustainable alternative to polluting chemicals used in industrial processes. Enzyme-catalyzed reactions yield fewer byproducts and have minimal toxicity. Enzymes play a key role in various industrial areas, including biofuels, leather, pulp and paper, textiles, animal feed, food and beverage, and detergents. Additionally, the use of enzymes is growing in fields such as diagnostics, research and development, and pharmaceuticals. The global enzyme market is expected to grow at a compound annual growth rate of 6.3% from 2020 to 2026, reaching $8.7 billion from an estimated $6.4 billion in 2021 (De Lise et al. 2023).

Industrial biotechnology is advancing further thanks to the ability of extremophilic Archaea to produce proteins and enzymes. These microorganisms are excellent candidates for biocatalytic functions because they have grown under extreme conditions. There are a lot of sectors benefiting from these important properties of Archaea, which are pharmaceuticals, textiles, leather, food, and paper. However, most commercially available enzymes are derived from bacteria or fungi; only a small number of these are extracted from Archaea. However, they are currently being studied for their ability to survive in environments where fungal and bacterial enzymes have been denatured. Also, the importance of these enzymes in many industrial sectors will gain attention with the help of biotechnological advancements. In this content, Archaeal enzymes, which are already in use in large-scale industrial applications are shown in Table 6.1 (Aparici-Carratalá et al. 2023; Littlechild 2015).

On the other hand, since 1987, the BRENDA database has been the primary database for categorizing enzymatic activity in literature, including taxonomic references and enzyme sequence identifiers. In the November 2022 BRENDA database, only 3938 Archaeal enzymes were found in all major enzyme classes, including oxidoreductases (EC 1), transferases (EC 2), hydrolases (EC 3), lyases (EC 4), isomerases (EC 5), ligases (EC 6), and translocases (EC 7) Specifically, the most prevalent classes identified in the Archaea domain are the Transferases (28.8%), Oxidoreductases (23.3%), and Hydrolases (22.3%) (Schomburg et al. 2017).

Table 6.1 Biotechnological applications of extremophilic Archaea enzymes. Adapted from Cabrera and Blamey (2018)

Extremophile type	Enzymes	Industrial usage
Acidophiles	Amylases, glucoamylases, xylanases, cellulases, proteases	Energy production, food and starch processing, mining
Alkaliphiles	Proteases, cellulases, amylases	Detergents, food and feed, beer brewing, paper industry
Halophiles	Proteases, dehydrogenases	Peptide synthesis, biocatalysis in organic media
Psychrophiles	Lipases, proteases, amylases, cellulases	Textile industry, detergents, laundry
Thermophiles	Proteases, lipases, glucoamylases, amylases, pullulanases, DNA polymerases, and dehydrogenases	Detergents, starch and cellulose processing, paper bleaching, molecular biology, and pharmaceuticals

6.3.1.1 Biotechnological Applications of Proteases

Proteases are important enzymes for protein hydrolysis into peptides or amino acids. The majority of extremophile proteases include serine amino acids, and many of them originated from hyperthermophilic Archaea. These genera are included: (i) *Pyrococcus*, (ii) *Thermococcus*, (iii) *Desulfurococcus*, (iv) *Pyrobaculum*, (v) *Staphylothermus*, and (vi) Thermoacidophilic archaeon *Sulfolobus* (Cabrera and Blamey 2018). Due to their numerous applications in various industrial sectors and their adaptable properties, proteases are more interesting than others. As a result, a significant portion of the worldwide enzyme market is made up of these enzymes (de Miguel Bouzas et al. 2006).

Hyperthermophilic Archaea like *Desulfurococcus, Thermococcus*, and *Pyrococcus* produce serine proteases and peptidases that are used for washing clothes at temperatures above 80°C, which denature other enzymes (Camacho et al. 2009). On the contrary, psychrophile-derived proteases are recommended for washing with cold water (Aparici-Carratalá et al. 2023). Moreover, they are used to remove proteinaceous stains from household laundry detergents. In this sector, proteases have been found to be resistant to detergents and alkaline denaturation. Therefore, proteases that show alkaline properties originated from halophilic Archaea are appropriate for this use (Antranikian et al. 2005).

In molecular biology, protein-based contaminants are broken down from DNA with the help of protease enzymes to elevate the quality of DNA amplification. As a result, thermostability is essential for PCR to function properly. Peptide synthesis has occurred in waterless environments. Moreover, peptides are employed as sweetener precursors, including aspartame. Thus, proteases that show alkaline properties are resistant to organic solvents, which are needed for this process (Satyanarayana et al. 2013). Additionally, proteases can shorten the time needed for dough fermentation in the bread industry and alter high-gluten blends by partially hydrolyzing them, which makes them stretchy and easy to cut off (Antranikian et al. 2005).

6.3.1.2 Biotechnological Applications of Esterases and Lipases

Lipases and esterases originated from extremophiles are better suited for industrial applications due to their ability to function in extreme conditions like low and high temperatures, acidic and alkaline pHs, low water activity, resistance to high radiation, high pressure, toxic metal concentrations, organic solvents, and combinations of these factors (Vivek et al. 2022).

Archaea lipases are often used as detergent-based molecules to eliminate oils and fats. As a result, they increase detergents' capacity to wash and facilitate the removal of stubborn stains, avoiding scaling (Cabrera and Blamey 2018). Lipases are used in the food industry to alter the structure of certain triglycerides to improve their flavor, physical characteristics, and nutritional value. The food industry can make healthful foods high in polyunsaturated fatty acids and monounsaturated fatty acids with the help of these enzymes (Reyes-Reyes et al. 2022). They are also utilized to produce human milk fat and cocoa butter substitutes, as well as to mature cheese (Aydemir et al. 2001).

In addition, Archaeal lipases are used in the production of new biopolymeric materials that are environmentally friendly and biodegradable (Sandoval et al. 2010). These ecologically friendly materials are also produced by Archaea as secondary metabolites. According to Li and Yu (2014), Archaeal lipase from *Haloarcula* sp. *G41* is used in the transesterification reactions of vegetable oils to produce biodiesel. They can also be utilized to produce nematicidal, pesticide, and insecticidal chemicals in the agrochemical industry (Meghwanshi et al. 2022).

6.3.1.3 Biotechnological Applications of Starch-Degrading Enzymes

Starch is a crucial storage product produced by several economically important crops, such as potatoes, tapioca, maize, rice, and wheat. Enzymatic hydrolysis of starch needs the action of numerous glycoside hydrolases, which are α-/β-amylases, also known as endo-amylases, producing oligosaccharides, exo-amylases, glucoamylases, and αglucosidases, producing glucose (Van Der Maarel et al. 2002). Nearly, 30% of the enzyme market is made up of thermostable amylases on a commercial basis (De Lise et al. 2023). The bioconversion of starch to sugars utilizing enzymes consists of three steps: gelatinization, liquefaction, and saccharification. Gelatinization at 110°C mostly uses thermostable α-amylases for starch enzymatic conversion due to their productivity and economic significance (Samanta et al. 2022).

Archaeal starch-converting enzymes such as pullulanases, amylopullulanases, and amylomaltases are commonly used in the food and feed industries to produce sweeteners, such as glucose, fructose, and maltose, as well as functional starch-based materials for use as texturizers, prebiotics, and aroma stabilizers. These enzymes aid in the production of syrups for candy and ice cream, improve the texture and shelf life of bread products, and produce gelatin alternatives. In pharmaceuticals, they allow for the production of maltose for intravenous feeding and branching cyclodextrins to stabilize medicinal compounds, as well as the development of slowly digestible starches with low glycemic impact (Cabrera and Blamey 2018).

6.3.1.4 Biotechnological Applications of Cellulose-Degrading Enzymes

Cellulose is a polysaccharide and a structural portion of plant cell walls. It consists of up to 15,000 glucose units linked together by β-1,4-glycosidic bonds. Cellulases are currently widely used in many industries, including biofuels, pulp and paper, detergents, textiles, animal feed, fruit juice, and biological treatment (Samanta et al. 2022). According to the 2022 global cellulase market research report, the global cellulase market is expected to reach a value of US$1621 million by 2032, with a compound annual growth rate of 6.9% from 2022 to 2032 (De Lise et al. 2023).

Cellulases have potential industrial applications, particularly in second-generation biorefineries. Cellulases convert lignocellulosic biomass into glucose, which can be used for bioethanol synthesis. Whenever ethanol made from lignocellulosic biomass becomes a significant transportation fuel, cellulases are predicted to become the most produced industrial enzyme. Together with proteases, cellulases show promise for application in laundry detergent due to their potent resistance to oxidizing chemicals, thermostability, and anionic surfactants. Cellulases may be utilized in bioremediation to eliminate cellulose from hypersaline wastewater (Khan et al. 2016).

On the other hand, β-1,4 bonds in cellulose are randomly restricted by endoglucanases to produce oligosaccharides such as cellobiose and glucose. Hyperthermophilic Archaea belonging to the genera *Metallosphaera, Pyrococcus, Ignisphaera*, and *Thermoproteus* have been found to have these enzymes. As a result, species belonging to these genera break down the β-1,4 bond easily to create their cellobiose and glucose for their metabolic requirements. The thermoacidophilic Archaea *Haloarcula, Sulfolobus solfataricus*, and *Acidilobus saccharovoran* have also been reported to have these enzymes (Cabrera and Blamey 2018). Industrial applications of these enzymes include increasing the quality, fermentation, and yields of beer and wine. Moreover, these enzymes, together with β-glucosidases, enhance color extraction, maceration, clarity, filtering, stability, and aroma of wine. In agriculture, cellulase-based treatments are used to manage plant diseases because they may destroy phytopathogen cell walls (Raddadi et al. 2015).

6.3.1.5 Biotechnological Applications of DNA Polymerases and DNA Ligases

DNA polymerases and DNA ligases are enzymes commonly used in molecular biology to carry out polymerase chain reaction (PCR) by forming phosphodiester bonds among nucleic acids and other analytical procedures related to molecular biology. For instance, thermostable DNA polymerases are responsible for DNA amplification, sequencing, and labeling (Cabrera and Blamey 2018).

The first thermostable DNA polymerase used in PCR, "Taq polymerase," was produced from the thermophilic bacterium *Thermus aquaticus*. This thermostable enzyme has revolutionized molecular biology research, especially PCR. Although its use was a significant milestone in the development of PCR, it has several disadvantages. The absence of 30–50 exonuclease of this enzyme is the most important disadvantages for analytical studies. Because this exonuclease activity has led to the use of more reliable substitutes, such as enzymes that exhibit both polymerization activity and correction properties. DNA polymerases "Pfu," "Pwo," "Vent," and "Tkod" have been isolated from the hyperthermophilic Archaea *Pyrococcus furiosus, Pyrococcus woesei, Thermococcus litoralis*, and *Thermococcus koda-karensis* (Zhang et al. 2015). Thanks to these enzymes, PCR takes place in a longer and more specific way due to their low error rate, high processing capacity, and high extension rate (Aparici-Carratalá et al. 2023).

Today, many thermostable DNA polymerases are used in PCR-based studies. However, thermostable enzymes used in molecular biology are not limited to DNA polymerases. In ligase-based reactions, DNA and RNA ligases are used because of their capability to form a phosphodiester bond. Also, these enzymes can be used as substitutes for Taq DNA or RNA ligase because of their properties of resistance to extremely high temperatures (Straub et al. 2018).

Additionally, Archaeal restriction endonucleases with unique recognition sites have been discovered and isolated. For molecular biology research, defined MaeI, MaeII, and MaeIII restriction enzymes are currently being produced for commercial sales. These defined restriction endonucleases have generally belonged to *Methanococcus aeolicus*. On the other hand, HcuI, HhII, and HsaI endonuclease enzymes are also used for commercial purposes. These enzymes originate from the *Hbt. cutirubrum, Hbt. halobium*, and *Hbt. salinarum*, respectively (Aparici-Carratalá et al. 2023).

6.3.2 Environmental and Energy Applications

Archaea make significant contributions to the circular economy by offering innovative waste management and resource recovery solutions. These resilient microorganisms can survive in harsh conditions and possess specialized metabolic skills that enable them to break down, purify, and transform a variety of waste products and contaminants. Archaea can be utilized for the bioremediation of polluted sites. In this process, Archaea convert organic waste into useful bioproducts and bioleach metals. As a result, Archaea's ability to operate in extreme environments makes them significant assets in closing the loop of resource use, reducing environmental consequences, and furthering circular economy principles (Rawat et al. 2024). Moreover, Archaea play an important role in the carbon cycle, primarily by producing methane in anoxic conditions and oxidizing released methane anaerobically. Methane emissions of one billion tons per year serve as a significant indicator of their role in the global carbon cycle. The ability of methanogenic Archaea to produce methane and contribute to global warming is widely established (Evans et al. 2019).

6.3.2.1 Energy Production

Methanogens or methanogenic Archaea produce methane as their primary metabolic product, which can take the place of fossil natural gas as a fuel. Methanogenic Archaea were previously exclusively identified in seven orders of the Euryarchaeota phylum: Methanobacteriales, Methanocellales, Methanococcales, Methanomicrobiales, Methanomassiliicoccales, Methano-dallales, and Methanopyrales (Enzmann et al. 2018). It is now understood that putative methanogenic Archaea exist outside of the Euryarchaeota. The categorization and diversity of methanogenic Archaea have changed dramatically as a result of in-depth genome sequencing. Currently, methanogenic Archaea are classified into four superphyla: Euryarchaeota, TACK, DPANN, and Asgard (Guerrero-Cruz et al. 2021). Especially, ANME (Anaerobic Methane Oxidizing Archaea) is the name given to the Archaea that oxidize methane in anaerobic conditions. They are divided into three groups: ANME-1, ANME-2, and ANME-3. They can exist as single cells or in syntrophic relationships with sulfate reducers. A single genome of an ANME-2a clade member's metatranscriptomic datasets revealed a collection of seven genes crucial for methanogenesis from CO_2 to methane. These genes dominated in an AOM enrichment setting and demonstrated a functional AOM pathway (Naitam and Kaushik 2021).

Anaerobic environments generally include Methanogens. These environments are deep oceans, wetlands, wastewater treatment plants, hydrothermal vents, plants, and the digestive systems of insects, herbivorous mammals, and humans. In these environments, methanogens catalyze methanogenesis. Specifically, methanogenesis or biomethanation is the production of methane by methanogens for energy conservation. It is the fourth and final stage of anaerobic digestion. The Carbon cycle is sustained through this process, and biobased methane is produced by these valuable microorganisms (Enzmann et al. 2018). Additionally, this methane metabolism, originating from Methanogens, is essential for producing methane from organic waste in anaerobic conditions. Especially, hydrogenotrophic methanogens have recently gained attention as biotechnological platform microorganisms for pure culture bioprocessing due to their ability to produce methane from H_2 and CO_2 in the biomethanation stage of a process called power-to-gas (Martin et al. 2013). Hydrogenotrophic methanogens are used in the power-to-gas conversion process, which is derived from renewable electricity sources and CO_2 point sources such as biogas and biomass plants (Contreras et al. 2022).

On the other hand, electroactive microorganisms can transfer electrons from cellular metabolism to an electrode, a process known as extracellular electron transfer. In bioelectrochemical systems, microorganisms catalyze oxidation and reduction reactions to generate or use electricity, utilizing their capacity to interact with electrodes (Logan et al. 2019). Methanogens, also known as electromethanogenesis, have been proposed as a means of producing methane in bioelectrochemical systems. CO_2 is converted to methane in these systems by using electrons from the cathode (Blasco-Gómez et al 2017).

6.3.2.2 Bioremediation

Bioremediation is the application of the metabolic capacities of microorganisms (including bacteria, certain algae, yeast, and fungi) and plants to break down and/or assimilate molecules that are hazardous to most living species or to change them into less toxic substances. The numerous benefits of bioremediation, including in situ treatment of soils or waters, high effectiveness, low cost in most situations, and no secondary pollution, make it one of the finest solutions for decontaminating polluted areas (Martínez-Espinosa 2024).

Archaea's remarkable tolerance and capacity to proliferate in contaminated conditions make them highly promising for use in bioremediation. The bioremediation process is also severely limited because many microorganisms cannot survive in situations with extreme temperatures, pH, and salinity. For this reason, Archaea's representatives are perfect for this role. They get carbon and energy by using contaminants as substrates (Aparici-Carratalá et al 2023).

Pollutant Removal

Haloarchaea have been characterized as hydrocarbon-decomposing organisms because they can thrive in environments with high salt. The salty wastewater produced by industries like the textile and oil sectors is one of these environments. They are frequently contaminated with harmful compounds, which represent a major threat to both human health and the environment. The salty, aerobic, and very halophilic archaeon *Haloferax* sp. *D1227* was isolated from oil-contaminated soil. This organism has the ability to break down and use a variety of aromatic compounds. Moreover, it has been claimed that halophilic Archaea, *Haloferax* sp., can be used to treat industrial wastewaters with high salt content or hypersaline environments contaminated with nitrates and nitrites (Singh and Singh 2017). Additionally, the breakdown of hydrocarbons by Archaeal microorganisms under anoxic conditions has been the object of various investigations. Archaean sulfate-reducing organisms have been identified for their ability to break down alkanes anaerobically (Schiraldi et al. 2002).

On the other hand, in the presence of nitrogen-rich hypersaline wastewater, haloarchaea need NO_3 and NO_2 as a nitrogen source in order to grow in an anaerobic condition. In these conditions, both ions operate as electron acceptors in the denitrification process. *Haloferax mediterranei* has been found to be a complete and model denitrifier when exposed to hypersaline wastewater that is rich in nitrogen (Martínez-Espinosa et al. 2007). Nevertheless, there aren't many appropriate research studies, and there isn't a report on how phosphorus is eliminated in haloarchaea.

Xenobiotics are compounds that are foreign to biological systems, such as medicines, insecticides, pollutants, carcinogens, volatile petrochemicals, food additives, and contaminated workplaces. It has been demonstrated that only a few numbers of Archaeal species are able to break down xenobiotics. *Methanosarcina thermophila* is a methanogenic and thermophilic archaeon identified from digester sludge. In addition to using methanol, methylamines, and acetate as possible growth substrates, this organism has been shown to break down chlorinated substances like carbon tetrachloride and chloroform through an anaerobic reduction route. *Sulfolobus solfataricus* is an aerobic, acidophilic, hyperthermophilic archaeon that metabolizes sulfur. Phenol can be employed as the only carbon source in the medium where this organism thrives (Ding and Lai 2010).

Methanogenic Archaea also play a role in oil spill cleanup and the removal of methanol and chlorinated chemicals from wastewater. Sludges, wastes of agriculture and industry have been broken down with the help of methanogenic Archaea via anaerobic treatment processes in the same way that ammonia-oxidizing Archaea may remove nitrogen from these waters, hyperthermophilic Archaea may also remove phosphate and arsenate from different locations. Additionally, acidophilic Archaea degrade sulfur compounds (Aparici-Carratalá et al. 2023).

Heavy Metal Removal

Metal bioremediation can take many different forms. To convert harmful redox states into harmless ones, it often requires the redox cycling of metals. The metal may precipitate or mobilize as a result of redox cycling, which can also change soluble metal redox states into insoluble ones or conversely. Furthermore, metals can be eliminated by sorption into biomass or by processes that allow heavy metals to volatilize (Krzmarzick et al. 2018).

Haloarchaea are capable of removing heavy metals from hypersaline wastewater and producing ecologically safe metallic nanoparticles (NPs). For example, Halobacterium absorbed Mn^{2+}, Zn^{2+}, As^{3+}, and Cd^{2+} through biosorption. *Haloferax volcanii* collected gold and silver intracellularly as AuNPs and AgNPs, but they were not adsorbed on external surfaces (Li et al. 2022).

On the other hand, Archaea play an important role in the microbial transformation of arsenic, particularly in the oxidation of dangerous arsenite (As III) to less hazardous arsenate (As V). Certain Archaeal strains, including *Sulfolobus acidocaldarius*, *Aeropyrum pernix*, *Pyrobaculum calidifontis*, and *Sulfolobus tokodaii*, have been verified or are predicted to contain arsenite oxidase genes (Krzmarzick et al. 2018). Furthermore, metagenomic investigations from extreme conditions such as Diamante Lake in Argentina show a high prevalence of arsenic metabolism genes in haloarchaea (Rascovan et al. 2016).

Archaea also help to detoxify heavy metals like mercury and uranium through redox reactions. Crenarchaeota and Euryarchaeota include mercury reductase genes, which convert poisonous mercuric mercury (Hg^{2+}) into less dangerous elemental mercury (Hg^{o}). This biological reduction mechanism was found in *Sulfolobus solfataricus*, where mercury reductase expression increased in response to mercury exposure (Schelert et al. 2013). Similar to this, hexavalent uranium (U^{6+}) can be reduced to tetravalent uranium (U^{4+}) by hyperthermophilic Archaea like *Pyrobaculum* species. This precipitation and immobilization of tetravalent uranium is essential for reducing uranium contamination in high-temperature sites of waste (Krzmarzick et al. 2018). Archaea's redox capabilities make them effective agents in the bioremediation of metal-polluted habitats.

Furthermore, Archaea are among the numerous prokaryotes that have the ability to change the oxidation state of metals, which can result in dissolution or biomineralization. Using hydrogen as an electron donor, the hyperthermophilic archaeon *Pyrobaculum islandicum* can grow at 100°C and change Fe^{3+} into Fe^{2+}. Utilizing the same electron donor source, this hyperthermophile has also demonstrated promising results in the reduction of U(VI), Tc(VII), Cr(VI), Co(III), and Mn(IV) all of which aid in the formation and expansion of mineral deposits. It has also been discovered that *P. furiosus*, another Fe^{3+} reducer archaeon, uses hydrogen as an electron donor to reduce gold Au(III) to Au(0) (Naitam and Kaushik 2021). These results show promise that the Archaea can be used in the treatment of wastewaters for heavy metal removal.

6.3.3 Medical and Pharmaceutical Applications

Along with their industrial enzymes and environmental uses, Archaea play a significant role in the pharmaceutical and medical industries through the production of a variety of small peptide molecules, other unique secondary metabolites. Moreover, extremophiles are currently causing confusion in the pharmaceutical industry thanks to their distinctive extremolytes. Scientists are currently looking into the potential use of extremolytes in human medicinal products, including cell cycle inhibitors, antioxidants, anticancer drugs, and cosmetic products (Gabani and Singh 2013). There is a lot of biotechnological promise in this field because of the diverse range of these products/metabolites generated. This section will examine the medical and pharmaceutical effects of Archaea in detail.

6.3.3.1 Drug Delivery Systems

Controlled drug release systems are a complex form of dosage used in the diagnosis, treatment, and prevention of disease. The goal of the design is to deliver a sufficient amount of drugs and/or diagnostic tools to the target areas at the appropriate moment in order to get the best possible treatment result with the fewest possible side effects and, preferably, no toxicity. In both pharmaceutical and medical research, lipid and phospholipid-based carriers are becoming more and more popular as drug delivery systems (Esfahani et al. 2024).

Liposomes are microscopic vesicular structures made up of one or more phospholipid bilayers surrounding an aqueous core. The aqueous core and the bilayers can both hold medicinal compounds. Most peptides and other hydrophilic compounds can be encapsulated and held in the aqueous core for long periods of time if they are big enough and charged enough to have slow leakage across the bilayers. By incorporating hydrophobic chemicals into the bilayers, their aqueous solubility can be enhanced (Jacobsen et al. 2017).

Vesosomes, solid lipid nanoparticles, nanostructured lipid carriers and archaeosomes are a few of the primary lipidic carriers. Polypeptides, nucleic acid molecules, and therapeutic, preventative, and diagnostic substances are all encapsulated or trapped in these carriers (Esfahani et al. 2024). The main drawback of traditional liposomes, however, is their instability in the biological environment as a result of pH, temperature, and enzyme variations. To lessen surface tension, the unstable liposomes often merge with the neighboring liposomes. These fused liposomes have very brief lifespans because the reticuloendothelial system (RES) rapidly captures them and removes them from the bloodstream. The term "biological half-life" refers to the time it takes for half of a pharmacological dose to be metabolized and eliminated from the bloodstream. Following their preferential accumulation in the RES's organs, the trapped liposomes are then broken down and removed by the target tissues. All things considered, this limits the therapeutic efficacy of the medication and reduces the effectiveness of traditional liposomes in delivering it to the targeted area (Santhosh and Genova 2022).

To overcome this problem, a new generation of liposomes with enhanced characteristics known as "archaeosomes" has been developed. Archaeosomes or Archaeolipid vesicles are particularly attractive among lipid-based drug delivery systems because of their immunomodulatory properties, long-term stability, resistance to enzyme degradation, and notable thermostability (Kejzar et al. 2023). In this way, Archaeosomes are used as vehicles to transfer a variety of pharmaceutical molecules, including anti-cancer drugs, vaccines, proteins, peptides, genes, and naturally occurring antioxidant chemicals within biological systems (Esfahani et al. 2024).

According to studies, archaeosomes made using synthetic, chemically modified, or natural Archaeal lipids can be effective drug carriers. The main advantages of using archaeosomes in drug delivery are increased shelf life, improved bioavailability, and protection against acidic or enzymatic degradation. Archaeosomes have the capability to serve as a drug carrier for breast cancer drug delivery, paclitaxel (PTX). According to research results, archaeosomal-PTX formulation cytotoxicity was dramatically lower than that of the standard, which only included the PTX formulation. The stability of Archaeal lipids under various conditions (temperature, pH, lipases, etc.) and the lipid coating enabled extended drug release, reduced side effects, and increased the therapeutic index. According to this study, drug loading into archaeosomes is a useful strategy for improving treatment response (Santhosh and Genova 2022).

Additionally, a variety of vaccines and immunomodulatory delivery systems can be effectively administered via archaeosomes. In contrast to conventional liposomes, Krishnan and colleagues have demonstrated that archaeosomes are highly effective. In this investigation, mice's humoral and cell-mediated immune responses were considerably improved by archaeosomes loaded with bovine serum albumin (BSA), which resulted in strong antibody production, spleen cell proliferation, and cytokine release (Krishnan et al 2000).

In another investigation, Attar et al. (2016) examined archaeosomes—which are composed of H. hispanica 2TK2 ether lipids—as vehicles for delivering plasmid DNA into mammalian cells (HEK293). Because of their anionic nature, the archaeosomes initially showed poor DNA binding and transfection efficiency. This was addressed by adding positively charged helper molecules to the formulations, such as calcium ions or 1,2-dioleoyl-3-(trimethylammonium) propane. This alteration greatly increased the archaeosomes' capacity to transport plasmid DNA encoding β-galactosidase or green fluorescent protein, which improved the efficiency of gene transfection in HEK293 cells.

On the other hand, researchers have looked into using Archaea's other distinctive characteristics as drug delivery systems. Low dissolved oxygen causes the halophilic

Archaea to create gas vesicles, which cause cells to float toward the air oxygen. Display vehicles with sustained release capabilities that demonstrate long-term immunological memory are recombinant gas vesicles linked with certain peptides (Singh and Singh 2017).

6.3.3.2 Anticancer Compounds

Cancer is one of the leading causes of death worldwide. Chemotherapy has problems such as inadequate drug absorption in tumor tissues and resistance to anti-cancer medicines. To address limitations, microorganisms have been intensively studied for cancer treatment. As crucial resources for investigating natural agents, microorganisms from the Archaea kingdom have drawn a lot of attention. Archaea can produce proteins, toxic substances, and metabolites that effectively suppress tumor cell development (Moopantakath et al. 2023).

The majority of haloarchaeal organisms produce pigments called carotenoids. By promoting gap-junctional communication, inhibiting the cell cycle, and triggering apoptosis, carotenoids have been shown to reduce the development of cancer cells (Abbes et al. 2013). Compared to 5-fluorouracil (5-FU), the carotenoids of *Natrialba* sp. *M6* demonstrated a much greater cytotoxic effect on cancer cells (Caco-2, MCF-7, HepG-2, and HeLa cells). Furthermore, these pigments caused more apoptosis in tumor cells than 5-FU. Interestingly, extracted pigment was proven to be far safer than 5-FU on normal cells. These findings support the idea that carotenoids derived from Archaea may offer a new avenue for cancer treatment that doesn't affect healthy cells (Moghimipour et al. 2021). Similarly, according to Hou and Cui (2018), the carotenoid pigment of *Halogeometricum limi* demonstrated around 23% anticancer activity against HepG2 cells.

On the other hand, bacterioruberin is more effective toward the liver cancer cells than β-carotene from the perspective of antihemolytic and cytotoxic properties (Hou and Cui 2018). C50 carotenoid was the main component produced by the Egyptian isolate *Natrialba* sp., *M6*, which belongs to the phylum Euryarchaeota (Hegazy et al. 2020). Compared to the common chemotherapy drug 5-fluorouracil, the pigment exhibited a better selectivity against various cancer-type cells. It works against breast, liver, and colon cancer cell types. Additionally, bacterioruberin combined with dexamethasone diminished inflammation by reducing TNF-α and IL-8 release, inducing macrophage morphological alterations, and operating as an intestinal barrier repair agent (Higa et al. 2020). The studies about anticancer properties of haloarchaea are listed in Table 6.2 (Moghimipour et al. 2021).

6.3.3.3 Antimicrobial and Antioxidant Compounds

Archaeocins are proteins produced by specific Archaea that have unique structures and potent antibacterial activity, in contrast to bacterial antibiotics. These are thought to compete with other Domains and archaeocin-producing strains of Archaea (Corral et al.

Table 6.2 Overview of the anticancer application of haloarchaea

Species	Compound	Anticancer activity	References
Halobacterium halobium	Carotenoids	Demonstrates antiproliferative activity	Abbes et al. (2013)
Halogeometricum limi	Carotenoids	Causes 23% inhibition of HepG2 cancer cells	Hou and Cui (2018)
Haloferax mediterranei	Carotenoids	Displays potential activity against HER2-positive and triple-negative breast cancer cell lines	Giani and Martínez-Espinosa (2020)
Natrialba sp. *M6*	Carotenoids	Maintains 50–100% viability in normal lung fibroblast cells. Causes ~50% cell death in colon cancer cells. Active against MCF7, HepG2, and HeLa cancer cell lines.	Hegazy et al. (2020)

2019). Currently, only two types of archaeocin are known to exist, halocins, which are extracted from haloarchaea belonging to the *Halobacteriales*, and *Sulfolobicins*, which are produced by thermoacidophilic bacteria belonging to the Sulfolobales order. For instance, KPS1 halocin has activity against certain Gram-positive and Gram-negative bacteria (Kavitha et al. 2011). *Sfb. Islandicus* has a peptide comparable to cationic antimicrobial peptides (CAMP), which is active against fungi, bacteria, and protozoa. It was demonstrated that this peptide possessed broad-spectrum antibacterial properties (Aparici-Carratalá et al. 2023). Moreover, *Haloferax larsenii HA1* supernatant contains halocin, which has significant antibiotic properties. According to earlier research on halocins from the haloarchaea, the antibacterial activity of halocins may be connected to processes of competition between various species for food, space, or niche. *Halobacterium salinarum ETD5*, for example, which was isolated from the sun saltern, had antagonistic activity against *Halorubrum chaoviator* sp. (Ghanmi et al. 2016; Moopantakath et al. 2023).

On the other hand, the microorganisms that cause infectious diseases are always changing and becoming more resistant to drugs. Drug resistance occurs when microorganisms are exposed to a chemical agent and develop the ability to resist it. Different antimicrobials are therefore needed to treat emerging health problems. Carotenoids are extracted from Haloarchaea for antimicrobial uses. Haloarchaeal carotenoids have been shown to possess antibacterial properties against many different diseases (Gómez-Villegas et al. 2020). According to recent studies, carotenoids obtained from *Halogeometricum* sp., have been shown to have antibacterial effects against *Vibrio anguillarum, Pseudomonas aeruginosa*, and *Pseudomonas anguilliseptica*, respectively (Sahli et al. 2022). Furthermore, the C50 carotenoid isolated from *Natrialba* sp., demonstrated remarkable potential in the neutralization of hepatitis C virus and hepatitis B virus from human blood mononuclear cells, indicating high antiviral activity (Hegazy et al. 2020).

The human body produces free radicals during metabolic processes, which cause oxidative stress, inflammation, and chronic diseases. At lower concentrations, haloarchaeal compounds have significantly more effective free radical scavenging properties than traditional reference compounds such as ascorbic acid, and are therefore attracting interest today. For biological applications, β-carotene and bacteriochlorophyll are the most sought-after carotenoids. As an antioxidant molecule, bacteriochlorophyll can absorb reactive oxygen species. It has stronger antioxidant activity than traditional ascorbic acid (Sahli et al. 2022). According to Giani et al. (2020), bacteriochrome synthesis in the *Haloferax mediterranei* strain R-4 increases by 78% when exposed to oxidative stress via H_2O_2. Furthermore, the antioxidant activity of halobacteria varies among different species. For example, *Haloferax* sp. *ME16* carotenoids exhibit greater antioxidant activity than *Haloarcula* and *Halogeometricum* (Sahli et al. 2022).

6.3.3.4 Cosmetics

Archaeosomes, as previously indicated, can be used in cosmetic applications because of their ability to store active substances and permeate the skin, similar to ordinary liposomes. This makes them suitable for usage as molecular vectorization systems in skin care treatments (Rastädter et al. 2020). Additionally, exopolysaccharides (EPS) produced by *Haloterrigena turmenica* may be utilized in the cosmetics industry because of their greater ability to hold onto moisture than hyaluronic acid, which is therefore used in this industry. Moreover, carotenoids and other secondary metabolites of Archaea, which protect radiophilic Archaea from UV light, could be used as sun creams because of their ability to prevent photooxidation harmful effects (Aparici-Carratalá et al. 2023).

For decades, ectoine, which is produced by *Nitrosopumilus* sp. has been a common ingredient in skincare products because of its ability to stabilize and protect human skin cell membranes from damaging environmental elements like wind, humidity, ultraviolet (UV) radiation, and extremely high or low temperatures (Bethlehem and van Echten-Deckert 2021). Previous research has shown that ectoine can minimize water loss from the skin barrier and keep the skin moisturized. As a result, Marini et al. (2014) developed an ectoine-containing cream as a local treatment for individuals who have atopic dermatitis (AD), which is frequently linked with impaired skin barrier function and dryness. Ectoine was found to be a unique and potent chemical for treating atopic dermatitis when compared to established reference products (Ng et al. 2023).

6.3.4 Other Biotechnological Applications

6.3.4.1 Biopolymers

The Archaea domain of microorganisms is significant in biotechnology because their biopolymers have unique features that promise insights into their biology and evolution. Two primary categories of biopolymers generated by these unusual microorganisms have received special attention: extracellular polysaccharides (EPSs) and endocellular polyhydroxyalkanoates (PHAs).

EPSs are extremely varied polymers that contain a variety of unique monosaccharides and species-specific non-carbohydrate substituents. Polysaccharide chains are often created by using an oligosaccharide as a repeating unit whose size varies depending on the degree of polymerization. Numerous uses for exopolysaccharides can be discovered in the food, pharmaceutical, and other industries. EPSs, which mainly belong to the thermophilic and halophilic groups, have been isolated from different Archaea. Therefore, certain haloarchaea are capable of producing large amounts of commercially available bioplastics, including polyhydroxyalkanoate (PHA), polyhydroxybutyrate (PHB), and polyhydroxyvalerate (PHV). The growing of such microorganisms on a pilot or industrial scale provides various benefits over other bioplastic producers in terms of culture sterilization, growth rate, and so on (Simó-Cabrera et al. 2021). Furthermore, *Thermococcus* and *Sulfolobus* are two of the thermophilic Archaeal genera that produce extracellular polymeric substances (EPSs). *Archaeoglobus fulgidus* and *Thermococcus litoralis* also produce large amounts of EPSs as biofilms, which are an assembly of microorganisms that are immobilized and confined within EPS, limiting the diffusion of substances and antimicrobial agents (Poli et al. 2011). Moreover, large amounts of EPS are produced by *Hfx. mediterranei* ATCC 33,500 when it is cultivated on a medium that contains KH2PO4, NH4Cl, and glucose (LitchWeld 2011).

On the other hand, plastics make up around 22–43% of the non-biodegradable waste that ends up in landfills globally. Petroleum depletion will likely lead to increased competition between oil-derived polymers and the energy-generation sector in the future. An estimated 25 million tonnes of environmental harm are caused annually by non-biodegradable plastics. As a result, it is crucial to identify, develop, and implement alternative methods for producing plastics from renewable resources to mitigate environmental impacts. Biopolymers are environmentally friendly alternatives to oil-based plastics, suitable for a wide range of applications, including biofuels, construction, medicine, and product packaging. One type of bio-based biodegradable polymer (biopolymer) is polyhydroxyalkanoates (PHAs) (Melanie et al. 2018).

The only genera described as PHA producers in the Archaea domain were those belonging to the Haloarchaea group. Because the cells require significant quantities of salt to remain alive, haloarchaeal cultivation on a control or large scale does not require rigid sterile conditions. Comparing this to eubacterial strains, culture is made simpler and more convenient. It is also simple to lyse the cell walls and membranes of haloarchaea without salt, especially with pure water. The isolation of PHA and PHB granules from crude extracts is made simpler and more cost-effective by this osmotic shock (Han et al. 2010; Simó-Cabrera et al. 2021).

Hfx. Mediterranei is likely the most popular PHA generator among all haloarchaeal strains due to its rapid growth rate, metabolic adaptability, genetic stability, and excellent transformation mechanism (Zuo et al. 2018). Numerous investigations have shown that *Hfx. Mediterranei* may produce PHA with substantial productivity using a variety of industrial and domestic wastes as carbon sources (Simó-Cabrera et al. 2021).

6.3.4.2 Surfactants

Biosurfactants are a class of surface-active chemicals produced naturally by diverse microorganisms, including bacteria and Archaea. Because of their remarkable hydrophilic and hydrophobic qualities, these chemicals can reduce the surface tension between various liquid phases (Alghamrawy et al. 2024). Surfactant generation by haloarchaea has received relatively less attention in comparison to biopolymer research. Just a few reports of biosurfactant generation were mentioned by extremophilic microorganisms (LitchWeld 2011). Because there are difficulties in producing biosurfactants by halophilic Archaea. The necessity for metabolic engineering and genetic modification, the scarcity of appropriate halophilic Archaeal strains, and challenges with extraction and purification all impede large-scale manufacturing. Overcoming these challenges is critical for optimizing biosurfactant manufacturing techniques and fully realizing the potential of these microorganisms in a variety of applications (Alghamrawy et al. 2024). On the other hand, biosurfactants have a wide range of industrial uses, frequently involving hypersaline solutions or suspensions, including bioremediation, mining, personal care products, the food industry, oil tank cleaning, oil recovery, and therapeutic and agricultural antimicrobial agents (Nikolova and Gutierrez 2021).

A recent study investigated the production, optimization, and biomedical potential of biosurfactants produced by two haloalkaliphilic Archaeal strains, which are *Natrialba* sp. *BG1* and N3 were isolated from Wadi El-Natrun, Egypt. These Archaea were able to produce biosurfactants under extreme saline and alkaline conditions. The biosurfactants were characterized chemically (via FTIR and NMR) and biologically for their functional properties. The produced biosurfactants demonstrated promising biomedical applications, exhibiting significant wound-healing acceleration, cytotoxic effects against various cancer cell lines, and notable anti-inflammatory and antioxidant activities. Overall, these findings highlight the potential use of haloarchaeal biosurfactants as natural, eco-friendly agents for medical and industrial applications (Alghamrawy et al. 2024).

In another investigation on biosurfactant activity, it was found that *Haloferax chudinovii HB1RANIA* has significant surface-active qualities, including a 64.28% emulsification index with soybean oil and a reduction in distilled water surface tension of approximately 15 mN/m. Importantly, the biosurfactant accelerated the breakdown of waste engine oil, with the HB1RANIA isolate utilizing 93.90% of the oil, indicating a high potential for bioremediation of hydrocarbon-contaminated salty settings (Ghaleb and Bhosale 2024).

6.4 Challenges and Future Directions of Archaeal Biotechnological Applications

Despite the numerous benefits of using Archaea for biotechnological applications, there are various obstacles and limitations to their full utilization. Archaea can benefit biotechnological systems; however, they must be procured as pure cultures for proper administration. Regardless of the advantages of metagenomic investigation, isolating Archaea in the laboratory is useful for gaining an in-depth awareness of their unique physiological and cellular features, metabolism, dietary requirements, genetic structure, and biotechnological potential. Isolating Archaea is problematic because the physicochemical and nutritional conditions for Archaea culture are difficult to achieve. As a result, traditional methods for Archaeal isolation, such as serial dilution and the pour plating method, are less effective in increasing the number of Archaeal isolates. Because of our poor understanding of Archaeal growth requirements, it is difficult to provide the optimal nutrients, salts, pH, temperature, and gases for Archaeal growth in the laboratory (Aparici-Carratalá et al. 2023).

Another drawback is that due to their smaller cell sizes and even smaller genomes, bacteria are. Therefore, some Archaeal groups lack vital metabolic genes and different metabolic pathways. However, because they are necessary for essential characteristics and cell division, a number of these pathways are very difficult for development (Rafiq et al. 2023). Moreover, very little information is known about the genome sequences of Archaeal species because fewer of them can be cultivated. This suggests that not enough genome sequences are available for annotation and metabolic pathway extraction. Because some Archaeal genes may have been mislabeled or linked to unknown and hypothetical proteins. As a result, there may be a genetic deficit for Archaeal cultivation (Karavaeva and Sousa 2024). According to the findings, up to 30% and in some cases up to 80% of the genomes of Archaea are annotated with hypothetical proteins, and they are composed of dark genomic matter. Thus, the functional profile and metabolic pathways for particular Archaeal species may be analyzed based on the precise genome annotation (Makarova et al. 2019).

Considering all biotechnological applications of Archaea covered in this chapter, it can be said that Archaea are a resource that is only now being utilized in biotechnology, and that there are still many unexplored applications for them. The current procedures must be replaced with improvements. Identified new Archaea with unexpected traits will probably be discovered because of expanding their field of study due to the development of bioinformatics-based technologies, the improvement of investigation techniques for extreme environments, and the enhancement of genome sequencing technology (Aparici-Carratalá et al. 2023). The application of multi-omics technologies with single-cell genomics has the potential to provide light on biotechnology advancements. Also, fostering international collaboration and alliances can help to maintain a viable database by allowing for data sharing, resource exchange, and cooperative research activities. International cooperation promotes scientific progress worldwide and increases the variety and scope of study on Archaea. As research into Archaea biotechnology increases, the design of bioreactors and biological processes will contribute to increased efficiency. In the extreme environmental conditions required for the development of extremophiles, pollution problems are very low. Therefore, the continuous culture approach is a point that is widely accepted by scientists. The use of continuous culture systems is also important for the development of new bioprocesses. As a result, scientists have focused their attention on the design of innovative bioreactors for the production of large amounts of biomass and metabolites. In addition, several strategies have been developed for the cloning of Archaeal enzymes into mesophilic organisms and the development of cell expression techniques, and these have begun to be implemented in various applications. The use of mesophilic microorganisms in the production of biomass and metabolites has been investigated in several studies (Rawat et al. 2024).

References

Abbes M, Baati H, Guermazi S, Messina C, Santulli A, Gharsallah N, Ammar E (2013) Biological properties of carotenoids extracted from Halobacterium halobium isolated from a Tunisian solar saltern. BMC Complement Altern Med 13(1):255

Alghamrawy BT, Hegazy GE, Sabry SA, Ghozlan H (2024) Production, characterization and biomedical potential of biosurfactants produced by haloalkaliphilic archaea from Wadi El-Natrun, Egypt. Microb Cell Fact 23(1):84

Antranikian G, Vorgias CE, Bertoldo C (2005) Extreme environments as a resource for microorganisms and novel biocatalysts. Marine Biotechnol 96:19–262

Aparici-Carratalá D, Esclapez J, Bautista V, Bonete MJ, Camacho M (2023) Archaea: current and potential biotechnological applications. Res Microbiol 174(7):104080

Attar A, Ogan A, Yucel S, Turan K (2016) The potential of archaeosomes as carriers of pDNA into mammalian cells. Artif Cells Nanomed Biotechnol 44(2):710–716

Aydemir S, Akin N, Koçak C (2001) Effect of lipase enzyme on the ripening of white pickled cheese. J Food Lipids 8(3):205–213

Baker BJ, De Anda V, Seitz KW, Dombrowski N, Santoro AE, Lloyd KG (2020) Diversity, ecology and evolution of Archaea. Nat Microbiol 5(7):887–900

Bethlehem L, van Echten-Deckert G (2021) Ectoines as novel anti-inflammatory and tissue protective lead compounds with special focus on inflammatory bowel disease and lung inflammation. Pharmacol Res 164:105389

Blasco-Gómez R, Batlle-Vilanova P, Villano M, Balaguer MD, Colprim J, Puig S (2017) On the edge of research and technological application: a critical review of electromethanogenesis. Int J Mol Sci 18(4):874

Cabrera MÁ, Blamey JM (2018) Biotechnological applications of archaeal enzymes from extreme environments. Biol Res 51(1):37

Camacho RM, Mateos JC, González-Reynoso O, Prado LA, Córdova J (2009) Production and characterization of esterase and lipase from Haloarcula marismortui. J Ind Microbiol Biotechnol 36(7):901–909

Coker JA (2023) 'All about' extremophiles. Faculty Rev 12:27

Contreras G, Thomsen J, Pfitzer M, Hafenbradl D, Kostner D, Holtmann D et al (2022) New perspectives for biotechnological applications of methanogens. Curr Res Biotechnol 4:468–474

Corral P, Amoozegar MA, Ventosa A (2019) Halophiles and their biomolecules: recent advances and future applications in biomedicine. Mar Drugs 18(1):33

Cowan DA, Albers SV, Antranikian G, Atomi H, Averhoff B, Basen M et al (2024) Extremophiles in a changing world. Extremophiles 28(2):26

De Lise F, Iacono R, Moracci M, Strazzulli A, Cobucci-Ponzano B (2023) Archaea as a model system for molecular biology and biotechnology. Biomolecules 13(1):114

de Miguel Bouzas T, Barros-Velázquez J, Gonzalez Villa T (2006) Industrial applications of hyperthermophilic enzymes: a review. Protein Pept Lett 13(7):645–651

Ding JY, Lai MC (2010) The biotechnological potential of the extreme halophilic archaea Haloterrigena sp. H13 in xenobiotic metabolism using a comparative genomics approach. Env Technol 31(8-9):905–914

Eichler J (2001) Biotechnological uses of archaeal extremozymes. Biotechnol Adv 19(4):261–278

Enzmann F, Mayer F, Rother M, Holtmann D (2018) Methanogens: biochemical background and biotechnological applications. AMB Express 8(1):1
Esfahani FN, Naghib SM, Atrooz O, Ataei S, Ak G, Zengin G, Alavi M, Mozafari MR, Taghavi AE (2024) Archaeal lipid vesicles as immune modulators, adjuvants and vaccine carriers: a review of the past and current trends. Preprints
Evans PN, Boyd JA, Leu AO, Woodcroft BJ, Parks DH, Hugenholtz P, Tyson GW (2019) An evolving view of methane metabolism in the Archaea. Nat Rev Microbiol 17(4):219–232
Gabani P, Singh OV (2013) Radiation-resistant extremophiles and their potential in biotechnology and therapeutics. Appl Microbiol Biotechnol 97(3):993–1004
Ghaleb RN, Bhosale HJ (2024) Characterization of biosurfactant and lipase produced by Haloferax chudinovii HB1RANIA from solar saltern of Mulund, Mumbai, India. Environ Sustain 7(2):197–207
Ghanmi F, Carré-Mlouka A, Vandervennet M, Boujelben I, Frikha D, Ayadi H et al (2016) Antagonistic interactions and production of halocin antimicrobial peptides among extremely halophilic prokaryotes isolated from the solar saltern of Sfax, Tunisia. Extremophiles 20(3):363–374
Giani M, Martínez-Espinosa RM (2020) Carotenoids as a protection mechanism against oxidative stress in Haloferax mediterranei. Antioxidants 9(11):1060
Gómez-Villegas P, Vigara J, Vila M, Varela J, Barreira L, Léon R (2020) Antioxidant, antimicrobial, and bioactive potential of two new haloarchaeal strains isolated from odiel salterns (Southwest Spain). Biology 9(9):298
Grivard A, Goubet I, Duarte Filho LMDS, Thiéry V, Chevalier S, de Oliveira-Junior RG et al (2022) Archaea carotenoids: natural pigments with unexplored innovative potential. Mar Drugs 20(8):524
Guerrero-Cruz S, Vaksmaa A, Horn MA, Niemann H, Pijuan M, Ho A (2021) Methanotrophs: discoveries, environmental relevance, and a perspective on current and future applications. Front Microbiol 12:678057
Han J, Hou J, Liu H, Cai S, Feng B, Zhou J, Xiang H (2010) Wide distribution among halophilic archaea of a novel polyhydroxyalkanoate synthase subtype with homology to bacterial type III synthases. Appl Env Microbiol 76(23):7811–7819
Hegazy GE, Abu-Serie MM, Abo-Elela GM, Ghozlan H, Sabry SA, Soliman NA, Abdel-Fattah YR (2020) In vitro dual (anticancer and antiviral) activity of the carotenoids produced by haloalkaliphilic archaeon Natrialba sp. M6. Sci Rep 10(1):5986
Higa LH, Schilrreff P, Briski AM, Jerez HE, de Farias MA, Portugal RV, Morilla MJ (2020) Bacterioruberin from Haloarchaea plus dexamethasone in ultra-small macrophage-targeted nanoparticles as potential intestinal repairing agent. Colloids Surf B Biointerfaces 191:110961
Hou J, Cui HL (2018) In vitro antioxidant, antihemolytic, and anticancer activity of the carotenoids from halophilic archaea. Curr Microbiol 75(3):266–271
Jacobsen AC, Jensen SM, Fricker G, Brandl M, Treusch AH (2017) Archaeal lipids in oral delivery of therapeutic peptides. Eur J Pharm Sci 108:101–110
Jain S, Caforio A, Driessen AJ (2014) Biosynthesis of archaeal membrane ether lipids. Front Microbiol 5:641
Karavaeva V, Sousa FL (2024) Navigating the archaeal frontier: insights and projections from bioinformatic pipelines. Front Microbiol 15:1433224
Kasirajan L, Maupin-Furlow JA (2021) Halophilic archaea and their potential to generate renewable fuels and chemicals. Biotechnol Bioeng 118(3):1066–1090
Kavitha P, Lipton AP, Sarika AR, Aishwarya MS (2011) Growth characteristics and halocin production by a new isolate, Haloferax volcanii KPS1 from Kovalam solar saltern (India). Res J Biol Sci 6:257–262
Kejzar J, Osojnik Črnivec IG, Poklar Ulrih N (2023) Advances in physicochemical and biochemical characterization of archaeosomes from polar lipids of Aeropyrum pernix K1 and stability in biological systems. ACS Omega 8(3):2861–2870

Khan MN, Luna IZ, Islam MM, Sharmeen S, Salem KS, Rashid TU et al (2016) Cellulase in waste management applications. In: New and future developments in microbial biotechnology and bioengineering. Elsevier, pp 237–256
Krishnan L, Dicaire CJ, Patel GB, Sprott GD (2000) Archaeosome vaccine adjuvants induce strong humoral, cell-mediated, and memory responses: comparison to conventional liposomes and alum. Infect Immun 68(1):54–63
Krzmarzick MJ, Taylor DK, Fu X, McCutchan AL (2018) Diversity and niche of archaea in bioremediation. Archaea 2018(1):3194108
Li J, Gao Y, Dong H, Sheng GP (2022) Haloarchaea, excellent candidates for removing pollutants from hypersaline wastewater. Trends Biotechnol 40(2):226–239
Li X, Yu HY (2014) Characterization of an organic solvent-tolerant lipase from Haloarcula sp. G41 and its application for biodiesel production. Folia Microbiol 59(6):455–463
Litchfield CD (2011) Potential for industrial products from the halophilic Archaea. J Ind Microbiol Biotechnol 38(10):1635
Littlechild JA (2015) Archaeal enzymes and applications in industrial biocatalysts. Archaea 2015(1):147671
Logan BE, Rossi R, Ragab AA, Saikaly PE (2019) Electroactive microorganisms in bioelectrochemical systems. Nat Rev Microbiol 17(5):307–319
Makarova KS, Wolf YI, Koonin EV (2019) Towards functional characterization of archaeal genomic dark matter. Biochem Soc Trans 47(1):389–398
Marini A, Reinelt K, Krutmann J, Bilstein A (2014) Ectoine-containing cream in the treatment of mild to moderate atopic dermatitis: a randomised, comparator-controlled, intra-individual double-blind, multi-center trial. Skin Pharmacol Physiol 27(2):57–65
Martin MR, Fornero JJ, Stark R, Mets L, Angenent LT (2013) A single-culture bioprocess of Methanothermobacter thermautotrophicus to upgrade digester biogas by CO2-to-CH4 conversion with H2. Archaea 2013(1):157529
Martínez-Espinosa RM (2024) Halophilic archaea as tools for bioremediation technologies. Appl Microbiol Biotechnol 108(1):401
Martínez-Espinosa RM, Lledó B, Marhuenda-Egea FC, Bonete MJ (2007) The effect of ammonium on assimilatory nitrate reduction in the haloarchaeon Haloferax mediterranei. Extremophiles 11(6):759–767
Meghwanshi GK, Verma S, Srivastava V, Kumar R (2022) Archaeal lipolytic enzymes: current developments and further prospects. Biotechnol Adv 61:108054
Melanie S, Winterburn JB, Devianto H (2018) Production of biopolymer polyhydroxyalkanoates (pha) by extreme halophilic marine archaea Haloferax mediterranei in medium with varying phosphorus concentration. J Eng Technol Sci 50(2):255–271
Moghimipour E, Abedishirehjin S, Baghbadorani MA, Handali S (2021) Bacteria and Archaea: a new era of cancer therapy. J Control Rel 338:1–7
Moopantakath J, Imchen M, Anju VT, Busi S, Dyavaiah M, Martínez-Espinosa RM, Kumavath R (2023) Bioactive molecules from haloarchaea: scope and prospects for industrial and therapeutic applications. Front Microbiol 14:1113540
Naitam MG, Kaushik R (2021) Archaea: an agro-ecological perspective. Curr Microbiol 78(7):2510–2521
Ng HS, Wan PK, Kondo A, Chang JS, Lan JCW (2023) Production and recovery of ectoine: a review of current state and future prospects. Processes 11(2):339
Nikolova C, Gutierrez T (2021) Biosurfactants and their applications in the oil and gas industry: current state of knowledge and future perspectives. Front Bioeng Biotechnol 9:626639
Orellana LH, Francis TB, Krüger K, Teeling H, Müller MC, Fuchs BM et al (2019) Niche differentiation among annually recurrent coastal Marine Group II Euryarchaeota. The. ISME J 13(12):3024–3036
Pfeifer K, Ergal İ, Koller M, Basen M, Schuster B, Rittmann SKM (2021) Archaea biotechnology. Biotechnol Adv 47:107668

Poli A, Di Donato P, Abbamondi GR, Nicolaus B (2011) Synthesis, production, and biotechnological applications of exopolysaccharides and polyhydroxyalkanoates by archaea. Archaea 2011(1):693253

Raddadi N, Cherif A, Daffonchio D, Neifar M, Fava F (2015) Biotechnological applications of extremophiles, extremozymes and extremolytes. Appl Microbiol Biotechnol 99(19):7907–7913

Rafiq M, Hassan N, Rehman M, Hayat M, Nadeem G, Hassan F et al (2023) Challenges and approaches of culturing the unculturable archaea. Biology 12(12):1499

Rascovan N, Maldonado J, Vazquez MP, Eugenia Farías M (2016) Metagenomic study of red biofilms from Diamante Lake reveals ancient arsenic bioenergetics in haloarchaea. ISME J 10(2):299–309

Rastädter K, Wurm DJ, Spadiut O, Quehenberger J (2020) The cell membrane of Sulfolobus spp.—homeoviscous adaption and biotechnological applications. Int J Mol Sci 21(11):3935

Rawat M, Chauhan M, Pandey A (2024) Extremophiles and their expanding biotechnological applications. Arch Microbiol 206(6):247

Reichart NJ, Bowers RM, Woyke T, Hatzenpichler R (2021) High potential for biomass-degrading enzymes revealed by hot spring metagenomics. Front Microbiol 12:668238

Rekadwad BN, Li WJ, Gonzalez JM, Punchappady Devasya R, Ananthapadmanabha Bhagwath A, Urana R, Parwez K (2023) Extremophiles: the species that evolve and survive under hostile conditions. 3 Biotech 13(9):316

Reyes-Reyes AL, Valero Barranco F, Sandoval G (2022) Recent advances in lipases and their applications in the food and nutraceutical industry. Catalysts 12(9):960

Sahli K, Gomri MA, Esclapez J, Gómez-Villegas P, Bonete MJ, León R, Kharroub K (2022) Characterization and biological activities of carotenoids produced by three haloarchaeal strains isolated from Algerian salt lakes. Arch Microbiol 204(1):6

Samanta D, Govil T, Saxena P, Thakur P, Narayanan A, Sani RK (2022) Extremozymes and their applications. In: Extremozymes and their industrial applications (pp 1-39). Academic Press

Sandoval G, Rivera I, Barrera-Rivera KA, Martínez-Richa A (2010) Biopolymer synthesis catalyzed by tailored lipases. Macromol Symp 289(1):135–139

Santhosh PB, Genova J (2022) Archaeosomes: new generation of liposomes based on archaeal lipids for drug delivery and biomedical applications. ACS Omega 8(1):1–9

Satyanarayana T, Littlechild J, Kawarabayasi Y (2013) Thermophilic microbes in environmental and industrial biotechnology. Biotechnol Thermophiles 3:1–954

Schelert J, Rudrappa D, Johnson T, Blum P (2013) Role of MerH in mercury resistance in the archaeon *Sulfolobus solfataricus*. Microbiology 159(Pt_6):1198–1208

Schiraldi C, Giuliano M, De Rosa M (2002) Perspectives on biotechnological applications of archaea. Archaea 1(2):75–86

Schomburg I, Jeske L, Ulbrich M, Placzek S, Chang A, Schomburg D (2017) The BRENDA enzyme information system–from a database to an expert system. J Biotechnol 261:194–206

Simó-Cabrera L, García-Chumillas S, Hagagy N, Saddiq A, Tag H, Selim S et al (2021) Haloarchaea as cell factories to produce bioplastics. Mar Drugs 19(3):159

Singh A, Singh AK (2017) Haloarchaea: worth exploring for their biotechnological potential. Biotechnol Lett 39(12):1793–1800

Straub CT, Counts JA, Nguyen DM, Wu CH, Zeldes BM, Crosby JR et al (2018) Biotechnology of extremely thermophilic archaea. FEMS Microbiology Rev 42(5):543–578

Tahon G, Geesink P, Ettema TJ (2021) Expanding archaeal diversity and phylogeny: past, present, and future. Annu Rev Microbiol 75(1):359–381

Van Der Maarel MJ, Van der Veen B, Uitdehaag JC, Leemhuis H, Dijkhuizen L (2002) Properties and applications of starch-converting enzymes of the α-amylase family. J Biotechnol 94(2):137–155

Vivek K, Sandhia GS, Subramaniyan SJBA (2022) Extremophilic lipases for industrial applications: a general review. Biotechnol Adv 60:108002

Woose CR, Kandler O, Wheelis ML (1990) Towards a natural system of organisms: proposal for the domains Archaea, Bacteria, and Eucarya. Proc Natl Acad Sci USA 87(12):4576–4579

Xie R, Wang Y, Huang D, Hou J, Li L, Hu H, et al (2022) Expanding Asgard members in the domain of Archaea sheds new light on the origin of eukaryotes. Sci China Life Sci 65(4):818–829
Zhang L, Kang M, Xu J, Huang Y (2015) Archaeal DNA polymerases in biotechnology. Appl Microbiol Biotechnol 99(16):6585–6597
Zuo ZQ, Xue Q, Zhou J, Zhao DH, Han J, Xiang H (2018) Engineering Haloferax mediterranei as an efficient platform for high level production of lycopene. Front Microbiol 9:2893

Chapter 7
Microbial Biotechnology in Medicine

Emily Addington, Alexa Gannon, Daniel E. Larcombe, Andreas C. Lawaetz, Rebecca E. McHugh, Silja Vahtokari and John T. Munnoch

Abstract The convergence of biology, technology, and medicine has revolutionised healthcare, with microbial biotechnology at the forefront. While many microbes are often considered solely for their infectious properties, many are major producers of natural products, including antimicrobials. Now, not only sources of clinically relevant drugs, they are also being directly engineered for advanced applications such as targeted drug delivery, immune modulation, and precision therapeutics. Microorganisms are key sources of novel antimicrobials, immunomodulatory, and anticancer agents, which synthetic biology and genomics mining can exploit. Bioengineering and exploration of underused microbial taxa offer promising solutions to the problem of rising antimicrobial resistance. Microbes also play crucial roles in modern vaccine development, from live attenuated to recombinant antigen production. The human microbiome has emerged as an interesting player in health, driving innovations in diagnostics and therapies that include next-generation probiotics and microbiota transplants. Furthermore, synthetic biology further empowers the design of 'smart' microbes for *in situ* therapeutic functions like imaging, biosensing, and targeted treatment. While transformative, these innovations also raise critical ethical and regulatory concerns, including biosafety, ecological impact, data privacy, and equitable access. This chapter explores the multifaceted roles of microbes in medical biotechnology—spanning therapeutics, vaccines, microbiome-based interventions, and engineered systems—underscoring their importance in the evolution of sustainable, personalised healthcare.

E. Addington · R. E. McHugh
School of Infection and Immunity, University of Glasgow, Glasgow, UK

A. Gannon · D. E. Larcombe · A. C. Lawaetz · S. Vahtokari · J. T. Munnoch (✉)
Strathclyde Institute of Pharmacy and Biomedical Sciences, University of Strathclyde, Glasgow, UK

J. T. Munnoch
e-mail: John.T.Munnoch@strath.ac.uk

M. Çalışkan and M. Arıkan (eds.), *Microbial World: Bacteria and Archaea*, Progress in Molecular and Subcellular Biology 62, https://doi.org/10.1007/978-3-032-11438-9_7

Keywords Antimicrobials · Vaccines · Microbiomes · Engineered microbes · Synthetic biology

7.1 Introduction

The twenty-first century has witnessed an unprecedented convergence of biology, technology, and medicine, leading to transformative innovations in healthcare. Central to this revolution is microbial biotechnology—the strategic application of microorganisms, including bacteria and archaea, to develop medical interventions that prevent, diagnose, or treat disease. While microbes were once solely viewed through the lens of their pathogenic potential, the last century has proved they are indispensable allies in the quest for human health and longevity.

Microorganisms have long played a foundational role in medicine, most famously in the production of antimicrobials such as penicillin (Fleming 1929) and streptomycin (Schatz et al. 1944). Following the historic discovery of these two antibiotics, bacteria and fungi have seen widespread use in industry, e.g., in large-scale industrial fermentations that produce these and many other antibiotics. However, modern microbial biotechnology expands far beyond traditional fermentation processes. Advances in genomics, synthetic biology, systems biology, and bioinformatics have opened new avenues for harnessing microbes as sophisticated tools in drug discovery, personalised medicine, vaccine development, and even as living therapeutics (Santos-Beneit 2024). Today, researchers are not only mining natural microbial diversity for bioactive compounds but are also engineering bacteria and archaea to perform customised medical functions—from delivering drugs at disease sites to modulating the immune system.

A major focus of microbial biotechnology is the use of bacteria, fungi, plants, and archaea in the discovery and production of novel drugs (considered in Sect. 7.2). These organisms are prolific producers of secondary metabolites, many of which possess potent antimicrobial, anticancer, or immunomodulatory properties (Barona-Gómez et al. 2023). Archaea, though historically less studied than bacteria, are gaining traction in bioprospecting due to their unique metabolic capabilities and resilience in extreme environments (Aparici-Carratalá et al. 2023a). As antimicrobial resistance (AMR) continues as a global health crisis (Murray et al. 2022), with a prediction of ~1.5 million deaths attributable per year (and 10 million per year predicted by 2050), microbial biotechnology offers new hope through the exploration of underutilised taxa and bioengineering approaches to develop next-generation therapeutics.

While historically one of the most pivotal developments in the clinical field, another advancing domain is microbial vaccine development (considered in Sect. 7.3). Live attenuated bacterial vectors, recombinant proteins produced in microbial hosts and engineered microbes expressing antigenic epitopes are redefining vaccine strategies (Cid and Bolívar 2021).

In parallel, the human microbiome—the vast and diverse microbial ecosystem residing in and on the human body—has emerged as a critical determinant of health and disease (considered in Sect. 7.4). Advances in sequencing technologies have enabled a deeper understanding of how shifts in microbiome composition influence metabolic disorders, immune responses, neurological function, and even cancer progression (Gilbert et al. 2018; Ma et al. 2024). This knowledge is giving rise to microbiome-based diagnostics and interventions, including faecal microbiota transplants, pro and prebiotics, and next-generation probiotics tailored to individual microbiome profiles.

Engineered microbes represent a particularly exciting frontier (considered in Sect. 7.5). Synthetic biology has empowered researchers to programme bacteria and archaea with circuits that sense environmental cues, perform logic operations and produce therapeutic outputs *in situ* (Chowdhury et al. 2019). These 'smart' microbes hold promise for applications such as targeted cancer therapy, sensing of disease biomarkers and precision modulation of the gut-brain axis. With CRISPR-based tools and genome-scale engineering, microbial chassis can be fine-tuned for safety, efficacy and regulatory compliance.

Yet, alongside these scientific advances lie complex regulatory and ethical considerations (considered in Sect. 7.6). The release of engineered microbes into the human body or environment raises important questions about biosafety, biosecurity, and long-term ecological impacts. Regulatory frameworks must advance to keep pace with innovation, ensuring that products are rigorously evaluated without stifling progress. Ethical issues also arise around data privacy in microbiome research, equitable access to microbial therapies, and the patenting of naturally occurring or engineered organisms.

In an era of increasing antimicrobial resistance (AMR), global pandemics and chronic disease burden, microbial biotechnology has proven itself an innovative and resilient field. By harnessing the remarkable capabilities of bacteria and archaea—both naturally occurring and engineered—we have the opportunity to reshape the medical landscape and deliver more sustainable, precise, and effective healthcare solutions.

7.2 Bacteria and Archaea in Antimicrobial Drug Discovery and Production

Antimicrobials are compounds that kill or inhibit the growth of microorganisms, including bacteria, fungi, algae, protozoa and, more controversially, viruses. Most of the clinically relevant antimicrobials are naturally produced by bacteria, fungi, and plants (Hutchings et al. 2019; Khameneh et al. 2019; Kumar et al. 2021), and are an important cornerstone of modern medicine. Bacteria have been used in natural product drug discovery and production since the early 1900s with the first discovery of the antibiotic Streptomycin in 1943, led by Selmon Waksman (Hutchings et al. 2019). The use of archaea in drug discovery is a significantly younger field; however, it has already provided interesting and novel results.

While antimicrobial research and development is a strategic priority globally, the economics surrounding industrial development have been marked with declining interest, resources, and success stories for decades (Blaskovich and Cooper 2025). Even with the discovery of new or novel compounds in development and an increase in antibiotic consumption, there has been limited re-engagement from industry and big pharma due to the substantial risk of investment and failure. While key incentive strategies including CARB-X (The Combating Antibiotic Resistant Bacteria Accelerator) and GARDP (Global Antibiotics Research and Development Partnership) are stimulating the field, it is clear that the future antimicrobial discovery is only one weapon in our arsenal.

This section will highlight and explore antimicrobial (antibacterials, antifungals, and antivirals) compounds produced by bacteria and archaea.

7.2.1 Antimicrobials

Some of the most prolific antimicrobial-producing bacteria come from soil and specifically from the order Actinomycetales. Perhaps the most famous genus from this order is *Streptomyces. Streptomyces* are responsible for the industrial production of over 2/3rds of all clinically relevant antimicrobials and not just their discovery (Hutchings et al. 2019). The study of *Streptomyces*, and specifically the study of the antibiotic actinorhodin, has, for example, led to the discovery of biosynthetic gene clusters (BGCs), groups of linked genes that include all the relevant genetic information for the production, export, and self-resistance of a particular product (Schlimpert and Elliot 2023). Genome mining has led to the discovery that many of these organisms encode 20–50 natural products, though only a few are produced under normal lab conditions (Hoskisson and Seipke 2020). Approaches to manipulate the host genome through genetic manipulation and synthetic biology are at the forefront of this research.

In addition to large-scale screening projects in the pre-genomic era, advances in genome sequencing, and bioinformatic tools for genome mining allow allow us to search for and interrogate BGCs in sequenced and assembled genomes (Randall and Davies 2021). One such tool is antiSMASH (antibiotics and Secondary Metabolite Analysis Shell). AntiSMASH works by predicting BGC based on a database of identified BGCs and providing tractable outputs and resources to aid in analysis (Medema et al. 2011).

Many bacteria produce antimicrobial peptides called bacteriocins. Similar products have been observed in a few archaea species, termed ‘archaeocins’. Thus far, two classes of archaeocins have been discovered: sulfolobicins produced by *Sulfolobales*, and halocins produced by *Halobacteriales*. Like bacteriocins, these peptides are being explored for their therapeutic potential as antimicrobials. For example, *Halobacteriales* produce the KPS1 halocin, which was effective against gram-positive and gram-negative bacteria (Kavitha et al. 2011). Halocins have the added benefit of being stable at high temperatures and variations in pH, which makes them attractive options for remote isolated communities where the distances and terrain can make it challenging to store drugs at low temperatures (Haque et al. 2019). For the remainder of this section, we will focus on antibacterial, antiviral and antifungal compounds.

7.2.2 Antibacterial Compounds: Classification and Function

Classification of antibacterials is performed according to their molecular structure and functional targets. Some of the most notable antibacterial classes (Table 7.1) include beta-lactams, tetracyclines, aminoglycosides, and macrolides (De Simeis

Table. 7.1 Example classes of antibacterials with natural origins. Adapted from De Simeis and Serra (2021)

Class	Target	Defining molecular feature	Famous examples and their producers
Beta-lactams	Cell wall biosynthesis	Four-member cyclic amide	Clavulanic acid (*Streptomyces clavuligerus*) Nocardicin A (*Nocardia uniformis subsp. tsuyamanensis*) Thienamycin (*Streptomyces cattleya*)
Tetracyclines	Protein synthesis through the 16S rRNA site	Naphthacene carboxamide nucleus	Tetracycline (*Streptomyces aureofaciens*) Oxytetracycline (Streptomyces rimosus)
Aminoglycosides	Protein synthesis through the 16S rRNA site	Amino-sugar core structure	Streptomycin, kanamycin, neomycin (Many *Streptomyces spp.)* Gentamycin (*Micromonospora spp.*)
Macrolides	Translocation of the ribosome through 50S subunit of the prokaryotic ribosome	Macrocyclic lactone with one or two sugar moieties	Erythromycin (*Aeromicrobium erythreum, Streptomyces erythreus*) Spiramycin (*Streptomyces ambofaciens*) Oleandomycin (*Streptomyces antibioticus*)

and Serra 2021). While the majority of antibacterials are natural products, there are an increasing number of semisynthetic and fully synthetic antibiotics (De Simeis and Serra 2021; Stojković et al. 2023).

Often, natural products can have poor physical and chemical properties for clinical use. For example, these can include solubility issues in water, or they are easily degraded in the human body, reducing their effectiveness. Semisynthetic antimicrobials can address this, modifying the natural products, often by adding specific chemical groups to address these issues. Many natural products are large, complex molecules that are impossible or too costly to make in the lab. Fully synthetic antimicrobials are often small, relatively simple molecules. Examples include salvarsan, also known as arsphenamine, the first modern antimicrobial agent, which was deployed against syphilis (Vernon Gervase 2019) and chloramphenicol, which is a natural product from *S. venezualae* but is now produced entirely synthetically at an industrial scale (Tevyashova 2021).

Functional targets of antibacterials are limited. To be effective, they rely on the differing fundamental biology of eukaryotes and prokaryotes (De Simeis and Serra 2021). Targets include synthesis of cell wall (human tissues have no cell wall), protein synthesis (different ribosome structure and subunits), the different enzyme pathways involved in DNA and RNA synthesis and the specific components of the cell membrane (differing lipid content). The fundamental differences in the cell walls/membranes of Gram-negative (inner and outer membrane and thin cell wall) and Gram-positive bacteria (single membrane and thick cell wall) define the target specificity of many compounds and are a specific challenge. Antibacterials have either bactericidal (kill) or bacteriostatic (inhibit growth) effects. These differences don't completely stop unwanted side-effects, and classes like aminoglycosides are known to cause nephro- and ototoxicity (Lang et al. 2023).

While antibacterials and antimicrobials in general have been pivotal in advancing clinical practices, major challenges face us now and in the future (Farha et al. 2025). Far fewer antibacterials are effective against Gram-negative bacteria than Gram-positive bacteria. Our limited understanding of the permeability of the Gram-negative bacterial cell membranes, in combination with the developing AMR crisis, along with limitingly generic in vitro screening methods for antimicrobial compounds, restricts our capacity to develop novel and impactful treatments. One area of research that aims to combat these issues is the use of antivirulence (AV) drugs. These compounds act differently from traditional antibiotics by targeting virulence mechanisms rather than affecting bacterial viability. AV drugs may represent an important tool for infections in which the use of antibiotics is contraindicated, for example, Shiga Toxin-producing *E. coli* infections in which toxin production can be activated by antibiotic therapy (Tucker et al. 2024). An example of an AV compound produced by bacteria is Aurodox, a specialised metabolite of *Streptomyces goldiniensis* (McHugh et al. 2019). Aurodox is a type III secretion system inhibitor, capable of blocking STEC attachment to epithelial cells, and is one of many AV compounds that show great potential in future use against bacterial infections (Mchugh et al. 2022). Another group of AV drugs is quorum-sensing inhibitors (QSI) (Rather et al. 2022). Many bacterial pathogens form biofilms, protective structures that can inhibit phage infection, antimicrobial action, and provide a defensive home to their hosts. Especially in the case *Pseudomonas aeruginosa*, an emerging global public-health threat, biofilm formation and other virulence factors are controlled by Quorum sensing (QS), an intercellular or cell-cell communication method. QSI offers an effective approach to combating virulence and AMR by reducing drug-resistant selection pressures.

7.2.3 Antiviral Compounds

Viruses are the most abundant biological entities on Earth, with an estimated 10^{31} particles and a virus-to-cell ratio of approximately 10:1 (Hendrix et al. 1999; Bar-On et al. 2018; Mushegian 2020). Viruses are unable to propagate autonomously and hence depend on the molecular machinery of host cells to reproduce (Louten 2016; Hobbs and Abedon 2016). Viruses have evolved to infect organisms across all domains of life—including bacteria (as bacteriophages), archaea, and eukaryotes (Broecker and Moelling 2019).

Historically, antiviral drug development has not progressed as quickly as antibacterial drugs (Swartz 2000), in part due to the unique intracellular life cycles of viruses and their relatively small genomes, which present fewer therapeutic targets compared to bacterial pathogens (Müller and Kräusslich 2009). Considering this, viral infections continue to cause significantmorbidity and mortality worldwide (Nathanson 2016). Compounding this challenge is the hypermutable nature of viruses (Sanjuán et al. 2010), which facilitates the rapid development of resistance to antiviral therapies (Müller and Kräusslich 2009). These challenges highlight the ongoing need for the discovery of novel antiviral agents. Recent discoveries of conserved antiviral defence systems (Ofir et al. 2021; Cury et al. 2024), along with broad-spectrum antiviral compounds (Martinez et al. 2015; Bernheim et al. 2021), demonstrating the potential of microbial genome mining to aid discovery of natural products with antiviral activity.

Bacterial and archaeal natural products have demonstrated antiviral activity against viruses infecting humans (Raihan et al. 2021). A recurrent theme of antiviral compounds includes chain terminators that inhibit viral RNA polymerase to halt transcription of viral genes (Müller and Kräusslich 2009). Notable examples include aciclovir, 3′-azido-3′-deoxythymidine (AZT), and sofosbuvir which are used to treat herpes virus, HIV, and hepatitis C, respectively (Müller and Kräusslich 2009; Feld et al. 2015).

In some cases, antiviral proteins (rather than small molecules) have been identified. The interferon induced human Viperin protein is another chain terminator with broad antiviral activity against numerous viruses including HIV and hepatitis C (Helbig and Beard 2014; Rivera-Serrano et al. 2020). It catalyses the conversion of CTP to ddhCTP which lacks the 3' hydroxyl group required for RNA chain elongation, thereby terminating viral transcription (Gizzi et al. 2018). Intriguingly, homologs of the viperin gene have been identified in bacteria and archaea, where they exhibit antiphage activity—for example, against phage T7 in *E. coli* (Bernheim et al. 2021).

Moreover, expression of the human viperin gene in *E. coli* confers similar antiphage effects (Bernheim et al. 2021). This functional conservation raises the possibility that microbial viperins might also possess antiviral activity against human-infecting viruses, although this remains to be experimentally tested.

Other antiviral strategies include inhibition of viral attachment and several natural products produced by bacteria have been shown to bind the viral envelope glycoprotein gp120 to inhibit viral entry and transmission between cells. Examples include cyanovirin-N (from *Nostoc ellipsosporum*), agglutinin (from *Oscillatoria agardhii*), and labyrinthopeptin A1 (from *Actinomadura namibiensis*) (Boyd et al. 1997; Sato et al. 2007; Koharudin et al. 2012; Férir et al. 2013).

7.2.4 Antifungal Compounds: Classification and Function

The global burden of fungal infections is currently increasing, with worrying rates of infection and antifungal resistance. Approximately a quarter of the human population has a dermatophytosis infection that can lead to tissue, organ and nerve damage. Furthermore, in rarer extreme cases where invasive fungal infections become systemic, they have an over 90% mortality rate leading to more than 1.5 million deaths annually (Zhang et al. 2023). Fungi are eukaryotes and hence have, relative to bacteria and viruses, fewer fundamental differences in cell biology, when compared to humans, which ultimately leads to even fewer differential molecular targets and relatively fewer identified treatment regimes. Due to the high rate of dermatophytosis, a clear distinction can be made between topical (i.e., surface level treatment, e.g., antifungal creams) and systemic treatments (i.e. blood stream and internal organs, e.g., intravenous or ingested drugs) where localised treatment can limit human tissue damage.

Targets include cell wall (glucan synthase and chitin), cell membrane (ergosterol synthesis pathway and direct binding), lipid metabolism, DNA/RNA functions, mitochondrial function, tubulin function, signalling pathways, and transport functions (Zhang et al. 2023). Of the 80+ natural product antifungal compounds with clinical relevance, only 16 have bacterial origins, while the remaining have mainly fungal or plant origins (Zhang et al. 2023).

While far more antibacterial compounds have been discovered/reported, it is unclear whether this is due to a relatively higher abundance of producing sources, bias in the search for and screening of antibacterials over antifungals, or the difficulty in finding new and novel antifungals that do not also cause harm to human health. The need for novel antifungal compounds is pressing, with the vast number of deaths worldwide alongside the increase in AMR among strains. As with bacterial and viral infections, fungal infections are a global priority with the search for new compounds ongoing (WHO fungal priority pathogens 2022).

7.2.5 *Conclusions and Future Perspectives*

As we face the growing challenge of antimicrobial resistance, the opportunities to discover novel natural products with antimicrobial activity have never been greater. Advances in genomics and the development of tools to interrogate BGCs offer remarkable potential to uncover a plethora of uncharacterised antimicrobials. Ongoing research into this field holds the promise of sparking a renaissance in antimicrobial development.

7.3 Bacteria and Archaea in Vaccine Development

7.3.1 *Introduction*

Throughout history, few medical breakthroughs involving microorganisms have revolutionised human health and survival as profoundly as the development of vaccines. The first vaccine is credited to Dr Edward Jenner's smallpox vaccine in 1796. Smallpox was a viral infection that is estimated to be responsible for 10% of all mortality in Europe during the eighteenth century. Today, thanks to a global vaccination effort, smallpox is the first human pathogen to be completely eradicated (Berche 2022).

Vaccines work by exposing the immune system to antigens specific to a pathogen. The antigens present in the vaccine trigger an immune response as the body attempts to neutralise the vaccine. In many vaccines, the humoral response mediated by the B cells is responsible for the long-term immunity as antigens are presented to B cells, which differentiate into long-lived memory cells (Pollard and Bijker 2020). Most vaccines will also include the involvement of the cellular immune system to both enhance the humoral response and directly mediate pathogen killing by T-cells, which is the case for the tetanus vaccination (Fryauff et al. 1999; van Riet et al. 2008). When the antigen, which originally triggered the immune system, is detected again, for example, during contact with the fully active pathogen, by memory B cells can rapidly differentiate into antibody-producing plasma cells. This rapid response, facilitated by the memory cells, allows the infection to be dealt with before it can become established and overwhelm the immune system (Ghattas et al. 2021).

Research into vaccines has focused on finding safe methods to trigger a strong, effective immune response to the antigens without risk of live infection or damage from overactivation of the immune system. Designing vaccines, therefore, requires in-depth knowledge of the biology of the disease-causing microorganisms.

7.3.2 *Overview of Traditional Vaccination Strategies*

Many diverse strategies have been developed for vaccinations. In the case of Edward Jenner's smallpox vaccine, live cowpox, a related but less virulent virus, was used.

This virus retains many of the recognisable molecular markers the immune system responds to, without the high level of virulence. Introduction of this live virus triggered a robust immune response due to activation by pathogen-associated molecular patterns (PAMPS) that closely mimicked those of the smallpox virus.

For pathogens without a closely related, less virulent organism, creation of a weakened form of the pathogen can provide a robust immune response with a significant reduction in the risk of causing disease. This class of vaccine is known as live attenuated. Strategies to attenuate bacteria for the development of vaccines can include genetic manipulation if virulence factors are known, or random mutagenesis (treatment with DNA/RNA mutagens) can be utilised to produce avirulent strains (by treating with mutagens, then selecting viable strains with reduced virulence identifiers). As these pathogens are still live and present antigens as live pathogens would, these types of vaccines often elicit a robust and long-lived protection. Some bacterial examples include typhoid T21a and cholera CVD 103-HgR (Detmer and Glenting 2006). Although live attenuated bacteria play an important role in more recently developed vaccination platforms described later, the possibility of reversion to virulence makes other platforms more appealing, especially for those with weakened immune systems (Lauring et al. 2010).

A similar technique to live attenuated vaccines is the use of inactivated pathogens, which are cells that have been killed or otherwise inactivated. Techniques used to prepare these include heat inactivation or chemical inactivation using, for example, formalin. Using inactivated cells is a straightforward approach that is often easy to produce, provided the bacteria can be cultured. Much of the early research into vaccines focused on this approach. Due to this ease, many bacterial infections have been trialled for inactivated vaccine, ranging from enteric pathogens such as *Campylobacter jejuni* and *Shigella spp.*, respiratory pathogens such as Streptococcus spp. and others, including urinary tract *Escherichia coli* (Pace et al. 1998). Inactivated vaccines are considerably safer than attenuated with no chance of replicating. Although many early studies focused on this immunisation strategy, researchers are now focusing on other techniques. This is primarily due to the weaker immune response generated by dead compared to live cells. The resulting protection is often short-lived, requiring the need for multiple booster shots to maintain immunity. Although cheaper to produce, the repeated immunisations required reduce the economic benefit of inactivated vaccines (Pace et al. 1998).

Previously mentioned vaccination strategies relied on using whole cells; however, in many cases, specific pathogenic antigens are enough to elicit an immune response. These acellular subunit vaccines include proteins (multiple or singular whole, subunit or peptide), polysaccharides, conjugated protein polysaccharide, or inactivated toxic proteins called toxoids. Examples of each of these include hepatitis B, *Neisseria meningitidis, Haemophilus influenzae* type B, and tetanus, respectively (Pollard 2020). The antigens present in subunit vaccines are selected to be specific to the pathogenic organisms and unable to cause major damage, for example, virus capsid proteins or polysaccharides from the bacterial cell wall. Antigens that are specific to a pathogen but are not strongly immunogenic by themselves can be joined with a strongly antigenic protein, for example, tetanus toxoid or meningococcal outer membrane protein.

Conjugation of weaker antigens with proteins allows a stronger immune response, primarily through the involvement of the cellular branch of immunity, which is often lacking with non-protein antigens (Pollard and Bijker 2020).

The final major class of vaccines that sees widespread use is nucleic acid vaccines, where specific DNA or mRNA molecules encoding pathogenic antigens are delivered into cells. The antigen is then produced by the host transcriptional machinery, and a strong immune response is elicited. The main vaccines of this class are the SARS-CoV-2 RNA vaccines developed during the COVID-19 pandemic (Pollard and Bijker 2020).

As well as vaccines containing the antigen components of the target pathogen listed above, adjuvants are also often included. Traditional examples of vaccine adjuvants include aluminium salts and oil–water emulsions. Adjuvants can increase the effectiveness of vaccinations via several mechanisms, including resembling co-stimulatory molecules such as PAMPs, facilitating antigen trafficking to lymph nodes, enhancing antigen presentation to antigen presenting cells, and prolonging the bioavailability of antigens (Zhao et al. 2023).

7.3.3 Role of Bacteria in Vaccine Development and Research

Bacteria play numerous essential roles in the research and development of almost all vaccines, even those designed to protect against viruses, fungi and cancer. Fundamental biological knowledge about bacterial pathogens is clearly key to the development of vaccines, allowing us to design adjuvants and vaccination strategies that help mimic infections as closely as possible. For attenuated and inactivated bacterial pathogens, the role of bacteria is clear. The bacteria must first be cultured to be able to harvest the vaccine. Development of methods of attenuating bacteria can also include passaging of bacterial cultures until virulence is lost (through selection pressures). Alternatively, directed genetic engineering of bacterial species utilising synthetic biology. Genetic engineering can not only include weakening the target pathogen but can also involve the expression of heterologous antigens on commensal or otherwise harmless 'carrier' bacteria (Galen et al. 2021). Clinical trials using attenuated *Salmonella enterica* TY2 expressing the virulent *Streptococcus pneumoniae* surface protein A (PspA) or *Pseudomonas aeruginosa* outer membrane proteins (OprF and OprI) have shown partial success at inducing a response to the target pathogen and both showed a promising safety profile (Galen et al. 2021).

In the case of protein subunit vaccines, production of the desired peptide can either be from the bacterial pathogen itself, followed by protein purification or the desired protein can be recombinantly expressed in a heterologous bacterial host. *Escherichia coli* is the most used host because it is easily transformed and has a thoroughly dissected genome, enabling many specialised protein-expressing strains to be developed, such as BL21 (Pouresmaeil and Azizi-Dargahlou 2023). Although *E. coli* is often the easiest and most effective system, alternative expression species such as yeast may be more suitable for the expression of large or glycosylated proteins, and this should be evaluated on a case-by-case basis. Despite the inability to post-translationally modify viral proteins, *E. coli* has successfully been used to produce a recombinant Hepatitis E virus-like particle vaccine named Hecolin, which is currently licensed for use in China (Zhang et al. 2014).

Some vaccine adjuvants are produced by bacterial systems. These include the bacterial cell wall components lipopolysaccharides (LPS), and lipid A or specific bacterial DNA sequences recognised as danger signals such as CpG motifs. While LPS and lipid A are potent activators of the immune system, there has historically been issues with safety due to high toxicity. Modified versions such as monophosphoryl lipid A show reduced toxicity, and certain bacterial species' LPS forms show reduced toxicity (Zhao et al. 2023).

As mentioned previously, bacterial toxoids or inactivated protein toxins are used as complete vaccines but are also fused with polysaccharides to enhance the immune response. The conjugation process is often traditionally performed chemically in vitro, but researchers have developed mutants that allow *E. coli* to produce glycosylated proteins. Recently, a cell-free system using dehydrated *E. coli* lysate has been demonstrated which represents a promising method of transporting conjugate vaccines without refrigeration (Warfel et al. 2023).

Although mRNA vaccines such as those used for the treatment of COVID-19 are usually produced by cell-free in vitro transcription reactions the process is still fundamentally reliant on the use of bacterial products. Many of the reagents for the reverse transcriptase reaction, such as the enzymes and plasmid template are produced by *E. coli* fermentation (Liu 2019). Additionally, bacterial systems have been evaluated as a delivery mechanism for nucleic acid-based vaccines, although none have been tested. Bactofection, a method of gene therapy where bacteria are used as vectors to deliver genetic material into eukaryotic cells, has some advantages over other methods of transforming mammalian cells, including intracellular delivery producing a strong immune response, delivery of very large cargo sequences, and low cost. However, the strong immune response may also be detrimental in some cases and as with all attenuated vaccine strategies, there is the risk of reversion to virulence for the transfecting bacteria (Heggie et al. 2025). The creation of the plasmids for mRNA and DNA vaccines is reliant on cloning processes facilitated by selection in *E. coli*. Indeed, this is true for any vaccine application requiring the recombination of heterologous DNA, including but not limited to recombinant protein expression, conjugate vaccines, and attenuation of pathogenic strains.

The final emerging technique for using bacteria in vaccination efforts is the use of bacterial Outer Membrane Vesicles (OMVs). OMVs are small lipid nanoparticles derived from the outer membrane of Gram-negative bacteria. They contain the same phospholipid and protein makeup as the bacterial membrane and are often highly immunogenic but cannot replicate themselves, limiting OMVs' capacity to cause disease. Two OMV-based vaccines have been licensed, VA-MENGOC-BC and Bexsero, both of which immunise against *N. meningitides* B (Kashyap et al. 2022). Heterologous gene expression can be combined with OMV formation to produce effective OMV vaccines against nonbacterial targets, such as viruses. In one study, a fusion protein of the Influenza H1N1 virus haemagglutinin and receptor binding domain of the Middle East Respiratory Syndrome Coronavirus was expressed on the surface of *E. coli*-derived OMVs, and these were found to successfully immunise mice against both viral infections (Shehata et al. 2019; Kashyap et al. 2022).

7.3.4 Role of Archaea in Vaccine Development and Research

Although untested in human clinical trials, archaea-derived vesicles have shown promise as delivery platforms and adjuvants for vaccines. Archaesomes are liposomes derived from the lipids of archaea cells. In comparison to bacterial vesicles, archaesomes are more thermally stable, making delivery of vaccines significantly cheaper. They are also more resistant to extremes of pH, chemical degradation, and oxidative damage, potentially extending shelf life. Gas vesicles are another type of vesicle produced by archaea naturally to aid with floatation, and these have also been investigated by researchers as vaccine delivery platforms. Both Archaesomes and Gas vesicles have shown promising initial results, activating both humoral and cellular arms of the immune system, being readily phagocytosed and acting as adjuvants themselves. Successful vaccinations using archaeal vesicles have been demonstrated in animal models against a host of pathogenic bacteria and viruses. Reviews by Natalia Adamiak et al. (2021) and Amy M Hill et al. (2020) provide a comprehensive summary of these recent results. Despite this, research on archaeal-produced vaccines and the genetic tools available is still in its infancy compared to the long history of bacterial exploitation.

7.3.5 Conclusions and Future Perspectives

While vaccines are an invaluable tool in medicine, their wider and more dynamic use is in question. Between rapid adaptation, immune avoidance of pathogenic organisms through short-term transient benefits of certain technologies, there are clear areas for vaccine improvement. The use of synthetic biology approaches in engineering bacteria to enhance and develop live attenuated strains and techniques for avirulent antigen presentation, to address this and archaeasomes with their increased stability could be a potential avenue for wider distribution, especially in storage for geographically remote locations, are encouraging avenues. Vaccines represent a valuable strategy for addressing AMR as well as wider public health priorities.

7.4 Microbiome, Phages and Synthetic Biology

7.4.1 Introduction

The dynamic, diverse, and vast community of symbiotic and pathobiontic microbes that inhabit a specific environment is known as a microbiome. Humans are populated by numerous, varied microbiomes, which include those found on the skin, in the oral cavity, gastrointestinal tract, lungs, and vaginal tract. Each of these microbiomes is composed of unique compositions of bacteria, archaea, fungi, protozoa, and viruses suited to their specific niche. Over millennia, these complex microbial ecosystems have evolved alongside their human hosts to form an inter-reliant, sophisticated, and vital relationship in which microbiomes have profound roles in human development, physiology, and health.

It is becoming increasingly apparent that human microbiomes have essential functions in metabolism, immune development, and even behavioural and psychological well-being, with a robust, diverse, and functional microbiome correlated with good health. Conversely, dysbiosis—a state of imbalance and disruption to the composition of the microbiome—is associated with poor health outcomes and disease (Aggarwal et al. 2023). Specifically, microbial dysbiosis and perturbation of microbiome homoeostasis are implicated in conditions such as inflammatory bowel disease (IBD), metabolic syndrome, type 2 diabetes, psoriasis, Parkinson's disease, reflux oesophagitis, cancer, and multiple other diseases (Manos 2022). Resultantly, the field of human microbiome research has advanced in recent years, and there is a growing scientific and public interest in therapies that manipulate the microbiome for health benefits and the treatment of disease.

7.4.2 A Healthy Microbiome

Defining a 'healthy microbiome' is a significant challenge due to the complex, dynamic, and intricate nature of microbiomes. Much remains to be elucidated about how human microbial ecosystems function in health and disease and what constitutes an 'optimal' microbiome, as composition is highly unique to individuals and infinitely varied across the human population (Hou et al. 2022).

Although humans possess distinct microbiomes on the skin and across the digestive, respiratory, urinary, and reproductive systems, it is the gut microbiome that is arguably considered the most significant in relation to host health (Hou et al. 2022). Trillions of microorganisms populate the human gut, with the colon containing the highest microbial density and diversity of any human-associated microbiome. This ecosystem is most predominantly influenced by host diet and can also be altered with age, antibiotic usage, and other medical interventions (Hasan and Yang 2019). It is generally understood that a healthy gut microbiome is highly diverse, stable, and resilient (meaning that it may return to homoeostasis following perturbation), abundantly colonised by bacterial species considered beneficial (e.g. *Bifidobacteria*) and with minimal presence of pathogenic microorganisms (e.g. *Fusobacterium*, *Clostridium*) (Van Hul et al. 2024).

Dysbiosis of the gut microbiome driven by disease, diet, or antibiotic usage can deplete beneficial species, expose the niche to heightened colonisation by harmful microorganisms and shift pathobiontic species towards pathogenicity, predisposing the host to negative health outcomes associated with a dysbiotic gut microbiome (Bidell et al. 2022). Antibiotics can disrupt the composition and abundance of the gut microbiome as they have largely indiscriminate antibacterial activity. For this reason, novel therapeutics are in development to target pathogens with precision, therefore limiting harm to the commensal microbiome. Many of these treatments involve utilising aspects of bacteria-killing viruses, rely on synthetic biology approaches, or are exploring genetic manipulation of the microbiome. Similarly, treatments that aim to improve or restore the condition of the microbiome, such as probiotics, prebiotics, and FMT, have garnered significant scientific and public interest (Fong et al. 2020; Nazir et al. 2024).

7.4.3 Probiotics and Prebiotics

Probiotics are living non-pathogenic microbes, typically bacteria or yeast, believed to confer health benefits to the host when administered orally in an appropriate dosage. Fermented foods such as kefir, cheese, sauerkraut, and kimchi are full of microbes that are considered probiotic and have been recognised for their various beneficial effects on human health for centuries (Soemarie et al. 2021). These probiotic microbes are postulated to exert a favourable effect on human health by improving the host immunological response, antagonising or out-competing pathogens for niche space, and alleviating negative side-effects resulting from antibiotic use (Bodke and Jogdand 2022).

Commercial probiotics added to food and drink or available as oral tablets are widely available and are often a blend of bacterial species. Predominantly used in commercial probiotics are *Lactobacillus, Bifidobacterium, Enterococcus*, and the yeast strain *Saccharomyces*, and these organisms may be further genetically modified to increase the desired effect. Such designer probiotics, or 'pharmabiotics', are engineered and administered alongside other modified strains in 'cocktails' designed to optimise probiotic delivery, survival, and effect (Gul and Durante-Mangoni 2024). The benefits of such strains are thought to derive, in part, from probiotic fermentation of dietary fibres to produce metabolites such as the short-chain fatty acids (SCFAs), acetate, propionate, and butyrate, which are beneficial for gut microbiome health, reducing inflammation, and increasing gut barrier integrity (Markowiak-Kopeć and Śliżewska 2020).

Some researchers have suggested that select species of archaea may be used for a similar function, termed 'archaeobiotics'. The most prominent example is found among the methane-producing archaea found in the human gastrointestinal tract.

Trimethylamine (TMA) is a naturally occurring compound that is produced by the normal gut bacteria and as a by-product of digesting compounds like choline and carnitine, which are abundant in red meat (Mei et al. 2021). Accumulation of TMA and its derivatives can cause damage to mucus membranes and contribute to cardiovascular diseases like atherosclerosis. A landmark study in 2013 found that one species of archaea, *Methanomassiliicoccus luminyensis B10*, was able to deplete TMA during methanogenesis (Brugère et al. 2013). This, coupled with the lack of obvious pathogenesis, led them to propose *Methanomassiliicoccus luminyensis* as a treatment for gut dysbiosis and even liver dysfunction conditions that cause accumulations of TMA.

Other experts say the use of archaeobiotics is premature—we still lack a lot of basic information about archaea biology, as well as the tools to learn more. Current methods only support the culture of a small percentage of species, which limits what can be studied in a laboratory setting compounding the difficulty to accurately isolate and identify the archaea of the human microbiome, or to say definitively whether they can cause disease (Barron 2024).

There is conflicting evidence as to the efficiency of probiotics in treating disease, with results likely depending significantly on how such probiotics are formulated, the diversity, type, and quantity of strains included, and their method of delivery. However, a growing body of evidence suggests that probiotics may be beneficial for patients with diabetes, autoimmune disorders, cancer, gastrointestinal disorders, neurological and psychiatric conditions, and in alleviating side effects associated with antibiotic use (Petrariu et al. 2023).

Prebiotics are dietary compounds that humans are incapable of digesting, but which can be fermented by probiotic microbes to allow for their growth in the gastrointestinal tract. Such dietary compounds are typically oligosaccharide carbohydrates, such as xylooligosaccharides, galacto-oligosaccharides, inulin, and fructose-oligosaccharides, but can also be phenols obtained from various fruits (You et al. 2022). These prebiotic compounds, when fermented by probiotic microbes in the large intestine, produce secondary metabolites such as the SCFA. These are beneficial to the host through various physiological mechanisms, such as the regulation of immunity, resistance to pathogens, fortification of the intestinal barrier, and lowering of blood lipid levels (You et al. 2022). As such, prebiotics are now commercially available in convenient formulations such as powdered inulin, which can be added to food and drink, or formulated alongside probiotic preparations to increase the efficiency and benefit of the probiotics (You et al. 2022).

7.4.4 Faecal Microbiota Transplantation

Faecal Microbiota Transplantation (FMT) is a technique rapidly gaining clinical evidence as a suitable and effective treatment for diseases associated with dysbiosis of the gut microbiome. FMT involves the transplantation of faecal microbiota from a healthy donor to an unwell individual to re-establish or improve the condition of the gut microbiome in the recipient. Typically, FMT techniques involve the preparation of fresh faecal material from a suitable donor by stool resuspension in sterile sodium chloride, followed by blending and filtering to remove large particulate matter. The prepared sample may then be frozen or freeze-dried and reconstituted directly before application to the donor (Porcari et al. 2023). This faecal slurry may be taken orally if prepared in lyophilised tablet-based form or directly administered to the large bowel via rectal enema, nasogastric tube, or colonoscopy (Wang et al. 2019).

Currently, FMT is most used in clinical practice to treat severe cases of dysbiosis seen when prolonged and aggressive antibiotic treatments have resulted in severe *Clostridiodes difficile* infections (CDI) (Porcari et al. 2023). However, there is emerging interest in FMT as a treatment for Ulcerative Colitis (UC), obesity and metabolic syndrome, which are conditions associated with gut microbiome dysbiosis (Wang et al. 2019; Feng et al. 2023).

Currently, the most significant challenge to establishing FMT as a clinical treatment is the identification of suitable faecal microbiome donors, who must be strictly screened to prevent the transfer of infectious organisms and disease from the donor to the recipient. As all individuals typically carry at least one potentially infectious or procarcinogenic organism in their gut microbiome, donors who pass the criteria may be extremely limited. Furthermore, donors are also excluded on the basis of chronic GI disorders, autoimmune disorders, cancer, neurological disorders, psychiatric disorders, obesity, and metabolic disorders, as well as recent exposure to antimicrobials, immunosuppressants, chemotherapeutics, and for many other conditions and lifestyle factors (Porcari et al. 2023).

Similarly, as there is such uncertainty about what constitutes a ‘healthy microbiome’, there are serious questions as to how FMT may shift the recipient microbiome and what the long-term risks of this may be. Communicable diseases are a prominent issue with the transfer of microbes from individual to individual via FMT, as it is increasingly apparent that the gut microbiome can impact physiological and psychological health. As such, and coupled with the current limited understanding of a ‘healthy microbiome’, it is possible there may be more extensive and unapparent risks to FMT.

Globally, there are vastly different regulations for FMT, despite the increasing interest in the technique in both clinical and commercial sectors (Wang et al. 2019). It is likely that FMT will become increasingly used in the future, with FMT formulation and technique undergoing refinement as understanding of the gut microbiome increases.

7.4.5 Phage Technology

Bacteriophages, known as phages, are viruses that infect bacteria and can enter the bacterial cell, where they may provoke lysis and bacterial cell death. The use of bacteriophages to treat bacterial infections is known as phage therapy and is of increasing clinical and commercial interest due to the global antimicrobial resistance crisis and growing ineffectiveness of last-resort antibiotics. Indeed, phage therapy has been proposed as a method to combat multidrug-resistant (MDR) bacteria, as a more precise and targeted alternative to antibiotics, and for the treatment of intracellular and biofilm-forming pathogens (Keen 2015).

Bacteriophages are the most abundant organisms on Earth and exist in every environment where bacteria may be found. Resultantly, there is an inexhaustible diversity of bacteriophages that may be utilised in phage therapy to target pathogenic bacteria. It is the current belief that there exists at least one phage type capable of infecting every bacterium in known existence (Keen 2015). Significantly, as distinct phages have specific host ranges, they present a method of eliminating bacterial species from the microbiome with precision, whilst conserving the overall diversity and health of the gut microbiota. Phage therapy, therefore, presents an alternative and more conservative approach than the broad-spectrum and general antibacterial activity of antibiotics.

The phage preparations used in clinical treatment may consist of specialised phages of a single type, or a 'cocktail' of different phages that may improve the effectiveness of the treatment. Early phage therapies were administered orally, topically, intravascularly, intrarectally, or inhaled, whereas modern phage therapy is usually delivered intravenously (Strathdee et al. 2023).

To date, there have been several cases where phage therapy has been used successfully to treat MDR infections, including endocarditis, septic shock, and bacteraemia (Gutiérrez and Domingo-Calap 2020). There is particular interest in utilising phage therapy for microbiome-based diseases, such as a potential treatment for adherent-invasive *E. coli* (AIEC), which are believed to contribute to IBD-associated dysbiosis and pathogenesis in the gut. Similarly, there have been recent investigations into phage therapy to target and remove potential pro-carcinogenic bacterial species such as *F. nucleatum*, to prevent colorectal cancer (CRC) (Strathdee et al. 2023). Further, there is interest in engineering new phages for specialised purposes, such as carrying payloads capable of degrading bacterial genomes to prevent the spread of antibiotic resistance genes amongst pathogens in host microbiomes. Engineered phages can be designed with increased ability to penetrate biofilms, target pathogens, or have enhanced pharmacokinetic and pharmacodynamic properties (Strathdee et al. 2023).

7.4.6 Synthetic Biology and Genetic Manipulation

Synthetic biology is a rapidly expanding field allowing for the development of approaches for modelling and studying the human microbiome, for treating dysbiosis, and for the manipulation of microbial communities through synthetic engineering.

There has been significant success utilising synthetic biology to treat many diseases, such as the use of genetically engineered bacteria to deliver targeted therapeutics for HIV, inflammatory bowel disease, diabetes, and cancer (Ezzamouri et al. 2021). An example of a microbiome-linked treatment is the probiotic *E. coli*, Nissle 1917, which has been engineered to detect tetrathionate produced in the inflamed gut and begin synthesis of a microcin which inhibits the inflammatory organism (Palmer et al. 2018).

Bacteriophages are of extensive interest in synthetic biology applications, as they can be utilised in precision microbiome editing in several ways. The ability of bacteriophages to target and insert material into bacterial cells can be commandeered to deliver payloads, which may be heterologous genes capable of eliminating pathogens, or 'toolkits', such as the CRISPR-Cas9 systems, for the engineering of specific strains in a population (Palmer et al. 2018).

Similar to using bacteriophages as targeted delivery vehicles, the specific binding of bacteriophages to their host bacterium is being used to design precision therapeutics such as lysibodies. These lysibodies are engineered molecules that combine bacteriophage receptors with antibody effector regions as a chimera capable of potentising the host immune response to perform phagocytosis of targeted pathogens (Nazir et al. 2024). Bacteriophage peptidoglycan hydrolases, known as lysins, have been bioengineered by fusion with bacterial receptors such as those found on *P. aeruginosa* pyocin bacteriocins to create 'lysocins' which have targeted, narrow-spectrum anti-Gram-negative activity, and are a promising approach for precision therapeutics in the microbiome (Heselpoth et al. 2019).

In the future, synthetic biology offers the ability to develop personalised precision medicine for the microbiome, and this will likely be a significant area of commercial and therapeutic interest. As each patient has a unique gut microbiome, treatments can be tailored to their specific needs, considering microbial imbalances, as well as genetic and metabolic profiles to enhance efficiency and minimise side-effects. Personal probiotics are one such area where bacterial strains may be engineered via synthetic biology and genetic manipulation to address specific microbiome concerns (Nazir et al. 2024). Such bioengineered microbial strains can be endowed with the ability to sense and respond to pathogens in the microbiome, such as endowing these probiotic strains with the ability to compete or kill specific disease-causing organisms, potentiate host immune systems, or reduce harmful inflammation responses (Cruz et al. 2022).

7.4.7 Conclusions and Future Perspectives

Microbiomes are increasingly being recognised as critical to human health, and biotechnology, which may preserve healthy microbial communities whilst eradicating pathogens is of significant and growing clinical, public, and commercial interest. Several approaches exist today, which seek to alter the human microbiome for the treatment of disease and improvement of patient health, with some of these involving the introduction of beneficial microbial strains and others the precision targeting of disease-causing organisms. In the future, these methods are likely to become increasingly personalised to individual patients, due to the unique composition of human microbiomes. As the understanding of human microbial ecosystems evolves, the development of more specific synthetic and bioengineered techniques will be facilitated.

7.5 Engineered Bacteria and Archaea in Medicine

7.5.1 Introduction

When one thinks of 'bacteria' in the context of medicine, antibiotics and infection are the first things that come to mind. In creating antibiotics to treat infections, we learned a great deal about what makes bacteria so dangerous: their motility, the ability to manipulate the host immune system, the excretion of toxins, and other cell-damaging agents. Now, researchers are beginning to leverage those same traits to our benefit.

With modern tools and techniques, it's relatively simple to rewire a bacterial genome, to remove harmful genes, add new ones, and even key gene expression to specific queues. We can isolate compounds that may have once caused tissue damage and repurpose them into anti-cancer drugs. We can generate fluorescent microbes that improve imaging techniques, and it isn't just limited to bacteria—organisms from the archaea domain are also being reimagined for medicinal purposes.

7.5.2 Bacteria in Medicine

7.5.2.1 Bacteria as Imaging Tools
We now have many tools that make it easy to manipulate the bacterial genome and change it to amazing effect. Bacteria are being successfully engineered for a wide range of tools to aid in imaging and diagnosis, including the use of fluorescence imaging, ultrasound imaging, MRIs, and PET scans (Jiang et al. 2023).

7.5.2.2 Biologically Engineered Bacterial Probes

In the early 1990's researchers created the first engineered fluorescent bacteria by inserting specific genes into the genome. These genes encoded for proteins that fluoresce when excited by light of specific wavelengths and are still widely used today; some common examples include Green Fluorescent Protein (GFP) and Red Fluorescent Protein (RFP), while a full spectrum of options is now available. In 2012, Tangney et al. engineered strains of *E. coli* that carried another set of fluorescent genes, *lux*ABCDE. These bacteria were inoculated into mice and became living probes that could be visualised as they moved through the body. They allowed researchers to use high-resolution imaging techniques like micro-computed tomography and three-dimensional diffuse optical tomography to create detailed, 3D maps of imaged areas (Cronin et al. 2012).

7.5.2.3 Bacteria in Metabolic Engineering

Along with genetic engineering, researchers are also creating imaging solutions that take advantage of bacteria's unique metabolism. In 2015, Weinstien et al. created a radioactive probe that could be used to detect gram-negative bacteria. They synthesised 2-deoxy-2-[18F] fluoro-D-sorbitol (18F-FDS), which did not interact with mammalian or gram-positive cells, but could integrate into the cell walls of gram-negative bacteria. They tested the compound in two mouse models: a thigh wound inoculated with *E. coli* (Gram-negative) and *S. aureus* (Gram-positive), and a brain infection with *Klebsiella pneumonia*. In the first model, only *E. coli* infections were visible using a PET scan. In the second model, the cell-wall binding was so specific that researchers were able to visualise *K. pneumonia* infected tissues in 3D and distinguish them from sterile but inflamed tissues. Both models illustrate the utility of using metabolic engineering to enhance existing imaging methods (Weinstein et al. 2014).

7.5.2.4 Bacteria as Cancer Therapy

It may seem counterintuitive to use bacteria—an agent of disease—to treat cancer, but the idea tracks back to the 1800s. In 1888, a German physician named Bruns noted a successful reduction in tumour size when patients were injected with streptococci (Munir et al. 2024). A few years later, this inspired William Coley—often referred to as the Father of Immunotherapy—to develop Coley's toxin. The toxin, a combination of heat-killed *Serratia marcescens* and *Streptomyces pyogens*, was administered to patients with inoperable tumours as a last resort. It led to reductions in tumour size and more than 10-year remission in approximately 30% of patients (Nguyen et al. 2023). Coley's work laid the foundations for the modern understanding of immunotherapy and paved the way for several bacterial-derived cancer treatments in use today.

7.5.2.5 Bacteria for Immune Stimulation

Tumours and their surroundings—often referred to as the 'microenvironment'—are immune privileged, meaning the immune system ignores this area. Immune privilege allows the tumour to proliferate, but it also allows bacteria to flourish. Bacteria can be genetically modified to be non-virulent but stimulate an immune response large enough to bring immune cells to the tumour microenvironment, where they begin to attack the cancer. This is how Coley's toxin works, and it demonstrated how effective the patient's immune system could be if given the chance (Nguyen et al. 2023). Today, bacteria like *E. coli* are being used in a similar manner. In 2024, Li et al successfully engineered a non-pathogenic strain of *E.coli* that carried the cytokine interferon gamma (an important regulator involved in the immune responses identification of tumour) directly into mouse tumours. When combined with immunotherapy drugs, tumour growth was impeded more than either therapy could produce on its own

(Li et al. 2024). That same year, Redenti et al engineered a promising bacterial cancer vaccine. They expressed tumour-specific antigens in *E.coli* cells and injected the cells into tumour mouse models. Researchers noticed that mice exposed to antigen-expressing *E. coli* had a dose-dependent reduction in tumour size compared to controls (Redenti et al. 2024).

7.5.2.6 Bacterial Engineered Toxins

In addition to immune privilege, the tumour microenvironment is more acidic and has less oxygen than physiologically normal tissues. Many drugs currently on the market require specific pH and oxygenation to function properly. These drugs are also toxic to cancerous and healthy tissues, which leads to many of the side effects commonly associated with cancer treatment, like nausea, weight loss, and pain. These limitations make treating cancerous tumours challenging. However, bacteria can be genetically engineered to grow in the tumour microenvironment and express anti-cancer genes only under these conditions. In 2012, Flentie et al. engineered an attenuated strain of *Salmonella* carrying the genes to express Shiga toxin, a potent exotoxin that disrupts a cell's ability to make protein. The engineered *Salmonella* was then injected into mouse models. Shiga toxin expression was controlled by gene promoters that could only be activated in the highly acidic conditions of the tumour microenvironment, thus preventing damage to healthy tissues and showing strong anti-tumour activity (Flentie et al. 2012).

7.5.2.7 Bacterial Drug Delivery

Bacteria can be engineered to deliver more than just toxins and cytokines to the tumour. Payloads can include genes that encode small interfering RNAs (siRNA) that interrupt oncogene expression and prodrug enzymes that can convert inert drugs into cytotoxic compounds under selective control. In 2013, Berger et al successfully eradicated neuroblastoma tumours in mice using attenuated *Salmonella typhimurium* harbouring siRNA that silences the expression of the Stat3 oncogene. They found that this enhanced the activity of the host T and natural killer cells, leading to tumour eradication and mouse survival (Berger et al. 2013). During a pilot trial in 2003, three patients whose cancer was not responding to treatment were given *S. typhimurium* that expressed a prodrug, cytosine deaminase (CD). When CD interacts with the nontoxic compound 5-fluorocytosine (5-FC), it is converted into 5-fluorouracil (5-FU), a highly toxic chemotherapy drug. *S. typhimurium* expressed CD in the tumour, leading to delivery of 5-FU directly to the cancerous tissue (Nemunaitis et al. 2003). While the colonisation was observed in 3/25 patients, no tumour regression was observed.

7.5.3 *Archaea in Medicine*

7.5.3.1 Archaea in Diagnostics

Like bacteria, archaea are featured in several novel diagnostic technologies. For example, the thermophilic *Sulfolobus solfataricus*, which is found in the high temperatures and low pH of sulfuric lakes, produces a glucose dehydrogenase that converts glucose to pyruvate. Due to the extreme conditions of its natural environment, this enzyme is very stable and has been commercialised and is now a component in laboratory test kits for measuring blood glucose in patient samples (Aparici-Carratalá et al. 2023b). Halophilic, or salt-loving, archaea are also being studied as tools to aid diagnostics. Researchers in Granada are working to develop a colorectal cancer screening tool using a protein produced by *H. halobium*. This protein elicits an unusual immune response in the GI tract— production of antibodies that recognise an oncogene associated with colorectal cancer (Pérez Davó 2014). This could be developed into a tool for both cancer screening and treatment. Some haloarchaeal can also produce gas vesicles (GVs), which are being explored as contrast agents for diagnostic imaging like MRIs and ultrasounds (Bulte 2018).

7.5.3.2 Archaea for Anti-Cancer Compounds

Select species of archaea produce compounds that have shown promising results for treating cancer. Unlike many approaches using live engineered bacteria, archaeal contributions to cancer treatment are exclusively secreted compounds. Two promising example compounds were isolated from halophilic archaea, which live in extremely high-salinity environments like salt lakes and underground salt deposits. A study in 2019 found small molecules produced by *H. salinarum* showed specific and potent killing of human prostate cancer cells, with no impact on normal cells in the control group (Safarpour et al. 2019). This specificity is attractive for a potential cancer drug, as traditional treatments like radiation and chemotherapy damage healthy tissues and cause patients to feel very unwell. Another two studies (Abbes et al. 2013; Hou and Cui 2018) identified the treatment potential of carotenoid pigment compounds in several halophilic archaea. They found that these carotenoids were effective in reducing the survival of HepG2 (hepatoma) cell line, a model liver carcinoma cell line, through antiproliferative activity. Both compounds have potential as the basis for novel drug design or creation of combination therapies that can supplement existing medications (Corral et al. 2020).

7.5.3.3 Conclusions and Future Perspectives

With a wide-reaching set of technologies, targets and challenges, biotechnology is at the forefront of many medical practices. Traditional cancer therapies like surgery, chemotherapy, and radiation, while often effective, have wide-ranging effects on patient outcomes. While saving many lives, they come with significant side effects and may fail to completely eradicate the cancerous cells. While the application of microbes as alternative treatment methods might not appear groundbreaking, the localisation and specificity it provides, while minimising healthy tissue damage, is exciting. Coming advances have the potential to use components of extremophiles for cost-effective and stable diagnostic devices in a wide array of areas, making day-to-day monitoring of conditions more manageable.

7.6 Regulatory, Ethical, and Future Considerations

7.6.1 Introduction

Throughout history, humans have utilised microorganisms in both food and medicine. In recent years, microbial biotechnology has expanded far beyond these traditional fermentation processes and is now playing a role in many areas of medicine, including vaccine development, microbiome modulation, and drug discovery (Hibbert et al. 2024). Ultimately, these technologies have improved disease prevention, diagnosis, and cure (Weiman and Fox 2015). However, alongside these innovations comes new ethical territory (Rhodes 2016). Whilst research ethics has always been an essential part of microbiology, ethical standards and policies must be adapted to reflect recent advances in microbial biotechnology in medicine (Ma et al. 2018).

7.6.2 Safety and Risk of Unintended Consequences

The use of microbes in medicine often necessitates the cultivation of large bacterial or fungal cultures. Examples include the large bioreactors used to cultivate *Streptomyces clavuligerus* for the production of clavulanic acid (Heerema et al. 1979) or the *Corynebacterium glutamicum* seed tankers required for industrial glutamate production (Slowinski and Charm 1973). If released, industrial organisms have the potential to disrupt the balance of the environmental microbiome, and therefore, extreme care must be taken to avoid the release of industrial microbes (Pantoja Angles et al. 2022). This is particularly important if the organisms are genetically modified, such as the transgenic *E. coli* used to produce human hormones such as insulin and human growth hormone (Grey et al. 1985; Riggs 2021). The practices that are designed to protect the environment from the accidental release of these industrial organisms are known as biocontainment. Conventional methods of containment include the use of disinfection of equipment, destruction of biological waste, and the use of personal protective equipment by relevant personnel (Pantoja Angles et al. 2022). Recently, there have been advances in 'intrinsic biocontainment', whereby organisms are genetically engineered to become reliant on the environment in which they are industrially cultured. One example of this is described in recent work by Chang et al., where 'synthetic autotrophy' was engineered in yeast, constraining their survival on an unnatural amino acid (Chang et al. 2023). Consequently, these organisms are unable to survive outside of laboratory conditions. In practice, intrinsic biocontainment has not been widely implemented in industrial biotechnology, but the concept may provide an additional strategy as the use of microbes in biotechnology is expanded (George et al. 2024).

Although the use of microbial natural products in food and medicine is now well established, the use of microbes and their nucleic acids directly as therapeutic interventions remains an emerging field. As such, potential unintended consequences of their use may be poorly understood and therefore must be considered (Rhodes 2016). For example, although phage therapy represents a promising treatment for precision antimicrobial treatment of multidrug-resistant organisms (Biswas et al. 2024), side effects and long-term outcomes of treatment are still being investigated (Sawa et al. 2024).

Unlike traditional antimicrobial therapies, phages are living entities with the capacity to mutate and evolve. This creates regulatory challenges, particularly for the standardisation and quality control of medicinal phage products. The genetic content of the chosen phage for therapy must also be scrutinised, ensuring these phages do not encode genes encoding potentially harmful phenotypes such as virulence (for unintended microbial targets) or antimicrobial resistance (inadvertently acquired through infection of resistant bacteria). Even with these controls in place, the intrinsic dynamic nature of phage genomes may still lead to issues with variations between batches that may affect efficacy or safety (Suleman et al. 2025).

The potential adverse effects of phage therapy on patients warrant careful consideration. The lytic phages commonly used in these treatments can trigger the systemic release of bacterial cell contents and lipopolysaccharides. While many of these cellular components activate the innate immune system and may facilitate rapid infection clearance, they could also potentially trigger an immune overreaction. When deploying phage therapy against bacterial infections, thorough characterisation of the causative organism is essential to ensure the treatment's appropriateness for the specific infection. Notably, for pathogens harbouring potent toxins, lysogenic phages may inadvertently release these toxins during bacterial lysis, resulting in cytotoxic effects for the host. These gaps in understanding alongside the complex regulatory procedures have complicated the approval of phage therapies in humans, and as such, regulatory approval remains limited to a small number of countries, including Russia, Poland and Georgia (Yang et al. 2023).

7.6.3 Informed Consent

Informed consent is a founding principle of medical ethics, which is traditionally described as the process whereby individuals are fully informed about the details of medical treatment, including potential benefits and risks, before agreeing to participate (Pietrzykowski and Smilowska 2021). As innovations such as microbiome-based therapies, probiotics, live biotherapeutic products (LBPs), faecal microbiota transplants (FMTs), and microbial gene therapy continue to advance, ensuring proper informed consent becomes not merely procedural but essential to upholding patient autonomy and trust (Pietrzykowski and Smilowska 2021).

When gaining consent to use microbial therapies, the traditional principles of informed consent should be applied, including discussing potential outcomes, the downstream use of biological samples and data, alternative treatment options and the right to withdraw. However, the informed consent process should also emphasise the novel and specialised nature of these treatments. Patients should be fully informed about the scientific rationale behind their use and the specific considerations for their chosen therapy (Ma et al. 2018; Shou et al. 2024). For example, patients undergoing FMTs should be informed about the donor selection process and the status of their treatment (Bokek-Cohen and Ravitsky 2017). In the UK, FMTs are approved only for patients with recurrent *Clostridioides difficile* infections (Mullish et al. 2018), and consequently, the use of this treatment outside of this remit would require discussion with the patient about the experimental nature of their treatment. It is imperative that potential adverse outcomes are discussed with patients. For FMTs, this could include unprecedented allergic reactions or the transfer of pathogenic or multidrug-resistant organisms (Merrick et al. 2020). Finally, the long-term consequences of treatment should be discussed with patients, which for FMTs may present a unique issue as the long-term effects of FMTs on the microbiomes are not yet understood. This is exacerbated by the complexity of the microbial community, and inherent variance from donor to donor, unlike with, for example, stable, defined small molecule treatments. A clear outline of potential benefits should also be presented, for example, the reduction in *C. difficile* recurrence in FMT participants, along with potential risks (Mikail et al. 2020).

7.6.4 Access and Equity

The use of microbes in medicine has the potential to revolutionise healthcare and provide treatments for diseases that currently have limited options. The World Health Organisation classifies antimicrobial resistance as a major threat to global health, and therefore, our efforts must be focussed on providing global solutions. When novel therapeutics emerge, it is crucial that their pricing model allows for global distribution. Mechanisms which facilitate this include global pricing models, tiered by GDP, or encouragement of generic manufacturing to lower the overall price, or drug donation programmes.

One case study in how these principles can enhance global access to a vital therapeutic is the antiparasitic drug, Ivermectin. The molecule is a derivative of avermectin, a natural product produced by *Streptomyces avermitilis* (Ikeda et al. 1987), first discovered by Satoshi Omura in 1973. Through collaboration with Merck, the compound was found to be a potent anthelminthic, and its optimised derivative, Ivermectin, was approved for veterinary use in 1981 (Õmura and Crump 2004). The observed effects were profound, and ivermectin was approved for human use by 1987. The drug was manufactured under the brand name Mectizan ©, and distributed for free by Merck as a treatment for Onchocerciasis (river blindness) (Crump and Omura 2011).

Currently, approximately 250 million people annually rely on ivermectin to combat two diseases: onchocerciasis (river blindness) and lymphatic filariasis. Most of these individuals live in remote, rural, and severely under-resourced areas of developing countries. The next goal is to continue to distribute Ivermectin—now available generically—o eradicate these diseases (Omura and Crump 2014). While controversy and ill-reputed science surround the use of Ivermectin for the treatment of other diseases/conditions, the importance of scientific integrity cannot be overstated.

Antimicrobial research is a global endeavour. As researchers explore new ecological niches to discover novel microbial natural products in new chemical space, it is essential that countries and territories that play a fundamental role in this research are adequately credited and rewarded (Katz and Baltz 2016). The Nagoya Protocol was established in 2014 and implements a legal framework that necessitates benefit-sharing agreements and informed consent implementation before accessing genetic resources from other countries (Soares 2011). This addresses the historical inequity where developing countries rich in biodiversity provided biological materials that were developed into profitable pharmaceuticals by companies in developed nations without fair compensation. The protocol promotes sustainable partnerships, encourages collaborative research with local institutions, and provides economic incentives for biodiversity conservation. As a result of the protocol, commitment towards the recognition of indigenous knowledge and local bioresources is now a fundamental principle in natural product discovery (Dedeurwaerdere et al. 2016; Comizzoli and Holt 2016).

7.6.5 Conclusions and Future Perspectives

Regulatory and ethical considerations are progressively becoming commonplace in research. The implementation of the Nagoya protocol was a milestone event, indicative of the global movement toward ethical practices in natural product drug discovery. The success has already had a butterfly effect on the research community as a whole, and its success heralds a more equitable future for all.

References

Abbes M, Baati H, Guermazi S et al (2013) Biological properties of carotenoids extracted from *Halobacterium halobium* isolated from a Tunisian solar saltern. BMC Complement Altern Med 13:1–8. https://doi.org/10.1186/1472-6882-13-255/FIGURES/7

Adamiak N, Krawczyk KT, Locht C, Kowalewicz-Kulbat M (2021) Archaeosomes and gas vesicles as tools for vaccine development. Front Immunol 12:746235. https://doi.org/10.3389/FIMMU.2021.746235

Aggarwal N, Kitano S, Puah GRY et al (2023) Microbiome and human health: current understanding, engineering, and enabling technologies. Chem Rev 123:31–72. https://doi.org/10.1021/ACS.CHEMREV.2C00431

Antimicrobial Resistance Division (AMR) Control of Neglected Tropical Diseases (NTD), Global Coordination and Partnership (GCP) (2022) WHO fungal priority pathogens list to guide research, development and public health action. In: World Health Organization. https://www.who.int/publications/i/item/9789240060241. Accessed 30 May 2025

Aparici-Carratalá D, Esclapez J, Bautista V et al (2023a) Archaea: current and potential biotechnological applications. Res Microbiol 174:104080. https://doi.org/10.1016/J.RESMIC.2023.104080

Aparici-Carratalá D, Esclapez J, Bautista V et al (2023b) Archaea: current and potential biotechnological applications. Res Microbiol 174:104080. https://doi.org/10.1016/J.RESMIC.2023.104080

Bar-On YM, Phillips R, Milo R (2018) The biomass distribution on Earth. Proc Natl Acad Sci USA 115:6506–6511. https://doi.org/10.1073/pnas.1711842115

Barona-Gómez, F, Chevrette, MG, Hoskisson, PA (2023). On the evolution of natural product biosynthesis. Elsevier

Barron M (2024) The Archaeome: an emerging player in health and disease. In: American Society for Microbiology. https://asm.org/articles/2024/july/the-archaeome-emerging-player-health-disease. Accessed 28 May 2025

Berche P (2022) Life and death of smallpox. Presse Med 51:104117. https://doi.org/10.1016/J.LPM.2022.104117

Berger E, Soldati R, Huebener N et al (2013) Salmonella SL7207 application is the most effective DNA vaccine delivery method for successful tumor eradication in a murine model for neuroblastoma. Cancer Lett 331:167–173. https://doi.org/10.1016/J.CANLET.2012.12.026

Bernheim A, Millman A, Ofir G et al (2021) Prokaryotic viperins produce diverse antiviral molecules. Nature 589:120–124. https://doi.org/10.1038/s41586-020-2762-2

Bidell MR, Hobbs ALV, Lodise TP (2022) Gut microbiome health and dysbiosis: a clinical primer. Pharmacotherapy 42:849–857. https://doi.org/10.1002/PHAR.2731

Biswas SK, Sumon MMH, Ahmed S et al (2024) Beyond antibiotics: exploring the potential of bacteriophages and phage therapy. Phage 5:186–202. https://doi.org/10.1089/PHAGE.2024.0005;WEBSITE:WEBSITE:MAL-SITE

Blaskovich MAT, Cooper MA (2025) Antibiotics re-booted—time to kick back against drug resistance. npj Antimicrob Resist 3(1):1–20. https://doi.org/10.1038/s44259-025-00096-1

Bodke H, Jogdand S (2022) Role of probiotics in human health. Cureus 14. https://doi.org/10.7759/CUREUS.31313

Bokek-Cohen Y, Ravitsky V (2017) Cultural and personal considerations in informed consent for fecal microbiota transplantation. Am J Bioeth 17:55–57. https://doi.org/10.1080/15265161.2017.1299241

Boyd MR, Gustafson KR, McMahon JB et al (1997) Discovery of cyanovirin-N, a novel human immunodeficiency virus-inactivating protein that binds viral surface envelope glycoprotein gp120: potential applications to microbicide development. Antimicrob Agents Chemother 41:1521–1530. https://doi.org/10.1128/AAC.41.7.1521

Broecker F, Moelling K (2019) What viruses tell us about evolution and immunity: beyond Darwin. Ann N Y Acad Sci 1447:53–68. https://doi.org/10.1111/nyas.14097

Brugère JF, Borrel G, Gaci N et al (2013) Archaebiotics: proposed therapeutic use of archaea to prevent trimethylaminuria and cardiovascular disease. Gut Microbes 5:5. https://doi.org/10.4161/GMIC.26749

Bulte JWM (2018) Gas vesicles as collapsible MRI contrast agents. Nat Mater 17(5):386–387. https://doi.org/10.1038/s41563-018-0073-x 2018

Chang T, Ding W, Yan S et al (2023) A robust yeast biocontainment system with two-layered regulation switch dependent on unnatural amino acid. Nat Commun 14(1):1–13. https://doi.org/10.1038/s41467-023-42358-4 2023

Chowdhury S, Castro S, Coker C et al (2019) Programmable bacteria induce durable tumor regression and systemic antitumor immunity. Nat Med 25(7):1057–1063. https://doi.org/10.1038/s41591-019-0498-z 2019

Cid R, Bolívar J (2021) Platforms for production of protein-based vaccines: from classical to next-generation strategies. Biomolecules 11:1072. https://doi.org/10.3390/BIOM11081072

Comizzoli P, Holt WV (2016) Implications of the Nagoya Protocol for genome resource banks composed of biomaterials from rare and endangered species. Reprod Fertil Dev 28:1145–1160. https://doi.org/10.1071/RD15429

Corral P, Amoozegar MA, Ventosa A (2020) Halophiles and their biomolecules: recent advances and future applications in biomedicine. Mar Drugs 18. https://doi.org/10.3390/MD18010033

Cronin M, Akin AR, Collins SA et al (2012) High resolution in vivo bioluminescent imaging for the study of bacterial tumour targeting. PLoS One 7:e30940. https://doi.org/10.1371/JOURNAL.PONE.0030940

Crump A, Omura S (2011) Ivermectin, 'Wonder drug' from Japan: the human use perspective. Proc Jpn Academy, Ser B 87:13–28. https://doi.org/10.2183/PJAB.87.13

Cruz KCP, Enekegho LO, Stuart DT (2022) Bioengineered probiotics: synthetic biology can provide live cell therapeutics for the treatment of foodborne diseases. Front Bioeng Biotechnol 10:890479. https://doi.org/10.3389/FBIOE.2022.890479/XML/NLM

Cury J, Haudiquet M, Hernandez Trejo V et al (2024) Conservation of antiviral systems across domains of life reveals immune genes in humans. Cell Host Microbe 32:1594–1607. https://doi.org/10.1016/j.chom.2024.08.002 e5

De Simeis D, Serra S (2021) Actinomycetes: anever-ending source of bioactive compounds—an overview on antibiotics production. Antibiotics 10. https://doi.org/10.3390/antibiotics10050483

Dedeurwaerdere T, Melindi-Ghidi P, Broggiato A (2016) Global scientific research commons under the Nagoya Protocol: Towards a collaborative economy model for the sharing of basic research assets. Env Sci Policy 55:1–10. https://doi.org/10.1016/J.ENVSCI.2015.08.006

Detmer A, Glenting J (2006) Live bacterial vaccines—a review and identification of potential hazards. Microb Cell Fact 5:23. https://doi.org/10.1186/1475-2859-5-23

Ezzamouri B, Shoaie S, Ledesma-Amaro R (2021) Synergies of systems biology and synthetic biology in human microbiome studies. Front Microbiol 12:681982. https://doi.org/10.3389/FMICB.2021.681982/XML/NLM

Farha MA, Tu MM, Brown ED (2025) Important challenges to finding new leads for new antibiotics. Curr Opin Microbiol 83:102562. https://doi.org/10.1016/J.MIB.2024.102562

Feld JJ, Jacobson IM, Hézode C et al (2015) Sofosbuvir and Velpatasvir for HCV genotype 1, 2, 4, 5, and 6 infection. N Engl J Med 373:2599–2607. https://doi.org/10.1056/NEJMoa1512610

Feng J, Chen Y, Liu Y et al (2023) Efficacy and safety of fecal microbiota transplantation in the treatment of ulcerative colitis: a systematic review and meta-analysis. Sci Rep 13. https://doi.org/10.1038/S41598-023-41182-6

Férir G, Petrova MI, Andrei G et al (2013) The lantibiotic peptide labyrinthopeptin A1 demonstrates broad anti-HIV andanti-HSV activity with potential for microbicidal applications. PLoS One 8:e64010. https://doi.org/10.1371/journal.pone.0064010

Fleming A (1929) On the antibacterial action of cultures of a penicillium, with special reference to their use in the isolation of *B. influenzæ*. Br J Exp Pathol 10:226

Flentie K, Kocher B, Gammon ST et al (2012) A bioluminescent transposon reporter-Trap identifies tumor-specific microenvironment-induced promoters in Salmonella for conditional bacterial-based tumor therapy. Cancer Discov 2:624. https://doi.org/10.1158/2159-8290.CD-11-0201

Fong W, Li Q, Yu J (2020) Gut microbiota modulation: a novel strategy for prevention and treatment of colorectal cancer. Oncogene 39(26):4925–4943. https://doi.org/10.1038/s41388-020-1341-1 2020

Fryauff DJ, Mouzin E, Preston Church LW et al (1999) Lymphocyte response to tetanus toxin T-cell epitopes: effects of tetanus vaccination and concurrent malaria prophylaxis. Vaccine 17:59–63. https://doi.org/10.1016/S0264-410X(98)00144-3

Galen JE, Wahid R, Buskirk AD (2021) Strategies for enhancement of live-attenuated salmonella-based carrier vaccine immunogenicity. Vaccines 9:162. https://doi.org/10.3390/VACCINES9020162

George DR, Danciu M, Davenport PW et al (2024) A bumpy road ahead for genetic biocontainment. Nat Commun 15:1–5. https://doi.org/10.1038/S41467-023-44531-1

Ghattas M, Dwivedi G, Lavertu M, Alameh MG (2021) Vaccine technologies and platforms for infectious diseases: current progress, challenges, and opportunities. Vaccines 9:1490. https://doi.org/10.3390/VACCINES9121490

Gilbert JA, Blaser MJ, Caporaso JG et al (2018) Current understanding of the human microbiome. Nat Med 24:392–400. https://doi.org/10.1038/NM.4517

Gizzi AS, Grove TL, Arnold JJ et al (2018) A naturally occurring antiviral ribonucleotide encoded by the human genome. Nature 558:610–614. https://doi.org/10.1038/s41586-018-0238-4

Gray GL, Baldridge JS, McKeown KS et al (1985) Periplasmic production of correctly processed human growth hormone in *Escherichia coli*: natural and bacterial signal sequences are interchangeable. Gene 39:247–254. https://doi.org/10.1016/0378-1119(85)90319-1

Gul S, Durante-Mangoni E (2024) Unraveling the puzzle: health benefits of probiotics—a comprehensive review. J Clin Med 13. https://doi.org/10.3390/JCM13051436

Gutiérrez B, Domingo-Calap P (2020) Phage therapy in gastrointestinal diseases. Microorganisms 8:1–11. https://doi.org/10.3390/MICROORGANISMS8091420

Haque RU, Paradisi F, Allers T (2019) Haloferax volcanii for biotechnology applications: challenges, current state and perspectives. Appl Microbiology Biotechnol 104(4):1371–1382. https://doi.org/10.1007/S00253-019-10314-2

Hasan N, Yang H (2019) Factors affecting the composition of the gut microbiota, and its modulation. PeerJ 2019. https://doi.org/10.7717/PEERJ.7502

Heerema MS, Musher DM, Williams TW (1979) Clavulanic acid and penicillin treatment of Staphylococcus aureus renal infection in mice. Antimicrob Agents Chemother 16:798–800. https://doi.org/10.1128/AAC.16.6.798;PAGE:STRING:ARTICLE/CHAPTER

Heggie A, Thurston TLM, Ellis T (2025) Microbial messengers: nucleic acid delivery by bacteria. Trends Biotechnol 43:145–161. https://doi.org/10.1016/J.TIBTECH.2024.07.010

Helbig KJ, Beard MR (2014) The role of viperin in the innate antiviral response. J Mol Biol 426:1210–1219. https://doi.org/10.1016/j.jmb.2013.10.019

Hendrix RW, Smith MCM, Burns RN et al (1999) Evolutionary relationships among diverse bacteriophages and prophages: all the world's a phage. Proc Natl Acad Sci USA 96:2192–2197. https://doi.org/10.1073/pnas.96.5.2192

Heselpoth RD, Euler CW, Schuch R, Fischetti VA (2019) Lysocins: Bioengineered antimicrobials that deliver lysins across the outer membrane of Gram-negative bacteria. Antimicrob Agents Chemother 63. https://doi.org/10.1128/AAC.00342-19

Hibbert T, Krpetic Z, Latimer J et al (2024) Antimicrobials: an update on new strategies to diversify treatment for bacterial infections. Adv Microb Physiol 84:135–241. https://doi.org/10.1016/BS.AMPBS.2023.12.002s

Hill AM, Salmond GPC (2020) Microbial gas vesicles as nanotechnology tools: exploiting intracellular organelles for translational utility in biotechnology, medicine and the environment. Microbiology 166:501. https://doi.org/10.1099/MIC.0.000912

Hobbs Z, Abedon ST (2016) Diversity of phage infection types and associated terminology: the problem with 'Lytic or lysogenic.'. FEMS Microbiol Lett 363:fnw047. https://doi.org/10.1093/femsle/fnw047

Hoskisson PA, Seipke RF (2020) Cryptic or silent? The known unknowns, unknown knowns, and unknown unknowns of secondary metabolism. mBio 11:1–5. https://doi.org/10.1128/MBIO.02642-20

Hou J, Cui HL (2018) In vitro antioxidant, antihemolytic, and anticancer activity of the carotenoids from halophilic Archaea. Curr Microbiol 75:266–271. https://doi.org/10.1007/S00284-017-1374-Z/FIGURES/4

Hou K, Wu ZX, Chen XY et al (2022) Microbiota in health and diseases. Signal Transduct Target Ther 7(1):1–28. https://doi.org/10.1038/s41392-022-00974-4

Hutchings M, Truman A, Wilkinson B (2019) Antibiotics: past, present and future. Curr Opin Microbiol 51:72–80. https://doi.org/10.1016/j.mib.2019.10.008
Ikeda H, Kotaki H, Omura S (1987) Genetic studies of avermectin biosythesis in *Streptomyces avermitilis*. J Bacteriol 169:5615–5621. https://doi.org/10.1128/JB.169.12.5615-5621.1987
Jiang H, Cao Z, Liu Y et al (2023) Bacteria-based living probes: preparation and the applications in bioimaging and diagnosis. Adv Sci 11:2306480. https://doi.org/10.1002/ADVS.202306480
Kashyap D, Panda M, Baral B et al (2022) Outermembrane vesicles: an emerging vaccine platform. Vaccines 10:1578. https://doi.org/10.3390/VACCINES10101578
Katz L, Baltz RH (2016) Natural product discovery: past, present, and future. J Ind Microbiol Biotechnol 43:155–176. https://doi.org/10.1007/S10295-015-1723-5
Kavitha P, Lipton AP, Sarika AR, Aishwarya MS (2011) Growth characteristics and halocin production by a new isolate, *Haloferax volcanii* KPS1 from Kovalam Solar Saltern (India). Res J Biol Sci. https://www.researchgate.net/publication/275518055_Growth_Characteristics_and_Halocin_Production_by_a_New_Isolate_Haloferax_volcanii_KPS1_from_Kovalam_Solar_Saltern_India#fullTextFileContent. Accessed 28 May 2025
Keen EC (2015) A century of phage research: bacteriophages and the shaping of modern biology. Bioessays 37:6. https://doi.org/10.1002/BIES.201400152
Khameneh B, Iranshahy M, Soheili V, Fazly Bazzaz BS (2019) Review on plant antimicrobials: a mechanistic viewpoint. Antimicrob Resist Infect Control 8(1):1–28. https://doi.org/10.1186/S13756-019-0559-6
Koharudin LMI, Kollipara S, Aiken C, Gronenborn AM (2012) Structural insights into the anti-HIV activity of the *Oscillatoria agardhii* agglutinin homolog lectin family. J Biol Chem 287:33796–33811. https://doi.org/10.1074/jbc.M112.388579
Kumar V, Singh B, van Belkum MJ et al (2021) Halocins, natural antimicrobials of Archaea: exotic or special or both. Biotechnol Adv 53:107834. https://doi.org/10.1016/j.biotechadv.2021.107834
Lang M, Carvalho A, Baharoglu Z, Mazel D (2023) Negative bacteria : new therapies with old molecules. Microbiol Mol Biol Rev 87:1–30
Lauring AS, Jones JO, Andino R (2010) Rationalizing the development of live attenuated virus vaccines. Nat Biotechnol 28:573. https://doi.org/10.1038/NBT.1635
Li F, Yang Z, Savage TM et al (2024) Programmable bacteria synergize with PD-1 blockade to overcome cancer cell-intrinsic immune resistance mechanisms. Sci Immunol 9:eadn9879. https://doi.org/10.1126/SCIIMMUNOL.ADN9879
Liu MA (2019) A comparison of plasmid DNA and mRNA as vaccine technologies. Vaccines (Basel) 7:37. https://doi.org/10.3390/VACCINES7020037
Louten J (2016) Virus replication. In: Essential human virology. Elsevier, pp 49–70. London, UK; San Diego, CA, Academic Press. https://onesearch.nihlibrary.ors.nih.gov/discovery/fulldisplay?docid=alma991001290579304686&context=L&vid=01NIH_INST:NIH&lang=en&adaptor=Local%20Search%20Engine
Ma Y, Chen H, Lan C, Ren J (2018) Help, hope and hype: ethical considerations of human microbiome research and applications. Protein Cell 9:404–415. https://doi.org/10.1007/S13238-018-0537-4
Ma Z, Zuo T, Frey N, Rangrez AY (2024) A systematic framework for understanding the microbiome in human health and disease: from basic principles to clinical translation. Signal Transduct Target Ther 9(1):1–36. https://doi.org/10.1038/s41392-024-01946-6
Manos J (2022) The human microbiome in disease and pathology. APMIS 130:690–705. https://doi.org/10.1111/APM.13225
Markowiak-Kopeć P, Śewska K (2020) The effect of probiotics on the production of short-chain fatty acids by human intestinal microbiome. Nutrients 12. https://doi.org/10.3390/NU12041107
Martinez JP, Sasse F, Brönstrup M et al (2015) Antiviral drug discovery: broad-spectrum drugs from nature. Nat Prod Rep 32:29–48. https://doi.org/10.1039/C4NP00085D
McHugh RE, Munnoch JT, Braes RE, McKean IJ, Giard J, Taladriz-Sender A, Hoskisson PA (2022) Biosynthesis of aurodox, a Type III secretion system inhibitor from *Streptomyces goldiniensis*. Appl Environ Microbiology 88(15):e00692–22

McHugh RE, O'Boyle N, Connolly JPR et al (2019) Characterization of the mode of action of aurodox, a type III secretion system inhibitor from *Streptomyces goldiniensis*. Infect Immun 87. https://doi.org/10.1128/IAI.00595-18/SUPPL_FILE/IAI.00595-18-S0001.PDF
Medema MH, Blin K, Cimermancic P et al (2011) antiSMASH: rapid identification, annotation and analysis of secondary metabolite biosynthesis gene clusters in bacterial and fungal genome sequences. Nucleic Acids Res 39:W339–W346. https://doi.org/10.1093/NAR/GKR466
Mei Z, Chen GC, Wang Z et al (2021) Dietary factors, gut microbiota, and serum trimethylamine-N-oxide associated with cardiovascular disease in the Hispanic Community Health Study/Study of Latinos. Am J Clin Nutr 113:1503. https://doi.org/10.1093/AJCN/NQAB001
Merrick B, Allen L, Masirah M, Zain N et al (2020) Regulation, risk and safety of faecal microbiota transplant. Infect Prev Pract 2:100069. https://doi.org/10.1016/J.INFPIP.2020.100069
Mikail M, O'Doherty KC, Poutanen SM, Hota SS (2020) Ethical implications of recruiting universal stool donors for faecal microbiota transplantation. Lancet Infect Dis 20:e44–e49. https://doi.org/10.1016/S1473-3099(19)30569-9
Müller B, Kräusslich H-G (2009) Antiviral strategies. In: Handbook of experimental pharmacology, vol 189, pp 1–24. Springer, Berlin, Heidelberg. https://doi.org/10.1007/978-3-540-79086-0_1
Mullish BH, Quraishi MN, Segal JP et al (2018) The use of faecal microbiota transplant as treatment for recurrent or refractory Clostridium difficile infection and other potential indications: joint British Society of Gastroenterology (BSG) and Healthcare Infection Society (HIS) guidelines. J Hospital Infect 100:S1–S31. https://doi.org/10.1016/J.JHIN.2018.07.037
Munir M, Cheema AY, Ogedegbe OJ et al (2024) William Coley: the pioneer and the father of immunotherapy. Cureus 16:e69113. https://doi.org/10.7759/CUREUS.69113
Murray CJ, Ikuta KS, Sharara F et al (2022) Global burden of bacterial antimicrobial resistance in 2019: a systematic analysis. Lancet 399:629–655. https://doi.org/10.1016/S0140-6736(21)02724-0
Mushegian AR (2020) Are there 10^{31} virus particles on earth, or more, or fewer? J Bacteriol 202. https://doi.org/10.1128/JB.00052-20
Nathanson N (2016) The human toll of viral diseases: past plagues and pending pandemics. In: Viral pathogenesis, pp 3–16. https://doi.org/10.1016/B978-0-12-800964-2.00001-X. Epub 2016 Feb 12. PMCID: PMC7149761.
Nazir A, Hussain FHN, Raza A (2024) Advancing microbiota therapeutics: the role of synthetic biology in engineering microbial communities for precision medicine. Front Bioeng Biotechnol 12:1511149. https://doi.org/10.3389/FBIOE.2024.1511149/XML/NLM
Nemunaitis J, Cunningham C, Senzer N et al (2003) Pilot trial of genetically modified, attenuated Salmonella expressing the *E. coli* cytosine deaminase gene in refractory cancer patients. Cancer Gene Ther 10:737–744. https://doi.org/10.1038/SJ.CGT.7700634
Nguyen DH, Chong A, Hong Y, Min JJ (2023) Bioengineering of bacteria for cancer immunotherapy. Nat Commun 14:1–5. https://doi.org/10.1038/S41467-023-39224-8
Ofir G, Herbst E, Baroz M et al (2021) Antiviral activity of bacterial TIR domains via immune signalling molecules. Nature 600:116–120. https://doi.org/10.1038/s41586-021-04098-7
Õmura S, Crump A (2004) The life and times of ivermectin—a success story. Nat Rev Microbiol 2:984–989. https://doi.org/10.1038/NRMICRO1048
Omura S, Crump A (2014) Ivermectin: panacea for resource-poor communities. Trends Parasitol 30:445–455. https://doi.org/10.1016/J.PT.2014.07.005
Pace JL, Rossi HA, Esposito VM et al (1998) Inactivated whole-cell bacterial vaccines: current status and novel strategies. Vaccine 16:1563–1574. https://doi.org/10.1016/S0264-410X(98)00046-2
Palmer JD, Piattelli E, McCormick BA et al (2018) Engineered probiotic for the inhibition of Salmonella via tetrathionate-induced production of microcin H47. ACS Infect Dis 4:39–45. https://doi.org/10.1021/ACSINFECDIS.7B00114
Pantoja Angles A, Valle-Pérez AU, Hauser C, Mahfouz MM (2022) Microbial biocontainment systems for clinical, agricultural, and industrial applications. Front Bioeng Biotechnol 10:830200. https://doi.org/10.3389/FBIOE.2022.830200/XML/NLM
Pérez Davó A (2014) *Microorganismos halófilos* en ambientes salinos de Andalucía: estudio taxonómico numérico y molecular. Thesis, Granada: Universidad de Granada, 2014. ISBN 978-84-9083-091-8

Petrariu OA, Barbu IC, Niculescu AG et al (2023) Role of probiotics in managing various human diseases, from oral pathology to cancer and gastrointestinal diseases. Front Microbiol 14. https://doi.org/10.3389/FMICB.2023.1296447

Pietrzykowski T, Smilowska K (2021) The reality of informed consent: empirical studies on patient comprehension—systematic review. Trials 22:1–8. https://doi.org/10.1186/S13063-020-04969-W/TABLES/1

Pollard A (2020) Types of vaccine. https://vaccineknowledge.ox.ac.uk/types-of-vaccine

Pollard AJ, Bijker EM (2020) A guide to vaccinology: from basic principles to new developments. Nat Rev Immunology 21(2):83–100. https://doi.org/10.1038/s41577-020-00479-7

Porcari S, Benech N, Valles-Colomer M et al (2023) Key determinants of success in fecal microbiota transplantation: From microbiome to clinic. Cell Host Microbe 31:712–733. https://doi.org/10.1016/J.CHOM.2023.03.020

Pouresmaeil M, Azizi-Dargahlou S (2023) Factors involved in heterologous expression of proteins in *E. coli* host. Arch Microbiology 205(5):1–18. https://doi.org/10.1007/S00203-023-03541-9

Raihan T, Rabbee MF, Roy P et al (2021) Microbial metabolites: the emerging hotspot of antiviral compounds as potential candidates to avert viral pandemic alike COVID-19. Front Mol Biosci 8. https://doi.org/10.3389/fmolb.2021.732256

Randall JR, Davies BW (2021) Mining for novel antibiotics. Curr Opin Microbiol 63:66–69. https://doi.org/10.1016/j.mib.2021.06.001

Rather MA, Saha D, Bhuyan S et al (2022) Quorum quenching: a drug discovery approach against *Pseudomonas aeruginosa*. Microbiol Res 264:127173. https://doi.org/10.1016/J.MICRES.2022.127173

Redenti A, Im J, Redenti B et al (2024) Probiotic neoantigen delivery vectors for precision cancer immunotherapy. Nature 635(8038):453–461. https://doi.org/10.1038/s41586-024-08033-4 2024

Rhodes R (2016) Ethical issues in microbiome research and medicine. BMC Med 14:1–4. https://doi.org/10.1186/S12916-016-0702-7/PEER-REVIEW

Riggs AD (2021) Making, cloning, and the expression of human insulin genes in bacteria: the path to humulin. Endocr Rev 42:374–380. https://doi.org/10.1210/ENDREV/BNAA029

Rivera-Serrano EE, Gizzi AS, Arnold JJ et al (2020) Viperin reveals its true function. Annu Rev Virol 7:421–446. https://doi.org/10.1146/annurev-virology-011720-095930

Safarpour A, Ebrahimi M, Fazeli SAS, Amoozegar MA (2019) Supernatant metabolites from halophilic Archaea to reduce tumorigenesis in prostate cancer in-vitro and in-vivo. Iran J Pharm Res 18:241

Sanjuán R, Nebot MR, Chirico N et al (2010) Viral mutation rates. J Virol 84:9733–9748. https://doi.org/10.1128/JVI.00694-10

Santos-Beneit F (2024) What is the role of microbial biotechnology and genetic engineering in medicine? Microbiologyopen 13. https://doi.org/10.1002/MBO3.1406

Sato Y, Okuyama S, Hori K (2007) Primary structure and carbohydrate binding specificity of a potent anti-HIV lectin isolated from the filamentous Cyanobacterium *Oscillatoria agardhii*. J Biol Chem 282:11021–11029. https://doi.org/10.1074/jbc.M701252200

Sawa T, Moriyama K, Kinoshita M (2024) Current status of bacteriophage therapy for severe bacterial infections. J Intensive Care 12:1–9. https://doi.org/10.1186/S40560-024-00759-7/TABLES/3

Schatz A, Bugle E, Waksman SA (1944) Streptomycin, a substance exhibiting antibiotic activity against Gram-positive and Gram-negative bacteria. Proc Soc Exp Biol Med 55:66–69. https://doi.org/10.3181/00379727-55-14461/PDF

Schlimpert S, Elliot MA (2023) The best of both worlds—*Streptomyces coelicolor* and *Streptomyces venezuelae* as model species for studying antibiotic production and bacterial multicellular development. J Bacteriol 205. https://doi.org/10.1128/JB.00153-23

Shehata MM, Mostafa A, Teubner L et al (2019) Bacterial outer membrane vesicles (OMVs)-based dual vaccine for influenza A H1N1 virus and MERS-CoV. Vaccines 7:46. https://doi.org/10.3390/VACCINES7020046

Shou Y, Yeo JE, Pang ASR et al (2024) Informed consent and risk communication challenges in antimicrobial clinical trials: a scoping review. BMJ Open 14:e082096. https://doi.org/10.1136/BMJOPEN-2023-082096

Slowinski W, Charm SE (1973) Glutamic acid production with gel-entrapped *Corynebacterium glutamicum*. Biotechnol Bioeng 15:973–979. https://doi.org/10.1002/BIT.260150514

Soares J (2011) The nagoya protocol and natural product-based research. ACS Chem Biol 6:289. https://doi.org/10.1021/CB200089W;SUBPAGE:STRING:FULL

Soemarie YB, Milanda T, Barliana MI (2021) Fermented foods as probiotics: a review. J Adv Pharm Technol Res 12:335. https://doi.org/10.4103/JAPTR.JAPTR_116_21

Stojković D, Petrović J, Carević T, et al (2023) Synthetic and semisynthetic compounds as antibacterials targeting virulence traits in resistant strains: a narrative updated review. Antibiotics 12. https://doi.org/10.3390/antibiotics12060963

Strathdee SA, Hatfull GF, Mutalik VK, Schooley RT (2023) Phage therapy: from biological mechanisms to future directions. Cell 186:17–31. https://doi.org/10.1016/j.cell.2022.11.017

Suleman M, Clark JR, Bull S, Jones JD (2025) Ethical argument for establishing good manufacturing practice for phage therapy in the UK. J Med Ethics 51:298–302. https://doi.org/10.1136/JME-2023-109423

Swartz MN (2000) Impact of antimicrobial agents and chemotherapy from 1972 to 1998. Antimicrob Agents Chemother 44:2009–2016. https://doi.org/10.1128/AAC.44.8.2009-2016.2000

Tevyashova AN (2021) Recent trends in synthesis of chloramphenicol new derivatives. Antibiotics 10:370. https://doi.org/10.3390/ANTIBIOTICS10040370

Tucker SK, McHugh RE, Roe AJ (2024) One problem, multiple potential targets: where are we now in the development of small molecule inhibitors against Shiga toxin. Cell Signal 121:111253. https://doi.org/10.1016/j.cellsig.2024.111253

Van Hul M, Cani PD, Petifils C et al (2024) What defines a healthy gut microbiome. Gut 73:1893–1908. https://doi.org/10.1136/GUTJNL-2024-333378

van Riet E, Retra K, Adegnika AA et al (2008) Cellular and humoral responses to tetanus vaccination in Gabonese children. Vaccine 26:3690–3695. https://doi.org/10.1016/J.VACCINE.2008.04.067

Vernon Gervase (2019) Life & times Syphilis and Salvarsan. Br J Gen Pract 36:2019

Wang JW, Kuo CH, Kuo FC et al (2019) Fecal microbiota transplantation: review and update. J Formos Med Assoc 118:S23–S31. https://doi.org/10.1016/j.jfma.2018.08.011

Warfel KF, Williams A, Wong DA et al (2023) A low-cost, thermostable, cell-free protein synthesis platform for on-demand production of conjugate vaccines. ACS Synth Biol 12:95–107. https://doi.org/10.1021/ACSSYNBIO.2C00392/SUPPL_FILE/SB2C00392_SI_001.PDF

Weiman S, Fox J (2015) Harnessing the power of microbes as therapeutics: bugs as drugs. https://doi.org/10.1128/AAMCOL.APR.2014

Weinstein EA, Ordonez AA, DeMarco VP et al (2014) Imaging Enterobacteriaceae infection in vivo with 18F-fluorodeoxysorbitol positron emission tomography. Sci Transl Med 6:259ra146. https://doi.org/10.1126/SCITRANSLMED.3009815

Yang Q, Le S, Zhu T, Wu N (2023) Regulations of phage therapy across the world. Front Microbiol 14:1250848. https://doi.org/10.3389/FMICB.2023.1250848/BIBTEX

You S, Ma Y, Yan B, et al (2022) The promotion mechanism of prebiotics for probiotics: a review. Front Nutr 9. https://doi.org/10.3389/FNUT.2022.1000517

Zhang CW, Zhong XJ, Zhao YS et al (2023) Antifungal natural products and their derivatives: a review of their activity and mechanism of actions. Pharmacol Res Mod Chin Med 7:100262. https://doi.org/10.1016/J.PRMCM.2023.100262

Zhang X, Wei M, Pan H et al (2014) Robust manufacturing and comprehensive characterization of recombinant hepatitis E virus-like particles in Hecolin®. Vaccine 32:4039–4050. https://doi.org/10.1016/j.vaccine.2014.05.064

Zhao T, Cai Y, Jiang Y et al (2023) Vaccine adjuvants: mechanisms and platforms. Signal Transduct Target Ther 8(1):1–24. https://doi.org/10.1038/s41392-023-01557-7

Chapter 8
Microbial Biotechnology in Agriculture

Mustafa Çilkiz

Abstract Global food security has become one of the greatest challenges of the twenty-first century due to the rapidly growing world population's food demands and environmental threats such as climate change, soil erosion, and the depletion of freshwater resources. The extensive use of chemical fertilizers and pesticides throughout conventional agriculture has increased productivity significantly, but it has additionally resulted in major ecological and socioeconomic problems, such as soil acidity, groundwater resource pollution, and decreased biodiversity. In this regard, microbial biotechnology is a particularly noteworthy technique that improves agricultural production while promoting environmental sustainability, maintaining ecological balance, and making effective use of resources. This application makes use of microorganisms to enhance soil health and structure, promote plant growth, and minimize both abiotic and biotic stresses. Microbial applications include nitrogen fixation, as well as biofertilizers that reduce the dependency on synthetic materials and biopesticides. Microbial consortia and biostimulants that improve plant physiology by producing phytohormones produce more dependable and durable consequences in the field. Metagenomics and metabolomics are the two types of omic technologies used in these areas of study that provide a thorough description of the variety and roles of microorganisms. Furthermore, the intentional production of microbes targeted at specific organisms has been made practical via synthetic biology and gene editing techniques. In-depth case studies performed in several countries reveal that microbial technologies significantly reduced expenses and improved soil production, advancing

M. Çilkiz (✉)
Transition Zone Agricultural Research Institute Directorate, Eskişehir, Türkiye
e-mail: mustafacilkiz@yahoo.com

M. Çalışkan and M. Arıkan (eds.), *Microbial World: Bacteria and Archaea*, Progress in Molecular and Subcellular Biology 62, https://doi.org/10.1007/978-3-032-11438-9_8

the sustainable development goals. Nevertheless, there are several barriers to the widespread use of microbial biotechnology in agriculture. These include unpredictable conditions in the fields, strict regulations, especially related to genetically modified organisms' problems with product quality, and farmers' insufficient understanding. Microbial biotechnology aims to accomplish its full potential as an advancement in technology and as an essential aspect of resource-efficient and environmentally friendly agricultural systems via responsible innovation, adaptable regulations, and worldwide cooperation.

Keywords Agricultural biotechnology · Soil · Food · Biofertilizers · Biostimulants · Microorganisms

8.1 Introduction

Global food security, or the demand to sustainably provide for the world's population's constantly increasing demand for food, is one of the twenty-first century's major challenges. By 2050, there will be approximately 9.7 billion people on the globe, estimated the Food and Agriculture Organization (FAO). According to Hamad and Tayel (2025), this scenario calls for a roughly 70% increase in food production above present levels. Agricultural systems are already under stress from a spectrum of environmental effects, particularly desertification, soil erosion, climate change, the depletion of clean water supplies, and the loss of biodiversity. This demand causes additional pressure on these ecosystems. All these components contribute to reducing the productivity of agriculture and endanger long-term resilience (Smith 2025; Lal et al. 2021; Wheeler and von Braun 2013). Through the widespread utilization of chemical pesticides, fertilizers, and high-yield crop varieties, the Green Revolution dramatically enhanced agricultural productivity; yet, these successes also resulted in considerable ecological and social costs. Soil acidification, organic matter depletion, contamination of groundwater, nitrate accumulation in renewable products, and excessive eutrophication in aquatic habitats are all results of agriculture's increased reliance on chemicals (John and Babu 2021; Kalogiannidis et al. 2022; Pretty and Bharucha 2014). Furthermore, widely used pesticides have caused damage to important soil microorganisms and promoted the emergence of insect species resistant to these chemicals, posing additional difficulties for crop protection (Singh 2024; Altman 2018). These concerns highlight attention to the drawbacks of conventional input-intensive agriculture systems and emphasize the urgent requirement for environmentally friendly alternative solutions that integrate environmental sustainability with agricultural products. In this connection, microbial biotechnology is becoming an increasingly promising approach, providing a range of strategies to reduce environmental damage, enhance agricultural yields, renew soil fertility, and enhance resilience. Given this, microbial biotechnology becomes a revolutionary discipline that provides resource-efficient and environmentally friendly solutions for modern agriculture (Mirsalami and Mirsalami 2025). The use of naturally occurring or genetically engineered microorganisms to improve ecosystem health and productivity in agriculture is commonly referred to as microbial biotechnology. By promoting plant development, enhancing soil fertility and structure, and minimizing issues associated with biotic stressors like pathogens, pests, and weeds, these microbes enhance agricultural production in many different aspects (Yadav et al. 2021; Verma et al. 2020). Additionally, microbial communities optimize nutrient cycling and bioavailability while decreasing environmental stresses including drought, salt, high temperatures, and heavy metal pollution, increasing plant resilience in challenging environments (Koza et al. 2022; Singh et al. 2023) Microbial biotechnology incorporates biological mechanisms to provide sustainable and renewable solutions for plants and soils in place of synthetic agricultural chemicals, which are extensively dependent on external stimuli regularly trigger catastrophic ecosystem damage (Folina et al. 2025). As a result, microorganisms in the soil microbiome are increasingly recognized to be vital players in the advancement of sustainable agriculture rather than passive organisms (O'Callaghan et al. 2022; Iqbal et al. 2025).

Microbial biotechnology is a discipline that combines a number of areas, including microbiology, soil science, plant biology, molecular biology, ecology, and environmental biotechnology (Sameen and Khalil 2025). This combination has been strengthened by its integration with emerging advancements in omics sciences, such as genomics, transcriptomics, proteomics, and metabolomics, which have drastically altered our knowledge of the workings of microbial communities in agricultural ecosystems (Wijayawardene et al. 2023). Metagenomic research has revealed the considerable taxonomic and functional diversity of rhizosphere microbiomes, which has contributed to the finding of genes associated with nutrient cycling, plant hormone synthesis, and stress management (El-Halim et al. 2025; Rajguru et al. 2025). Similarly, metabolomic research has provided opportunities for the discovery of a range of bioactive substances produced from organisms that interact with plants, including siderophores, exopolysaccharides, and volatile organic compounds. These compounds have key functions in improving nutrient absorption and improving tolerance to abiotic and biotic stresses, thereby contributing to plant growth (Ullah et al. 2025; Etesami et al. 2025). Transcriptomic and proteomic techniques have offered an alternative perspective on plant–microorganism–soil interactions by providing a better understanding of the responses of communities where microorganisms interact to environmental variations and plant signals (Jain et al. 2024; Sharma et al. 2020). When taken together, these omics-based data provide a holistic perspective of agricultural ecosystems and facilitate the design and engineering of microbial-derived products, such as biofertilizers, biostimulants, and biopesticides, with higher accuracy (Jansson and Hofmockel 2018).

Microbial biotechnology applications in agriculture are divided into four major categories, with each one addressing a distinctive but interrelated aspect of agricultural output and sustainability. First, biofertilizers utilize the capabilities of beneficial microbial organisms, including *Rhizobium*, *Azospirillum*, and arbuscular mycorrhizal fungi, which perform essential functions including nitrogen fixation, phosphate solubilization, and nutrient cycling. This promotes soil fertility and decreases the dependence on chemical fertilizers (Bhardwaj et al. 2014; Adesemoye et al. 2009).

Second, biopesticides consist of microbial agents, including *Bacillus thuringiensis* (Bt), Trichoderma species, and entomopathogenic fungi. These chemicals control plant diseases and damaging insects while minimizing undesirable effects and pesticide resistance, providing ecologically advantageous alternatives to chemical pesticides (Chaudhary 2024; Cai and Dimopoulos 2025). Third, biostimulants known as plant growth-promoting rhizobacteria (PGPR), which usually originate from bacteria such as *Pseudomonas fluorescens* and *Azospirillum* spp., support plant development through a variety of strategies. These strategies include producing phytohormones, improving nutrient absorption, and establishing systemic resistance to abiotic and biotic stresses (de Andrade et al. 2023; Rouphael and Colla 2020; Sun et al. 2024). Finally, microbial consortia are a pioneering innovation that employs formulations including multiple strains or species joined together to perform efficiently in combination. This enables more consistent and strong effects under various soil and agricultural ecological conditions (Bhattacharyya and Jha 2012; Malusà et al. 2016).

Microbial biotechnology has a profound social and economic impact on agricultural systems, especially in resource-limited regions. Data from countries such as India, Brazil, and China show that using microbial biological fertilizers and biopesticides could decrease input costs by 20% while improving soil quality, nutrient efficiency, and sustainability. (Adesemoye et al. 2009; da Costa et al. 2018). This dual effect, which supports environmental sustainability by increasing farmers' profits, is extremely important for small farmers who are at higher risk due to rising agricultural chemical prices, contaminated soil conditions, and climate change (Pretty et al. 2018; Sharma and Singh 2019). Microbial biotechnology not only addresses farmers' economic expectations by reducing economic barriers to sustainable agricultural practices but also contributes to rural development and food security issues from a broader perspective. Furthermore, the application of microbial technologies is closely aligned with international sustainability goals such as the United Nations Sustainable Development Goals. The benefits provided by these technologies directly support SDG 2 (Zero Hunger) by increasing agricultural productivity, SDG 12 (Responsible Consumption and Production) by reducing the use of synthetic agricultural chemicals, and SDG 13 (Climate Action) by reducing fertilizer-related greenhouse gas emissions (FAO 2017; United Nations 2015). Thus, microbial biotechnology is not only considered a technical innovation but also serves as a strategic policy tool that can combine agricultural intensification with ecological management and social equity. It is increasingly being recognized by policymakers, development agencies, and international organizations, highlighting its potential to transform agricultural systems toward a sustainable and inclusive future.

The role of microbial biotechnology in agriculture extends far beyond the technical realm of crop management. It represents a new approach in agricultural science, where ecological management is increasingly integrated with technological innovations to address dual requirements such as yield and sustainability (Tilman et al. 2002; Pretty 2008). While traditional methods rely heavily on synthetic inputs, microbial-based strategies harness the natural abilities of living organisms to improve soil health, increase nutrient use efficiency, and provide resistance to both biotic and abiotic stresses, thereby aligning agricultural practices with long-term ecological balance

(Foley et al. 2011; Hartmann et al. 2015). Microbial biotechnology symbolizes the convergence of traditional agricultural knowledge with elements such as compost, preparations, and organic fertilizers, along with the latest innovations in molecular biology, genomics, and synthetic biology (Bender et al. 2016; Finkel et al. 2017). This convergence has increased the importance of microbial solutions in addressing global food security issues such as soil erosion and pollution, climate change, and the growing need for sustainable agricultural practices (Godfray et al. 2010; Rockström et al. 2017). This has led to its evolution from being a complementary application to being accepted as a fundamental element of sustainable agriculture. In the following sections of this review, the scientific foundations of microbial biotechnology, its various applications, agricultural and socioeconomic benefits, technological innovations, current challenges, and emerging trends are analyzed from different perspectives.

8.2 Scientific Foundations of Microbial Biotechnology in Agriculture

Microbial biotechnology in agriculture involves uncovering and harnessing the complex relationships between plants, microorganisms, and soil. Microorganisms have developed biochemical, physiological, and ecological adaptations that vary depending on the environment in which they live, and these adaptations play an important role in regulating nutrient cycles, maintaining soil fertility, and ensuring plant health (van der Heijden et al. 2008; Schloter et al. 2018). Microbial communities do not merely play a passive role in agricultural ecosystems; they also actively influence plant growth and resistance by directing important processes such as nitrogen fixation, phosphate solubilization, siderophore production, hormone modulation, and pathogen suppression (Glick 2012; Compant et al. 2019). For these reasons, a thorough understanding of the scientific foundations of microbial biotechnology requires addressing three interrelated key topics. First, examining the ecological functions of microorganisms in agricultural systems should aim to reveal their contributions to soil structure formation, nutrient cycling, and plant–microorganism symbioses such as mycorrhizal relationships and rhizobium–legume interactions (Smith and Read 2010; Oldroyd 2013). Second, the technical aspects of supporting plant growth and reducing stress factors require the investigation of various microbial functions, such as the regulation of phytohormones (e.g., indole-3-acetic acid, cytokinins, and gibberellins), increasing systemic resistance to biotic and abiotic stress factors, and structuring the root system (Vessey 2003; Backer et al. 2018). Finally, due to the rapid advancement of molecular and genomic technologies, the identification, investigation, and integration of beneficial microbial strains into agro-engineering studies should be pursued. Innovations in next-generation sequencing, metagenomics, and synthetic biology enable researchers not only to characterize microbial diversity but also to optimize microbial functions for specific agricultural objectives (Berg 2020; Timmis et al. 2017). Bringing these three areas together will enable the integrated use of microbial biotechnology in agriculture to create sustainable and resilient food systems based on beneficial plant–microorganism–soil interactions.

8.2.1 Microorganisms as Key Factors in Agroecosystem Function

Microorganisms play a significant role in the sustainability of soil health and productivity, directing the functioning of fundamental biochemical cycles such as nitrogen, phosphorus, sulfur, and carbon in agricultural ecosystems (Timmis et al. 2017; Fierer 2017). Biological nitrogen fixation, one of these processes, forms the basis of interactions between plants and microorganisms. Symbiont bacteria such as *Rhizobium* spp. and related diazotrophs such as *Azospirillum* spp. convert atmospheric dinitrogen (N_2) into ammonia (NH_3) form that can be used by plants, reducing the need for synthetic nitrogen fertilizers (Hungria and Mendes 2015; Mus et al. 2016). The contributions of these microorganisms not only increase crop yields but also mitigate environmental problems such as nitrate leaching and greenhouse gas emissions caused by excessive fertilizer use. Phosphorus, another important macronutrient, is often found in insoluble forms that are unavailable to plants in soils. Phosphate-solubilizing microorganisms (PSMs), such as *Pseudomonas*, *Bacillus*, and *Aspergillus*, significantly improve plant nutrition and growth by producing organic acids, phosphatases, and other metabolites that convert bound phosphates into biologically available forms (Richardson and Simpson 2011; Kuhad 2011). These microorganisms, which facilitate phosphorus uptake, help overcome one of the major challenges in many agricultural soils and reduce dependence on limited mineral phosphate resources. However, arbuscular mycorrhizal fungi (AMF) are also of great importance. These fungi form symbiotic relationships with over 80% of terrestrial plant species. AMF spreads its hyphal networks beyond the roots into the soil, providing an effective root surface area that enhances nutrient and water uptake. In this way, AMF increase phosphorus, nitrogen, and micronutrient uptake while also strengthening plants' ability to cope with abiotic stresses such as salinity, drought, and heavy metal toxicity (Smith and Read 2008; van der Heijden et al. 2015). In addition to the direct effects of AMF on plant nutrition, they contribute to soil structure and aggregation by producing soil proteins associated with glomalin, which stabilizes soil particles and increases water retention capacity. Overall, these microbial processes demonstrate the indispensable role of soil microorganisms in regulating nutrient availability, maintaining soil fertility, and enhancing plant resilience. Microorganisms serve as natural biofertilizers and bioregulators by directing the biochemical cycles of essential nutrients, offering ecologically sustainable alternatives to chemical inputs, and forming the basis of microbial biotechnology in agriculture.

The rhizosphere, a narrow area directly exposed to the effects of root exudates, is an important area for plant, soil, and microorganism interactions. Plant roots actively secrete various organic molecules such as sugars, amino acids, fatty acids, phenolic compounds, and organic acids from root exudates, and these compounds serve as both substrates and signals for microorganisms in the environment (Philippot et al. 2013; Sasse et al. 2018). These components are not only related to the taxonomic structure of microbial communities but also affect metabolic processes and ecological functions in the rhizosphere. In turn, microorganisms produce phytohormones (e.g., auxins, cytokinins, and gibberellins), antibiotic components, siderophores, volatile organic compounds, and signaling molecules that directly or indirectly regulate factors such as plant growth, nutrient uptake, immunity, and stress tolerance (Berendsen et al. 2012; Vacheron et al. 2013). This mutual interaction between roots and microorganisms makes the rhizosphere a dynamic center of biochemical and ecological interactions, giving rise to a specific rhizosphere microbiome. This microbiome is increasingly considered an extension of the plant's function due to its critical role in supporting the plant's metabolic and physiological capabilities and is often referred to as the plant's second genome (Bulgarelli et al. 2013; Mendes et al. 2013). Through processes such as nutrient utilization, pathogen control, and regulation of plant defense pathways, the rhizosphere microbiome is seen as a central determinant of crop productivity, resilience to environmental stress, and long-term soil health (Tkacz et al. 2015; Compant et al. 2019).

8.2.2 Mechanisms for Promoting Plant Growth and Reducing Stress

Microbial effects on plant growth and health can generally be divided into two categories: direct and indirect mechanisms (Table 8.1). Both mechanisms function through a wide variety of biochemical, physiological, and ecological processes that significantly influence the productivity and resilience of agricultural crops. Direct mechanisms affect plant growth, particularly by increasing the availability of nutrients. Such processes include biological nitrogen fixation, phosphate solubilization, siderophore synthesis, and phytohormone production. Nitrogen-fixing bacteria such as *Rhizobium*, *Azospirillum*, and *Bradyrhizobium* reduce the need for synthetic nitrogen fertilizers by converting atmospheric nitrogen into biologically available ammonia (Hungria et al. 2006). Phosphate-solubilizing bacteria and fungi, such as *Pseudomonas*, *Bacillus*, and *Aspergillus*, produce organic acids and phosphatase enzymes that mobilize insoluble phosphate components, thereby increasing phosphorus uptake efficiency (Richardson and Simpson 2011). Siderophore-producing bacteria of the *Pseudomonas fluorescens* species chelate ferric iron (Fe^{3+}) found in the soil, providing this important micronutrient to plants while also limiting its use by competing pathogens (Saha et al. 2016). In addition, many plant growth-promoting rhizobacteria (PGPR), such as *Azospirillum brasilense* and *Pseudomonas fluorescens*, produce phytohormones—auxins, cytokinins, and gibberellins—that affect the root system, promote lateral root formation, and increase nutrient uptake (Glick 2012; Vacheron et al. 2013). The combined effect of these processes leads to an expansion of root surface area, improved nutrient uptake, and ultimately high crop yields. Indirect mechanisms function through pathogen suppression, enhancement of plant immune responses, and reduction of abiotic stresses. Many beneficial microorganisms synthesize antibiotics, lytic enzymes, and volatile

Table 8.1 Direct and indirect microbial mechanisms in promoting plant growth

Category	Mechanism	Representative microorganisms	Function/benefit	Key references
Direct	Nitrogen fixation	*Rhizobium*, *Azospirillum*, *Bradyrhizobium*	Conversion of N^2 to ammonia for plant uptake	Hungria et al. (2006)
Direct	Phosphate solubility	*Pseudomonas*, *Bacillus*, *Aspergillus*	Mobilization of insoluble phosphate	Richardson and Simpson (2011)
Direct	Siderophore production	*Pseudomonas fluorescens*, *Enterobacter*	Fe^{3+} chelation, enhanced iron uptake, pathogen suppression	Saha et al. (2016)
Direct	Phytohormone synthesis	*Azospirillum brasilense*, *Pseudomonas fluorescens*	Auxins, cytokinins, and gibberellins regulate root structure	Vacheron et al. (2013)
Indirect	Pathogen suppression	*Trichoderma* spp., *Bacillus subtilis*	Enzyme secretion, antimicrobial metabolites, and biofilms	Harman (2011), Ongena and Jacques (2008)
Indirect	Induced systemic resistance	*Pseudomonas*, *Bacillus* spp.	Preparation of plant immune pathways	Pieterse et al. (2014)
Indirect	Abiotic stress tolerance	*Pseudomonas*, *Enterobacter*, mycorrhizal fungi	ACC deaminase, exopolysaccharides, osmotic adjustment	Glick (2014), Smith and Read (2008)

organic compounds that suppress phytopathogens. For example, Trichoderma spp. fungi secrete hydrolytic enzymes such as chitinases and glucanases that break down cell walls, while *Bacillus subtilis* produces antimicrobial lipopeptides and creates competition against pathogens in the rhizosphere by forming protective biofilms (Harman 2011; Ongena and Jacques 2008). Furthermore, microbial effectors can induce systemic resistance (ISR) in plants; this refers to a state in which microbial signals prepare the host to mount a faster and stronger defense response when dealing with situations such as pathogen attack or abiotic stress (Pieterse et al. 2014).

Microorganisms play a critical role in enhancing plants' tolerance to abiotic stresses such as drought, salinity, and heavy metal toxicity, as well as possessing a pathogen suppression function. Exopolysaccharide-producing bacteria improve plant access to water during drought periods by supporting the aggregation of soil aggregates and increasing water retention capacity (Sandhya et al. 2009). Arbuscular mycorrhizal fungi increase access to water and nutrients through the hyphal networks they spread to roots, while also strengthening the osmotic balance of host plants (Smith and Read 2008). Some PGPR, such as *Pseudomonas* and *Enterobacter* strains, produce the enzyme 1-aminocyclopropane-1-carboxylate (ACC) deaminase, which reduces stress-induced ethylene levels in plants and thus alleviates growth pressure under adverse conditions

(Saleem et al. 2007; Glick 2014). When considered together, these microorganisms demonstrate that microbial processes offer numerous benefits beyond traditional agricultural chemicals in microbial biotechnology. While synthetic inputs typically focus on a specific problem (e.g., nutrient deficiencies or pest control), microbial effectors also improve nutrient cycles, promote growth, suppress pathogens, and increase plants' resilience to environmental stress. This versatility makes microbial biotechnology a fundamental part of sustainable and climate-resilient agricultural practices.

8.2.3 Omics Technologies and Systems Biology in Microbial Biotechnology

Over the past 20 years, the development of high-throughput omics technologies has brought about a fundamental shift in microbial biotechnology, transforming it from a largely descriptive field into a predictive, quantitative, and engineering-focused discipline (Table 8.2). Methods such as genomics, transcriptomics, proteomics, metabolomics, and systems biology have significantly deepened our understanding of microbial diversity while providing applicable data on the ecological and functional roles of microorganisms within agricultural ecosystems (Jamil et al. 2022; Dubey et al. 2019). These advances have laid the scientific groundwork for the rational design of microbial inoculants and the application of synthetic biology techniques in agriculture.

Metagenomics has been quite successful in illuminating the vast taxonomic and functional diversity of soil and rhizosphere microbiomes. A large portion of these microbiomes cannot be revealed using traditional laboratory techniques. Thanks to high-throughput sequencing technologies, researchers have been able to identify genes associated with important agricultural processes such as nitrogen fixation, phosphate solubilization, siderophore biosynthesis, and abiotic stress tolerance (Fierer 2017; Mendes et al. 2019). These datasets not only determine the composition of microbial communities but also make functional predictions linking microbial presence to ecosystem services, allowing researchers to better understand the contributions of soil microbiomes to crop productivity and resilience. While metagenomics reveals genetic potential, metatranscriptomic and metaproteomic studies provide the opportunity to identify which genes are actively expressed and which proteins are produced in situ. This distinction is crucial because microbial potential is linked to the ecological functions performed. For example, transcriptomic studies have shown that rhizobacteria supporting plant growth increase stress response genes under drought and salinity conditions, thereby enhancing plant tolerance (Carrión 2019). Similarly, metaproteomic studies have revealed the production of enzymes related to the nutrient cycle, such as phosphatase secretion for phosphorus mobilization and ACC deaminase activity for stress reduction (Wilmes et al. 2015). These methods go beyond merely static inventories of microbial diversity, providing a functional perspective on microbial interactions in agricultural soils. Metabolomics adds another dimension by profiling small molecules produced by microorganisms that are directly involved in

Table 8.2 Omics approaches and their contributions to microbial biotechnology in agriculture

Omics approach	Main focus	Key findings	Agricultural applications	References
Metagenomics	DNA sequencing of all microbial communities	Determines taxonomic diversity; predicts functional genes (e.g., N fixation, P solubilization, stress tolerance)	Discovery of new biofertilizer candidates; functional markers for soil health monitoring	Fierer (2017), Mendes et al. (2019)
Metatranscriptomics	RNA sequencing to evaluate gene expression profiles	Links microbial potential to active functions; identifies stress response genes	Understanding microbial adaptation under drought/salinity conditions; improving biostimulant formulations	Carrión et al. (2019)
Metaproteomics	Large-scale protein identification and quantification	Detects functionally expressed enzymes and proteins in situ	Discovery of enzymes for the nutrient cycle (e.g., phosphatases, ACC deaminases); studies on microbial stress physiology	Wilmes et al. (2015)
Metabolomics	Profiling of microbial metabolites and signaling molecules	Identifies bioactive compounds (phytohormones, siderophores, antibiotics, VOCs)	Development of microbial biostimulants and biocontrol products	Schulz-Bohm et al. (2017)
Systems biology	Integration of omics datasets into network models	Models microorganism–microorganism and plant–microorganism interactions; predicts functional stability	Rational design of microbial consortia; prediction of responses to climate and soil management	Sessitsch et al. (2019), Toju et al. (2018)
Synthetic biology	Rational engineering of microbial genomes and functions	CRISPR/Cas gene editing; redesign of metabolic pathways	Enhanced N-fixation, stress tolerance, biological control traits; programmable microbial inoculants	Wang et al. (2016), Ali et al. (2025)

plant–microorganism interactions and stress regulation. Components such as phytohormones (e.g., indole-3-acetic acid), siderophores, antibiotics, and volatile organic compounds have been identified as important metabolites for plant growth, nutrient uptake, pathogen resistance, and stress tolerance (Schulz-Bohm et al. 2017). Most of these bioactive compounds are currently considered pioneers in the development of new-generation microbial biostimulants and biocontrol agents.

The integration of these various omics datasets using systems biology has enabled the development of network models that reveal the complex interactions between microorganisms, plants, and soils. These models are vital for the logical design of microbial groups with complementary functions, such as nitrogen-fixing bacteria, phosphate-solubilizing fungi, and biological control agents, because these groups produce consistent and resilient outcomes under variable environmental conditions (Sessitsch et al. 2019; Toju et al. 2018). This type of information also helps predict how microbial communities will respond to agricultural applications and climate change, making systems biology an indispensable tool for sustainable agriculture.

Finally, while the convergence of omics and systems biology with synthetic biology has opened new opportunities for the deliberate engineering of microbial genomes, genome editing techniques, particularly CRISPR/Cas systems, are increasingly being applied to enhance nitrogen fixation efficiency, alter symbiotic capacities, and improve stress tolerance or pathogen suppression (Wang et al. 2016; Wang 2017). Synthetic approaches can be designed to enable microorganisms to overproduce bioactive metabolites such as phytohormones or antibiotics, thereby increasing agricultural benefits (Liu and Stewart 2015). This "microbial factory" concept may enable a significant shift in agricultural biotechnology from the use of naturally occurring strains to the deliberate design of bioinoculants specifically tailored for particular products and environments. These tools, which enable researchers to move from defining microbial diversity to effectively modifying microbial functions, have advanced precision microbial management in agriculture. This transformation is consistent with broader goals such as sustainable agriculture, ecological management, and global food security, in addition to increasing crop productivity and resilience.

8.2.4 Microbial Biotechnology as an Ecological and Evolutionary Discipline

Although microbial biotechnology is generally approached from a technical perspective, it is equally important to evaluate it within broader ecological and evolutionary frameworks. Soil microbial communities are in a state of constant change and are shaped by selective pressures such as nutrient availability, climate variability, and agricultural practices (Fierer 2017; Prosser 2020). The sustainability of microbial biotechnology depends on engineered or modified strains being ecologically compatible with established microbial communities, thereby protecting against adverse effects that threaten ecosystem functionality. The ecological complexity of soils presents significant challenges, such as the variable performance of microbial inoculants due to competition or suppression by microorganisms (Schmidt et al. 2019; Kaminsky et al. 2019). These ecological possibilities reveal the limitations of simplified approaches that disregard community-level interactions. Consequently, system-level strategies that integrate microbial ecology, evolutionary processes, and biotechnological innovations are essential for creating robust and sustainable microbial solutions in agriculture (de Vries and Wallenstein 2017; Toju et al. 2018).

8.2.5 Summary

Fundamentally, the scientific basis of microbial biotechnology in agriculture is an integration encompassing microbial physiology, plant–microorganism interactions, ecosystem dynamics, and various layers of molecular biology. Microorganisms provide essential ecosystem services that sustain the function of agricultural ecosystems. These functions include nutrient cycling, organic matter decomposition, abiotic stress reduction, and phytopathogen control (van der Heijden et al. 2008; Timmis et al. 2019). They also form the basis of sustainable agricultural productivity. In high-throughput omics technologies, innovations in the fields of genomics, transcriptomics, proteomics, and metabolomics have revolutionized the identification, characterization, and functional elucidation of microbial communities in agricultural soils (Fierer 2017; Dubey et al. 2019). These approaches correlate microbial taxonomic diversity with functional potentials, providing deeper mechanistic insights into the effects of soil microbiomes on plant growth and resilience. Furthermore, systems biology methods integrate these datasets to develop predictive models of microbial–plant–soil interactions, thereby enabling the design of microbial inoculants and consortia that offer greater stability and efficacy (Sessitsch et al. 2019; Toju et al. 2018).

The combination of ecological approaches and technological innovations positions microbial biotechnology not only as efficient and resilient but also as a powerful scientific framework for developing environmentally sustainable agricultural systems. By utilizing natural microbial processes in conjunction with molecular engineering and synthetic biology, this field has the potential to reduce the negative effects of chemical inputs, improve soil health, and directly contribute to climate-friendly agricultural practices (Lal 2020; Jacoby et al. 2017).

8.3 Applications of Microbial Biotechnology in Agriculture

The use of microbial biotechnology in agriculture plays an important role in supporting ecosystem sustainability by providing increasingly more benefits. By using natural and engineered microorganisms, microbial products offer more environmentally friendly alternatives to chemical fertilizers and pesticides, thereby addressing the need to strike the necessary balance between agricultural productivity and ecosystem health

(Adesemoye et al. 2009). These innovations work in many ways, such as increasing the availability of nutrients, controlling soil-borne pests and diseases, supporting plant growth, and increasing resistance to abiotic stresses such as drought, salt, and heavy metal toxicity (Glick 2012; Backer et al. 2018).

The practical applications of microbial biotechnology can be divided into four main categories: biofertilizers, biopesticides, biostimulants, and microbial consortia. Biofertilizers include species such as nitrogen-fixing bacteria and phosphate-solubilizing microorganisms and contribute directly to the nutrient cycle, reducing dependence on chemical substances (Richardson and Simpson 2011; Bhattacharyya and Jha 2012). Biopesticides include species such as *Bacillus thuringiensis* and *Trichoderma* spp., and reduce the ecological threats associated with traditional pesticides by controlling pests specific to the target content (Glare et al. 2012). Biostimulants, often based on plant growth-promoting rhizobacteria such as *Pseudomonas fluorescens*, enhance crop health by increasing phytohormone production (Vessey 2003; Calvo et al. 2014). Ultimately, microbial consortia formed by combining multiple strains establish a new technological standard, offering more resilient and consistent synergistic benefits under various environmental conditions (Bashan et al. 2014; Malusá et al. 2016).

Each category represents a theoretical advancement in agricultural science that encompasses not only applications in plant cultivation but also a deeper understanding of plant–microorganism–soil interactions. Furthermore, these innovations provided by microbial biotechnology also contribute to reducing greenhouse gas emissions, increasing soil biodiversity, and aligning with global goals such as the United Nations Sustainable Development Goals (Timmis et al. 2019). Therefore, the uses of microbial biotechnology go beyond limited technical solutions and offer systematic innovations in the development of environmentally friendly and resource-efficient agricultural systems.

8.3.1 Biofertilizers: Microbial Agents That Increase Nutrient Availability

Bio-fertilizers are microorganisms that enhance plant nutrient availability by capturing atmospheric nitrogen, converting insoluble phosphorus into a usable form, activating potassium and zinc, or transforming organic matter into a biologically active state. Unlike chemical fertilizers, which often leach into water sources, cause eutrophication, and disrupt microbial balance in the soil, biofertilizers work in harmony with natural soil processes, thereby contributing to sustainable nutrient cycles and long-term soil fertility (Vessey 2003; Adesemoye et al. 2009). Their application methods exemplify the transition from input-based agriculture to biologically directed nutrient management, guided by the principles of increasing ecological intensity and sustainable food production (Bashan et al. 2014).

One of the most researched groups of biofertilizers is nitrogen-fixing microorganisms. Symbiotic bacteria such as Rhizobium spp. form nodules on the roots of leguminous plants, where they convert atmospheric nitrogen (N_2) into ammonia, providing the host plant with directly usable nitrogen (Hungria and Mendes 2015). Free-living diazotrophs such as Azotobacter and Azospirillum independently fix nitrogen and promote root development by producing phytohormones such as indole-3-acetic acid (Bhattacharyya and Jha 2012). Similarly, cyanobacteria such as Anabaena and Nostoc species are of great importance in rice-based ecosystems by enriching nitrogen and providing organic carbon (Singh et al. 2011). Phosphate-solubilizing microorganisms (PSMs) properly represent the second important group of biofertilizers. Bacteria such as Pseudomonas and Bacillus, along with fungi like Aspergillus and Penicillium, produce organic acids and phosphatases that render insoluble phosphates usable and convert them into bioavailable forms for plant uptake (Rodríguez and Fraga 1999; Sharma et al. 2013). These microorganisms play a critical role in tropical and subtropical soils, where phosphorus is often bound in insoluble complexes that limit crop yields. Microorganisms that solubilize potassium and zinc further enhance soil fertility by mobilizing essential macro- and micronutrients. For example, *Bacillus mucilaginosus* and *Frateuria aurantia* solubilize potassium from minerals, supporting plant growth and productivity (Meena et al. 2014). Additionally, some Pseudomonas species have the ability to increase the bioavailability of zinc, a vital micronutrient for enzyme activity, chlorophyll production, and grain nutritional quality (Saravanan et al. 2007).

In addition to bacteria, fungi also play an important role in biofertilizer applications. Arbuscular mycorrhizal fungi (AMF) establish a symbiotic relationship with plant roots that provides mutual benefits and increases access to water and nutrients, especially phosphorus and micronutrients, by sending hyphae into the soil (Smith and Read 2008). In addition to improving nutrient uptake, AMF increase plant resistance to salinity, drought, and heavy metal toxicity, while also contributing to soil aggregation processes by producing a glycoprotein called glomalin, which strengthens soil structure (Rillig 2004).

The advantages provided by biofertilizers are not limited to nutrient supply. Increased use of biofertilizers reduces farmers' production costs by decreasing chemical fertilizer use and contributes to environmental protection goals by reducing nutrient leaching and greenhouse gas emissions (Jain et al. 2020). Field studies in India and Brazil have shown that when biofertilizers are used in conjunction with reduced amounts of chemical fertilizers, yield increases range from 10%–30%, demonstrating the application potential of these fertilizers in various agricultural ecosystems (Adesemoye et al. 2009; Hungria et al. 2010). Importantly, biofertilizers restore soil fertility and increase microbial diversity, making them particularly critical for smallholder farmers with limited resources and for sustainable agriculture in the face of climate change.

8.3.2 Biopesticides: Utilizing Microorganisms for Pest and Disease Control

Biopesticides are microbial substances that combat organisms and pathogens harmful to plants, achieving this through parasitism, antibiosis, resource competition, and increased host resistance (Chandler et al. 2011; Glare et al. 2012). Instead of synthetic pesticides, which often cause chemical residues, have a broad toxic effect, and accelerate the emergence of resistant pest species, microbial biopesticides are generally species-specific, naturally degradable, and environmentally friendly. These characteristics make them attractive elements for sustainable agriculture, especially when integrated into Integrated Pest Management (IPM) approaches that prioritize ecological balance, economic viability, and long-term pest control (Jamil et al. 2022; Areej et al. 2024).

Bacterial biopesticides constitute one of the most well-known categories. *Bacillus thuringiensis* (Bt) is known as the most widely used microbial pesticide, producing Cry and Cyt protein crystals that selectively affect the digestive system of susceptible insect larvae, causing them to die without harming non-target organisms (Sanahuja et al. 2011; Bravo et al. 2017). Similarly, *Bacillus subtilis* has become commercially important due to its production of antifungal lipopeptides such as iturins, surfactins, and fengisins, which inhibit many phytopathogens and increase systemic resistance in host plants (Ongena and Jacques 2008; Dunlap et al. 2019).

Fungal biopesticides are extremely important for sustainable pest control and disease management (Table 8.3). Trichoderma species have been extensively studied as biological control agents due to their ability to infect pathogenic fungi, produce hydrolytic enzymes such as chitinases and glucanases, and protect plants systemically (Harman 2011; Mukherjee et al. 2013). Entomopathogenic fungi such as *Beauveria bassiana* and *Metarhizium anisopliae* are widely used in the control of insect pests. These fungi infect their hosts by penetrating their body structure, reproducing within the hemocoel, and producing secondary metabolites that cause the rapid death of pests (Shah and Pell 2003; Lovett and Leger, 2017). These fungi offer a promising alternative to chemical insecticides in agricultural fields and greenhouses.

Viral biopesticides, particularly baculoviruses, constitute another distinct category. Viruses such as the Helicoverpa armigera nucleopolyhedrovirus (HearNPV) and Spodoptera frugiperda nucleopolyhedrovirus (SfMNPV) have been effectively used against lepidopteran pests in agriculture and forestry (Ayilara et al. 2023). The high specificity of these viruses on their hosts minimizes the impact on beneficial insects and vertebrates, offering a sustainable alternative for pest control in ecologically sensitive environments.

The benefits of biopesticides go beyond simply eliminating pests. Their use reduces the use of harmful synthetic chemicals, minimizes pesticide residues in food and the environment, protects beneficial organisms such as pollinators and natural enemies, and slows the development of resistance in pest populations (Sharma and Ortiz 2002; Glare et al. 2012). Furthermore, they contribute to climate-friendly agricultural practices by reducing the carbon footprint associated with the production and use of agricultural chemicals (Chandler et al. 2011). However, despite these advantages, certain limitations hinder the widespread adoption of biopesticides. The field performance of microbial biopesticides is often variable due to abiotic factors such as ultraviolet radiation, extreme temperatures, and desiccation (Jaronski 2010; Glare et al. 2012). Furthermore, formulation stability and the challenges associated with large-scale production and distribution pose significant barriers ( ). To overcome these obstacles and enhance the efficacy of biopesticides under various agricultural conditions,

Table 8.3 Categories of microbial biopesticides: representative organisms, mechanisms, and target pests/pathogens

Category	Representative organisms	Mode of action	Target pests/ pathogens	Key references
Bacterial biopesticides	*Bacillus thuringiensis* (Bt)	Production of Cry and Cyt protein crystals that disrupt the midgut epithelial cells of insects	Lepidopteran, coleopteran, and dipteran larvae	Sanahuja et al. (2011), Bravo et al. (2017)
	Bacillus subtilis	Antifungal lipopeptides (iturins, surfactins, fengycins); induction of systemic resistance	Root and leaf fungal diseases (e.g., Fusarium, Rhizoctonia)	Ongena and Jacques (2008), Dunlap et al. (2015)
Fungal biopesticides	Trichoderma spp.	Mycoparasitism; secretion of hydrolytic enzymes (chitinases, glucanases); secondary metabolites; ISR induction	Soil-borne fungal pathogens (Fusarium, Pythium, Sclerotinia)	Harman (2011), Mukherjee et al. (2013)
	Beauveria bassiana, *Metarhizium anisopliae*	Direct cuticle penetration; production of insecticidal secondary metabolites	Insect pests (whiteflies, aphids, beetles, caterpillars)	Shah and Pell (2003), Lovett and Leger (2017)
Viral biopesticides	Baculoviruses (HearNPV, SfMNPV)	Infection and replication in insect host cells; host-specific lethality	Lepidopteran pests (Helicoverpa armigera, Spodoptera frugiperda)	Moscardi (1999), Sun et al. (2019)

innovations in formulation science, such as encapsulation, UV-protective carriers, and nanotechnology, are being explored (Kumar 2019; de Oliveira et al. 2021).

8.3.3 Biostimulants: Microorganisms that Improve Plant Physiology

Biostimulants are microbial inoculant products that promote plant growth, increase nutrient utilization, and enhance resistance to environmental stress factors, and the effects of these substances are not limited to nutrient supply alone. Their effects are mainly evident through their regulation of plant physiological processes, particularly through phytohormone production and systemic resistance pathways (du Jardin 2015; Rouphael and Colla 2020). Plant growth-promoting rhizobacteria (PGPR) constitute one of the most studied types of biostimulants. For example, Pseudomonas fluorescens produces indole-3-acetic acid (IAA), which promotes branching in roots and thereby increases nutrient uptake (Backer et al. 2018). Similarly, *Azospirillum brasilense* produces gibberellin and cytokinin, which increase shoot elongation and overall vitality in plants (Cassán 2014). Endophytic microorganisms, including bacteria and fungi, represent another important class of microbial biostimulants. These organisms multiply within the plant without showing symptoms, benefiting the plant and increasing its tolerance to abiotic stress conditions such as drought and salinity (Santoyo et al. 2016). The mechanisms of action of biostimulants are quite diverse. These mechanisms include the synthesis of phytohormones such as auxins, gibberellins, and cytokinins; the activity of the enzyme 1-aminocyclopropane-1-carboxylate (ACC) deaminase enzyme, which reduces ethylene levels in plants in response to stress; and the activation of induced systemic resistance (ISR), which enables plants to respond more strongly to biotic and abiotic challenges (Glick 2012; Pieterse et al. 2014). From a climate-friendly agriculture perspective, biostimulants appear quite promising. By enhancing physiological resistance, these microbial compounds support the sustainable continuation of crop yields under conditions of water scarcity, high temperatures, or soil salinity. Their multifaceted nature highlights that they offer sustainable options for increasing agricultural productivity under variable environmental conditions while reducing dependence on chemical inputs (Calvo et al. 2014; Rouphael and Colla 2020).

8.3.4 Microbial Consortia: Synergistic Communities for Sustainable Agriculture

Although single-strain microbial applications offer agricultural advantages, their success in application areas often varies due to competition with microbial communities and environmental variability (Schmidt et al. 2019). To overcome these limitations, microbial consortia have been designed, combining multiple microbial strains or species that offer complementary functions such as nitrogen fixation, phosphate solubilization, pathogen suppression, and abiotic stress reduction (Sessitsch et al. 2019; Pankievicz et al. 2020). Consortia offer a more integrated and resilient approach to microbial biotechnology in agriculture by leveraging functional diversity. Numerous case studies demonstrate the effectiveness of microbial consortia. For example, the combined use of Rhizobium, phosphate-solubilizing Pseudomonas, and arbuscular mycorrhizal fungi (AMF) has been proven to increase both crop yield and soil health in legume-based agricultural systems (Verma et al. 2018). Similarly, large-scale trials conducted in China have shown that multi-strain bioinoculants increase rice yields by approximately 25%, while also improving soil organic carbon levels, contributing to food security and soil health

(Just et al. 2024). The benefits of microbial consortia extend beyond yield increases. They provide greater resilience to different soil and climate conditions, increase functional diversity compared to single-strain inoculants, and develop the ecological balance of new microorganisms through cooperative and synergistic interactions (Toju et al. 2018; Raklami et al. 2019). Overall, these qualities make microbial consortia a promising new-generation strategy for the sustainable intensification of agriculture.

8.3.5 Summary

The various applications of microbial biotechnology demonstrate its important role as both a sustainable alternative and a complementary approach to traditional chemical inputs in agriculture (Table 8.4). Bio-fertilizers contribute to the nutrient cycle and soil fertility by mobilizing essential macronutrients such as nitrogen, phosphorus, and potassium; thus, they reduce the use of synthetic fertilizers, which often have negative environmental consequences (Vessey 2003; Adesemoye et al. 2009). Similarly, microbial biopesticides effectively control harmful insects and plant pathogens while minimizing ecological hazards such as unwanted toxicity and the development of pesticide-resistant populations, which are often associated with chemical pesticides (Chandler et al. 2011; Santos et al. 2013). Biostimulants enhance these benefits by increasing the physical performance of plants through mechanisms such as phytohormone modulation, nutrient uptake, and systemic resistance induction, thereby improving their ability to withstand drought, salinity, and nutrient deficiency conditions (du Jardin 2015; Rouphael and Colla 2020). Additionally, microbial consortia consisting of nitrogen fixers, phosphate solubilizers, and biological control agents, whether engineered or naturally occurring communities, are increasingly recognized for their potential to deliver highly reliable results under diverse soil and climate conditions (Sessitsch et al. 2019; Raklami et al. 2019). Overall, these microbial interventions increase crop productivity, reduce input costs, and improve environmental sustainability by balancing agricultural intensification with ecological conservation. However, their widespread acceptance is not guaranteed; success largely depends on optimized formulation strategies, compatible regulatory frameworks, and farmers' willingness to adopt microbial solutions as reliable components of integrated crop management systems (Bashan et al. 2014; Malusá et al. 2016).

8.4 Case Studies and Practical Applications of Microbial Biotechnology in Agriculture

The applicability of microbial biotechnology in the agricultural sector is demonstrated through concrete examples where biofertilizers, biopesticides, biostimulants, and microbial consortia are successfully used on a large scale. For example, large-scale

Table 8.4 Comparative overview of microbial biotechnology applications in agriculture

Category	Representative microorganisms	Primary mechanisms	Key advantages	Selected references
Bio-fertilizers	Rhizobium, Azospirillum, Azotobacter, Pseudomonas, Bacillus, Cyanobacteria	Nitrogen fixation; phosphate solubilization; mobilization of potassium and micronutrients; organic matter decomposition	Reduced dependence on synthetic fertilizers; improved nutrient cycling; increased yield	Vessey (2003), Adesemoye et al. (2009)
Biopesticides	*Bacillus thuringiensis*, *Bacillus subtilis*, Trichoderma spp., *Beauveria bassiana*, *Metarhizium anisopliae*, Baculoviruses	Antibiosis; parasitism; competition; toxin production; induction of plant defense responses	Targeted pest/pathogen suppression; environmentally friendly; reduced pesticide resistance	Chandler et al. (2011), Santos et al. (2013)
Biostimulants	*Pseudomonas fluorescens*, *Azospirillum brasilense*, endophytic bacteria and fungi	Phytohormone production (auxins, cytokinins, gibberellins); ACC deaminase activity; induction of systemic resistance	Enhanced root/shoot growth; increased stress tolerance; stabilized yield under adverse conditions	du Jardin (2015), Rouphael and Colla (2020)
Microbial Consortia	Multi-strain formulations (e.g., Rhizobium + Pseudomonas + AMF; PGPR mixtures)	Synergistic functions such as nitrogen fixation, phosphate solubilization, stress reduction, and pathogen suppression	Greater field resilience; broader functional coverage; enhanced soil fertility and crop productivity	Sessitsch et al. 2019; Raklami et al. (2019)

field trials conducted in India and Brazil have shown that biofertilizers can increase yields in cereal and legume crops by 10–30% while also reducing the need for chemical fertilizers. Similarly, the widespread use of *Bacillus thuringiensis* (Bt)-based biopesticides in certain regions of North America, Europe, and Asia demonstrates that these pesticides are effective in reducing insect pests with minimal harm to the environment. Biostimulants such as *Azospirillum brasilense* and Arbuscular mycorrhizal fungi (AMF) have been used in the Mediterranean and semi-arid regions to increase plant resistance under drought and salt stress, contributing to more climate-resilient farming methods. Furthermore, microbial consortia combining nitrogen-fixing bacteria, phosphate-solubilizing microorganisms, and AMF have been adopted in agricultural practices in Africa and Asia, helping to increase soil fertility and ensure sustainable crop productivity. These situations demonstrate that microbial methods are not only scientifically valid but also economically viable, environmentally beneficial, and politically important. However, they also point to broader challenges hindering wider adoption, such as regulatory barriers to bioinoculant approval processes due to soil and climate differences, and farmers' lack of awareness or inadequate training on the subject. Overcoming these barriers is critical to ensuring the widespread use of microbial biotechnology as a fundamental element of sustainable agriculture.

8.4.1 India: Bio-fertilizers as a Tool for Soil Health and Cost Reduction

India has led the world in adopting biofertilizers due to its deep reliance on nitrogen fertilizers and the need to reduce costs for small farmers with limited resources (Tilak et al. 2005). A good example involves inoculating chickpeas (*Cicer arietinum*) and wheat with Rhizobium strains and *Azotobacter chroococcum*. Field trials have frequently demonstrated that this type of inoculation increases biological nitrogen fixation, yields increase of 15–25%, and reduces nitrogen fertilizer requirements by up to 30 kg per hectare (Wani et al. 1995; Adesemoye et al. 2009). Economically, these strategies have saved approximately 20% in fertilizer costs without yield loss, highlighting their importance for the sustainability of smallholder farmers. In rice-based systems, the integration of biofertilizers such as blue-green algae (Anabaena and Nostoc) has been particularly beneficial. Adding these cyanobacteria to rice field soils increases the direct availability of nitrogen, reducing the need for chemical nitrogen by up to 20% and enhancing soil microbial diversity (Prasanna et al. 2015; Prasanna et al. 2008). Importantly, the use of algal biofertilizers has been associated with reduced nitrate leaching into irrigation water, providing additional benefits to environmental quality (Çakirsoy et al. 2022). Building on these successes, government-supported programs promoting biofertilizer use have been launched in various Indian states, such as Tamil Nadu and Maharashtra; these have been implemented through subsidies and farmer education projects (Jain et al. 2013). However, obstacles persist, particularly the inconsistent quality of commercial formulations and insufficient training for farmers on appropriate application methods (Malusá and Vassilev 2014). Overcoming such obstacles is critical to expanding the use of biofertilizers and achieving lasting impacts on agricultural productivity and environmental sustainability.

8.4.2 Brazil: Utilizing Microbial Consortia in Soybean Cultivation

Brazil has become one of the world's leading countries in the application of agricultural microbial biotechnology, with soybean production being the most effective example of its widespread use. The widespread use of microorganism inoculation in soybean systems has made biological nitrogen fixation (BNF) a fundamental element of the country's agricultural productivity. Specifically, inoculation with *Bradyrhizobium japonicum* and similar strains enables soybeans (Glycine max) to store up to 200 kg of nitrogen per hectare, significantly reducing the need for artificial nitrogen fertilizers (Hungria and Mendes, 2015; Kaschuk et al. 2010). This practice not only helps Brazil maintain its position as the world's largest soybean supplier but also provides significant economic benefits. Estimates indicate that the adoption of rhizobial inoculants saves the country more than US$13 billion annually in fertilizer imports, highlighting the scalability and cost-effectiveness of integrating microbial technologies into conventional agriculture (Hungria and Mendes 2015). Brazil has also made progress in the use of microbial consortia in other agricultural products besides soybeans, particularly sugarcane (*Saccharum officinarum*). Multi-strain formulations containing Gluconacetobacter diazotrophicus, which has nitrogen fixation properties, bacteria with phosphate solubilization ability, and rhizobacteria (PGPR) that promote plant growth have increased sugar content and juice quality while providing a 20–30% increase in yield (Oliveira et al. 2002; Reis et al. 2017). Reducing the use of nitrogen fertilizers in sugarcane farming has made agricultural production compatible with climate change mitigation strategies by further contributing to the reduction of greenhouse gas emissions (Thorburn and Biggs 2010B). Brazil's success story demonstrates that the large-scale implementation of microbial biotechnology is possible when supported by robust scientific research, favorable government policies, and the active contributions of industry stakeholders. The integration of microbial inoculants into national agricultural practices serves as an example for other nations seeking to combine economic growth, food security, and environmental sustainability (Hungria et al. 2010; Bashan et al. 2014).

8.4.3 China: Microbial Solutions to Boost Rice and Corn Production

China faces persistent challenges such as soil erosion, loss of organic matter, and excessive reliance on chemical fertilizers due to agricultural practices. In response to these challenges, microbial biotechnology has been strategically incorporated into national soil fertility enhancement and sustainability projects. A notable example is the application of multi-bioinoculants in rice production, achieved through the combined use of Azospirillum, phosphate-solubilizing bacteria, and arbuscular mycorrhizal fungi. Field trials have reported significant improvements in soil organic carbon and nutrient cycle efficiency, with yield increases ranging from 18% to 25%. These bioinoculants not only increase yields but also offer significant advantages for smallholder farmers by reducing fertilizer costs by approximately 15–20%. Another positive example is the application of biopesticides using Bacillus thuringiensis and Trichoderma spp. formulations in vegetable and corn production, particularly to control lepidopteran pests and soil-borne pathogens. The use of these microbial products has not only provided effective plant protection but also reduced pesticide residues in food and water systems, resulting in a 30%–40% reduction in synthetic pesticide use. The widespread adoption of these technologies has been significantly supported by government initiatives, including the "Zero Growth in Fertilizer and Pesticide Use by 2020" policy launched in 2015, which promotes biofertilizers and biopesticides as sustainable alternatives. These examples demonstrate that China is a global leader in promoting microbial biotechnology for agricultural productivity and environmental sustainability.

8.4.4 Europe: Biopesticides and Biostimulants in Sustainable Agriculture

Strict environmental regulations and growing consumer interest in organic products have influenced agricultural practices in Europe, and microbial biotechnology has become an indispensable component of sustainable agricultural systems (European Commission 2020; Malusà and Vassilev 2014). A notable example is the use of Trichoderma-based biopesticides in greenhouses. In Italy, *Trichoderma harzianum* formulations are widely preferred among tomato and cucumber growers and exert effective pressure on soil-borne pathogens such as Fusarium oxysporum and Pythium ultimum (Woo et al. 2014; Vitti et al. 2015). This application has not only reduced dependence on chemical fungicides but also increased compliance with organic farming standards and provided significant cost savings for farmers (Manczinger et al. 2002). Another example comes from the use of PGPR-based biostimulants in horticulture. Here, *Pseudomonas fluorescens* and Azospirillum spp. are used as seed coatings in lettuce and carrot cultivation. These microorganisms have increased germination rates, enhanced drought tolerance, and led to an increase in market value, primarily through increased nutrient uptake and strengthening of the root system (Calvo et al. 2014; Rouphael and Colla 2018). The increase in water use efficiency achieved through microbial biostimulants has also contributed to a reduction in irrigation needs. This offers a significant advantage under increasingly prevalent drought conditions in Europe (Berg et al. 2020). Despite all these successes, the European Union's strict regulatory framework for microbial products has slowed down the commercialization process of these products compared to other regions. (Frederiks and Wesseler, 2019). However, this detailed approval process has ensured that high-quality and effective products reach the market, thereby increasing farmers' confidence and supporting the long-term adoption of microbial technologies in European agriculture (European Food Safety Authority [EFSA] 2018).

8.4.5 Lessons Learned from Global Applications

These case studies present several important findings regarding the multiple applications of microbial biotechnology in agriculture. Economic viability, particularly for small farmers with limited resources, is the biggest motivating factor for adoption, as reduced fertilizer and pesticide requirements lead to lower costs, which directly increase profitability (Adesemoye et al. 2009). Policy support has also played a critical role; state support similar to Brazil's national inoculation program and China's "Zero Growth in Fertilizer and Pesticide Use" policy, along with extension and certification programs, have greatly accelerated farmers' adoption of this technology (Hungria and Mendes 2015; Li 2021). Another important finding is the importance of quality control and regulatory oversight. Fluctuating field performance, often associated with low-quality or contaminated products, remains a persistent problem, especially where regulatory structures are underdeveloped (Bashan et al. 2014; Malusà and Vassilev 2014). To increase reliability, microbial technologies should be integrated with modern agricultural technologies such as precision agriculture, soil health monitoring, and digital decision support systems that target applications and increase productivity (Berg et al. 2020; Singh et al. 2020). Ultimately, farmer awareness and education are also crucial for successful implementation. Research shows that knowledge sharing through advisory services, farmer cooperatives, and participatory research programs significantly increases the adoption and proper use of microbial inoculants, thereby maximizing agricultural and environmental gains (Choudhary et al. 2017; Wezel et al. 2018). These findings reveal that the success of microbial biotechnology depends not only on scientific innovation but also on socioeconomic, institutional, and educational factors that influence real-world adoption.

8.4.6 Summary

Case studies from India, Brazil, China, and Europe clearly demonstrate that microbial biotechnology is not only a laboratory innovation but also a field-ready solution that provides measurable economic and environmental benefits. In India, the use of nitrogen-fixing biofertilizers such as Rhizobium and Azotobacter in legumes and cereals has increased yields while reducing chemical nitrogen inputs (Hungria et al. 2013; Sharma and Singh 2017). Similarly, Brazil has pioneered the large-scale use of Bradyrhizobium inoculants in soybean cultivation. Here, biological nitrogen fixation has almost entirely replaced synthetic nitrogen fertilizers and has saved the country an estimated $13 billion annually in fertilizer imports (Hungria and Mendes 2015). In China, multi-strain microbial consortia applied to rice and other staple crops have directly contributed to the national "Zero Growth" fertilizer and pesticide policies by reducing fertilizer use while increasing yields and soil organic matter (Diedhiou et al. 2016; Etesami 2021; Elekhtyar et al. 2022). Meanwhile, in Europe, biopesticides such as Trichoderma and Bacillus formulations, along with PGPR-based biostimulants, are being widely integrated into organic and conventional greenhouse systems, improving crop protection and stress tolerance in compliance with strict EU environmental regulations (Woo et al. 2014; Pertot et al. 2017). These examples demonstrate the potential of microbial biotechnology to reduce chemical dependency, increase soil fertility, mitigate climate change, and improve farmers' livelihoods. However, the widespread adoption of such technologies requires compatible regulatory frameworks, significant investments in quality control, and continuous farmer education and participation to ensure reliability, acceptance, and long-term adoption (Bashan et al. 2014; Malusà and Vassilev, 2014).

8.5 Recent Developments in Microbial Biotechnology

The field of microbial biotechnology has undergone profound changes over the past two decades, driven by innovations in molecular biology, systems biology, and synthetic biology. Methods that previously focused on the isolation and experimental testing of natural microorganisms are now increasingly integrated with high-resolution, engineering-focused techniques that enable scientists to examine, regulate, and design microbial communities with unprecedented precision, often replacing these older methods (Fierer 2017; Dubey et al. 2019). These developments have deepened our scientific understanding of the interactions between plants, microorganisms, and soil, and have also increased the possibilities for scalable agricultural applications. Some of the most effective technological drivers include metagenomics, which reveals the taxonomic and functional diversity of uncultivable soil microbiomes (Mendes 2019); and metatranscriptomics and metaproteomics, which combine the genetic potential with microbial functions applied in situ (Carrión et al. 2019; Wilmes et al. 2015); and metabolomics, which identifies active components such as phytohormones and siderophores that regulate plant development and stress tolerance (Schulz-Bohm et al. 2017). A higher level of integration enables the rational design of balanced and functionally compatible inoculants by presenting microbial consortia and their interactions with plants and soil through network-based models via systems biology (Sessitsch et al. 2019; Toju et al. 2018). In genome editing, recent advances, particularly in the CRISPR/Cas toolkit, have further accelerated the engineering of beneficial traits in plant-associated microorganisms, such as nitrogen fixation, stress tolerance, or biological control properties (Wang et al. 2016; Shelake 2019). Simultaneously, synthetic biology methods have enabled the development of engineered microbial strains and synthetic consortia that function as "microbial factories" to intensively produce phytohormones, antibiotics, or stress-reducing enzymes (Liu and Stewart 2015; Mishra 2022). Taken together, these developments are giving rise to new precision microbial biotechnology paradigms where agricultural sustainability can be achieved through predictive, data-driven, and designer interventions.

8.5.1 Omics Approaches and Their Effects

8.5.1.1 Metagenomics and Soil Microbiomes

Metagenomics has fundamentally transformed the study of microbial communities in soil and the rhizosphere by enabling the analysis of genetic material in a culture-independent manner. Using methods such as shotgun sequencing and amplicon-based 16S rRNA profiling, scientists can identify microbial taxa, estimate functional gene pools, and track community dynamics under various environmental conditions (Scholz 2012; Fierer 2017). These methods offer an unparalleled level of detail in linking microbial diversity to ecosystem functions. For example, metagenomic studies of wheat and corn rhizospheres have revealed key taxa associated with important processes such as nitrogen cycling, siderophore production, and abiotic stress tolerance, demonstrating the importance of specific microbial groups' environmental functions on crop productivity (Peiffer 2013; Mendes et al. 2019). This information provides direct insight into the scientific design of microbial consortia assembled to perform functions such as nutrient mobilization, pathogen control, and growth promotion, thereby contributing to the development of more resilient and sustainable agricultural systems (Toju et al. 2018; Sessitsch and Sessitsch et al. 2015).

8.5.1.2 Transcriptomics and Gene Expression

Transcriptomic analyses provide important insights into the expression dynamics of plant and microbial genes during environmental stress conditions and beneficial plant–microorganism interactions. Compared to static genomic data, transcriptomics provides a practical perspective on host–microorganism relationships by showing the real-time activation or suppression of gene networks (Kukurba and Montgomery 2015; Pinu et al. 2019). For example, when examining interactions between plant growth-promoting rhizobacteria (PGPR) and host plants, an increase in root system structure, auxin signaling, and defense-related genes has been observed. This demonstrates that microbial inoculation can reprogram the transcriptional responses of plants (Zamioudis et al. 2013; Verbon and Liberman 2016). Such transcriptomic data not only enable effective rhizosphere colonization but also facilitate the strategic selection of microbial strains that enhance induced systemic resistance (ISR), thereby improving crop resilience against abiotic stress factors such as drought and salinity, as well as biotic threats (Niu et al. 2016; Eida et al. 2018). By correlating microbial inoculation with host transcriptome outcomes, transcriptomics has become an effective tool in the design of next-generation bioinoculants.

8.5.1.3 Metabolomics and Functional Profiling

Metabolomics plays an important complementary role in genomic and transcriptomic studies by elucidating small molecules such as phytohormones, siderophores, antibiotics, and signaling metabolites that drive interactions between plants and microorganisms. Unlike DNA- or RNA-based methods, metabolomics detects biochemical phenotypes and provides direct evidence of functional effects in the rhizosphere (Fiehn 2002; Patti et al. 2012). Advances in analytical methods such as liquid chromatography–mass spectrometry (LC–MS) and nuclear magnetic resonance (NMR) spectroscopy have enabled the identification and characterization of various microbial metabolites that have significant effects on plant health and nutrient cycling. For example, siderophore-producing rhizobacteria have the ability to increase iron uptake in plants while suppressing pathogens by reducing iron availability (Aznar and Dellagi 2015). Similarly, metabolites related to organic acid secretion improve crop nutrient uptake by contributing to phosphorus solubility (Richardson and Simpson 2011). These functional data contribute to the development of biostimulant formulations enriched with microbial metabolites or engineered strains capable of overproducing beneficial components, creating a promising field for precision microbial biotechnology in agriculture (Schirawski and Perlin 2018; Rouphael and Colla 2020).

8.5.2 Systems Biology and Predictive Modeling

The complexity of plant, microorganism, and soil interactions requires systems-level approaches; these approaches enable the integration of genomic, transcriptomic, proteomic, and metabolomic data. Systems biology provides mathematical and computational tools to model the dynamics of microbial communities, nutrient flows, and plant responses at multiple scales (Konopka 2009; Shanmugam and Kannan 2020). For example, flux balance analysis has been applied to predict the metabolic resource allocation of nitrogen-fixing bacteria under varying soil and environmental conditions, contributing to the rational design of more effective biofertilizers (Orth et al. 2010; Herring et al. 2015). Another important advancement is the use of network analysis to reveal the structures and functions of microbial consortia. Such studies reveal the cooperative and competitive interactions between microbial taxa, enabling the creation of new synthetic communities with the potential to reduce antagonism and increase cooperation (Faust and Raes 2012; Toju et al. 2018). Such information is particularly important for agricultural applications where the stability and durability of microbial inoculants remain an ongoing challenge. Furthermore, predictive modeling frameworks increasingly support the development of precision agriculture practices that involve tailoring microbial formulations to specific soil physicochemical properties, crop genotypes, and climatic conditions (Papp et al. 2018; Choudhary et al. 2022). Systems biology, which combines ecological theory with biotechnological innovations, contributes to the development of effective, reproducible, and environmentally sustainable next-generation microbial inoculants.

8.5.3 Genome Editing and CRISPR/Cas Technologies

The development of CRISPR/Cas gene editing technology has created a historic shift in the field of microorganism-based biotechnology and offers unparalleled precision compared to traditional mutagenesis or recombinant DNA strategies (Doudna and Charpentier 2014; Pickar-Oliver and Gersbach 2019). This innovation enables specific modifications to be made in the DNA of microorganisms, allowing for the systematic design of microorganisms with improved agricultural traits (Table 8.5). Regarding biofertilizers, it has been demonstrated that the CRISPR method can be used to modify the nitrogenase enzyme complex in *Rhizobium* and *Azotobacter* species to optimize nitrogen-fixing genes, thereby increasing the efficiency and productivity of nitrogenase under variable oxygen conditions (Wang et al. 2016; Mus et al. 2016). Similarly, the CRISPR-based regulation of genes effective in organic acid production in Bacillus strains has contributed to increased phosphate solubility, which in turn has enhanced the biological availability of phosphorus in soils (Zhou et al. 2021). Applications of this technology also include biopesticides and biostimulants. CRISPR has modified the genetic makeup of *Trichoderma* spp. strains, increasing the production of antifungal enzymes and secondary metabolites, thereby strengthening defense mechanisms against soil-borne pathogens (Mishra 2021). Furthermore, plant growth-promoting rhizobacteria such as *Pseudomonas fluorescens* have been engineered to increase resistance to salt and drought stress, thereby ensuring sustainable colonization and functionality under adverse environmental conditions (Ali 2025; Khan et al. 2021; Shelake et al. 2022; Al-Turki et al. 2023).

CRISPR technology, which offers more than just single-feature improvements, enables the addition of beneficial genes ("knock-in") across species, allowing for the production of engineered microorganisms capable of performing multiple functions simultaneously, such as nutrient mobility, pathogen control, and stress reduction (Liu and Stewart 2015). Despite these advances, challenges remain in achieving commercial success, particularly in areas such as the European Union, which regulates the release of genetically modified organisms (GMOs) with strict biosafety and regulatory frameworks (Nap et al. 2003; Eriksson et al. 2018). These situations highlight the tension between the opportunities offered by the technology and societal acceptance, emphasizing the need for coherent global policies.

8.5.4 Synthetic Biology and the "Microbial Factory"

Synthetic biology expands the boundaries of microbial biotechnology by treating microorganisms as programmed biological structures capable of performing specific agricultural functions in a precise and stable manner. By combining modular genetic circuits and variables, synthetic biologists can develop microorganisms that consistently perform tasks such as food production, pest control, and stress and pressure reduction (Way et al. 2014; Liu and Stewart 2015). This approach is often referred to as the "microbial factories" model, in which engineered microorganisms act as production units living in soil ecosystems. For example, synthetic consortia have been developed that are designed to continuously fix atmospheric nitrogen and release it in forms that plants can use, thereby reducing dependence on chemical fertilizers (Mancinelli 2001; Temme et al. 2012). Engineered Bacillus species have been developed to offer a rich repertoire of insecticidal proteins, increasing the potential to control pests beyond the effects of natural biopesticides (Wang et al. 2012; Baum et al. 1996). Similarly, engineered microorganisms capable of simultaneously producing various phytohormones

Table 8.5 CRISPR applications in agricultural microbial biotechnology

Application area	Target trait	Microbial system	CRISPR result	Key references
Bio-fertilizers	Enhanced nitrogen fixation	Rhizobium, Azotobacter	Optimized nitrogenase complex activity under variable oxygen conditions	Wang et al. (2016), Mus et al. (2016)
Bio-fertilizers	Phosphate solubility	*Bacillus subtilis*, *Bacillus megaterium*	Regulated organic acid secretion genes → improved P mobilization	Zhou et al. (2021)
Biopesticides	Increased antagonistic activity	*Trichoderma harzianum*	Enhanced antifungal enzyme and metabolite production	Mishra et al. (2021)
Biostimulants	Stress tolerance (salinity, drought)	*Pseudomonas fluorescens*, PGPR strains	Engineered tolerance genes → improved root colonization under stress	Zhang et al. (2020), Kibret et al. (2024)
Multifunctional microorganisms	Designed strains with multiple beneficial traits	Synthetic microbial consortia	Gen knock-ins for combined N-fixation, P-solubilization, and pathogen suppression	Liu and Stewart (2015), Pickar-Oliver and Gersbach (2019)

such as auxins, gibberellins, and cytokinins are being investigated as multifunctional biostimulants to improve plant growth and resilience under abiotic stress conditions (Pardo et al. 2014; Liu et al. 2018). All these innovations demonstrate how synthetic biology enables the planned design of micro-bio systems with specialized and scalable functions, thereby transforming agricultural biotechnology.

Recent advances in synthetic biology have opened up the opportunity to create microbial species that offer highly controllable and predictable agricultural functions. For example, synthetically produced *Escherichia coli* strains have been programmed to release ammonia at specific rates, thereby increasing nitrogen use efficiency and minimizing environmental nitrogen losses due to evaporation or leaching (Rogers and Oldroyd 2014; Temme et al. 2012). Similarly, engineered Pseudomonas strains with high affinity for iron have been developed that help plants absorb micronutrients more efficiently while also contributing to disease control by inhibiting iron uptake by pathogens (Schmid 2011; Cornelis 2010). Synthetic biology enables the design of genetically stable microorganisms optimized for the addition and expression of beneficial genes (Renda et al. 2014; Martinez-Garcia and de Lorenzo 2017). These strains offer greater predictability and resilience for field applications compared to natural isolates, whose performance varies more under different soil and climate conditions. By standardizing their genetic structure and facilitating regulatory control, they represent an important step toward the reliable and scalable use of microbial products.

8.5.5 Artificial Intelligence and Data-Driven Microbial Design

The integration of artificial intelligence and machine learning into the discipline of microbial biotechnology is creating a new field with great potential for sustainable agriculture. By leveraging large-scale omics databases, AI-based tools enable the prediction of microbial traits best suited to specific plants, soil types, and climatic conditions (Arango-Argoty et al. 2018; Libbrecht and Noble 2015). In the field of strain discovery, machine learning methods have been used to access metagenomic data and identify potential strains with functional properties such as nitrogen fixation, phosphate solubilization, or pathogen suppression. In the formulation improvement stage, artificial intelligence models can simulate microbial interactions and predict synergistic or antagonistic interactions between the consortia involved, enabling the faster design of multi-strain inoculants with enhanced stability and performance (Nguyen 2019; de Souza et al. 2020). Furthermore, through remote sensing, soil microbiome profiling, and climate simulation, AI-supported precision agriculture enables the monitoring of field applications, real-time assessment of microbial activity, and the preparation of customized management strategies (Liakos et al. 2018; Mahlein 2016). Thus, the interaction between artificial intelligence and microbial biotechnology provides a strong foundation for developing reliable, customizable, and scalable next-generation microbial products in various agricultural ecosystems.

8.5.6 Challenges and Future Directions

While innovations in genomics, CRISPR/Cas gene editing, and synthetic biology offer great potential for agricultural advancement, there are also some challenges that hinder the widespread use of these innovations (Table 8.6). Engineered microorganisms that work successfully in the laboratory often fail to establish themselves or function consistently under different soil and climate conditions; this situation makes field variability a major obstacle (Bashan et al. 2014; Kaminsky et al. 2019). Legal barriers also pose a significant obstacle: The approval process for genetically modified microorganisms can be particularly difficult and time-consuming in regions with the strictest biosafety regulations, such as the European Union (Hartung and Schiemann 2014). Biosafety concerns complicate the commercialization of such products and require comprehensive risk assessments and monitoring protocols, as uncertainties remain regarding the long-term ecological impacts of releasing engineered strains into soil ecosystems (Kumar et al. 2019; Ray et al. 2020). Furthermore, farmers'

Table 8.6 Key challenges and potential solutions in the field application of microbial biotechnology

Challenge	Explanation/example	Impact	Possible solutions	References
Field variability	Engineered products or beneficial microorganisms perform well under laboratory/greenhouse conditions, but cannot survive or function consistently in different soils and climates.	Limits the reproducibility and scalability of microbial products.	Development of stress-resistant strains; use of encapsulation and seed coatings; region-specific microbial consortia.	Bashan et al. (2014), Kaminsky et al. (2019)
Legal barriers	Approval processes for genetically modified microorganisms are lengthy and stringent, particularly in the EU.	Delays commercialization; increases development costs.	Simplified risk assessment frameworks; international harmonization of biosafety guidelines.	Hartung and Schiemann (2014)
Biosafety Concerns	Uncertainty about the long-term ecological effects of releasing engineered microorganisms (e.g., horizontal gene transfer, unintended interactions).	Public/environmental risks; stricter monitoring requirements.	Comprehensive ecological risk assessments; use of kill switches and biological containment strategies.	Jamil et al. (2022), Ray et al. (2020)
Farmer acceptance	Negative perceptions of GMOs; limited awareness of microbial products.	Low adoption rates despite proven benefits.	Education campaigns; participatory trials with farmers; government/NGO dissemination support.	Tensi et al. (2022)
Product consistency	Naturally occurring isolates may vary in performance due to genetic drift or environmental stress.	Inconsistent crop responses.	Using synthetic biology to create stable microbial chassis with predictable performance.	Malusá et al. (2016)
Distribution efficiency	Low survival rate of microorganisms during storage and after application to fields.	Decreased efficacy of inoculants.	Advances in encapsulation, nano carriers, and polymer-based formulations.	John et al. (2011), Malusá et al. (2016)
Information integration	Lack of integration between omics, predictive modeling, and field-level agronomy.	Limits the precise application of microbial inputs.	AI-driven strain selection, predictive modeling, and integration with precision agriculture platforms.	Arango-Argoty et al. (2018), Zhou et al. (2021)

acceptance of these technologies is not guaranteed; perceptions about GMOs and limited awareness of microbial technologies may create challenges in adopting these technologies without effective extension services and knowledge-sharing programs (Tensi et al. 2022).

Future developments will necessitate interdisciplinary collaboration between microbiologists, synthetic biologists, agriculturalists, and data analysts to bridge the gaps between technological innovations and application performance. The combination of CRISPR-based engineering with AI-supported predictive modeling will accelerate the development of microbial products with greater resilience and stability in different environments (Arango-Argoty et al. 2018; Zhou et al. 2021). At the same time, innovations in delivery systems such as seed coatings, polymer-based encapsulation, and nanocarrier formulations are expected to increase microbial survival, colonization efficiency, and efficacy under field conditions (John et al. 2011; Malusá et al. 2016). Overcoming these challenges will be critical to realizing the full potential of microbial biotechnology in sustainable agriculture.

8.5.7 Summary

Recent advances in microbial biotechnology have transformed this field from a discipline limited to the identification of natural isolates into a high-tech area capable of precisely designing and engineering microbial systems. High-throughput omics techniques, such as metagenomics and transcriptomics, have provided significant depth in understanding microbial diversity, functional potential, and plant–microorganism–soil interactions (Philippot et al. 2013; Fierer 2017). Systems biology has provided predictive models by modeling the dynamics of microbial communities and developing microbial consortia under variable field conditions (Zhou et al. 2021). At the same time, gene editing methods such as CRISPR/Cas and innovations in synthetic biology have provided new opportunities for creating customized "engineered microorganisms" for nutrient mobility, biological control, and stress tolerance (Jinek et al. 2012; Schmidt-Dannert and López-García 2016). Recently, artificial intelligence (AI) and machine learning (ML) have accelerated both strain discovery and formulation development by combining large omics datasets with field performance data (Arango-Argoty et al. 2018; Ray et al. 2020). All these innovations demonstrate progress toward precision microbial management, offering unique solutions for sustainable agriculture in terms of plants and soil. However, to fully realize this potential, scientific advances alone will not suffice; careful consideration of biosafety, compatible regulatory frameworks, and appropriate adoption strategies for farmers will also be necessary (Kaminsky et al. 2019; Malusá et al. 2016).

8.6 Challenges, Limitations, and Policy Perspectives

Despite the remarkable advances made in microbial biotechnology and its proven potential to increase agricultural productivity, sustainability, and resilience, several critical challenges continue to hinder the widespread adoption of this technology. From a scientific perspective, the inconsistent performance of microbial preparations under different soil and climate conditions limits their reliability in field conditions (Kaminsky et al. 2019; Timmusk et al. 2017). Ecologically, concerns persist regarding the undesirable effects of introduced strains on native microbial communities and long-term soil health (Backer et al. 2018). Technological barriers, such as the scalability of microbial formulations and the development of stable delivery systems, further constrain their application (Sessitsch et al. 2019). From a socioeconomic perspective, low awareness among farmers, limited extension support, and the higher upfront costs of microbial products compared to traditional inputs reduce adoption rates, particularly in developing regions (Malusá et al. 2016). Legal barriers also play a significant role; the lack of harmonized international standards for registration, quality control, and biosafety creates uncertainty for both producers and users (Calvo et al. 2014; Mishra et al. 2021). Overcoming these challenges requires not only continuous scientific innovation but also consistent policy frameworks, capacity-building initiatives for farmers, effective regulations, and international cooperation to increase confidence in microbial technologies.

8.6.1 Scientific and Technical Limitations

8.6.1.1 Variability in Field Performance

One of the biggest challenges in microbial biotechnology is the inconsistent performance of preparations under field conditions. Microbial strains that exhibit high efficacy in laboratory or greenhouse environments often fail to provide the same level of benefit in open agricultural fields. This situation stems from the complexity of soil microbiomes; these microbiomes are shaped by the need for introduced strains to compete with local microbial communities and adapt to changing abiotic conditions such as temperature, humidity, and pH (Kaminsky et al. 2019; Timmusk et al. 2017). For example, phosphate-solubilizing bacteria that can effectively release bound phosphorus in neutral soils may be ineffective in alkaline areas where phosphorus is largely immobilized as calcium compounds (Janati 2023; Qetrani et al. 2024). Similarly, nitrogen-fixing bacteria generally exhibit reduced activity under saline or waterlogged soil conditions, which negatively affects the function of the nitrogenase enzyme complex (Vessey 2003). These ecosystem and environmental factors highlight the need for site-specific preparation and development and demonstrate the importance of integrating microbial ecology into biotechnology development for sustainable agriculture.

8.6.1.2 Short Shelf Life and Storage Restrictions

Many microorganism-based products are naturally sensitive to environmental factors such as temperature changes, ultraviolet (UV) rays, and humidity, and this often leads to a shortened shelf life and reduced effectiveness during storage and transportation. In addition to chemically stable synthetic fertilizers and pesticides, microbial preparations require careful formulation and often special packaging or cold chain transportation systems to maintain their viability (Mishra and Arora 2016; Herrmann and Lesueur 2013). These requirements pose significant challenges for large-scale distribution, particularly in isolated areas with weak infrastructure or limited resources (Malusá et al. 2016). To overcome these limitations, research has been conducted on the use of protective carrier matrices such as encapsulation methods, nanomaterials and alginate beads, biochar, and polymer-based coatings to increase microbial resistance and controlled release (John et al. 2011; Vassilev et al. 2020). Although promising, these innovations are still in the development stage, and practical issues such as formulation stability and cost efficiency remain the biggest obstacles to the large-scale commercialization of microbial biotechnology.

8.6.1.3 Knowledge Gaps in Microbial Ecology

Although metagenomics and other omics-based methods have significantly increased our knowledge of microbial diversity and functional vitality, the interactions between microbial communities in soil and the rhizosphere are still only partially understood. There are still significant gaps in our knowledge about the cumulative effects of long-lived microbial strains on ecosystem functions such as nutrient cycling, disease suppression, and carbon storage, due to their competitive and cooperative relationships with local microbial communities (van der Heijden and Hartmann 2016; Fierer 2017). For example, rhizobacteria (PGPR) that promote the growth of inoculated plants may initially colonize roots effectively, but over time, their survival and functionality often decline due to competition with the local microbiota and changing soil conditions (Schmidt 2019; Compant et al. 2019). These uncertainties make it difficult to design stable, resilient, and predictable microbial products for agricultural applications. To overcome this challenge, integrated ecological research combining field-based monitoring and system-level modeling is required. This will enable a deeper understanding of the complexity of plant–soil–microorganism interactions across space and time (Trivedi 2020).

8.6.2 Ecological and Biosafety Concerns

8.6.2.1 Risks of Ecological Degradation

The deliberate introduction of microorganisms that are not native to soil ecosystems or that are genetically modified raises significant concerns about unexpected environmental impacts. Added microbial inoculants may compete with native species, alter nutrient cycling pathways, or disrupt existing symbiotic relationships between plants and native microbiota (Hart et al. 2018; Finkel et al. 2017). Traditional biological fertilizers and biopesticides are generally considered safe due to their host specificity and limited persistence in soils (Herrmann and Lesueur 2013), but the use of genetically modified species carries greater uncertainty. Engineered microorganisms may possess enhanced adaptation traits, potential for horizontal gene transfer, or broad metabolic capabilities, increasing the likelihood of unpredictable long-term ecological consequences (Sessitsch et al. 2019; Snow et al. 2005). These points highlight the need for detailed ecological risk assessments and long-term monitoring requirements prior to large-scale field applications of genetically modified microbial products.

8.6.2.2 Horizontal Gene Transfer

Another significant biosafety concern in the field of microbial biotechnology is the possibility of horizontal gene transfer (HGT), defined as the potential for engineered genes to transfer from the host organism to unwanted microbial communities in the soil or rhizosphere. Such transfer events can lead to the unintended spread of abilities such as antibiotic resistance or enhanced metabolic traits. This may increase the likelihood of opportunistic pathogens acquiring new abilities or disrupting the dynamics of natural microbial communities (Heuer and Smalla, 2012; Van Elsas et al. 2014). Although it is difficult to accurately measure the frequency of HGT in soil environments, available data suggest that mobile genetic elements, plasmids, and transposons can facilitate such changes under field conditions (Hall et al. 2017, 2014; Domingues et al. 2012). These threats, along with rigorous post-release monitoring activities and ecological risk assessments, highlight the importance of integrating genetic safety measures such as kill switch systems, dependency circuits, and auxotrophy-based constraints (Wright et al. 2013; Schmidt and de Lorenzo 2016).

8.6.2.3 Regulatory and Public Perception Challenges

Although the reliability and effectiveness of engineered microbial products are supported by solid scientific evidence, public skepticism regarding genetically modified organisms (GMOs) remains a major obstacle to the adoption of these products. In places such as the European Union, strict biosafety regulations and lengthy approval processes, reflecting both precautionary approaches and societal concerns, make it even more difficult to bring genetically modified microorganisms to market (Schmidt et al. 2009; Hartung and Matthiessen 2013). Producers and consumers generally exhibit a cautious attitude towards "biotechnologically engineered microorganisms" compared to preparations produced using natural resources, associating them with genetically modified products and broader debates on biotechnology (Frewer et al.

2013). This conflict between technological development and public acceptance highlights the importance of effective risk communication, stakeholder engagement, and developing regulatory frameworks that aim to strike a balance between precaution and innovation (Marris 2015; Stirling 2008).

8.6.3 Socioeconomic Barriers

8.6.3.1 Farmers' Awareness and Education

In many places, particularly in developing countries, the use of microbial biotechnology remains limited due to farmers' limited knowledge, insufficient technical expertise, and concerns about the effectiveness of the products. Traditional chemical resources, such as synthetic fertilizers and pesticides, still play a strong role in agricultural applications and pose both behavioral and structural barriers to the transition to bio-based alternatives (Herrera-Estrella and López-Arredondo 2016; Choudhary et al. 2017). The effective adoption of microbial preparations requires not only clearly demonstrating their cost-effectiveness and productivity advantages but also robust extension services and farmer education programs that will increase confidence in these innovations (Ghosh 2004). Furthermore, participatory approaches that involve farmers in local trials and adapt microbial technologies to regional agroecological conditions have been observed to significantly increase adoption rates (Sanginga and Woomer 2009). Therefore, education, capacity building, and knowledge dissemination activities targeting farmers play a vital role in the dissemination of microbial biotechnology in small-scale farming systems.

8.6.3.2 Economic Sustainability and Market Competition

Microbial products have long faced significant challenges competing in markets dominated by traditional agricultural chemicals. Synthetic fertilizers and pesticides have benefited from years of industrial optimization, well-organized global distribution networks, and significant government subsidies in many countries that reduce farmers' costs and encourage continued dependence on chemical inputs (Tilman et al. 2002; Pingali 2012). In contrast, microbial biotechnology companies are typically smaller businesses with limited financial and infrastructure resources to increase production volume, standardize formulations, and employ effective marketing methods (Malusá and Vassilev 2014; Herrmann and Lesueur 2013). This structural imbalance often limits the integration of microbial products as common agricultural inputs, confining them to niche markets instead. Therefore, policy incentives such as subsidies for bio-based products, certification systems, and investments in production infrastructure are necessary to level the playing field and support the transition to more sustainable agricultural systems (Choudhary et al. 2017; Malusá et al. 2016).

8.6.3.3 Intellectual Property and Access Issues

The patenting of microbial strains and genetic engineering techniques presents both opportunities and challenges for the advancement and widespread adoption of microbial biotechnology. Intellectual property rights play an important role in supporting innovation, encouraging private sector investment, and bringing new microbial products to market (Graff et al. 2003; Jefferson et al. 2015). However, restrictive patent systems can inadvertently hinder access for small farmers and public research institutions, particularly in low- and middle-income countries where affordability and accessibility are critical (Stiglitz 2008). Concerns have also been raised about the concentration of ownership among certain multinational corporations, which could further exacerbate inequalities in global agricultural innovation (Louwaars et al. 2009). For this reason, striking a balance between proprietary technologies and microbial resources made available through open access has become an urgent policy issue, particularly in terms of ensuring global food security and achieving the United Nations Sustainable Development Goals (Pingali 2012; FAO 2020).

8.6.4 Regulatory Frameworks and Policy Perspectives

8.6.4.1 Current Regulatory Environment

Legal regulations related to microbial biotechnology reveal significant regional differences in risk perception, institutional capabilities, and policy objectives (Table 8.7). In the United States, oversight is shared among federal agencies: the Environmental Protection Agency (EPA) is responsible for regulating microbial pesticides under the Federal Insecticide, Fungicide, and Rodenticide Act (FIFRA), while the U.S. Department of Agriculture (USDA) is responsible for genetically modified microbial products that affect plant health (Wend et al. 2024; USDA-APHIS 2020). Approval processes typically require detailed environmental reviews and field testing to ensure biosafety and efficacy. Regulatory legislation in the European Union (EU) is quite strict on genetically modified organisms (GMOs), particularly with Directive 2001/18/EC and Regulation 1829/2003. This cautious regulatory stance has slowed the market entry of CRISPR-based microbial products, despite robust scientific advances (Hartung and Schiemann 2014; Eriksson et al. 2018). On the other hand, rapidly growing economies such as India, Brazil, and China are establishing new regulatory frameworks to meet the growing demand for microbial preparation components. These countries have reported significant progress in promoting biofertilizers and biopesticides as environmentally friendly alternatives to agricultural chemicals, but they still face many challenges in standardizing product quality, ensuring compliance, and providing adequate training to farmers (Malusá and Vassilev 2014; Jahan et al. 2016). Establishing a science-based and harmonized regulatory framework remains a vital step in ensuring the innovative use and safe integration of microbial technologies in agriculture.

Table 8.7 Comparative regulatory frameworks for microbial biotechnology

Region	Regulatory authorities	Focus of regulation	Key features	Challenges	References
United States	- Environmental Protection Agency (EPA) - U.S. Department of Agriculture (USDA-APHIS)	- Microbial pesticides (EPA) - Genetically modified microorganisms affecting plant health (USDA)	- Regulated under FIFRA (EPA) and the Plant Protection Act (USDA) - Requires environmental risk assessment and field trials	- Long approval times - Complex multi-agency oversight	EPA (2020), USDA-APHIS (2020)
European Union	- European Food Safety Authority (EFSA) - European Commission under Directive 2001/18/EC and Regulation 1829/2003	- GMOs, including engineered microorganisms - Risk assessment and post-market monitoring	- Strict GMO regulation - The precautionary principle prevails in the decision-making process	- Slows down the adoption of CRISPR-based and synthetic biology products - Public skepticism remains high	Hartung and Schiemann (2014), Eriksson et al. (2018)
India	- Ministry of Agriculture - Department of Biotechnology (DBT) - Fertilizer Control Ordinance (FCO) for bio-fertilizers	- Quality standards for bio-fertilizers and bio-pesticides - Genetically modified microorganisms regulated under the 1986 EPA	- Expansion of biofertilizer promotion for sustainable agriculture - Supportive government programs	- Weak implementation of quality control - Farmers' lack of awareness	Malusá and Vassilev (2014), Jahan et al. (2016)
Brazil	- National Technical Commission on Biosafety (CTNBio) - Ministry of Agriculture, Livestock, and Food Supply	- Biosafety of genetically modified microorganisms - Registration of microbial preparations for agriculture	- Flexible and adaptable biosafety policies - Strong focus on agricultural biotechnology	- Monitoring and compliance - Dissemination and adoption among small farmers	Bomfim et al. (2021)
China	- Ministry of Agriculture and Rural Affairs (MARA) - Biosafety Committee for GMOs	- Registration of bio-fertilizers and bio-pesticides - GMO risk assessments and field trials	- Aggressive investment in microbial biotechnology - Rapid commercialization of bio-fertilizers	- Quality control and regulatory harmonization remain weak	Pandey and Chandra (2016), Sun et al. (2025)

8.6.4.2 Towards Risk-Based and Adaptive Regulation

Many experts state that regulations in the field of microbial biotechnology should evolve into flexible, risk-based systems that balance innovation with biosafety concerns. Instead of blanket bans on genetically modified microorganisms, it is recommended that regulations examine products on a case-by-case basis, taking into account information about the organism's specific characteristics, its intended agricultural use, and the ecosystem in which it is introduced. This approach has been proposed as a solution for supporting responsible innovation while protecting the environment (Eriksson et al. 2018; Hartung and Schiemann 2014). By aligning regulations with scientific risk assessments, governments can promote the safe market introduction of new microbial technologies while preserving ecosystem integrity (Raybould and Poppy 2012).

8.6.4.3 Policy Incentives for Adoption

Considering permanent market and institutional barriers, policy interventions are needed to accelerate the adoption of microbial biotechnology. Financial incentives, such as subsidies for microbial inputs, are similar to incentives applied to chemical fertilizers in the past and thus have the potential to reduce cost differences and increase competitiveness (Adesemoye et al. 2009; Malusá and Vassilev 2014). Including microbial biotechnology in climate-friendly agriculture programs will enable these technologies to be used as a tool to provide resilience, productivity, and sustainability in the face of changing climate conditions (FAO 2013). Public–private partnerships are another important mechanism that encourages collaboration between research institutions, industry, and government agencies to support large-scale demonstration projects and establish viable supply chains (Larson 2021). Furthermore, farmer participation, through extension services and training, is critical for effective knowledge dissemination and enables farmers to safely incorporate microbial products into their cropping systems (Prasad 2019). At the global level, international frameworks such as the United Nations Sustainable Development Goals (SDGs) establish a strategic relationship between microbial biotechnology and global development goals. In particular, microbial innovations directly contribute to SDG 2 (Zero Hunger), SDG 12 (Sustainable Consumption and Production), SDG 13 (Climate Action), and SDG 15 (Life on Land), highlighting their importance in promoting food security, sustainable resource management, and environmental protection (United Nations 2015).

8.6.5 Future Directions and Strategic Considerations

Future plans to overcome the current barriers hindering the widespread acceptance of microbial biotechnology must encompass scientific, political, and social elements. Strengthening the interaction between research and policy. Policymakers must have access to robust, evidence-based data that includes long-term ecological studies, cost-benefit analyses, and social impact assessments. Increased collaboration between researchers and regulators will enable science-based decision-making processes and accelerate the responsible application of microbial technologies (Borrelli et al. 2017; Kamilaris et al. 2019). Policy frameworks supported by scientific data can facilitate the integration of microbial biotechnology into broader sustainability and food safety goals. Establishing infrastructure for quality control. Ensuring the consistency and reliability of microbial products is particularly important in regions where variability in product quality affects farmer confidence. Developing microbial culture collections, standard test protocols, and certification systems can enhance both quality assurance and traceability (Malusá and Vassilev 2014). Public investments in infrastructure and regulatory oversight play a critical role in building confidence and promoting adoption, particularly in developing countries. Enhancing global cooperation. Agricultural issues such as soil degradation, climate change, and food insecurity have an international dimension, making global cooperation essential. Platforms such as the Food and Agriculture Organization (FAO) and the Consultative Group on International Agricultural Research (CGIAR) can promote knowledge sharing, harmonize regulatory standards, and build capacity in low-income countries (FAO 2017; Pretty et al. 2018). Such initiatives will ensure that microbial biotechnology contributes to global sustainability goals. Promoting inclusive innovation. Equal access must remain at the center of future innovation processes. Open-source platforms for microbial genome data, stronger involvement of farmer cooperatives and the public sector interested in local production can democratize access to microbial technologies, thereby preventing the dominance of large agricultural companies (Bezner Kerr et al. 2019; De Schutter 2011). Inclusive innovation not only promotes fairness but also increases resilience by diversifying participation among stakeholders.

8.6.6 Summary

Microbial biotechnology lies at the intersection of scientific innovation and societal needs, offering promising alternatives for enhancing the sustainability, productivity, and resilience of agriculture. The transformation potential of agricultural systems is clearly observed in applications such as biofertilizers, biopesticides, biostimulants, and microbial consortia (Bashan et al. 2014; Malusá et al. 2016). However, several significant challenges exist, including inconsistent performance under variable field conditions, uncertainties related to biosafety, stringent and diverse regulatory frameworks, competition with established agricultural chemical industries, and limited societal acceptance of biotechnology products (Schmidt et al. 2019; Kaminsky et al. 2019; Marrone 2008). To overcome these barriers, a comprehensive strategy is needed that combines the latest scientific data with flexible regulatory structures, targeted farmer education, and inclusive innovative models. When supported by enabling policy frameworks and international collaborations, microbial biotechnology can evolve from a promising innovation into a fundamental pillar of sustainable agriculture. Integrating scientific advances with goals of ecological sustainability, food security, and socioeconomic development is critical to realizing its full potential. By bringing together science, policy, and practice, microbial biotechnology has demonstrated its capacity to play an important role in transforming the global food system to address twenty-first century challenges such as climate change, soil erosion, and chemical input dependency (FAO 2017; Pretty et al. 2018).

8.7 Conclusion and Future Expectations

8.7.1 Reshaping Agriculture in the Twenty-First Century

In the twenty-first century, agriculture faces numerous interconnected crises, including rapid population growth, climate change, soil erosion, declining biodiversity, and freshwater scarcity, creating a critical threshold for agriculture (Foley et al. 2011; IPCC 2022). In particular, the intensive use of synthetic fertilizers and pesticides has significantly contributed to increased yields through traditional farming methods, but it has also brought about high environmental and social costs. These costs include problems such as depletion of soil nutrients, water pollution, greenhouse gas emissions, and pesticide-resistant pests (Tilman et al. 2002; Matson et al. 1997; Carvalho 2017). In this context, microbial biotechnology presents itself not merely as a marginal innovation but as a transformative paradigm with the potential to reshape global food production.

Microbial biotechnology views soil not as a static environment requiring only chemical nutrient supplements, as in traditional methods, but as a dynamic ecosystem regulated by microbial diversity and complex interactions (Schmidt et al. 2019). Leveraging the abilities of microorganisms to fix atmospheric nitrogen, solubilize phosphorus, prevent plant diseases, and increase resilience against abiotic stresses is emerging as a way to balance agricultural productivity with ecosystem integrity (Bhattacharyya and Jha 2012; Singh and Trivedi 2017). In this context, microbial biotechnology represents more than just a technical advancement; it symbolizes a broader philosophical and practical shift toward system-focused, regenerative, and sustainable agricultural approaches (Pretty et al. 2018; Bender et al. 2016).

8.7.2 Consolidation of Existing Gains

The four main branches of microbial biotechnology—biofertilizers, biopesticides, biostimulants, and microbial consortia—show significant potential in various agricultural systems. Biofertilizers increase the efficient use of nutrients while reducing dependence on synthetic materials (Adesemoye et al. 2009; Vessey 2003). Biopesticides offer environmentally sensitive and targeted alternatives to chemical pesticides, minimizing ecological damage and resistance development (Mishra et al. 2015; Glare et al. 2012). Biostimulants enhance plants' resistance to abiotic and biotic stresses;

this property is particularly important in situations of climate change and limited resources (Rouphael and Colla 2020). Microbial consortia represent todays newest technology, offering unique benefits by combining numerous microbial functions suitable for various soils and cropping systems (Hu et al. 2022; Toju et al. 2018). Overall, these microbial innovations form an effective set of intervention options that increase productivity, reduce input costs, and support environmental sustainability.

Practical data from developing economies demonstrate the tangible benefits of these technologies. Case studies from India, Brazil, and China highlight both economic and environmental benefits, showing that microbial inputs can reduce fertilizer costs by 20–30%, increase soil fertility, and reduce greenhouse gas emissions (Hungria et al. 2013; Sessitsch et al. 2021). These results confirm that microbial biotechnology is not merely a possibility but a practical application that produces tangible results in agricultural systems. However, fully realizing its potential will require the standardization of quality control measures, the widespread adoption of supportive policy frameworks, and the implementation of farmer-focused education and extension activities to ensure consistent benefits (Herrmann and Lesueur 2013; Kaminsky et al. 2019).

8.7.3 Future Scientific Directions

Advances in synthetic biology and CRISPR-based genome editing offer unique opportunities for the intelligent design of microorganisms with specific functions. Engineered strains can now be programmed to release nutrients only under specific soil or plant conditions, produce phytohormones in response to host signals, or increase productivity by breaking down soil contaminants (Jiang and Doudna, 2017). This shift in the concept of the "microbial factory" redefines microorganisms as programmable partners, expanding their roles beyond the nutrient cycle to include dynamic environmental management within agricultural systems.

Simultaneous advances in multi-omics and systems biology have enabled a better understanding of the complex interactions between plants, microorganisms, and soils. Metagenomics, transcriptomics, metabolomics, and proteomics provide complementary information about microbial diversity, gene expression, metabolic activity, and protein function (Bashiardes et al. 2016; Fierer 2017). By combining these data sets, systems biology enables the modeling of rhizosphere dynamics and the prediction of community-level behavior under changing environmental conditions (Zhalnina 2018). Such data support the methodical design of microbial consortia optimized for specific crop types, soil varieties, or climate zones, thereby better integrating microbial biotechnology into precision agriculture.

Artificial intelligence and machine learning are accelerating these developments by providing tools for analyzing large microbiome datasets and identifying microbial signatures of soil health. Predictive modeling can forecast the field performance of microbial inoculants and provide additional support for creating compatible, functional consortia (Arango-Argoty et al. 2020; Trivedi et al. 2020). Digital agriculture platforms that combine microbiome data with soil chemistry, climate data, and crop needs are emerging and offer decision support systems that guide farmers in selecting and applying the most suitable microbial products for their fields (Mahlein 2016).

Climate change, drought, salinity, and heat stress are increasing in frequency, while microbial biotechnology is also being deployed to enhance resilience. The development of engineered or selected microbial strains capable of providing stress tolerance is a critical innovation for maintaining productivity in vulnerable regions (Naylor et al. 2021; Mikiciuk et al. 2024). These climate-responsive microbial solutions are of great importance for smallholder farmers in Africa and South Asia, where climate variability poses the greatest threat to food security. Microbial biotechnology, which combines precision engineering, systems biology, AI-assisted analysis, and climate adaptation strategies, is poised to play a transformative role in establishing sustainable and resilient agricultural systems.

8.7.4 Socioeconomic and Policy Outlook

For microbial biotechnology to effectively transition from research to widespread agricultural applications, significant investments in extension services and participatory research are essential. Farmers must play an active role as co-creators in this process, testing microbial products under different field conditions, providing context-specific feedback, and adapting applications to the local agroecological structure (Vanlauwe et al. 2019; Hellin et al. 2020). Without such participation, even the most technologically advanced microbial innovations may not be used sufficiently and may fail to produce the desired effect.

Equal access is one of the fundamental elements for the sustainable expansion of microbial biotechnology. Small holder farmers, who account for over 70% of global food production, should benefit from this technology alongside industrial agriculture (FAO 2014; Samberg et al. 2016). To ensure inclusivity, public investments, open-access microbial culture banks, and the establishment of local or community-based production units are necessary. Such strategies can reduce dependence on multinational corporations, lower costs, and increase farmers' confidence in bio-based solutions. Therefore, policy guidelines should prioritize accessibility to ensure that microbial technologies reduce rather than exacerbate inequalities in the global food system (Scoones et al. 2020).

Finally, the development of regulatory frameworks continues to be a decisive factor in shaping the direction of microbial biotechnology. Current regulations, particularly within the EU, continue to impose strict restrictions on genetically modified microorganisms, hindering the commercialization of microbial products based on CRISPR and synthetic biology (Purnhagen et al. 2021; Gelinsky and Hilbeck 2018). A transition to adaptable, risk-based regulations is necessary to unlock the potential of precision microbial biotechnology. These frameworks should be based on and evaluate the characteristics of the products and their interactions with the ecosystem, rather than the methods used in their development (Eriksson et al. 2019). Ensuring transparency in biosafety monitoring and proactive, science-based communication with the public will be crucial for building public trust and regulatory acceptance (Marris 2015).

8.7.5 Towards Global Food Security and Sustainability

The opportunities offered by microbial biotechnology encompass not only increased yields but also the potential to reduce chemical dependency, improve soil health, and enhance agricultural resilience against climate stresses. These microbial developments contribute by directly benefiting the three main components of sustainability: economy, environment, and society (Tiwari and Park 2024; Singh et al. 2020). Furthermore, these findings are in line with the United Nations Sustainable Development Goals (SDGs). For example, microbial technologies support SDG 2 (Zero Hunger) by increasing crop productivity and resilience. They also contribute to SDG 12 (Responsible Consumption and Production) by reducing dependence on synthetic agricultural chemicals. Their ability to reduce greenhouse gas emissions and develop adaptation strategies supports SDG 13 (Climate Action), while efforts to increase soil fertility and restore biodiversity align with SDG 15 (Life on Land) (United Nations 2015; Adesemoye et al. 2009). Based on this, microbial biotechnology should be considered not only as an agricultural innovation but also as a fundamental element of global sustainability strategies. Integration with international climate initiatives, food security programs, and biodiversity conservation approaches can accelerate the transition to regenerative food systems, particularly in regions most vulnerable to ecological collapse and climate change (FAO 2017). By combining scientific discoveries with global policy goals, microbial biotechnology has the capacity to reshape agricultural food systems in the twenty-first century.

8.7.6 Conclusion

Microbial biotechnology is a two-way concept; it refers to processes that are constantly ongoing and evolving. Fundamental processes such as fermentation and nitrogen fixation demonstrate humanity's long-standing dependence on microorganisms (Smid and Lacroix 2013; Vessey, 2003) and, at the same time, synthetic biology, gene editing, and multi-omic approaches are promoting this evolution (de Lorenzo 2018; Lorenz and Eckert 2022). Thus, microbial biotechnology blends past ecological knowledge with the latest technological advances, defining microorganisms as programmable partners in sustainable agriculture.

However, the success of microbial biotechnology depends not only on scientific innovation but also on social, political, and ethical choices. Policies must encourage the transition from chemical dependency to sustainable resources (Tilman 2011; Glick 2012) and ensure that small farmers, who account for more than 70% of global food production, have access to microbial resources (FAO 2017; Singh et al. 2020). Intellectual property practices and corporate concentration pose a risk of restricting equitable access and raise concerns about monopolization and exclusion (Louwaars et al. 2009; Aerni 2013). Furthermore, regulatory frameworks, particularly in the European Union, continue to take a cautious approach to genetically modified microorganisms and emphasize the need for flexible, risk-based approaches that balance innovation and biosafety (Kawall 2019; Trump 2017).

To overcome these barriers, a responsible innovation approach is needed, where inclusivity, transparency, and intellectual flexibility guide technological development (Stilgoe et al. 2013). Public trust is linked to effective science communication, participatory governance, and post-market ecological monitoring (Marris 2015). Microbial biotechnology can only overcome skepticism, ensure biosafety, and deliver on its promises at scale through such an integrative approach. When applied inclusively and responsibly, microbial biotechnology has the potential to play a significant role both as a technical solution to urgent agricultural problems and as the foundation of sustainable food systems. By reducing chemical dependency, improving soil health, and strengthening resistance to climate stress, it contributes directly to the United Nations Sustainable Development Goals (FAO 2017; UN 2015). In this broader perspective, microorganisms are more than just elements present in the soil: they are partners in survival, architects of regenerative agriculture, and critical supporters for humanity in its efforts to feed a growing population while protecting the planet's limits (Berg et al. 2014; Bezner Kerr et al. 2021; Boddey et al. 1995; Borriss 2014; Bravo et al. 2011; Cawoy et al. 2011; Choudhary et al. 2016; Choudhary et al. 2011; Compant et al. 2010; Costa et al. 2021; e Lorenzo et al. 2015; Edwards et al. 2015; Eriksson et al. 2020; FAO 2018; Herridge et al. 2008; Herring et al. 2006; Hu et al. 2016; Intergovernmental Panel on Climate Change 2023; Ju et al. 2007; Kamilaris et al. 2017; Kuntz 2012; Lacey et al. 2015; Liu et al. 2017; Liu et al. 2013; Mishra et al. 2014; Nadeem et al. 2014; Nguyen et al. 2016; Ofek-Lalzar et al. 2014; Oliveira et al. 2003; Pankievicz et al. 2019; Rillig et al. 2015; Rockström et al. 2009; Schütz et al. 2018; Shokralla et al. 2012; United States Department of Agriculture, Animal and Plant Health Inspection Service USDA-APHIS 2020; United States Department of Agriculture, Animal and Plant Health Inspection Service USDA-APHIS 2020; Wezel et al. 2014; Zhang et al. 2013; Zhang et al. 2019; Zhang et al. 2013; Fanzo et al. 2022; Aktar et al. 2009; Nuruzzaman et al. 2025; Sudheer et al. 2020; Iqbal et al. 2023; Daniel et al. 2022; Pandiyan et al. 2024; Aguiar-Pulido et al. 2016; Dudeja et al. 2025; Dhawan and Gupta 2024; Lensch et al. 2024; Karalis et al. 2020).

References

Adesemoye AO, Torbert HA, Kloepper JW (2009) Plant growth-promoting rhizobacteria allow reduced application rates of chemical fertilizers. Microb Ecol 58(4):921–929. https://doi.org/10.1007/s00248-009-9531-y

Aerni P (2013) Resistance to agricultural biotechnology: the importance of distinguishing between weak and strong public attitudes. Biotechnol J 8(10):1129–1132. https://doi.org/10.1002/biot.201300188

Aguiar-Pulido V, Huang W, Suarez-Ulloa V, Cickovski T, Mathee K, Narasimhan G (2016) Metagenomics, metatranscriptomics, and metabolomics approaches for microbiome analysis. Evolut Bioinforma Online 12(Suppl 1):5–16. https://doi.org/10.4137/EBO.S36436

Aktar MW, Sengupta D, Chowdhury A (2009) Impact of pesticides use in agriculture: their benefits and hazards. Interdiscip Toxicol 2(1):1–12. https://doi.org/10.2478/v10102-009-0001-7

Altman J (2018) Pesticide interactions in crop production. In: Pesticide interactions in crop production. CRC Press, pp 3–9

Arango-Argoty G, Garner E, Pruden A, Heath LS, Vikesland P, Zhang L (2018) DeepARG: a deep learning approach for predicting antibiotic resistance genes from metagenomic data. Microbiome 6(1):23. https://doi.org/10.1186/s40168-018-0401-z

Aznar A, Dellagi A (2015) New insights into the role of siderophores as triggers of plant immunity: what can we learn from animals. J Exp Bot 66(11):3001–3010. https://doi.org/10.1093/jxb/erv155

Backer R, Rokem JS, Ilangumaran G, Lamont J, Praslickova D, Ricci E, Smith DL (2018) Plant growth-promoting rhizobacteria: context, mechanisms of action, and roadmap to commercialization of biostimulants for sustainable agriculture. Front Plant Sci 9:1473. https://doi.org/10.3389/fpls.2018.01473

Bashan Y, de-Bashan LE, Prabhu SR, Hernandez JP (2014) Advances in plant growth-promoting bacterial inoculant technology: formulations and practical perspectives (1998–2013). Plant Soil 378(1–2):1–33. https://doi.org/10.1007/s11104-013-1956-x

Bashiardes S, Zilberman-Schapira G, Elinav E (2016) Use of metatranscriptomics in microbiome research. Bioinforma Biol Insights 10:19–25. https://doi.org/10.4137/BBI.S34610

Bender SF, Wagg C, van der Heijden MG (2016) An underground revolution: biodiversity and soil ecological engineering for agricultural sustainability. Trends Ecol Evol 31(6):440–452. https://doi.org/10.1016/j.tree.2016.02.016

Berendsen RL, Pieterse CM, Bakker PA (2012) The rhizosphere microbiome and plant health. Trends Plant Sci 17(8):478–486. https://doi.org/10.1016/j.tplants.2012.04.001

Berg G, Grube M, Schloter M, Smalla K (2014) Unraveling the plant microbiome: looking back and future perspectives. Front Microbiol 5:148. https://doi.org/10.3389/fmicb.2014.00148

Bezner Kerr R, Madsen S, Stüber M, Liebert J, Enloe S, Borghino N et al (2021) Can agroecology improve food security and nutrition? Glob Food Secur 29:100540. https://doi.org/10.1016/j.gfs.2021.100540

Bhardwaj D, Ansari MW, Sahoo RK, Tuteja N (2014) Biofertilizers function as key player in sustainable agriculture by improving soil fertility, plant tolerance and crop productivity. Microb Cell Fact 13: 66. https://doi.org/10.1186/1475-2859-13-66

Bhattacharyya PN, Jha DK (2012) Plant growth-promoting rhizobacteria (PGPR): emergence in agriculture. World J Microbiol Biotechnol 28(4):1327–1350. https://doi.org/10.1007/s11274-011-0979-9

Boddey RM, de Oliveira OC, Urquiaga S, Reis VM, de Olivares FL, Baldani VLD, Döbereiner J (1995) Biological nitrogen fixation associated with sugar cane and rice: contributions and prospects for improvement. Plant Soil 174(1–2):195–209. https://doi.org/10.1007/BF00032247

Borrelli P, Robinson DA, Fleischer LR, Lugato E, Ballabio C, Alewell C et al (2017) An assessment of the global impact of 21st century land use change on soil erosion. Nat Commun 8(1):2013. https://doi.org/10.1038/s41467-017-02142-7

Borriss R (2014) Bacillus, a plant-beneficial bacterium https://doi.org/10.13140/2.1.3220.8642

Bravo A, Likitvivatanavong S, Gill SS, Soberón M (2011) Bacillus thuringiensis: a story of a successful bioinsecticide. Insect Biochem Mol Biol 41(7):423–431. https://doi.org/10.1016/j.ibmb.2011.02.006

Bulgarelli D, Schlaeppi K, Spaepen S, van Themaat EVL, Schulze-Lefert P (2013) Structure and functions of the bacterial microbiota of plants. Annu Rev Plant Biol 64:807–838. https://doi.org/10.1146/annurev-arplant-050312-120106

Cai P, Dimopoulos G (2025) Microbial biopesticides:a one health perspective on benefits and risks. One Health 20:100962

Çakirsoy I, Miyamoto T, Ohtake N (2022) Physiology of microalgae and their application to sustainable agriculture: a mini-review. Front Plant Sci 13:1005991

Calvo P, Nelson L, Kloepper JW (2014) Agricultural uses of plant biostimulants. Plant Soil 383(1–2):3–41. https://doi.org/10.1007/s11104-014-2131-8

Carvalho FP (2017) Pesticides, environment, and food safety. Food Energy Secur 6(2):48–60. https://doi.org/10.1002/fes3.108

Cassán F, Vanderleyden J, Spaepen S (2014) Physiological and agronomical aspects of phytohormone production by model plant-growth-promoting rhizobacteria (Azospirillum spp.). Plant Soil 356(1–2):1–20. https://doi.org/10.1007/s00344-013-9362-4

Cawoy H, Bettiol W, Fickers P, Onge M (2011) Bacillus-based biological control of plant diseases. In: Pesticides in the modern world—pesticides use and management. InTech, https://doi.org/10.5772/17184

Chandler D, Bailey AS, Tatchell GM, Davidson G, Greaves J, Grant WP (2011) The development, regulation and use of biopesticides for integrated pest management. Philos Trans R Soc B 366(1573):1987–1998. https://doi.org/10.1098/rstb.2010.0390

Chaudhary R, Nawaz A, Khattak Z, Butt MA, Fouillaud M, Dufossé L, Mukhtar H (2024) Microbial bio-control agents: a comprehensive analysis on sustainable pest management in agriculture. J Agric Food Res 18:101421

Choudhary DK, Sharma KP, Gaur R (2011) Biotechnological perspectives of microbes in agro-ecosystems. Biotechnol Lett 33(10):1905–1910. https://doi.org/10.1007/s10529-011-0662-0

Choudhary DK, Kasotia A, Jain S, Vaishnav A, Kumari S, Sharma KP (2016) Bacterial-mediated tolerance and resistance in plants under abiotic and biotic stresses. J Plant Growth Regul 35(2):276–300. https://doi.org/10.1007/s00344-015-9521-x

Compant S, Clément C, Sessitsch A (2010) Plant growth-promoting bacteria in the rhizo- and endosphere of plants: their role, colonization, mechanisms involved and prospects for utilization. Soil Biol Biochem 42(5):669–678. https://doi.org/10.1016/j.soilbio.2009.11.024

Compant S, Samad A, Faist H, Sessitsch A (2019) A review on the plant microbiome: ecology, functions, and emerging trends in microbial application. J Adv Res 19:29–37. https://doi.org/10.1016/j.jare.2019.03.004

Cornelis P (2010) Iron uptake and metabolism in pseudomonads. Appl Microbiol Biotechnol 86(6):1637–1645. https://doi.org/10.1007/s00253-010-2550-2

Costa F, da S, de Moraes AS, Ferreira NS, Borges WL (2021) Technical viability of improving soil chemical characteristics by using biofertilizers. Rev Brasi Ciênc Agrárias 16(3):e331. https://doi.org/10.5039/agraria.v16i3a331

Daniel AI, Fadaka AO, Gokul A, Bakare OO, Aina O, Fisher S, Burt AF, Mavumengwana V, Keyster M, Klein A (2022) Biofertilizer: the future of food security and food safety. Microorganisms 10(6):1220. https://doi.org/10.3390/microorganisms10061220

de Andrade LA, Santos CHB, Frezarin ET, Sales LR, Rigobelo EC (2023) Plant growth-promoting rhizobacteria for sustainable agricultural production. Microorganisms 11(4):1088. https://doi.org/10.3390/microorganisms11041088

De Schutter O (2011) Agroecology and the right to food (A/HRC/16/49). United Nations Human Rights Council.

de Souza RSC, Armanhi JSL, Arruda P (2020) From microbiome to traits: designing synthetic microbial communities for improved crop resiliency. Front Plant Sci 11:1179. https://doi.org/10.3389/fpls.2020.01179

de Vries FT, Wallenstein MD (2017) Below-ground connections underlying above-ground food production: a framework for optimising ecological connections in the rhizosphere. J Ecol 105(4):913–920. https://doi.org/10.1111/1365-2745.12783

Dhawan P, Gupta V (2024) Role of microbiology for sustainable development in India. Role of science and technology for sustainable future: volume 1: sustainable development: a primary goal. Springer Nature Singapore, Singapore, pp 119–132

Domingues S, Harms K, Fricke WF, Johnsen PJ, da Silva GJ, Nielsen KM (2012) Natural transformation facilitates transfer of transposons, integrons and gene cassettes between bacterial species. PLoS Pathog 8(8):e1002837. https://doi.org/10.1371/journal.ppat.1002837

Doudna JA, Charpentier E (2014) The new frontier of genome engineering with CRISPR-Cas9. Science 346(6213):1258096. https://doi.org/10.1126/science.1258096

Dubey A, Malla MA, Khan F, Chowdhary K, Yadav S, Kumar A et al (2019) Soil microbiome: a key player for conservation of soil health under changing climate. Biodivers Conserv 28 (8–9):2405–2429. https://doi.org/10.1007/s10531-019-01760-5

du Jardin P (2015) Plant biostimulants: definition, concept, main categories, and regulation. Sci Horticult 196:3–14. https://doi.org/10.1016/j.scienta.2015.09.021
Dudeja C, Mishra A, Ali A, Singh PP, Jaiswal AK (2025) Microbial Genome Editing with CRISPR–Cas9: recent advances and emerging applications across sectors. Fermentation 11(7):410
e Lorenzo V, Sekowska A, Danchin A (2015) Chemical reactivity drives spatiotemporal organisation of bacterial metabolism. FEMS Microbiol Rev 39(1):96–119
Edwards J, Johnson C, Santos-Medellín C, Lurie E, Podishetty NK, Bhatnagar S et al (2015) Structure, variation, and assembly of the root-associated microbiomes of rice. Proc Natl Acad Sci USA 112(8):E911–E920. https://doi.org/10.1073/pnas.1414592112
Eida AA, Ziegler M, Lafi FF, Michell CT, Voolstra CR, Hirt H, Saad MM (2018) Desert plant bacteria reveal host influence and beneficial plant growth properties. PLoS One 13(12):e0208223. https://doi.org/10.1371/journal.pone.0208223
El-Halim HMA, El-Hadidi M, Fouad N, Hamed RR, Megid IA, Taha MH, Radwan KH (2025) Metagenomic insight into drought-induced changes in the Egyptian wheat rhizosphere microbiome. World J Microbiol Biotechnol 41(8):310. https://doi.org/10.1007/s11274-025-04518-0
Eriksson D, Harwood W, Hofvander P, Jones H, Rogowsky P, Stöger E, Visser RGF (2018) A welcome proposal to amend the GMO legislation of the EU. Trends Biotechnol 36(11):1100–1103. https://doi.org/10.1016/j.tibtech.2018.05.001
Eriksson D, Kershen D, Nepomuceno A, Pogson BJ, Prieto H, Romano A, Whelan AI (2019) A comparison of the EU regulatory approach to directed mutagenesis with that of other jurisdictions, consequences for international trade and potential steps forward. N Phytol 222(4):1673–1684. https://doi.org/10.1111/nph.15627
Eriksson D, Custers R, Edvardsson Björnberg K, Hansson SO, Purnhagen K, Qaim M, Romeis J, Schiemann J, Schleissing S, Tosun J, Visser RGF (2020) Options to reform the European Union legislation on GMOs: scope and definitions. Trends Biotechnol 38(3):231–234. https://doi.org/10.1016/j.tibtech.2019.12.002
Etesami H (2025) The dual nature of plant growth-promoting bacteria: benefits, risks, and pathways to sustainable deployment. Curr Res Microb Sci https://doi.org/10.1016/j.crmicr.2025.100421
European Commission (2020) Farm to Fork strategy: for a fair, healthy and environmentally-friendly food system. https://food.ec.europa.eu/horizontal-topics/farm-fork-strategy_en
European Food Safety Authority (EFSA) (2018) Guidance on the characterisation of microorganisms used as plant protection products. EFSA J 16(7):e05206. https://doi.org/10.2903/j.efsa.2018.5206
Fanzo J, Rudie C, Sigman I, Grinspoon S, Benton TG, Brown ME, Covic N, Fitch K, Golden CD, Grace D, Hivert MF, Huybers P, Jaacks LM, Masters WA, Nisbett N, Richardson RA, Singleton CR, Webb P, Willett WC (2022) Sustainable food systems and nutrition in the 21st century: a report from the 22nd annual Harvard Nutrition Obesity Symposium. Am J Clin Nutr 115(1):18–33. https://doi.org/10.1093/ajcn/nqab315
FAO (2013) Climate-Smart Agriculture Sourcebook. Food and Agriculture Organization of the United Nations. https://www.fao.org/3/i3325e/i3325e.pdf
FAO (2014) The State of Food and Agriculture 2014: innovation in family farming. https://www.fao.org/3/i4040e/i4040e.pdf
FAO (2017) The future of food and agriculture: trends and challenges. http://www.fao.org/3/i6583e/i6583e.pdf
FAO (2018) The future of food and agriculture: alternative pathways to 2050. https://www.fao.org/documents/card/en/c/CA1553EN
FAO, IFAD, UNICEF, WFP & WHO (2020) The State of Food Security and Nutrition in the World 2020. https://doi.org/10.4060/ca9692en
Faust K, Raes J (2012) Microbial interactions: from networks to models. Nat Rev Microbiol 10(8):538–550. https://doi.org/10.1038/nrmicro2832
Fiehn O (2002) Metabolomics—the link between genotypes and phenotypes. Plant Mol Biol 48(1–2):155–171. https://doi.org/10.1023/A:1013713905833
Fierer N (2017) Embracing the unknown: disentangling the complexities of the soil microbiome. Nat Rev Microbiol 15(10):579–590. https://doi.org/10.1038/nrmicro.2017.87

Finkel OM, Castrillo G, Herrera Paredes S, Salas González I, Dangl JL (2017) Understanding and exploiting plant beneficial microbes. Curr OpPlant Biol 38:155–163. https://doi.org/10.1016/j.pbi.2017.04.018
Foley JA, Ramankutty N, Brauman KA, Cassidy ES, Gerber JS, Johnston M et al (2011) Solutions for a cultivated planet. Nature 478(7369):337–342. https://doi.org/10.1038/nature10452
Folina A, Kakabouki I, Baginetas K, Bilalis D (2025) Integration of bioresources for sustainable development in organic. Farming 14(7):102
Frewer LJ, van der Lans IA, Fischer AR, Reinders MJ, Menozzi D, Zhang X et al (2013) Public perceptions of agri-food applications of genetic modification: a systematic review and meta-analysis. Trends Food Sci Technol 30(2):142–152. https://doi.org/10.1016/j.tifs.2013.01.003
Gelinsky E, Hilbeck A (2018) European Court of Justice ruling regarding new genetic engineering methods scientifically justified: a commentary on the biased reporting about the recent ruling. Environ Sci Europe 30(1):52. https://doi.org/10.1186/s12302-018-0182-9
Ghosh N (2004) Promoting bio-fertilisers in Indian agriculture. Econ Polit Wkly 39(52):5617–5625. https://www.epw.in/journal/2004/52/review-agriculture-review-issues-specials/promoting-biofertilisers-indian
Glare T, Caradus J, Gelernter W, Jackson T, Keyhani N, Köhl J, Marrone P, Morin L, Stewart A (2012) Have biopesticides come of age. Trends Biotechnol 30(5):250–258. https://doi.org/10.1016/j.tibtech.2012.01.003
Glick BR (2012) Plant growth-promoting bacteria: mechanisms and applications. Scientifica 2012:963401
Glick BR (2014) Bacteria with ACC deaminase can promote plant growth and help to feed the world. Microbiol Res 169(1):30–39. https://doi.org/10.1016/j.micres.2013.09.009
Godfray HCJ, Beddington JR, Crute IR, Haddad L, Lawrence D, Muir JF et al (2010) Food security: the challenge of feeding 9 billion people. Science 327(5967):812–818. https://doi.org/10.1126/science.1185383
Graff GD, Cullen SE, Bradford KJ (2003) The public-private structure of intellectual property ownership in agricultural biotechnology. Nat Biotechnol 21(9):989–995
Hamad A, Tayel A (2025) Food 2050 concept: trends that shape the future of food. J Future Foods
Harman GE (2011) Multifunctional fungal plant symbionts: new tools to enhance plant growth and productivity. N Phytol 189(3):647–649. https://doi.org/10.1111/j.1469-8137.2010.03614.x
Hart MM, Antunes PM, Chaudhary VB, Abbott LK (2018) Fungal inoculants in the field: is the reward greater than the risk. Funct Ecol 32(1):126–135. https://doi.org/10.1111/1365-2435.12976
Hartmann M, Frey B, Mayer J, Mäder P, Widmer F (2015) Distinct soil microbial diversity under long-term organic and conventional farming. ISME J 9(5):1177–1194. https://doi.org/10.1038/ismej.2014.210
Hartung F, Schiemann J (2014) Precise plant breeding using new genome editing techniques: opportunities, safety and regulation in the EU. Plant J 78(5):742–752. https://doi.org/10.1111/tpj.12413
Herrera-Estrella L, López-Arredondo D (2016) Phosphorus: the underrated element for feeding the world. Trends Plant Sci 21(6):461–463. https://doi.org/10.1016/j.tplants.2016.04.010
Herridge DF, Peoples MB, Boddey RM (2008) Global inputs of biological nitrogen fixation in agricultural systems. Plant Soi 311(1–2):1–18. https://doi.org/10.1007/s11104-008-9668-3
Herring CD, Raghunathan A, Honisch C, Patel T, Applebee MK, Joyce AR et al (2006) Comparative genome sequencing of *Escherichia coli* provides insights into the adaptive evolution of bacterial populations. Nat Genet 38(12):1406–1412. https://doi.org/10.1038/ng1906
Herrmann L, Lesueur D (2013) Challenges of formulation and quality of biofertilizers for successful inoculation. Appl Microbiol Biotechnol 97(20):8859–8873. https://doi.org/10.1007/s00253-013-5228-8
Heuer H, Smalla K (2012) Plasmids foster diversification and adaptation of bacterial populations in soil. FEMS Microbiol Rev 36(6):1083–1104. https://doi.org/10.1111/j.1574-6976.2012.00337.x

Hu J, Wei Z, Friman VP, Gu SH, Wang XF, Eisenhauer N, Jousset A (2016) Probiotic diversity enhances rhizosphere microbiome function and plant disease suppression. mBio 13(1):e03520–21. https://doi.org/10.1128/mbio.01790-16

Hungria M, Nogueira MA, Araujo RS (2013) Co-inoculation of soybeans and common beans with rhizobia and azospirilla: strategies to improve sustainability. Biol Fertil Soils 49(7):791–801. https://doi.org/10.1007/s00374-012-0771-5

Hungria M, Campo RJ, Souza EM, Pedrosa FO (2010) Inoculation with selected strains of Azospirillum brasilense and Burkholderia sp. improves yields of maize and wheat in Brazil. Plant Soil 331(1–2):413–425. https://doi.org/10.1007/s11104-009-0262-0

Hungria M, Franchini JC, Campo RJ, Crispino CC, Moraes JZ, Sibaldelli RNR, Mendes IC, Arihara J (2006) Nitrogen nutrition of soybean in Brazil: contributions of biological N_2 fixation and N fertilizer to grain yield. Can J Plant Sci 86(4):927–939. https://doi.org/10.4141/P05-098

Hungria M, Mendes IC (2015) Nitrogen fixation with soybean: the perfect symbiosis? In: Werner D, Newton WE (eds) Nitrogen fixation in agriculture forestry, ecology, and the environment. Springer, Cham, pp 100–114. https://doi.org/10.1002/9781119053095.ch99

Intergovernmental Panel on Climate Change (2023) Climate change 2022—Impacts, adaptation and vulnerability. Cambridge University Press, Cambridge. https://doi.org/10.1017/9781009325844

Iqbal B, Li G, Alabbosh KF, Hussain H, Khan I, Tariq M, Ahmad N (2023) Advancing environmental sustainability through microbial reprogramming in growth improvement, stress alleviation, and phytoremediation. Plant Stress 10:100283

Iqbal S, Begum F, Nguchu BA, Claver UP, Shaw P (2025) The invisible architects: microbial communities and their transformative role in soil health and global climate changes. Environ Microbiome 20(1):36. https://doi.org/10.1186/s40793-025-00694-6

Jacoby R, Peukert M, Succurro A, Koprivova A, Kopriva S (2017) The role of soil microorganisms in plant mineral nutrition—current knowledge and future directions. Front Plant Sci 8:1617. https://doi.org/10.3389/fpls.2017.01617

Jahan MAHS, Hossain A, Waqas M, Khan A, Ashrafuzzaman M, Rashid MH (2016) Biofertilizers and their role in sustainable agriculture. J Anim Plant Sci 26(3):656–669

Jain A, Singh D, Singh AK, Singh S, Sparks DL (2020) Microbial biofertilizers: recent trends and future outlook. Adv Agron 161:51–95

Jain A, Sarsaiya S, Singh R, Gong Q, Wu Q, Shi J (2024) Omics approaches in understanding the benefits of plant-microbe interactions. Front Microbiol 15:1391059. https://doi.org/10.3389/fmicb.2024.1391059

Jamil F, Mukhtar H, Fouillaud M, Dufossé L (2022) Rhizosphere signaling: insights into plant–rhizomicrobiome interactions for sustainable agronomy. Microorganisms 10(5):899. https://doi.org/10.3390/microorganisms10050899

Jansson JK, Hofmockel KS (2018) The soil microbiome—from metagenomics to metaphenomics. Curr Op Microbiol 43:162–168. https://doi.org/10.1016/j.mib.2018.01.013

Jansson JK, Hofmockel KS (2018) The soil microbiome-from metagenomics to metaphenomics. Curr Opmicrobiol 43:162–168. https://doi.org/10.1016/j.mib.2018.01.013

Jefferson OA, Köllhofer D, Ehrich TH, Jefferson RA (2015) The ownership question of plant gene and genome intellectual properties. Nat Biotechnol 33(11):1138–1143. https://doi.org/10.1038/nbt.3393

Jiang F, Doudna JA (2017) CRISPR-Cas9 structures and mechanisms. Annu Rev Biophys 46:505–529. https://doi.org/10.1146/annurev-biophys-062215-010822

Jinek M, Chylinski K, Fonfara I, Hauer M, Doudna JA, Charpentier E (2012) A programmable dual-RNA-guided DNA endonuclease in adaptive bacterial immunity. Science 337(6096):816–821. https://doi.org/10.1126/science.1225829

John DA, Babu GR (2021) Lessons from the aftermaths of green revolution on food system and health. Front Sustafood Syst 5:644559. https://doi.org/10.3389/fsufs.2021.644559

John RP, Tyagi RD, Brar SK, Surampalli RY, Prévost D (2011) Bio-encapsulation of microbial cells for targeted agricultural delivery. Crit Rev Biotechnol 31(3):211–226. https://doi.org/10.3109/07388551.2010.513327

Ju XT, Kou CL, Christie P, Dou ZX, Zhang FS (2007) Changes in the soil environment from excessive application of fertilizers and manures to two contrasting intensive cropping systems on the North China Plain. Environ Pollut 145(2):497–506. https://doi.org/10.1016/j.envpol.2006.04.017

Kalogiannidis S, Kalfas D, Chatzitheodoridis F, Papaevangelou O (2022) Role of crop-protection technologies in sustainable agricultural productivity and management. Land 11(10):1680

Kamilaris A, Kartakoullis A, Prenafeta-Boldú FX (2017) A review on the practice of big data analysis in agriculture. Comput Electron Agric 143:23–37. https://doi.org/10.1016/j.compag.2017.09.037

Kaminsky LM, Trexler RV, Malik RJ, Hockett KL, Bell TH (2019) The inherent conflicts in developing soil microbial inoculants. Trends Biotechnol 37(2):140–151. https://doi.org/10.1016/j.tibtech.2018.11.011

Karalis DT, Karalis T, Karalis S, Kleisiari AS (2020) Genetically modified products, perspectives and challenges. Cureus 12(3):e7306. https://doi.org/10.7759/cureus.7306

Kaschuk G, Alberton O, Hungria M (2010) Three decades of soil microbial biomass studies in Brazilian ecosystems: lessons learned about soil quality and indications for improving sustainability. Soil Biol Biochem 42(1):1–13. https://doi.org/10.1016/j.soilbio.2009.08.020

Kawall K (2019) New possibilities on the horizon: genome editing makes the whole genome accessible for changes. Front Plant Sci 10:525. https://doi.org/10.3389/fpls.2019.00525

Koza NA, Adedayo AA, Babalola OO, Kappo AP (2022) Microorganisms in plant growth and development: roles in abiotic stress tolerance and secondary metabolites secretion. Microorganisms 10(8):1528. https://doi.org/10.3390/microorganisms10081528

Kukurba KR, Montgomery SB (2015) RNA sequencing and analysis. Cold Spring Harb Protoc 2015(11):951–969. https://doi.org/10.1101/pdb.top084970

Kuntz M (2012) The postmodern assault on science. EMBO Rep 13(10):885–889. https://doi.org/10.1038/embor.2012.130

Lacey LA, Grzywacz D, Shapiro-Ilan DI, Frutos R, Brownbridge M, Goettel MS (2015) Insect pathogens as biological control agents: back to the future. J Invertebrate Pathol 132:1–41. https://doi.org/10.1016/j.jip.2015.07.009

Lal R (2020) Regenerative agriculture for food and climate. J Soil Water Conserv 75(5):123A–124A. https://doi.org/10.2489/jswc.2020.0620A

Lal R, Monger C, Nave L, Smith P (2021) The role of soil in regulation of climate. Philos Trans R Soc Lond Ser B Biol Sci 376(1834):20210084. https://doi.org/10.1098/rstb.2021.0084

Larson PD (2021) Relationships between logistics performance and aspects of sustainability: a cross-country analysis. Sustainability 13(2):623. https://doi.org/10.3390/su13020623

Lensch A, Lindfors HA, Duwenig E, Fleischmann T, Hjort C, Kärenlampi SO, van Kranenburg R (2024) Safety aspects of microorganisms deliberately released into the environment. EFB Bioecon J 4:100061

Liakos KG, Busato P, Moshou D, Pearson S, Bochtis D (2018) Machine learning in agriculture: a review. Sensors 18(8):2674. https://doi.org/10.3390/s18082674

Libbrecht MW, Noble WS (2015) Machine learning applications in genetics and genomics. Nat Rev Genet 16(6):321–332. https://doi.org/10.1038/nrg3920

Liu H, Carvalhais LC, Crawford M, Singh E, Dennis PG, Pieterse CMJ, Schenk PM (2017) Inner plant values: diversity, colonization and benefits from endophytic bacteria. Front Microbiol 8:2552. https://doi.org/10.3389/fmicb.2017.02552

Liu W, Stewart CN (2015) Plant synthetic biology. Trends Plant Sci 20(5):309–317. https://doi.org/10.1016/j.tplants.2015.02.004

Liu XJ, Zhang Y, Han WX, Tang AH, Shen JL, Cui ZL, Zhang FS (2013) Enhanced nitrogen deposition over China. Nature 494(7438):459–462. https://doi.org/10.1038/nature11917

Louwaars NP, Dons H, Van Overwalle GRL, Raven H, Arundel A, Eaton D, Nelis A (2009) Breeding business: the future of plant breeding in the light of developments in patent rights and plant breeder's rights (CGN Report No. 14). Wageningen UR Centre for Genetic Resources

Lovett B, StLeger RJ (2017) The insect pathogens: entomopathogenic fungi and oomycetes. In: Heitman J, Howlett BJ, Crous PW, Stukenbrock EH, James TY, Gow NAR (eds) The Fungal Kingdom. ASM Press, 923–943. https://doi.org/10.1128/9781555819583.ch45

Mahlein AK (2016) Plant disease detection by imaging sensors—parallels and specific demands for precision agriculture. Plant Dis 100(2):241–251. https://doi.org/10.1094/PDIS-03-15-0340-FE
Malusá E, Vassilev N (2014) A contribution to set a legal framework for biofertilisers. Appl Microbiol Biotechnol 98(15):6599–6607. https://doi.org/10.1007/s00253-014-5828-y
Malusá E, Pinzari F, Canfora L (2016) Efficacy of biofertilizers: challenges to improve crop production. In: Singh DP, Singh HB, Prabha R (eds) Microbial inoculants in sustainable agricultural productivity. Springer, India, 17–40. https://doi.org/10.1007/978-81-322-2644-4_2
Malusà E, Pinzari F, Canfora L (2016) Efficacy of biofertilizers: challenges to improve crop production. In: Microbial inoculants in sustainable agricultural productivity: vol. 2: functional applications. New Delhi, pp 17–40
Manczinger L, Antal Z, Kredics L (2002) Ecophysiology and breeding of mycoparasitic Trichoderma strains. Acta Microbiol Immunol Hungarica 49(1):1–14. https://doi.org/10.1556/AMicr.49.2002.1.1
Marris C (2015) The construction of imaginaries of the public as a threat to synthetic biology. Sci Cult 24(1):83–98. https://doi.org/10.1080/09505431.2014.986320
Martinez-Garcia E, de Lorenzo V (2017) Molecular tools and emerging strategies for deep genetic/genomic refactoring of Pseudomonas. Curr OpBiotechnol 47:120–132. https://doi.org/10.1016/j.copbio.2017.06.013
Matson PA, Parton WJ, Power AG, Swift MJ (1997) Agricultural intensification and ecosystem properties. Science 277(5325):504–509. https://doi.org/10.1126/science.277.5325.504
Meena VS, Maurya BR, Verma JP (2014) Does a rhizospheric microorganism enhance K^+ availability in agricultural soils?. Microbiol Res 169(5–6):337–347. https://doi.org/10.1016/j.micres.2013.09.003
Mendes R, Garbeva P, Raaijmakers JM (2013) The rhizosphere microbiome: significance of plant beneficial, plant pathogenic, and human pathogenic microorganisms. FEMS Microbiol Rev 37(5):634–663. https://doi.org/10.1111/1574-6976.12028
Mirsalami SM, Mirsalami M (2025) Advances in genetically engineered microorganisms: Transforming food production through precision fermentation and synthetic biology. Future Foods 11:100601
Mishra J, Tewari S, Singh S, Arora NK (2014) Biopesticides: where we stand. Front Plant Sci 12:664020. https://doi.org/10.3389/fpls.2021.664020
Mishra J, Tewari S, Arora NK (2015) Biopesticides: where we stand? In: Plant-microbe interaction: an approach to sustainable agriculture. Springer, Cham, pp 37–75. https://doi.org/10.1007/978-81-322-2068-8_2
Moscardi F (1999) Assessment of the application of baculoviruses for control of Lepidoptera. Annu Rev Entomology 44(1):257–289. https://doi.org/10.1146/annurev.ento.44.1.257
Mukherjee PK, Horwitz BA, Herrera-Estrella A, Schmoll M, Kenerley CM (2013) Trichoderma research in the genome era. Annu Rev Phytopathol 51(1):105–129. https://doi.org/10.1146/annurev-phyto-082712-102353
Mus F, Crook MB, Garcia K, Garcia Costas A, Geddes BA, Kouri ED et al (2016) Symbiotic nitrogen fixation and the challenges to its extension to non-legumes. Appl Environ Microbiol 82(13):3698–3710. https://doi.org/10.1128/AEM.01055-16
Nadeem SM, Ahmad M, Zahir ZA, Javaid A, Ashraf M (2014) The role of mycorrhizae and plant growth promoting rhizobacteria (PGPR) in improving crop productivity under stressful environments. Biotechnol Adv 32(2):429–448. https://doi.org/10.1016/j.biotechadv.2013.12.005
Nap JP, Metz PL, Escaler M, Conner AJ (2003) The release of genetically modified crops into the environment. Part I: overview of current status and regulations. Plant J 33(1):1–18. https://doi.org/10.1046/j.0960-7412.2002.001607.x
Nguyen NH, Song Z, Bates ST, Branco S, Tedersoo L, Menke J, Kennedy PG (2016) FUNGuild: an open annotation tool for parsing fungal community datasets by ecological guild. Fungal Ecol 20:241–248. https://doi.org/10.1016/j.funeco.2015.06.006
Niu B, Paulson JN, Zheng X, Kolter R (2016) Simplified and representative bacterial community of maize roots. Proc Natl Acad Sci USA 114(12):E2450–E2459. https://doi.org/10.1073/pnas.1616148114

Nuruzzaman M, Bahar MM, Naidu R (2025) Diffuse soil pollution from agriculture: Impacts and remediation. Sci Total Environ 962:178398

O'Callaghan M, Ballard RA, Wright D (2022) Soil microbial inoculants for sustainable agriculture: limitations and opportunities. Soil Use Manag 38(3):1340–1369

Ofek-Lalzar M, Sela N, Goldman-Voronov M, Green SJ, Hadar Y, Minz D (2014) Niche and host-associated functional signatures of the root surface microbiome. Nat Commun 5: 4950. https://doi.org/10.1038/ncomms5950

Oldroyd GED (2013) Speak, friend, and enter: signalling systems that promote beneficial symbiotic associations in plants. Nat Rev Microbiol 11(4):252–263. https://doi.org/10.1038/nrmicro2990

Oliveira ALM, Canuto EL, Reis VM, Baldani JI (2003) Response of micropropagated sugarcane varieties to inoculation with endophytic diazotrophic bacteria. Braz J Microbiol 34:59–61. https://doi.org/10.1590/S1517-83822003000500020

Ongena M, Jacques P (2008) Bacillus lipopeptides: Versatile weapons for plant disease biocontrol. Trends Microbiol 16(3):115–125. https://doi.org/10.1016/j.tim.2007.12.009

Orth JD, Thiele I, Palsson BØ (2010) What is flux balance analysis. Nat Biotechnol 28(3):245–248. https://doi.org/10.1038/nbt.1614

Pandiyan A, Sarsan S, Durga GGS, Ravikumar H (2024) Biofertilizers and biopesticides as microbial inoculants in integrated pest management for sustainable agriculture. In: Microbial essentialism. Academic, pp 485–518

Pankievicz VCS, Irving TB, Maia LGS, Ané J-M (2019) Are we there yet? The long walk towards the development of efficient symbiotic associations between nitrogen-fixing bacteria and non-leguminous crops. BMC Biol 17: 99. https://doi.org/10.1186/s12915-019-0710-0

Patti GJ, Yanes O, Siuzdak G (2012) Innovation: metabolomics: the apogee of the omics trilogy. Nat Rev Mol Cell Biol 13(4):263–269. https://doi.org/10.1038/nrm3314

Peiffer JA, Spor A, Koren O, Jin Z, Tringe SG, Dangl JL, Buckler ES, Ley RE (2013) Diversity and heritability of the maize rhizosphere microbiome under field conditions. Proc Natl Acad Sci USA 110(16):6548–6553. https://doi.org/10.1073/pnas.1302837110

Philippot L, Raaijmakers JM, Lemanceau P, van der Putten WH (2013) Going back to the roots: the microbial ecology of the rhizosphere. Nat Rev Microbiol 11(11):789–799. https://doi.org/10.1038/nrmicro3109

Pickar-Oliver A, Gersbach CA (2019) The next generation of CRISPR-Cas technologies and applications. Nat Rev Mol Cell Biol 20(8):490–507. https://doi.org/10.1038/s41580-019-0131-5

Pieterse CMJ, Zamioudis C, Berendsen RL, Weller DM, Van Wees SCM, Bakker PAHM (2014) Induced systemic resistance by beneficial microbes. Annu Rev Phytopathol 52:347–375. https://doi.org/10.1146/annurev-phyto-082712-102340

Pingali PL (2012) Green revolution: impacts, limits, and the path ahead. Proc Natl Acad Sci USA 109(31):12302–12308. https://doi.org/10.1073/pnas.0912953109

Pinu FR, Beale DJ, Paten AM, Kouremenos KA, Swarup S, Schirra HJ, Wishart D (2019) Systems biology and multi-omics integration: viewpoints from the metabolomics research community. Metabolites 9(4):76. https://doi.org/10.3390/metabo9040076

Prasanna R, Jaiswal P, Kaushik BD (2008) Cyanobacteria as potential options for environmental sustainability—promises and challenges. Indian J Microbiol 48(1):89–94. https://doi.org/10.1007/s12088-008-0009-2

Pretty J (2008) Agricultural sustainability: concepts, principles and evidence. Philos Trans R Soc B 363(1491):447–465. https://doi.org/10.1098/rstb.2007.2163

Pretty J, Bharucha ZP (2014) Sustainable intensification in agricultural systems. Ann Botany 114(8):1571–1596. https://doi.org/10.1093/aob/mcu205

Pretty J, Benton TG, Bharucha ZP, Dicks LV, Flora CB, Godfray HCJ et al (2018) Global assessment of agricultural system redesign for sustainable intensification. Nat Sustain 1(8):441–446. https://doi.org/10.1038/s41893-018-0114-0

Rajguru B, Shri M, Bhatt B, Bhatt VD (2025) Comparative metagenomic analysis of rhizospheric soil of *Mangifera indica* (mango) and *Carica papaya* (papaya). Discov Plants 2(1):1–16

Raklami A, Bechtaoui N, Tahiri AI, Anli M, Meddich A, Oufdou K (2019) Use of rhizobacteria and mycorrhizae consortium in the open field as a strategy for improving crop nutrition, productivity, and soil fertility. Front Microbiol 10:1106. https://doi.org/10.3389/fmicb.2019.01106

Ray P, Lakshmanan V, Labbé J, Craven KD (2020) Microbe to microbiome: a paradigm shift in the application of microorganisms for sustainable agriculture. Front Microbiol 11:622926. https://doi.org/10.3389/fmicb.2020.622926

Raybould A, Poppy GM (2012) Commercializing genetically modified crops under EU regulations: objectives and barriers. GM Crop Food 3(1):9–20. https://doi.org/10.4161/gmcr.18961

Renda BA, Hammerling MJ, Barrick JE (2014) Engineering reduced evolutionary potential for synthetic biology. Mol Biosyst 10(7):1668–1678. https://doi.org/10.1039/C3MB70606K

Richardson AE, Simpson RJ (2011) Soil microorganisms mediating phosphorus availability. Plant Physiol 156(3):989–996. https://doi.org/10.1104/pp.111.175448

Rillig MC (2004) Arbuscular mycorrhizae, glomalin, and soil aggregation. Can J Soil Sci 84(4):355–363. https://doi.org/10.4141/S04-003

Rillig MC, Aguilar-Trigueros CA, Bergmann J, Verbruggen E, Veresoglou SD, Lehmann A (2015) Plant root and mycorrhizal fungal traits for understanding soil aggregation. N Phytol 205(4):1385–1388. https://doi.org/10.1111/nph.13045

Rockström J, Steffen W, Noone K, Persson Å, Chapin FS, Lambin EF, Foley JA (2009) A safe operating space for humanity. Nature 461(7263):472–475. https://doi.org/10.1038/461472a

Rockström J, Williams J, Daily G, Noble A, Matthews N, Gordon L, Smith J (2017) Sustainable intensification of agriculture for human prosperity and global sustainability. Ambio 46(1):4–17. https://doi.org/10.1007/s13280-016-0793-6

Rodríguez H, Fraga R (1999) Phosphate-solubilizing bacteria and their role in plant growth promotion. Biotechnol Adv 17(4-5):319–339. https://doi.org/10.1016/S0734-9750(99)00014-2

Rogers C, Oldroyd GED (2014) Synthetic biology approaches to engineering the nitrogen symbiosis in cereals. J Exp Botany 65(8):1939–1946. https://doi.org/10.1093/jxb/eru098

Rothschild LJ, Mancinelli RL (2001) Life in extreme environments. Nature 409(6823):1092–1101. https://doi.org/10.1038/35059215

Rouphael Y, Colla G (2018) Synergistic biostimulatory action: designing the next generation of plant biostimulants for sustainable agriculture. Front Plant Sci 9:1655. https://doi.org/10.3389/fpls.2018.01655

Rouphael Y, Colla G (2020) Editorial: biostimulants in agriculture. Front Plant Sci 11:40. https://doi.org/10.3389/fpls.2020.00040

Saha M, Sarkar S, Sarkar B, Sharma BK, Bhattacharjee S, Tribedi P (2016) Microbial siderophores and their potential applications: a review. Environ Sci Pollut Res 23(5):3984–3999. https://doi.org/10.1007/s11356-015-4294-0

Saleem M, Arshad M, Hussain S, Bhatti AS (2007) Perspective of plant growth promoting rhizobacteria (PGPR) containing ACC deaminase in stress agriculture. J Ind Microbiol Biotechnol 34(10):635–648. https://doi.org/10.1007/s10295-007-0240-6

Samberg LH, Gerber JS, Ramankutty N, Herrero M, West PC (2016) Subnational distribution of average farm size and smallholder contributions to global food production. Environ Res Lett 11(12):124010. https://doi.org/10.1088/1748-9326/11/12/124010

Sameen S, Khalil MB (2025) Integrating biotechnology into plant biology: challenges and future directions. J Plant Biol 3:100014

Sanahuja G, Banakar R, Twyman RM, Capell T, Christou P (2011) Bacillus thuringiensis: a century of research, development and commercial applications. Plant Biotechnol J 9(3):283–300. https://doi.org/10.1111/j.1467-7652.2011.00595.x

Santos DKF, Rufino RD, Luna JM, Santos VA, Sarubbo LA (2016) Biosurfactants: multifunctional biomolecules of the 21st century. Int J Mol Sci 17(3):401. https://doi.org/10.3390/ijms17030401

Santoyo G, Moreno-Hagelsieb G, Orozco-Mosqueda MDC, Glick BR (2016) Plant growth-promoting bacterial endophytes. Microbiol Res 183:92–99. https://doi.org/10.1016/j.micres.2015.11.008

Sasse J, Martinoia E, Northen T (2018) Feed your friends: do plant exudates shape the root microbiome. Trends Plant Sci 23(1):25–41. https://doi.org/10.1016/j.tplants.2017.09.003

Schirawski J, Perlin MH (2018) Plant–microbe interaction 2017—the good, the bad and the diverse. Int J Mol Sci 19(5):1374. https://doi.org/10.3390/ijms19051374

Schloter M, Nannipieri P, Sørensen SJ, van Elsas JD (2018) Microbial indicators for soil quality. Biol Fertil Soils 54(1):1–10. https://doi.org/10.1007/s00374-017-1248-3
Schmidt R, Ulanova D, Wick LY, Bode HB, Garbeva P (2019) Microbe-driven chemical ecology: past, present and future. ISME J 13(11):2656–2663. https://doi.org/10.1038/s41396-019-0469-x
Schmidt M, Kelle A, Ganguli-Mitra A, de Vriend H (2009) Synthetic biology: the technoscience and its societal consequences. Springer. https://doi.org/10.1007/978-90-481-2678-1
Scholz MB, Lo CC, Chain PS (2012) Next generation sequencing and bioinformatic bottlenecks: the current state of metagenomic data analysis. Curr OpBiotechnol 23(1):9–15. https://doi.org/10.1016/j.copbio.2011.11.013
Schulz-Bohm K, Martín-Sánchez L, Garbeva P (2017) Microbial volatiles: small molecules with an important role in intra- and inter-kingdom interactions. Front Microbiol 8:2484. https://doi.org/10.3389/fmicb.2017.02484
Schütz L, Gattinger A, Meier M, Müller A, Boller T, Mäder P, Mathimaran N (2018) Improving crop yield and nutrient use efficiency via biofertilization—a global meta-analysis. Front Plant Sci 8:2204. https://doi.org/10.3389/fpls.2017.02204
Scoones I, Stirling A, Abrol D, Atela J, Charli-Joseph L, Eakin H, Yang L (2020) Transformations to sustainability: combining structural, systemic and enabling approaches. Curr OpEnviron Sustain 42:65–75. https://doi.org/10.1016/j.cosust.2019.12.004
Sessitsch A, Mitter B (2015) 21st century agriculture: integration of plant microbiomes for improved crop production and food security. Microb Biotechnol 8(1):32–39. https://doi.org/10.1111/1751-7915.12180
Sessitsch A, Pfaffenbichler N, Mitter B (2019) Microbiome applications from lab to field: facing complexity. Trends Plant Sci 24(3):194–198. https://doi.org/10.1016/j.tplants.2018.12.004
Shah PA, Pell JK (2003) Entomopathogenic fungi as biological control agents. Appl Microbiol Biotechnol 61(5–6):413–423. https://doi.org/10.1007/s00253-003-1240-8
Sharma A, Shukla A, Attri K, Kumar M, Kumar P, Suttee A, Varma A (2020) Global trends in pesticides: a looming threat and viable alternatives. Ecotoxicol Environ Saf 201:110812. https://doi.org/10.1016/j.ecoenv.2020.110812
Sharma M, Sudheer S, Usmani Z, Rani R, Gupta P (2020) Deciphering the Omics of Plant-Microbe Interaction: perspectives and new insights. Curr Genomics 21(5):343–362. https://doi.org/10.2174/1389202921999200515140420
Sharma SB, Sayyed RZ, Trivedi MH, Gobi TA (2013) Phosphate-solubilizing microbes: sustainable approach for managing phosphorus deficiency in agricultural soils. SpringerPlus 2(1):587. https://doi.org/10.1186/2193-1801-2-587
Shokralla S, Spall JL, Gibson JF, Hajibabaei M (2012) Next-generation sequencing technologies for environmental DNA research. Mol Ecol 21(8):1794–1805. https://doi.org/10.1111/j.1365-294X.2012.05538.x
Singh A, Mazahar S, Chapadgaonkar SS, Giri P, Shourie A (2023) Phyto-microbiome to mitigate abiotic stress in crop plants. Front Microbiol 14:1210890. https://doi.org/10.3389/fmicb.2023.1210890
Singh BK, Trivedi P (2017) Microbiome and the future for food and nutrient security. Microb Biotechnol 10(1):50–53. https://doi.org/10.1111/1751-7915.12592
Singh BK, Trivedi P, Egidi E, Macdonald CA, Delgado-Baquerizo M (2020) Crop microbiome and sustainable agriculture. Nat Rev Microbiol 18(11):601–602. https://doi.org/10.1038/s41579-020-00446-y
Singh JS, Pandey VC, Singh DP (2011) Efficient soil microorganisms: a new dimension for sustainable agriculture and environmental development. Agric Ecosyst Environ 140(3–4):339–353. https://doi.org/10.1016/j.agee.2011.01.017
Singh M (2024) Pesticide predicament: exploring environmental and health impacts, and possible eco-friendly solutions. Int J Environ Health Sci 6:38–47
Smid EJ, Lacroix C (2013) Microbe–microbe interactions in mixed culture food fermentations. Curr OpBiotechnol 24(2):148–154. https://doi.org/10.1016/j.copbio.2012.11.007
Smith SE, Read DJ (2008). Mycorrhizal symbiosis, 3rd ed. Academic
Smith SE, Read DJ (2010). Mycorrhizal symbiosis, 3rd ed. Academic

Stilgoe J, Owen R, Macnaghten P (2013) Developing a framework for responsible innovation. Res Policy 42(9):1568–1580. https://doi.org/10.1016/j.respol.2013.05.008

Sudheer S, Bai RG, Usmani Z, Sharma M (2020) Insights on engineered microbes in sustainable agriculture: biotechnological developments and future prospects. Curr Genomics 21(5):321–333. https://doi.org/10.2174/1389202921999200603165934

Sun W, Shahrajabian MH, Soleymani A (2024) The roles of plant-growth-promoting rhizobacteria (PGPR)-based biostimulants for agricultural production systems. Plants (Basel, Switz) 13(5):613. https://doi.org/10.3390/plants13050613

Temme K, Zhao D, Voigt CA (2012) Refactoring the nitrogen fixation gene cluster from Klebsiella oxytoca. Proc Natl Acad Sci USA 109(18):7085–7090. https://doi.org/10.1073/pnas.1120788109

Thorburn PJ, Biggs JS (2010) Opportunities for and potential consequences of reducing nitrous oxide emissions from sugarcane crops. Proceedings of the International Society of Sugar Cane Technologists (ISSCT), 27th Congress

Tilman D, Balzer C, Hill J, Befort BL (2011) Global food demand and the sustainable intensification of agriculture. Proc Natl Acad Sci USA 108(50):20260–20264. https://doi.org/10.1073/pnas.1116437108

Tilman D, Cassman KG, Matson PA, Naylor R, Polasky S (2002) Agricultural sustainability and intensive production practices. Nature 418(6898):671–677. https://doi.org/10.1038/nature01014

Timmis K, de Vos WM, Ramos JL, Vlaeminck SE, Prieto A, Danchin A, Verstraete W, de Lorenzo V, Lee SY, Brüssow H, Timmis JK, Singh BK (2017) The contribution of microbial biotechnology to sustainable development goals. Microb Biotechnol 10(5):984–987. https://doi.org/10.1111/1751-7915.12818

Timmusk S, Behers L, Muthoni J, Muraya A, Aronsson A-C (2017) Perspectives and challenges of microbial application for crop improvement. Front Plant Sci 8:49. https://doi.org/10.3389/fpls.2017.00049

Tkacz A, Cheema J, Chandra G, Grant A, Poole PS (2015) Stability and succession of the rhizosphere microbiota depends upon plant type and soil composition. ISME J 9(11):2349–2359. https://doi.org/10.1038/ismej.2015.41

Toju H, Peay KG, Yamamichi M, Narisawa K, Hiruma K, Naito K, Sato H (2018) Core microbiomes for sustainable agroecosystems. Nat Plants 4(5):247–257. https://doi.org/10.1038/s41477-018-0139-4

Trivedi P, Leach JE, Tringe SG, Sa T, Singh BK (2020) Plant–microbiome interactions: from community assembly to plant health. Nat Rev Microbiol 18(11):607–621. https://doi.org/10.1038/s41579-020-0412-1

Trump BD (2017) Synthetic biology regulation and governance: lessons from TAPIC for the United States, European Union, and Singapore. Health Policy 121(11):1139–1146. https://doi.org/10.1016/j.healthpol.2017.07.010

Ullah F, Ali S, Siraj M, Akhtar MS, Zaman W (2025) Plant microbiomes alleviate abiotic stress-associated damage in crops and enhance climate-resilient agriculture. Plants 14(12):1890

United Nations (2015) Transforming our world: the 2030 Agenda for Sustainable Development. https://sdgs.un.org/2030agenda

USDA-APHIS (2020) Biotechnology regulatory services. https://www.aphis.usda.gov/aphis/ourfocus/biotechnology

Vacheron J, Desbrosses G, Bouffaud ML, Touraine B, Moënne-Loccoz Y, Muller D et al (2013) Plant growth-promoting rhizobacteria and root system functioning. Front Plant Sci 4:356. https://doi.org/10.3389/fpls.2013.00356

Vanlauwe B, Hungria M, Kanampiu F, Giller KE (2019) The role of legumes in the sustainable intensification of African smallholder agriculture: lessons learned and challenges for the future. Agric Ecosyst Environ 284:106583. https://doi.org/10.1016/j.agee.2019.106583

van der Heijden MG, Martin FM, Selosse MA, Sanders IR (2015) Mycorrhizal ecology and evolution: the past, the present, and the future. N Phytol 205(4):1406–1423. https://doi.org/10.1111/nph.13288

van der Heijden MGA, Hartmann M (2016) Networking in the plant microbiome. PLoS Biol 14(2):e1002378. https://doi.org/10.1371/journal.pbio.1002378
van der Heijden MGA, Bardgett RD, van Straalen NM (2008) The unseen majority: soil microbes as drivers of plant diversity and productivity in terrestrial ecosystems. Ecol Lett 11(3):296–310. https://doi.org/10.1111/j.1461-0248.2007.01139.xVerbon EH, Liberman LM (2016) Beneficial microbes affect endogenous mechanisms controlling root development. Trends Plant Sci 21(3):218–229. https://doi.org/10.1016/j.tplants.2016.01.013
Verma JP, Gupta VK, Macdonald C, Podile AR (2020) New and future developments in microbial biotechnology and bioengineering. Elsevier, Amsterdam, The Netherlands, pp 19–30
Vessey JK (2003) Plant growth promoting rhizobacteria as biofertilizers. Plant Soil 255(2):571–586. https://doi.org/10.1023/A:1026037216893
Wang HH, Kim H, Cong L, Jeong J, Bang D, Church GM (2012) Genome-scale promoter engineering by coselection MAGE. Nat Methods 9(6):591–593. https://doi.org/10.1038/nmeth.1971
Wani SP, Rupela OP, Lee KK (1995) Sustainable agriculture in the semi-arid tropics through biological nitrogen fixation in grain legumes. Plant Soil 174(1–2):29–49
Way JC, Collins JJ, Keasling JD, Silver PA (2014) Integrating biological redesign: where synthetic biology came from and where it needs to go. Cell 157(1):151–161. https://doi.org/10.1016/j.cell.2014.02.039
Wend K, Zorrilla L, Freimoser FM, Gallet A (2024) Microbial pesticides: challenges and future perspectives for testing and safety assessment with respect to human health. Environ Health 23(1):49. https://doi.org/10.1186/s12940-024-01090-2
Wezel A, Casagrande M, Celette F, Vian JF, Ferrer A, Peigné J (2014) Agroecological practices for sustainable agriculture: a review. Agron Sustain Dev 34(1):1–20. https://doi.org/10.1007/s13593-013-0180-7
Wheeler T, von Braun J (2013) Climate change impacts on global food security. Science 341(6145):508–513. https://doi.org/10.1126/science.1239402
Wijayawardene NN, Boonyuen N, Ranaweera CB, de Zoysa HK, Padmathilake RE, Nifla F et al (2023) OMICS and other advanced technologies in mycological applications. J Fungi 9(6):688
Wilmes P, Heintz-Buschart A, Bond PL (2015) A decade of metaproteomics: where we stand and what the future holds. Proteomics 15(20):3409–3417. https://doi.org/10.1002/pmic.201500183
Woo SL, Ruocco M, Vinale F, Nigro M, Marra R, Lombardi N et al (2014) Trichoderma-based products and their widespread use in agriculture. Open Mycol J 8:71–126. https://doi.org/10.2174/1874437001408010071
Wright O, Stan GB, Ellis T (2013) Building-in biosafety for synthetic biology. Microbiology 159(7):1221–1235. https://doi.org/10.1099/mic.0.066308-0
Yadav AN, Singh J, Singh, C, Yadav N (2021) Current trends in microbial biotechnology for sustainable agriculture. Springer, Singapore
Zamioudis C, Mastranesti P, Dhonukshe P, Blilou I, Pieterse CMJ (2013) Unraveling root developmental programs initiated by beneficial Pseudomonas spp. bacteria. Plant Physiol 162(1):304–318. https://doi.org/10.1104/pp.112.212597
Zhalnina K, Louie KB, Hao Z, Mansoori N, da Rocha UN, Shi S et al (2018) Dynamic root exudate chemistry and microbial substrate preferences drive patterns in rhizosphere microbial community assembly. Nat Microbiol 3(4):470–480. https://doi.org/10.1038/s41564-018-0129-3
Zhang F, Chen X, Vitousek P (2013) An experiment for the world. Nature 497(7447):33–35. https://doi.org/10.1038/497033a
Zhang J, Liu YX, Zhang N, Hu B, Jin T, Xu H et al (2019) NRT1.1B is associated with root microbiota composition and nitrogen use in rice. Nat Biotechnol 37(6):676–684. https://doi.org/10.1038/s41587-019-0104-4
Zhang W, Dou Z, He P, Ju XT, Powlson D, Chadwick D, Zhang F (2013) New technologies reduce greenhouse gas emissions from nitrogenous fertilizer in China. Proc Natl Acad Sci USA 110(21):8375–8380. https://doi.org/10.1073/pnas.1210447110

Chapter 9
Microbiota and Microbiome

Ozgenur Hacioglu

Abstract The human microbiota represents a complex and dynamic ecosystem composed of microorganisms from various taxonomic groups, including bacteria, viruses, fungi, archaea, and protozoa. These microorganisms inhabit different anatomical regions of the human body, such as the genitourinary system, the gastrointestinal tract, the oral cavity, the skin, and the respiratory tract, exhibiting distinct densities, compositions, and functional characteristics, and interact reciprocally with the host organism. The term microbiota not only defines the diversity and abundance of microorganisms but also encompasses their functional influence on host physiology. At this point, the concept of the microbiome becomes relevant. The microbiome refers to the collective genomic content of all microorganisms comprising the microbiota, that is, their genetic material and the potential biological functions encoded by their genes. Therefore, microbiome analysis enables not only the assessment of microbial diversity, but also of metabolic capacity, signal transduction, immune regulation, and other host–microbe interactions. The microbiota and microbiome play important roles in preserving human health and homeostatic balance. A healthy microbial composition promotes immune system development, aids digestion and nutrient absorption, reduces pathogenic microorganism colonization, and contributes to the integrity of the mucosal barrier. In contrast, dysbiosis, or disruption of microbial equilibrium, has been linked to a variety of pathophysiological illnesses, including inflammatory diseases, metabolic disorders, neurodegenerative diseases, and some neoplasms.

O. Hacioglu (✉)
Kirklareli University, Faculty of Health Sciences, Kirklareli, Türkiye
e-mail: ozgenuryilmaz@klu.edu.tr

M. Çalışkan and M. Arıkan (eds.), *Microbial World: Bacteria and Archaea*, Progress in Molecular and Subcellular Biology 62, https://doi.org/10.1007/978-3-032-11438-9_9

Today, microbiome research is not only essential for understanding health and disease mechanisms but also forms the foundation for innovative future medical applications.

Keywords Human microbiota · Human microbiome · Microbial diversity · Dysbiosis · Host–microbe interaction · Microbiome-based therapeutics

9.1 Introduction

Microbiota refers to the ecosystem composed of microorganisms such as bacteria, viruses, fungi, and protozoa, whereas microbiome is the term that encompasses the genetic material of these microorganisms and their interactions with the environment (Ma et al. 2024). The concept of microbiota broadly refers to the community of microorganisms associated with any living organism (plants, animals, and humans) or a specific environmental niche (soil, water, air, marine ecosystems, etc.). Microbiota is not confined to the human body; from an ecological perspective, it encompasses various examples such as soil, marine, plant, and animal microbiota (Abdi et al. 2024; Kang et al. 2024). Accordingly, microbiota constitutes a multifaceted phenomenon that requires examination at both biological and ecological levels. This section predominantly addresses the human microbiota and microbiome.

The human body is today recognized as an ecosystem inhabited by trillions of microbes, rather than a collection of only human cells. The human microbiota is a complex ecosystem made up of microorganisms such as bacteria, viruses, fungi, archaea, and protozoa (Hou et al. 2022). The microbiota colonizes various anatomical regions such as the gut, skin, oral cavity, respiratory tract, and urogenital system, each hosting its own unique community of microbes. These microbial communities differ significantly in terms of density, diversity, and function, making each niche a universe waiting to be discovered. What was once referred to as "normal flora" is now more accurately and inclusively described by the term "microbiota" (Afzaal et al. 2022).

Although molecular techniques are increasingly allowing us to identify and understand the functions of these microbial members, fully grasping the broader picture is still a formidable challenge. A look back at the historical trajectory of microbiome research shows that what began with the discovery of microorganisms has now evolved into an in-depth exploration of disease etiology (Kim et al. 2023; El Mazouri et al. 2024).

Recent studies involving microbiome analyses have shown the distribution of microorganisms resident in most human bodies according to their locations (Hou et al. 2022; Kang et al. 2024). However, studies clearly show that determining the composition maps of these members will be possible with each study conducted. Therefore, there are many more microorganisms to be discovered. Studies report that the members of the microbiota must be in balance, and that if the balance is disrupted, it may be linked to the pathophysiological conditions of various diseases. Various diseases, especially inflammatory diseases, metabolic disorders, neurological degenerations and some cancer processes, have been associated with the number and composition of these microorganisms in the microbiota (Kim et al. 2023; El Mazouri et al. 2024; Kang et al. 2024). Therefore, today's microbiome research has become a pioneering field of research that not only explains the biological mechanisms underlying healthy and diseased conditions but also shapes individual-specific treatment approaches and innovative medical practices (Cryan et al. 2020; Di Domenico et al. 2022; Abdi et al. 2024; Garcia et al. 2024).

This section deals with the microbiota and microbiome, which we have heard frequently in recent years in the etiology of most diseases due to dysbiosis of almost all microbial members. First, the concepts of human microbiota and microbiome and their historical development are discussed in detail. The following three sections address the microbial ecosystems in the human body, microbiota diversity and the factors affecting this diversity. The fifth section discusses the effects of microbiota on health by giving examples of various diseases. The sixth section, the conclusion, summarizes new approaches in microbiota and microbiome research and draws attention to current deficiencies.

9.2 Microbiota: The Invisible Regulator of Health

9.2.1 Definitions of the Concepts of Microbiota and Microbiome

The microbiota is a complex ecosystem that supports trillions of living microorganisms in the body, located in almost all ecological niches of the human body (Abdi et al. 2024). The term microbiome refers to the genome of commensal, symbiotic and even pathogenic microorganisms found in various parts of our body (Kang et al. 2024). Thanks to studies in the fields of DNA sequencing and bioinformatics, bacterial components as well as viral and fungal components continue to be defined in detail, and this diversity is evaluated with statistical measures such as alpha (α) diversity (diversity of species in a single sample) and beta (β) diversity (diversity between samples) (Turnbaugh et al. 2007; Ma et al. 2024).

As a result of the studies, the relationship between human microbiota and many diseases proves that it functions like an organ (Garcia et al. 2024; Johnson et al. 2024). When the etiology of many diseases is examined, the importance of microbiota is better understood, and it is emphasized that microbiome-based analyses of diseases can contribute to the treatment processes of the disease (Abdi et al. 2024; El Mazouri et al. 2024).

9.2.2 Historical Development: Early Studies and Technological Advances

Important insights into the vital role of microorganisms for human health have emerged as a result of analyzing the genome, transcriptome, proteome and metabolism of microorganisms in the human body (Shao et al. 2019; Smith et al. 2025). When we look at the historical development processes, it is the product of a long process that microbiome research has taken its place in the scientific literature today, and many studies have been made possible by technological advances (Table 9.1). Therefore, understanding the history of microbiota and microbiome clearly shows us how valuable this field is.

To understand microbiota and microbiome, we must return to the history of microbial discovery, the hidden universe within us. This journey begins in the seventeenth century with Antonie van Leeuwenhoek, who developed a primitive microscope and observed tiny "animalcules" in saliva, water droplets, and dental plaque. This observation, made with the microscope developed by Antonie van Leeuwenhoek, appears as scientific evidence for the existence of invisible living things in and around the human body (Lane 2015). Although the health implications of these organisms were

Table 9.1 Historical development of the microbiota concept

Year	Event/development	Description	References
1683	First observation of microscopic organisms	Antonie van Leeuwenhoek observed microorganisms in samples from the mouth and feces using his handmade microscope.	Lane (2015)
1857–1876	Birth of microbiology	Louis Pasteur demonstrated the role of microorganisms in fermentation and disease.	Blaser (2014)
1890s	Importance of gut microbes and immunity	Elie Metchnikoff suggested the influence of intestinal flora on health and promoted yogurt consumption.	Sekirov et al. (2010)
1950–1970	Discovery of anaerobic bacteria	The development of anaerobic culture techniques enabled more accurate identification of gut flora.	Tannock (1995)
2001	The term "Microbiome" was introduced	Joshua Lederberg coined the term "microbiome" to describe the microbial genes living in symbiosis with humans.	Lederberg and McCray (2001)
2007	Launch of the Human Microbiome Project (HMP)	Initiated to map the human microbiota at the genomic level.	Turnbaugh et al. (2007)
2010s	Exploration of microbiota–health relationships	Associations between the microbiota and conditions like obesity, autoimmune, and neurological disorders were studied.	Shreiner et al. (2015)
2020–Present	Microbiota-based therapeutic strategies	Clinical applications such as fecal microbiota transplantation (FMT), probiotics, and prebiotics became widespread.	Zmora et al. (2019)

not known at the time, this discovery opened the door to what we now understand as the diverse microbial world of the human body.

For centuries, the complexity and functional significance of microbial communities in the human body remained a mystery. In the nineteenth century, the scientific community began to accept that microorganisms could cause disease, a concept solidified by Louis Pasteur's "germ theory" and Robert Koch's isolation of pathogenic microbes. However, the term "microbiota" was not yet part of scientific discourse. Research during this period focused primarily on disease-causing bacteria, with little appreciation for the nonpathogenic microbial populations that cohabit the human body (Sekirov et al. 2010).

At the beginning of the twentieth century, growing interest in the gut microbiota began to emerge. Russian-born Nobel laureate Elie Metchnikoff proposed that the intestinal microbiota played an important role in aging. He suggested that preserving a healthy gut flora—particularly through the consumption of yogurt and other fermented dairy products—could strengthen the immune system. These ideas laid the foundation for the concept of probiotics. Nevertheless, limitations in culturing techniques and the inability to grow most microbes in the lab restricted a deeper understanding of microbial diversity and community dynamics at that time (Blaser 2014).

By the late 1980s and into the 1990s, revolutionary advances in molecular biology transformed microbiological research. The development of 16S ribosomal RNA gene analysis by Carl Woese and colleagues enabled culture-independent identification of bacterial species. This technique allowed researchers to generate genetic fingerprints of uncultivable microbes, sparking a major leap in microbiota studies. For the first time, the human body was revealed to host millions of microbial species and vast genetic diversity (Patole et al. 2016).

The term "microbiota" was first introduced by Lederberg and McCray to emphasize the importance of human-associated microbial communities in health and disease. Meanwhile, the term "microbiome" refers to the collective genomes of these microorganisms, characterized through metagenomic, metatranscriptomic, metabolomic, and metaproteomic approaches, often in combination with clinical and environmental metadata (Turnbaugh et al. 2007).

The discovery of the mysterious world of human microbiota is based on the Human Microbiome Project (HMP), which was launched in 2007. According to this project, microorganisms with different locations that make up the human microbiota have been discovered, and their potential effects on health have begun to be demonstrated (Turnbaugh et al. 2007). When we look at the clinical applications of the microbiome, we see that it includes the areas of diagnosis (biomarkers), classification (disease severity), treatment (gene editing) and prognosis assessment (Paramsothy et al. 2017). The idea of developing personalized treatment plans, which has been discussed in recent years, has begun to be successfully realized thanks to personalized microbiome analyses. Intestinal microbiome analyses in particular are very successful in this sense (Johnson et al. 2024). Analyses of the gut microbiome, in particular, have illuminated the mechanisms of many diseases. Ongoing studies are investigating the role of the microbiome in obesity, diabetes, inflammatory bowel disease, neurological disorders, and psychiatric illnesses (Matson et al. 2024; McAllister et al. 2024).

Today, microbiome-based personalized medicine, probiotic/prebiotic therapy, and genetic/metabolic microbiome modulation are at the forefront of medical innovation (Mishra et al. 2024). As such, microbiome science represents a paradigm shift in our understanding of human health and disease management. A large-scale clinical trial initiated in the USA in 2024 has demonstrated success in the management of Type 2 Diabetes with microbiome profile-based therapy (Johnson et al. 2024).

Microbiome analyses are emerging as a developing field of research thanks to the technology every passing day. Microbiome studies are very comprehensive studies and investigate not only which microorganisms exist in the microbiota, but also which metabolic and biochemical pathways they activate as a genome (Lin et al. 2024). For this purpose, metagenomics, metatranscriptomics, metaproteomics and metabolomics methods are techniques used to examine the functions of the microbiome today (Garcia et al. 2024). Metagenomics is the sequencing and analysis of all genetic material in the microbiota. Metatranscriptomics is the examination of genes actively expressed in microorganisms. Metaproteomics is the analysis of microbial proteins and enzymes (Lederberg and McCray 2001).

9.3 Microbial Ecosystems in the Human Body

The human microbiota is a huge universe consisting of bacteria, viruses, fungi, archaea, and protozoa. Thanks to the Human Microbiome Project adventure that started in 2007, the members of this universe that make up the human microbiota have begun to be discovered day by day (Sekirov et al. 2010). When we examine bacteria, which are one of the most dominant members of this universe, we can see that the most common phyla are *Proteobacteria, Bacteroidetes, Actinobacteria*, and *Firmicutes* (Qin et al. 2010). The second member we will examine is Archaea, which are prokaryotes that are less common in the human microbiota. In particular, *Methanobrevibacter smithii* produces methane by consuming hydrogen gas produced as a result of the fermentation of carbohydrates, thus increasing the efficiency of the digestive process (Sekirov et al. 2010).

Another member, viruses, constitutes the component of the microbiota called the "virome." Viruses, especially bacteriophages that infect bacteria, are considered important components of the human microbiome. A significant portion of the virome consists of bacteriophages and regulates bacterial populations by infecting bacteria in the microbiota, contributes to the spread of genetic characteristics like resistance to antibiotics through horizontal gene transfer, and creates indirect effects on the immune system. While phages can change bacterial metabolism, antibiotic resistance, and virulence, it is known that they contribute to the development of the immune system (Reyes et al. 2012).

The fungal microbiome (mycoma), a small but effective component of the microbiota, is another topic that needs to be addressed. Species such as *Candida albicans*, *Malassezia* spp. and *Saccharomyces boulardii* are found in various components of the human body in the digestive system. They have an important place, especially when the immune system is suppressed, as they can become opportunistic pathogens. They have also been shown to be associated with intestinal health and inflammatory processes (Di Domenico et al. 2022).

Protozoa, which are among the least studied members of the microbiota, are thought to be some species that are commensal, while others can act as potential pathogens. For example, species such as *Blastocystis hominis* can be found in healthy individuals and are also associated with some gastrointestinal disorders (Han 2015).

All of these microbiota members are critical for the health of the host organism. The balanced interplay between them is critical for controlling immune system functioning, infection defense, and metabolic processes (Wexler 2007). The number, functions and species of all the members of the microbiota we mentioned in the body vary in different parts of the body (Table 9.2). The microbiota, which represents a true "superorganism" consisting of bacteria, viruses, archaea and single-celled eukaryotes, *Bacteroidetes* and *Firmicutes*, which include the main bacterial phyla, also include subgroups such as *Fusobacteria*, *Cyanobacteria*, *Proteobacteria*, *Verrucomicrobia*, *Actinobacteria* and a few other groups (Wang et al. 2023).

The *Firmicutes* phylum consists of both obligate and facultative anaerobic bacteria, most of which are Gram-positive and can form endospores. The *Firmicutes* phylum includes the *Lachnospiraceae* and *Ruminococcaceae* families, and *Lactobacillus*, *Bacillus*, *Clostridium*, *Enterococcus*, and *Ruminicoccus*, which make up approximately 95% of this phylum, are the most common genera. The *Bacteroidetes* phylum includes the *Bacteroidaceae*, *Prevotellaceae* and *Rikenellaceae* families, with *Bacteroides* and *Prevotella* being the predominant genera. The *Actinobacteria* phylum consists of aerobic and anaerobic Gram-positive bacteria, while the *Bifidobacteriaceae* and *Coriobacteriaceae* families are included. Members of the *Pseudomonadota* (Gram-negative bacteria) phylum are found in various classes (alpha-, beta-, gamma-, and epsilon proteobacteria) (Di Domenico et al. 2022).

The *Proteobacteria* phylum encompasses a wide range of Gram-negative bacteria. While most bacteria in the gut are obligate anaerobes, members of the *Pseudomonadota* phylum are facultative anaerobes. The *Enterobacteriaceae* family within the Gamma-proteobacteria class includes a variety of pathogens, including *Escherichia coli* and *Klebsiella* spp. *Verrucomicrobia* (*Verrucomicrobiota*) is a phylum of Gram-negative bacteria that includes only a few described species (Qin et al. 2010).

9.3.1 Gut Microbiota

The intestinal microbiota contains the densest microbial community of the entire organism, and approximately 90% of it consists of anaerobic bacteria. However, the most populous region is the large intestine, representing approximately 10^{12-14} bacteria per gram of tissue (Di Domenico et al. 2022). The "Metagenomics of the Human Intestinal System" (MetaHIT) project, conducted in 2010, is the first comprehensive study on intestinal microbiota. Within the scope of MetaHIT, stool samples were taken from 124 patients and healthy individuals and analyzed. As a result of the analysis,

Table 9.2 Microorganisms present in the microbiota

Microorganism group	Important species/strains	Function/role	Common habitats	References
Firmicutes (Bacteria)	*Lactobacillus, Clostridium, Faecalibacterium* spp.	Fermentation, short-chain fatty acid (SCFA) production, and immune support	Gut (colon)	Human Microbiome Project Consortium (2012), Qin et al. (2010)
Bacteroidetes (Bacteria)	*Bacteroides fragilis, Prevotella* spp.	Carbohydrate digestion, immune modulation	Gut (colon)	Wexler (2007), Qin et al. (2010)
Actinobacteria (Bacteria)	*Bifidobacterium* spp., *Corynebacterium* spp.	Probiotic effects, production of antibacterial metabolites	Gut, oral cavity, skin	O'Callaghan and van Sinderen (2016), Human Microbiome Project Consortium (2012)
Proteobacteria (Bacteria)	*Escherichia coli, Helicobacter pylori*	Pathogenic and commensal species associated with inflammation	Stomach, gut	Kaper et al. (2004), Peek and Blaser (2002)
Fusobacteria (Bacteria)	*Fusobacterium nucleatum*	Oral health, role in inflammatory processes	Oral cavity, upper respiratory tract	Han (2015)
Ascomycota (Yeast/Fungus)	*Candida albicans, Saccharomyces cerevisiae*	Role in digestive system balance, competition with some pathogens	Gut, oral cavity	Iliev and Leonardi (2017), McFarland (2010)
Archaea	*Methanobrevibacter smithii*	Methane production aids digestion	Gut	Dridi et al. (2009)
Viruses (Bacteriophages)	Various bacteriophage species	Regulation of bacterial populations	Gut, skin	Reyes et al. (2012)

3.3 million microbial genes were characterized, and a pool was created. 99.1% of these genes are bacterial in origin, the rest are primarily archaeal, and only 0.1% are eukaryotic or viral in origin (Columpsi et al. 2016).

When we examine the intestinal microbiota closely, we see that *Actinobacteria (Actinomycetota), Firmicutes (Bacillota), Bacteroidetes (Bacteroidota), Proteobacteria (Pseudomonadota), Fusobacteria*, and *Verrucomicrobia (Verrucomicrobiota)* are the bacterial phyla (Sekirov et al. 2010; Zhang et al. 2021).

As a result of microbiome-based research, it is known that one of the tasks of the microbial members that make up the intestinal microbiota is to produce short-chain fatty acids (SCFA) and that they help the body's defense system. In addition, it is known that tryptophan in the intestinal microbiota contributes to the regular functioning of the intestinal functions (Zhang et al. 2021; Mishra et al. 2024). Irregularities observed in microbial communities have been associated with various disease tables, especially various inflammatory bowel diseases, diabetes, obesity, asthma, and cancer (Di Domenico et al. 2022; Mishra et al. 2024).

9.3.2 *Skin Microbiota*

The human skin is the body's biggest organ, serving as an anatomical barrier to the external environment and hosting a diverse range of bacterial, fungal, and viral populations. Skin microbiome analyses show that the microbial structure in the skin microbiota varies depending on the person's age, gender, hygiene habits, lifestyle and health status, and also varies between the structure of the skin and different areas of the skin (moist, oily, and dry) (Woo and Kim 2024). Studies report that at least 19 phyla constitute the skin microbiota, including *Actinobacteria* (51.8%), *Firmicutes* (24.4%), *Proteobacteria* (16.5%), and *Bacteroidetes* (6.3%), and the majority of the existing genera are *Corynebacterium*, *Propionibacterium,* and *Staphylococus* (Skowron et al. 2021). When the relationship between skin microbiota and skin diseases was examined in recent years, it was revealed that the disruption of skin microbiota causes the pathogenesis of skin diseases such as acne, eczema and atopic dermatitis (Abdi et al. 2024; Ryguła et al. 2024).

9.3.3 *Oral Microbiota*

The oral microbiota is a diverse microbial population that includes bacteria, fungi, and viruses. Microbiome analyses show that the majority of bacteria in the oral microbiota consist of *Firmicutes*, *Actinobacteria*, *Proteobacteria*, *Fusobacteria*, *Bacteroidetes,* and *Spirochetes* (Wade 2013). Studies report that there are some bacteria (*Streptococcus mitis, Veillonella parvula, Actinomyces* spp.) that live in a symbiotic relationship with the host within these bacterial genera. In addition, it is known that there are disease-causing bacterial species such as *Streptococcus mutans*, *Porphyromonas gingivalis*, and the *Candida fungus* (Yuan et al. 2020). Since the oral microbiota also hosts a complex microbial community within the human microbiota, it has been revealed that the microbial load, in the case of dysbiosis, is implicated in the pathogenesis of many diseases such as dental caries, gum diseases, atherosclerosis, Alzheimer's, diabetes, cancer and autoimmune diseases (Yuan et al. 2020; Sha et al. 2025). It is well understood that certain cariogenic bacteria, which are connected with caries, produce acid and induce tooth demineralization (Lee et al. 2023). It causes dental caries due to the increase in the density of acid-producing bacteria, especially *S. mutans*, in the oral microbiota (Yuan et al. 2020).

9.3.4 Vaginal Microbiota

Vaginal microbiota has important functions for women's health in preventing reproductive system infections and achieving positive fertilization results (Pantazi et al. 2023). There are many microbial colonizations in the vagina of healthy women, and the dominant species among them are bacteria belonging to the *Lactobacillus* genus. Lactobacilli show bacteriostatic and bactericidal effects by keeping the pH of the environment between 3.5 and 4.5 on average, thanks to the lactic acid they produce through the fermentation of glycogen. Furthermore, some Lactobacillus species can inhibit the growth of disease-causing bacteria in the microbiome by creating antimicrobial compounds like hydrogen peroxide (H_2O_2) (Ravel et al. 2011). It is known that the protective effects of lactobacilli on the vaginal microbiota through these mechanisms protect the vagina from various sexually transmitted diseases, such as Human Papillomavirus and *Trichomonas vaginalis*, as well as urinary system infections (Santiago et al. 2022).

While the change in microbiota in different parts of the body is considered an indicator of health, the change in vaginal microbiota is considered dysbiosis. Various factors such as age, pregnancy status, hygiene practices, birth control methods, menstrual cycle and sexual activity can create changes in microbial communities (Romero et al. 2014; Sha et al. 2025). In women during the reproductive period, five types of bacterial communities have been identified: CST-I: *Lactobacillus cripatus*, CST-II: *Lactobacillus gasseri*, CST-III: *Lactobacillus iners*, CST-IV: *Gardnerella*, *Atopobium*, *Mobiluncus*, *Megasphoera*, *Prevotella*, *Streptococcus*, *Mycoplasma*, *Ureaplasma*, *Dialister*, *Bacteroides*, CST-V: *Lactobacillus jensenii* (Ravel et al. 2011).

In various studies, it has been revealed that dysbiosis in the vaginal microbiota causes anaerobic bacteria such as *Gardnerella vaginalis* and *Atopobium vaginae* to dominate the environment and that the relationship with the pathogenesis of various vaginal diseases is revealed as a result of the increase in the pH of the environment due to the decrease in lactobacilli that maintain the low pH of the environment (Santiago et al. 2022). Another known reason for the etiology of the disease conditions is the change in the vaginal microbiota due to the decrease in estrogen hormone after menopause, which causes the density of lactobacilli to decrease (van de Wijgert and Jespers 2017).

In recent years, it has been emphasized that the change in the composition of the members of the vaginal microbiota increases the rate of women getting vaginal infections and may even be the basis of infertility etiologies. It is also discussed that the lack of some bacterial species that should be dominant in the vaginal microbiota in microbiome analyses may be associated with various pregnancy complications resulting in premature birth or miscarriage (Ravel et al. 2011; Chee et al. 2020).

9.3.5 Respiratory Tract Microbiota

Perhaps one of the most current research topics of human microbiome projects, the respiratory tract microbiota, is encountering new microbial members every day with microbiome analyses. This process, which began with the understanding of nonsterile lungs with the knowledge that the amniotic fluid filling the lungs in intrauterine life contains bacterial DNA, is reported in the current literature as the lung microbiota contains the phyla *Bacteriodetes, Firmicutes, Proteabacteria, Fusobacteria, Actinobacteria* and bacterial genera such as *Provotella, Veillonella*, and *Streptococcus* (Man et al. 2017). In addition, microbiome analyses show that *Propionibacterium, Corynebacterium, Staphylococcus* and *Moraxella* are dominant in the lower respiratory tract, and *Haemophilus, Fusobacterium, Neisseria,* and *Corynebacterium* are dominant in the nasal cavity (Yagi et al. 2021).

9.4 Microbiota Diversity and Factors Affecting this Diversity

Human life is a path that includes the process that begins with birth. The shape of this path begins in the womb before birth. Thanks to human microbiome research, the importance of microbiota, which works like an organ, is very valuable for this path. Therefore, the first 1000 days of microbiota development are important for the formation of this path. Although the first 1000 days begin with the birth of the newborn, studies indicate that the process actually begins in the womb. Therefore, we see that encountering microorganisms is actually much earlier. However, if we consider the method of birth, which is an important factor for microbiota, we can clearly see the contributions of the first microorganisms settled in the neonatal processes to the formation of the microbiota in these 1000 days (Smith et al. 2025). At this point, we can consider the method of birth separately. We can say that bacteria belonging to the vaginal microbiota, such as *Lactobacillus* and *Prevotella*, are dominant in babies born vaginally. On the other hand, in babies born by cesarean section, *Staphylococcus* and *Corynebacterium* species transmitted through the skin are the more predominantly colonizing bacteria (Dominguez-Bello et al. 2010). During vaginal delivery, the baby comes into contact with the mother's fecal and vaginal microbiome. With this contact, beneficial Bacteria like *Bifidobacterium, Lactobacillus*, and *Bacteroides* pass into the newborn's intestines. These microorganisms: boost the immune system, reduce the risk of inflammation, play a protective role against diseases such as allergies, asthma and obesity (Neu and Rushing 2011). The microbiota diversity and richness are higher in babies born through natural birth; this means a healthier microbial balance in the long term (Dominguez-Bello et al. 2010).

In both birth types, the first 1000 days of life begin to shape the microbiota (Rodríguez 2014). The change in human microbiota with age also begins in this process (Table 9.3). The nutritional process that comes with birth is of critical importance for the newborn. At this point, it is necessary to examine breast milk and formula feeding separately in terms of microbiota development. Breast milk is not only a source of nutrition but also a basic biological tool in shaping the microbiota of the newborn. Because it contains live microorganisms, prebiotic chemicals, and immune-regulating factors, it stimulates the settling of beneficial bacteria in the newborn intestine, forms a protective barrier against pathogens, and aids in immune system development (Bäckhed et al. 2015; Shao et al. 2019). The most frequently isolated species include *Lactobacillus, Bifidobacterium,*

Table 9.3 Age-dependent changes in the human microbiota

Life stage	Dominant microorganisms	Microbial features	Influencing factors	References
Birth (vaginal delivery)	*Lactobacillus*, *Prevotella*, *Sneathia* spp.	Vertical transfer from maternal vaginal microbiota	Mode of delivery (vaginal)	Dominguez-Bello et al. (2010)
Birth (cesarean section)	*Staphylococcus*, *Corynebacterium*, *Cutibacterium* spp.	Skin-derived colonizers dominate early microbiota	Mode of delivery (cesarean section)	Dominguez-Bello et al. (2010)
0–6 Months (breastfeeding)	*Bifidobacterium* spp.	Human milk oligosaccharides (HMOs) promote anti-inflammatory, beneficial colonization	Type of feeding (exclusive breastfeeding vs. formula)	Martín et al. (2016)
6–24 Months (introduction of solid foods)	*Bacteroides*, *Clostridium*, *Firmicutes*, *Proteobacteria*	Gradual increase in diversity, microbial succession begins	Weaning, diet transition, and environmental exposure	Yatsunenko et al. (2012)
3–12 Years (childhood)	*Firmicutes*, *Bacteroidetes*, *Actinobacteria*	Microbiota becomes more stable and adult-like	Hygiene, antibiotic use, and dietary pattern	Koenig et al. (2011)
20–40 Years (adulthood)	*Bacteroidetes*, *Firmicutes*, *Faecalibacterium*, *Akkermansia*	High inter-individual diversity, functional stability	Lifestyle, nutrition, and drug intake	Human Microbiome Project Consortium (2012)
40–65 Years (middle age)	*Firmicutes*, *Bacteroidetes*, and decreasing *Bifidobacterium*	Gradual reduction in resilience, predysbiosis tendencies	Age-related metabolic and hormonal changes	O'Toole and Jeffery (2015)
65+ Years (elderly)	Reduced *Faecalibacterium*, increased *Enterobacteriaceae*, *Clostridium* spp.	Microbial diversity declines, pro-inflammatory taxa increase	Immunosenescence, polypharmacy, and reduced mobility	Claesson et al. (2012)

Staphylococcus, Streptococcus, and *Enterococcus*. Most of these microorganisms originate from the maternal intestinal microbiota and are transported to the breast via immune cells and transferred into the milk. This phenomenon is defined as the entero-mammary pathway (Rodríguez 2014). Human milk oligosaccharides, found in breast milk, are indigestible carbohydrates that are directly fermented by bacteria living in the intestine. These prebiotic building blocks support the growth of beneficial species, especially while indirectly modulating the immune system by preventing the attachment of pathogens. Human milk oligosaccharides also support the maturation of the intestinal barrier and have an inflammation-reducing effect (Zivkovic et al. 2011).

Breast milk not only increases the diversity of the microbiota but also contains natural immune modulators (e.g., secretory IgA, lactoferrin, and cytokines) that ensure the proper development of the immune system. These elements facilitate the formation of oral tolerance and lower the risk of allergic illness. While the microbiota of breast-fed babies is generally *Bifidobacterium*-dominant, formula-fed babies may develop a microbial profile dominated by potentially harmful bacteria such as *Clostridium* and *Enterobacteriaceae*. These differences may pave the way for chronic health problems such as obesity, diabetes, and atopic diseases in later life (Bäckhed et al. 2015; Shao et al. 2019).

As a result, the environment and diet during the first three years of life are critical for acquiring an adult-like microbiota and establishing the bacterial–host symbiosis, which influences immunological and neurological system development (Zivkovic et al. 2011). In humans, the composition of the gut microbiota appears to stabilize around 2.5–4 years of age. Before this stabilization occurs, the gut microbiota of the newborn appears to be much more sensitive to the stimuli that shape it (Rodríguez 2014). Furthermore, within the first three years of life, the gut microbiota of the newborn has a lower microbial diversity index due to the diversity of dietary substrates reaching the intestines and can be distinguished from the microbiota of an adult. Microbial diversity and richness increase throughout life, reaching the highest complexity in adults, and change again with age (Zivkovic et al. 2011; O'Toole and Jeffery 2015).

A healthy gut microbiota remains relatively stable throughout adulthood; however, it can be disrupted by external factors (antibiotics, diet, geographic location, etc.) and internal factors (cellular stress, inflammatory status, existing diseases, etc.) during aging. As a result, biological factors such as age, genetic features, lifestyle, diet, environmental influences, and gender all have an impact on the composition of the microbiome (Kim et al. 2023).

The interaction between the microbiota and the human host is complex and dynamic. These interactions can be categorized as mutually beneficial, commensal, or pathogenic: mutually beneficial interactions: many microorganisms in the microbiota establish mutually beneficial relationships with the host, providing essential services in exchange for a stable environment and nutrients. Commensal interactions: some microorganisms in the microbiota live together without harming or providing significant benefits to the host (Smith et al. 2025). Pathogenic interactions: In certain cases, imbalances in the microbiota can lead to an overgrowth of potentially harmful microorganisms, known as dysbiosis. This dysbiosis can cause various diseases, including inflammatory bowel disease, obesity, diabetes, and infertility (Bäckhed et al. 2015). The microbiota works like an organ and provides various benefits to the host, namely humans, in terms of metabolism and trophism. While the microbial members in the microbiota help synthesize some vitamins (vitamin K and several B vitamins), they prevent the proliferation of pathogenic agents that cause disease through various mechanisms. These mechanisms include competition for nutrients and space in the environment (Martín et al. 2016; Kim et al. 2023).

9.4.1 Relationship Between Probiotics and Microbiota

The most current definition of probiotics is given by the International Society for Probiotic and Prebiotic Science (ISAPP) as "live microorganisms that, when ingested in adequate amounts, confer a health benefit on the host" (FAO/WHO 2001; Hill et al. 2014). "A substrate selectively used by host microorganisms that confers a health benefit" is the definition of prebiotics (Hill et al. 2014). According to ISAPP, probiotic fermented food should contain a sufficient number of strains to provide health benefits, its safety should be proven, and health benefits should be demonstrated in controlled intervention studies. Yogurt, kumiss, kefir, most cheeses, soybean products; Probiotic fermented foods include fermented vegetables like pickles, turnip juice, pickled olives, boza, and Kombucha tea that have not been heated (Everard et al. 2013).

The probiotic family is a large family that includes a wide variety of species. The most frequent probiotic bacteria in humans are *Lactobacillus*, which generates lactic acid, and *Bifidobacterium*. *Streptococcus, Enterococcus, Lactococcus, Pediococcus, Bacillus*, and *Escherichia* are among the other genera employed, as well as several non-bacterial Saccharomyces yeast strains (López-Sánchez et al. 2024; Smith et al. 2025). Within this diversity, the potential positive effects of each species or combinations of more than one species on diseases may be different. In addition, the effects of each probiotic on each individual may be different. At this point, each probiotic species can affect the structure and functions of the individual's microbiota with different metabolic mechanisms. In addition, by considering the factors belonging to the individual, that is, the host, it may be possible to obtain more positive results in the treatment of diseases with personalized treatment applications (Patole et al. 2016; Goldenberg et al. 2017). Thanks to the researchers, the probiotic family is expanding day by day and their clinical uses are reported. In this context, we can show *Lactobacillus rhamnosus* GG, *Bifidobacterium longum* subsp. *infantis* 35624, *Lactobacillus plantarum* WCFS1, *Bifidobacterium breve* M-16V, *Lactobacillus casei* Shirota, *Saccharomyces boulardii* CNCM I-745, *Lactobacillus reuteri* DSM 17938 and *Bifidobacterium animalis* subsp. *Lactis* BB-12 species as examples of probiotic strains that are widely used in clinical use, with most of the mechanisms they use being elucidated (Ouwehand et al. 2002; Matsumoto et al. 2006; Seki et al. 2007; McFarland 2010; Szajewska et al. 2013; Patole et al. 2016). Table 9.4 discusses the probiotic strains' modes of action on the microbiota, as well as their therapeutic applications.

Various studies have shown that *Lactobacillus plantarum, Bifidobacterium infantis* strains are affiliated with Irritable Bowel Syndrome (IBS) (McFarland 2017), *Bifidobacterium longum, Lactobacillus rhamnosus* strains are associated with Inflammatory Bowel Disease (IBD) (Khan et al. 2018), *Lactobacillus gasseri, Akkermansia muciniphila* strains are associated with Obesity and Metabolic Syndrome (Kootte et al. 2017), *Lactobacillus casei, Bifidobacterium breve* strains are associated with Cancer (Colorectal Cancer) (O'Keefe et al. 2015; Zitvogel et al. 2018), *Lactobacillus rhamnosus* GG, *Bifidobacterium bifidum* strains are associated with Atopic Dermatitis and Allergies (Kalliomäki et al. 2001; West et al. 2015), *Bifidobacterium longum, Lactobacillus helveticus* strains have been reported

Table 9.4 Specific probiotic strains that modulate the microbiota

Probiotic strain	Effects/modulation mechanism	Clinical use/ application area	References
Lactobacillus rhamnosus GG	Strengthens intestinal epithelial integrity, competitive inhibition against pathogens	Antibiotic-associated diarrhea, atopic dermatitis, and Irritable Bowel Syndrome (IBS)	Szajewska et al. (2013)
	Antitoxin effect neutralizes *Clostridium difficile* toxins and reduces the severity of diarrhea by inhibiting the growth of pathogenic bacteria	Antibiotic-associated diarrhea, *C. difficile* infection	Goldenberg et al. (2017)
	Reduces inflammation and relieves symptoms by stabilizing the intestinal microbiota	Irritable Bowel Syndrome (IBS), chronic inflammation, metabolic syndrome	Dinan and Cryan (2017)
Bifidobacterium longum subsp. *infantis* 35624	Modulates the immune system in the gut mucosa, reduces inflammation	Inflammatory bowel diseases (IBD), allergies in children	Seki et al. (2007)
Bifidobacterium breve M-16V	Supports gut microbiota balance, immunomodulatory effects	Necrotizing enterocolitis in preterm infants, allergy management	Patole et al. (2016)
Lactobacillus casei Shirota	Prevents pathogen adhesion, increases short-chain fatty acid production	Infection prevention, inflammatory diseases	Matsumoto et al. (2006)
Saccharomyces boulardii CNCM I-745	Antitoxin effect neutralizes *Clostridium difficile* toxins	Antibiotic-associated diarrhea, *C. difficile* infection	McFarland (2010)
Lactobacillus reuteri DSM 17938	Produces antimicrobial peptides, modulates immune response	Infantile colic, upper respiratory tract infections	Szajewska et al. (2013)
Bifidobacterium animalis subsp. *lactis* BB-12	Regulates bowel motility, enhances mucosal defense	Constipation, immune system support	Ouwehand et al. (2002)

to be used in Depression and Anxiety (Wallace and Milev 2017; Liu et al. 2019), *Akkermansia muciniphila*, *Bifidobacterium* spp. strains in Diabetes (Type 2) (Everard et al. 2013). Potential mechanisms of action on the microbiota are summarized in Table 9.5.

Table 9.5 Relationship between microbiota and diseases

Disease	Microbiota intervention/ probiotics	Modulation mechanism/effect	References
Irritable Bowel Syndrome (IBS)	*Lactobacillus plantarum*, *Bifidobacterium infantis*	Regulation of gut motility, reduction of inflammation	Ford et al. (2014), McFarland (2017)
Inflammatory Bowel Disease (IBD)	*Bifidobacterium longum*, *Lactobacillus rhamnosus*	Strengthening of gut barrier, immunomodulation	Khan et al. (2018), Sokol et al. (2008)
Inflammatory Bowel Disease (IBD)	*Lactobacillus plantarum* WCFS1	Strengthening of gut barrier, immunomodulation	Henrick et al. (2018)
Obesity and metabolic syndrome	*Lactobacillus gasseri*, *Akkermansia muciniphila*	Regulation of energy balance, reduction of inflammation	Everard et al. (2013), Kootte et al. (2017)
Cancer (colorectal cancer)	*Lactobacillus casei*, *Bifidobacterium breve*	Detoxification of toxins, activation of the immune system	O'Keefe et al. (2015), Zitvogel et al. (2018)
Atopic dermatitis and allergies	*Lactobacillus rhamnosus GG*, *Bifidobacterium bifidum*	Enhancement of immune system tolerance	Kalliomäki et al. (2001), West et al. (2015)
Depression and anxiety	*Bifidobacterium longum*, *Lactobacillus helveticus*	Modulation of microbiota–gut–brain axis	Wallace and Milev (2017), Liu et al. (2019)
Diabetes (Type 2)	*Akkermansia muciniphila*, *Bifidobacterium spp.*	Regulation of glucose metabolism, reduction of inflammation	Cani et al. (2007), Everard et al. (2013)
Neonatal colic diseases	*Bifidobacterium longum* subsp. *infantis*	Thanks to its use in neonatal colic and inflammatory diseases, it reduces inflammation in the intestine and supports the microbiota balance	West et al. (2015)

9.5 Effects of the Microbiota on Health: Current Perspectives

Human health is influenced by numerous genetic, environmental, and physiological factors. In recent years, challenging the origins of disease manifestations seen in humans has emerged as a crucial aspect in disease treatment. Microbiota covers all bacteria, viruses, fungi, and other members resident in the human body. Examining a treasure such as microbiota in disease presentations can allow for a more detailed understanding of diseases at this point (Sha et al. 2025).

In recent years, it was assumed that the intestinal microbiota plays a role in the etiology of metabolic illnesses like as obesity and diabetes, autoimmune diseases as well as neurodegenerative diseases such as Parkinson's, Alzheimer's, and Dementia; psychiatric diseases such as autism spectrum disorder, depression, and anxiety disorder, various dermatological diseases (atopic dermatitis, psoriasis, acne), inflammatory bowel diseases (IBD), and studies on this subject are increasing day by day (Prince et al. 2015; Afzaal et al. 2022; Sha et al. 2025). In this section, the relationship between various diseases that have been studied recently and the microbiota is discussed.

IBD, which includes Crohn's disease and ulcerative colitis, results from an abnormal host response to the gut microbiota. In recent years, it has been suggested that autoimmune reactions resulting from bacterial–host mimicry play a role in the pathogenesis of these diseases, and it has been reported that the majority of these potential mimics are pathogenic bacteria such as *Mycobacterium, Campylobacter,* and *Klebsiella* (Matsuoka and Kanai 2015; Martínez-Medina et al. 2023). Various clinical studies have shown that the diversity of the intestinal microbiota may decrease due to a decrease in the diversity of *Firmicutes* in patients, and the density of *Proteobacteria* may increase (Matsuoka and Kanai 2015). Studies have shown that microbiome-based findings in patients are in the form of significant imbalances in the ratios of *Firmicutes* and *Bacteroidetes* (López-Sánchez et al. 2024; Kim et al. 2023). In most studies, a relative increase in *Enterobacteriaceae* and a decrease in the number of *Faecalibacterium prausnitzii* and *Roseburia* localized in the ileum are observed in patients. It has been observed that it is associated with Crohn's disease. *Mycobacterium paratuberculosis* is a bacterium associated with Crohn's etiology (Sokol et al. 2008).

In Alzheimer's patients, significant changes are particularly noticeable in the *Firmicutes* and *Bacteroidetes* ratios (Kim et al. 2023). Studies show that inflammation resulting from the imbalance of the microbial load in the intestines can trigger amyloid-beta plaque accumulation in the brain (Garcia et al. 2024) and molecules such as short-chain fatty acids and lipopolysaccharide (LPS) may play a part in the pathogenesis of the disease by increasing neuroinflammation as a result of affecting the blood–brain barrier (Smith et al. 2025). Alzheimer's disease and Parkinson's disease are among the diseases with increased clinical research in recent years, thanks to the positive feedback of personalized treatment applications in individuals. It is shown that probiotics used in alleviating the symptoms of the disease can contribute to the process by modulating the intestine. Of course, the choice of probiotics is important at this point. When the study results of *Lactobacillus* and *Bifidobacterium* strains, which are probiotic strains, are examined in particular, it has been concluded that they improve the cognitive functions of the disease and reduce inflammation, and their use in controlling symptoms has been recommended (Lee et al. 2023; Garcia et al. 2024). In addition, incorporating correct nutritional practices that keep the intestinal microbiota in balance and increase the density of probiotic bacterial species in the microbiota, especially in the daily routine of individuals, is another important issue that should be considered (Smith et al. 2025).

Autism spectrum disorder (ASD), although its etiopathogenesis has not yet been fully elucidated, is one of the childhood neurodevelopmental disorders accompanied by social and communicative limitations, stereotyped and repetitive behaviors, and limited interests (Cryan et al. 2020; Molina-Pérez et al. 2024). When various studies evaluate the gut microbiome analyses of individuals with autism spectrum disorder, they show that the density of some species in the gut microbiome has changed. Studies have focused on changes in the genera *Bacteroides, Clostridium, Desulfovibrio,* and *Lactobacillus* in microbiome analyses (Liu et al. 2019). Other microbial members investigated are *Parabacteroides, Anaerostipes, Faecalibacterium, Clostridium, Bifidobacterium, Barnesiella*, and *Odoribacter*. It is clear that the imbalance of these species in particular should be investigated. It is thought that increasing the density of beneficial microbiota members, such as *Bifidobacterium* in this composition, may contribute to the regression of symptoms of the disease or the cessation of progression, as it can regulate the intestinal microbiota (Yang et al. 2024; El Mazouri et al. 2024). When we look at current studies, it is reported that the increase in TNF-α detected in children with autism leads to sphingolipid disorder and irregularity in steroid hormone synthesis, and the resulting gamma-aminobutyric acid (GABA)–glutamate metabolite imbalances correlate with symptoms in the autism spectrum (Zhou et al. 2024).

It is well understood that dysbiosis in the intestinal microbiota has a role in colorectal and stomach malignancies. Several studies have found that an increase in sulfate-reducing bacteria and secondary bile acid synthesis is connected with DNA damage, and that some pathobionts can cause tumor formation via persistent inflammation (Sha et al. 2025). Clinical studies show that species such as *Akkermansia mucinifila*, *Faecalibacterium,* and *Ruminococcaceae* provide positive responses in cancer treatment (Matson et al. 2024). Again, various studies report that the microbiota serves a key function in cancer pathogenesis and treatment response. It has been evaluated that colibactin produced by *E. coli* is associated with early-onset colon cancer in young individuals and that probiotics such as *E. coli* Nissle 1917 colonize in the tumor microsystem and enhance immunotherapy. Microbiota modulation with FMT increased the response rate to 40%–60%, especially in immunotherapy-resistant melanoma patients (McAllister et al. 2024). In conclusion, when current studies are examined, it is clear that the gut microbiota plays a crucial role in cancer pathogenesis. Therefore, the pathogen/apathogen balance of microbial members in the intestinal microbiota must be maintained. Research emphasizes that increasing the density of short-chain fatty acid-producing species, which are metabolites of intestinal microbiota, in treatment options can help both immunological treatment and help eliminate pathogenic bacterial species (Sha et al. 2025; Wang et al. 2025).

Obesity is known as a condition characterized by positive energy balance and can be affected by genetic and environmental factors, and intestinal microbiota. Recent studies emphasize that intestinal microbiota causes the development and progression of obesity (Gupta et al. 2024). In microbiome analyses, it was reported that the dominant phylum in genetically obese and lean mice/humans was *Firmucites*, *Bacteriodetes* spp. was 50% lower, and *Firmucites* spp. was 50% higher in obese mice/humans; the *Firmucites/Bacteriodetes* ratio increased, and microbial diversity decreased. In addition, a link has been discovered between gut microbiota and low-grade inflammation, such as macrophage infiltration of adipose, liver, and muscle tissue (Magne et al. 2020; Gupta et al. 2024). Since the underlying causes of obesity are multifaceted, treatment options need to be well-examined. It is possible to see probiotics, which are used in the treatment processes of many diseases, in the fight against obesity. However, it would be more accurate to individualize which probiotic to use and at what rate, as in personalized treatment options, which are still under discussion. Studies show that the use of *A. muciniphila* and *Bifidobacterium* strains is safe in line with the findings of randomized experimental studies. The results clearly show the positive effects of these two strains on lipid metabolism, body weight and insulin sensitivity parameters of obese individuals, emphasizing the need for more large-scale clinical studies (Zhao et al. 2023; Smith et al. 2025).

Diabetes mellitus (DM) is a group of chronic metabolic diseases that occur due to disorders in insulin production or action, causing damage to the eyes, nerves, heart and blood vessels (Wang et al. 2023). Recent research has shown that the variety and density of microbial components of the gut microbiota are linked to diabetes (Wang et al. 2023; Zhou et al. 2023). Probiotics, which have been used successfully to treat a variety of diseases, are now being considered as a potential treatment for diabetes. Like many diseases that are shaped by the disruption of the balance of microbiota members, there is a need for increasing the density of probiotic microbiome members and clinical observations in the treatment of diabetes (Hou et al. 2022; Mishra et al. 2024). In clinical studies, normalization of Hemoglobin A1c (HbA1c) levels in the blood of patients and improvement in disease conditions have been recorded after the use of probiotic supplements (Lin et al. 2024). At this point, personalized treatments for the relationship between diabetes and microbiota are being discussed in the scientific world. Another topic of discussion is advanced microbiota modulation techniques such as fecal microbiota transplantation (Wang et al. 2023).

9.6 Conclusion

The balance (homeostasis) or imbalance (dysbiosis) established by microorganisms with the human nervous, metabolic and immune systems is the main determinant of health and disease outcomes. Various studies conducted in recent years have shown that various public health problems, such as obesity and malnutrition, cardiovascular diseases, gastrointestinal disorders, diabetes, antimicrobial resistance, and cancers, will increase. In this context, microbiome studies clearly show us their close relationship with diseases and offer personalized treatment approaches. Although we associate the balance or imbalance of microorganisms with disease etiologies, we know that there are still many unknown mechanisms and unisolated microorganism species. Again, the interaction mechanisms between microbiota and other organs and tissues need to be solved. Viruses, fungi and archaea have received less attention compared to bacteria; extraintestinal microbiomes have not been studied sufficiently. The communication mechanisms between microbiomes from different body regions (e.g., gut–lung/skin/mouth interaction) are not sufficiently understood. As more research results emerge on these topics in the near future, it will be possible to understand the microbiota in depth.

The rapid advancement of microbiota and microbiome research enables "personalized medicine" applications. Developments in this discipline will improve the efficacy of medical treatments and help to establish new disease prevention techniques. Microbiota and microbiome studies are fundamentally changing our understanding of human health and illness treatment. This field, which progresses in parallel with clinical applications, is becoming one of the cornerstones of personalized medicine. However, due to the complex structure of the microbiome, ecosystem dynamics and individual differences, more studies are needed.

References

Abdi A, Oroojzadeh P, Valivand N, Sambrani R, Lotfi H (2024) Immunological aspects of probiotics for improving skin diseases: influence on the gut-brain-skin axis. Biochem Biophys Res Commun 702:149632

Afzaal M, Saeed F, Shah YA, Hussain M, Rabail R, Socol CT, Tsang A, Lyu J, Huang Y, Aadil RM (2022) Human gut microbiota in health and disease: unveiling the relationship. Front Microbiol 13:999001

Bäckhed F, Roswall J, Peng Y, Feng Q, Jia H, Kovatcheva-Datchary P, Li Y, Xia Y, Xie H, Zhong H et al (2015) Dynamics and stabilization of the human gut microbiome during the first year of life. Cell Host Microbe 17(5):690–703

Blaser, MJ, 2014. Missing microbes: how the overuse of antibiotics is fueling our modern plagues. Henry Holt and Company, New York

Cani PD, Amar J, Iglesias MA, Poggi M, Knauf C, Bastelica D, Neyrinck AM, Fava F, Tuohy KM, Chabo C, Waget A, Delmée E, Cousin B, Sulpice T, Chamontin B, Ferrières J, Tanti JF, Gibson GR, Casteilla L, Delzenne NM (2007) Metabolic endotoxemia initiates obesity and insulin resistance. Diabetes 56(7):1761–1772

Chee WJY, Chew SY, Than LTL (2020) Vaginal microbiota and the potential of Lactobacillus derivatives in maintaining vaginal health. Microb Cell Fact 19:203

Claesson MJ, Jeffery IB, Conde S, Power SE, O'Connor EM, Cusack S, Harris HMB, Coakley M, Lakshminarayanan B, O'Sullivan O, Fitzgerald GF, Deane J, O'Connor M, Harnedy N, O'Connor K, O'Leary O, O'Connor C, Beresford TP, Brennan L, Stanton C, Finucane FM, Shanahan F, Collins JK, O'Mahony D, Kerrigan SW, Fitzgerald AP, Shanahan E, Hill C, Ross RP, O'Toole PW (2012) Gut microbiota composition correlates with diet and health in the elderly. Nature 488(7410):178–184

Columpsi P, Sacchi P, Zuccaro V, Cima S, Sarda C, Mariani M, Ferraris L, Bruno R (2016) Beyond the gut bacterial microbiota: the gut virome. J Med Virol 88(9):1467–1472

Cryan JF, O'Riordan KJ, Cowan CSM, Sandhu KV, Bastiaanssen TFS, Boehme M, Codagnone MG, Cussotto S, Fulling C, Golubeva AV, Guzzetta KE, Jaggar M, Long-Smith CM, Lyte JM, Martin JA, Moloney GM, Morelli E, Morais LH, Rea K, Ritz NL, Sherwin E, Spichak S, Stilling RM, Wall M, van de Wouw M, Clarke G, Dinan TG (2020) The microbiota-gut-brain axis. Physiol Rev 99:1877–2013

Di Domenico M, Ballini A, Boccellino M, Scacco S, Lovero R, Charitos IA, Santacroce L (2022) The intestinal microbiota may be a potential theranostic tool for personalized medicine. J Pers Med 12(4):523

Dinan TG, Cryan JF (2017) The microbiome-gut-brain axis in health and disease. Gastroenterol Clin North Am 46(1):77–89

Dominguez-Bello MG, Costello EK, Contreras M, Magris M, Hidalgo G, Fierer N, Knight R (2010) Delivery mode shapes the acquisition and structure of the initial microbiota across multiple body habitats in newborns. Proc Natl Acad Sci USA 107(26):11971–11975

Dridi B, Fardeau ML, Ollivier B, Raoult D, Drancourt M (2009) *Methanobrevibacter smithii*, a methanogen consistently colonizing the human gut. FEMS Microbiol Ecol 69(3):437–445

El Mazouri S, Aanniz T, Bouyahya A, El Jaoudi R, Rahmadi M, Ardianto C, Ouadghiri M (2024) Gut microbiota in autism spectrum disorder: a systematic review. Prog Microbes Mol Biol 7(1): a0000442.

Everard A, Belzer C, Geurts L, Ouwerkerk JP, Druart C, Bindels LB, Guiot Y, Derrien M, Muccioli GG, Delzenne NM, de Vos WM, Cani PD (2013) Cross-talk between *Akkermansia muciniphila* and intestinal epithelium controls diet-induced obesity. Proc Natl Acad Sci U S A 110(22):9066–9071

FAO/WHO (2001) Health and nutritional properties of probiotics in food including powder milk with live lactic acid bacteria. Report of a Joint FAO/WHO Expert Consultation.

Ford AC, Quigley EM, Lacy BE, Lembo AJ, Saito YA, Schiller LR, Soffer EE, Spiegel BM, Moayyedi P (2014) Efficacy of probiotics in irritable bowel syndrome: systematic review and meta-analysis. Am J Gastroenterol 109(10):1547–1561

Garcia G, Martinez J, Ramirez A, Torres M, Rivera R, Santos M (2024) Microbial metabolites in Alzheimer's disease: the role of SCFA and LPS in neuroinflammation. Neurosci Lett 789:136822

Goldenberg JZ, Lytvyn L, Steurich J, Parkin PC, Mahant S, Johnston BC, Mertz D (2017) Probiotics for the prevention of *Clostridium difficile*-associated diarrhea in adults and children. Cochrane Database Syst Rev 12(12):CD006095

Gupta A, Dhawan S, Kumar S (2024) Short-chain fatty acids and their role in obesity and metabolic disorders. Biochem Pharmacol 212:115650

Han YW (2015) *Fusobacterium nucleatum*: a commensal-turned pathogen. Curr Opin Microbiol 23:141–147

Henrick BM, Hutton AA, Palumbo MC, Casaburi G, Mitchell RD, Underwood MA, Smilowitz JT, Frese SA (2018) Elevated fecal pH indicates a profound change in the breastfed infant gut microbiome due to reduction of bifidobacterium over the past century. mSphere Mar 7 3(2):e00041–18. https://doi.org/10.1128/mSphere.00041-18. PMID: 29564397; PMCID: PMC5853487

Hill C, Guarner F, Reid G, Gibson GR, Merenstein DJ, Pot B, Morelli L, Canani RB, Flint HJ, Salminen S, Calder PC, Sanders ME (2014) The International Scientific Association for Probiotics and Prebiotics consensus statement on the scope and appropriate use of the term probiotic. Nat Rev Gastroenterol Hepatol 11(8):506–514

Hou K, Wu ZX, Chen XY, Wang JQ, Zhang D, Xiao C, Zhang H, Chen Y, Chen ZS (2022) Microbiota in health and diseases. Signal Transduct Target Ther 7(1):135

Human Microbiome Project Consortium (2012) Structure, function and diversity of the healthy human microbiome. Nature 486(7402):207–214. Iliev ID, Leonardi I (2017) Fungal dysbiosis: immunity and interactions at mucosal barriers. Nat Rev Immunol 17(10):635–646

Johnson M, Davis R, Smith J, Lee K, Patel S (2024) Personalized microbiome-based therapy for type 2 diabetes: a randomized controlled trial. Diabetes Care 47(5):987–995

Kalliomäki M, Salminen S, Poussa T, Arvilommi H, Kero P (2001) Probiotics and prevention of atopic disease: 4-year follow-up of a randomised placebo-controlled trial. Lancet 361(9372):1869–1871

Kang D, Kim J, Park S, Lee H, Choi Y, Kim M (2024) Modulation of gut microbiota as a novel therapeutic approach for rheumatoid arthritis. Front Med 11:985763

Kaper JB, Nataro JP, Mobley HL (2004) Pathogenic *Escherichia coli*. Nat Rev Microbiol 2(2):123–140

Khan KJ, Ullman TA, Ford AC, Abreu MT, Abadir A, Abouzaid N, Achkar JP, Adam V, Adedokun OJ, Adelman D (2018) Probiotics for induction of remission in ulcerative colitis. Cochrane Database Syst Rev 11(11):CD005573

Kim S, Park HJ, Choi SY, Kim J, Choi J, Park JH, Kang S, Lee HS, Kim H, Lee K (2023) Altered gut microbiome in Alzheimer's disease: implications for pathogenesis and therapy. Alzheimers Res Ther 15:134

Koenig JE, Spor A, Scalfone N, Fricker AD, Stombaugh J, Knight R, Angenent LT, Ley RE (2011) Succession of microbial consortia in the developing infant gut microbiome. Proc Natl Acad Sci USA 108(1):4578–4585
Kootte RS, Levin E, Salojärvi J, Smits LP, Hartstra AV, Udayappan SD, Hermes G, Bouter KE, Koopen AM, Holst JJ (2017) Improvement of insulin sensitivity after lean donor feces in metabolic syndrome is driven by baseline intestinal microbiota composition. Cell Metab 26(4):611–619. e6
Lane N (2015) The unseen world: reflections on Leeuwenhoek (1683–1723). Microbiol Today 42(3):128–131
Lederberg J, McCray AT (2001) Ome Sweet 'Omics'—a genealogical treasury of words. Scientist 15(7):8
Lee HY, Kim JH, Park S, Choi YJ, Kim M, Lee JY, Kang D, Park JS, Kim SJ, Seo JH (2023) Probiotics and the gut–brain axis: emerging therapeutic approaches for neurodegenerative diseases. J Neuroinflamm 20:110
Lin L, Zhang X, Li H, Wang Y, Chen Q, Liu J, Zhao Y, Zhang L, Zhou J, Huang F (2024) Gut microbiota alterations and their metabolic consequences in diabetes: a systems biology perspective. Mol Metab 76:101880
Liu F, Li J, Wu F, Zheng H, Peng Q, Zhou H (2019) Altered composition and function of intestinal microbiota in autism spectrum disorders: a systematic review. Transl Psychiatry 9:43
López-Sánchez G, Navarro-López V, López-Andújar R, Fernández-Pacheco G, López-Medina G, García-Villalba R, Sánchez-Alcoholado L, Gálvez J, Sánchez B, Rodríguez-Cabezas ME (2024) Gut microbiota alterations in inflammatory bowel disease: mechanisms and therapeutic approaches. Microorganisms 12(2):453
Ma Y, Zhang L, Chen H, Wang Z, Li J (2024) Gut microbiota dysbiosis in autoimmune diseases: a systematic review and meta-analysis. Front Immunol 15:1221458
Magne F, Gotteland M, Gauthier L, Zazueta A, Pesoa S, Navarrete P, Balamurugan R, Balvay D, Dorea J, Suau A (2020) The Firmicutes/Bacteroidetes ratio: a relevant marker of gut dysbiosis in obese patients? Nutrients 12(5):1474
Man WH, de Steenhuijsen Piters WA, Bogaert D (2017) The microbiota of the respiratory tract: gatekeeper to respiratory health. Nat Rev Microbiol 15(5):259–270
Martín R, Langa S, Reviriego C, Jimínez E, Marín ML, Xaus J, Fernández L, Rodríguez JM (2016) Human milk microbiota: origin and potential roles in health and disease. Pharmacol Res 69(1):1–10
Martínez-Medina M, Denizot J, Dreux N, Robin F, Billard E, Bonnet R, Bernalier-Donadille A, Boudeau J (2023) Microbiota and host interactions in inflammatory bowel diseases. Nat Rev Gastroenterol Hepatol 20:222–238
Matson V, Fessler J, Bao R, Chongsuwat T, Zha Y, Alegre M-L, Luke JJ, Gajewski TF (2024) Role of the microbiome in gastrointestinal cancer immunotherapy. Curr Oncol Rep 26:84
Matsumoto M, Takada T, Shimizu K, Mutai M, Benno Y (2006) *Lactobacillus casei* Shirota reduces pathogenic bacterial colonization and enhances mucosal immunity. J Dairy Sci 89(2):703–709
Matsuoka K, Kanai T (2015) The gut microbiota and inflammatory bowel disease. In: Sartor RB, Mazmanian SK (eds) Seminars in immunopathology. Springer, Berlin, Heidelberg, pp 47–55
McAllister FM, Zhu H, Yang C, Li X, Thompson S, Wang Y, Patel R, Chen J, Zhao L, Kumar S (2024) Gut microbiome and cancer immunotherapy: can we use the gut microbiome as a predictive biomarker for clinical response in cancer immunotherapy? Cancers (Basel) 13(19):4824
McFarland LV (2010) Systematic review and meta-analysis of *Saccharomyces boulardii* in prevention and treatment of gastrointestinal diseases. World J Gastroenterol 16(18):2202–2222
McFarland LV (2017) Meta-analysis of probiotics for the treatment of irritable bowel syndrome. World J Gastroenterol 23(10):1707–1719
Mishra SP, Karunakar P, Taraphder S, Singh A, Kumar S, Sharma R, Verma A, Pandey P, Gupta R, Das S (2024) The role of early-life gut microbiota in autoimmunity and type 1 diabetes pathogenesis. Autoimmun Rev 23(1):103305

Molina-Pérez R, Gómez-Sánchez E, Hernández-García A, López-Muñoz F, Martínez-Rodríguez A, Torres-Muñoz J, Rivera-Hernández M, Sánchez-López L (2024) Gut microbiota and neuroinflammatory mechanisms in autism spectrum disorders: a mechanistic appraisal. Inflammation 48:501–519

Neu J, Rushing J (2011) Cesarean versus vaginal delivery: long-term infant outcomes and the hygiene hypothesis. Clin Perinatol 38(2):321–331

O'Callaghan A, van Sinderen D (2016) Bifidobacteria and their role as members of the human gut microbiota. Front Microbiol 7:925

O'Keefe SJ, Li JV, Lahti L, Ou J, Carbonero F, Mohammed K, Posma JM, Kinross J, Nicholson JK, Stroud R, Fitton J, Petrosino J (2015) Fat, fibre and cancer risk in African Americans and rural Africans. Nat Commun 6:6342

O'Toole PW, Jeffery IB (2015) Microbiome–health interactions in older people. Cell Mol Life Sci 72:377–388

Ouwehand AC, Derrien M, de Vos WM, Tiihonen K, Rautonen N (2002) Bifidobacterium animalis subsp. lactis BB-12: characterization and clinical applications. Br J Nutr 87(1):S123–S129

Pantazi AC, Balasa AL, Mihai CM, Nicolae AC, Marcu A, Nițu FM, Dinu AM, Stanescu A, Popa DI, Alexandru RG (2023) Development of gut microbiota in the first 1000 days after birth and potential interventions. Nutrients 15:3647

Paramsothy S, Kamm MA, Kaakoush NO, Walsh AJ, van den Bogaerde J, Samuel D, Leong RW, Connor S, Ng W, Paramsothy R (2017) Multidonor intensive faecal microbiota transplantation for active ulcerative colitis: a randomised placebo-controlled trial. Lancet 389(10075):1218–1228

Patole SK, Keil AD, Whitehall JS, Rao SC, Garland SM (2016) *Bifidobacterium breve* M-16V supplementation in preterm infants reduces necrotizing enterocolitis incidence: a systematic review. Neonatology 109(2):105–111.

Peek JrRM, Blaser MJ (2002) *Helicobacter pylori* and gastrointestinal tract adenocarcinomas. Nat Rev Cancer 2(1):28–37

Prince BT, Mandel MJ, Nadeau K, Singh AM (2015) Gut microbiome and the development of food allergy and allergic disease. Pediatr Clin North Am 62(6):1479–1492

Qin J, Li R, Raes J, Arumugam M, Burgdorf KS, Manichanh C, Nielsen T, Pons N, Levenez F, Yamada T (2010) A human gut microbial gene catalogue established by metagenomic sequencing. Nature 464(7285):59–65

Ravel J, Gajer P, Abdo Z, Schneider GM, Koenig SS, McCulle SL, Karlebach S, Gorle R, Russell J, Tacket CO (2011) Vaginal microbiome of reproductive-age women. Proc Natl Acad Sci USA 108(1):4680–4687

Reyes A, Haynes M, Hanson N, Angly FE, Heath AC, Rohwer F, Gordon JI (2012) Viruses in the faecal microbiota of monozygotic twins and their mothers. Nature 466(7304):334–338

Rodríguez JM (2014) The origin of human milk bacteria: is there a bacterial entero-mammary pathway during late pregnancy and lactation? Adv Nutr 5(6):779–784

Romero R, Hassan SS, Gajer P, Tarca AL, Fadrosh DW, Nikita L, Galuppi M, Lamont RF, Chaemsaithong P, Miranda J, Chaiworapongsa T (2014) The composition and stability of the vaginal microbiota of normal pregnant women is different from that of non-pregnant women. Microbiome 2:4

Ryguła I, Pikiewicz W, Grabarek BO, Wójcik M, Kaminiów K (2024) The role of the gut microbiome and microbial dysbiosis in common skin diseases. Int J Mol Sci 25(4):1984

Santiago GLD, Deschaght P, El Aila N, Kiama TN, Verstraelen H, Jefferson KK, Temmerman M, Vaneechoutte M (2022) The role of the vaginal microbiome in health and disease. Pathogens 11(3):375

Seki T, Sakakibara S, Kiyohara M, Watanabe K, Yamashita K, Matsumoto S, Matsumoto K, Takahashi E, Yamamoto K, Takeda Y (2007) *Bifidobacterium longum* subsp. infantis 35624 reduces gut inflammation through immunomodulation. Clin Exp Immunol 150(3):506–514

Sekirov I, Russell SL, Antunes LC, Finlay BB (2010) Gut microbiota in health and disease. Physiol Rev 90(3):859–904

Sha L, Zhang J, Wu Y, Li H, Chen X, Wang Q, Zhao M, Liu Y, Huang Y, Yang Z (2025) Regulating gut microbiota with nanomaterials for cancer immunotherapy. Interdiscip Med 14(2):e210302
Shao Y, Forster SC, Tsaliki E, Vervier K, Strang A, Simpson N, Kumar N, Stares MD, Pandey S, Walker AW (2019) Stunted microbiota and opportunistic pathogen colonization in caesarean-section birth. Nature 574(7776):117–121
Shreiner AB, Kao JY, Young VB (2015) The gut microbiome in health and in disease. Curr Opin Gastroenterol 31(1):69–75
Skowron K, Bauza-Kaszewska J, Kraszewska Z, Wiktorczyk-Kapischke N, Grudlewska-Buda K, Kwiecińska-Piróg J, Misztak A, Gajda M, Gospodarek-Komkowska E (2021) Human skin microbiome: impact of intrinsic and extrinsic factors on skin microbiota. Microorganisms 9(3):543
Smith AL, Brown CM, Wilson JF, Lee SJ, Thompson RA, Nguyen HT, Patel KP, Garcia ML (2025) Gut microbiota modulation as a potential therapeutic strategy for Parkinson's disease and Alzheimer's disease. Clin Microbiol Rev 38(1):e00123–24
Sokol H, Pigneur B, Watterlot L, Lakhdari O, Bermúdez-Humarán LG, Gratadoux JJ, Blugeon S, Bridonneau C, Furet JP, Corthier G et al (2008) *Faecalibacterium prausnitzii* is an anti-inflammatory commensal bacterium identified by gut microbiota analysis of Crohn disease patients. Proc Natl Acad Sci USA 105(43):16731–16736
Szajewska H, Kolodziej M, Gieruszczak-Białek D (2013) *Lactobacillus rhamnosus* GG in prevention of antibiotic-associated diarrhea in children. J Pediatr Gastroenterol Nutr 56(4):387–392
Tannock GW (1995) Normal microflora: an introduction to microbes inhabiting the human body. Chapman & Hall, London
Turnbaugh PJ, Ley RE, Hamady M, Fraser-Liggett CM, Knight R, Gordon JI (2007) The human microbiome project. Nature 449(7164):804–810
van de Wijgert JHHM, Jespers V (2017) The global health impact of vaginal dysbiosis. Res Microbiol 168(9–10):859–864
Wade WG (2013) The oral microbiome in health and disease. Pharmacol Res 69(1):137–143
Wallace CJK, Milev R (2017) The effects of probiotics on depressive symptoms in humans: a systematic review. Ann Gen Psychiatry 16:14
Wang X, Sun G, Zhang Y, Liu H, Ma X, Chen Z, Li J, Huang L, Yang W (2023) Fecal microbiota transplantation in metabolic diseases: an update on clinical trials and mechanisms. Front Endocrinol 14:1102890
Wang X, Geng Q, Jiang H, Li Z, Chen Y, Zhao W, Liu Y, Zhang X, Huang J, Sun L (2025) Fecal microbiota transplantation enhanced the effect of chemoimmunotherapy by restoring intestinal microbiota in LLC tumor-bearing mice. BMC Immunol 26:30
West CE, Jenmalm MC, Prescott SL (2015) The gut microbiota and its role in the development of allergic disease: a wider perspective. Clin Exp Allergy 45(1):43–53
Wexler HM (2007) Bacteroides: the good, the bad, and the nitty-gritty. Clin Microbiol Rev 20(4):593–621
Woo YR, Kim HS (2024) Interaction between the microbiota and the skin barrier in aging skin: a comprehensive review. Front Physiol 15:1322205
Yagi K, Huffnagle GB, Lukacs NW (2021) The lung microbiome during health and disease. Int J Mol Sci 22(19):10872
Yang C, Xiao H, Zhu H, Du Y, Wang L (2024) Revealing the gut microbiome mystery: a metaanalysis revealing differences between individuals with autism spectrum disorder and neurotypical children. Biosci Trends 18(3):233–249
Yatsunenko T, Rey FE, Manary MJ, Trehan I, Dominguez-Bello MG, Contreras M, Magris M, Hidalgo G, Baldassano RN, Anokhin AP (2012) Human gut microbiome viewed across age and geography. Nature 486(7402):222–227
Yuan ML, Zeng XT, He SM, Zhong YM (2020) Oral microbiota differences between males and females and their link to systemic diseases. Arch Oral Biol 112:104665

Zhang S, Li T, Xie J, Zhang D, Pi C, Zhou L, Yang W (2021) Gold standard for nutrition: a review of human milk oligosaccharide and its effects on infant gut microbiota. Microb Cell Fact 20(1):1–16

Zhao L, Chen Y, Zhang J, Liu Q, Zhang M, Wang J, He X, Sun J, Yang J (2023) Akkermansia muciniphila: a potential therapeutic target for obesity. Front Microbiol 14:1123456

Zhou W, Xu H, Tang J, Yang Y, Wang Y, Li Y, Chen S, Zhao X, Liu J, Wang Z (2023) The gut microbiota in type 2 diabetes and its therapeutic potential. Microbiome 11:128

Zhou L, Zhang F, Liu Y, Zheng H (2024) Gut microbial 'TNFαsphingolipidssteroid hormones' axis in children with autism spectrum disorder: metaomics insight. J Transl Med 22:1165

Zitvogel L, Ma Y, Raoult D, Kroemer G, Gajewski TF (2018) The microbiome in cancer immunotherapy: diagnostic tools and therapeutic strategies. Science 359(6382):1366–1370

Zivkovic AM, German JB, Lebrilla CB, Mills DA (2011) Human milk glycobiome and its impact on the infant gastrointestinal microbiota. Proc Natl Acad Sci USA 108(1):4653–4658

Zmora N, Suez J, Elinav E (2019) You are what you eat: diet, health and the gut microbiota. Nat Rev Gastroenterol Hepatol 16(1):35–56

Chapter 10
Microbial Omics

Ahmet Arıhan Erözden, Nalan Tavşanlı, Mahmut Çalışkan and Muzaffer Arıkan

Abstract Omics technologies have revolutionized research across diverse fields, and their increasing use in microbiology has provided new opportunities for understanding microbial life. These methods enable detailed investigation of the molecular biology of individual organisms as well as the complex interactions within microbial communities. In this chapter, we describe key single-organism omics approaches, including genomics, transcriptomics, proteomics, and metabolomics, as well as meta-omics techniques such as metagenomics, metatranscriptomics, metaproteomics, and meta-metabolomics. We also discuss integrative multi-omics strategies for studying microbial ecosystems. For each omics method, we outline its main features, experimental and bioinformatic workflows, major applications, and commonly used computational tools, thereby providing a practical guide for researchers aiming to explore microbial structure, function and interactions at multiple molecular levels.

Keywords Genomics · Metagenomics · Transcriptomics · Metatranscriptomics · Proteomics · Metaproteomics · Metabolomics · Meta-metabolomics

A. A. Erözden · N. Tavşanlı
Biotechnology Division, Department of Biology, Faculty of Science, Istanbul University, Vezneciler, Istanbul, Türkiye

Biotechnology Program, Department of Biology, Institute of Graduate Studies in Sciences, Istanbul University, Vezneciler, Istanbul, Türkiye

M. Çalışkan · M. Arıkan (✉)
Biotechnology Program, Department of Biology, Institute of Graduate Studies in Sciences, Istanbul University, Vezneciler, Istanbul, Türkiye

M. Arıkan
e-mail: muzafferarikan@istanbul.edu.tr

M. Çalışkan and M. Arıkan (eds.), *Microbial World: Bacteria and Archaea*, Progress in Molecular and Subcellular Biology 62, https://doi.rg/10.1007/978-3-032-11438-9_10

10.1 Introduction to Microbial Omics

Omics technologies encompass a variety of high-throughput analysis methods, including genomics, transcriptomics, proteomics, metabolomics, and other recent approaches like lipidomics (Pieters et al. 2021; Bhardwaj et al. 2013). These methods provide researchers with tools to discover the molecular composition, functional properties, and interactions of organisms with great detail. Consequently, the increasing use of omics methods has allowed new insights into disease mechanisms, gene-environment interactions, evolutionary biology, and systems-level cellular responses (Kim et al. 2017; Dai and Shen 2022).

High-throughput microbial omics methods can be used to study microorganisms, both at the species level and at the community level. Single-organism omics focuses on elucidating the genomic makeup, expression pathways, and functional capacity of individual microorganisms. While these approaches are often applied to well-established model organisms, such as *Escherichia coli* and *Saccharomyces cerevisiae* (Kwong et al. 2015; Baker et al. 2018), they can also be used to investigate any species that can be isolated and cultivated in the laboratory. Within this context, single-cell omics is a rapidly advancing technology that employs cutting-edge methods to enable researchers to narrow the focus of the study down to a single cell. The ability to detect biomolecules that are present inside just a single cell greatly enhances the efficiency of the analyses and enables more precise detection of target molecules. Thereby, it provides the researchers with a deeper knowledge about the molecular composition, functional properties, and interactions of microorganisms (Blainey 2013; Kelly 2020; Ali et al. 2019; Gaisser et al. 2024).

Studying a single microorganism using omics approaches can provide detailed information on its genetics and metabolism; however, this approach may not be optimal for all research questions or all microorganisms. For example, only a limited proportion of microbial species can be cultured efficiently in laboratory conditions (Steen et al. 2019). Moreover, culturing microorganisms removes them from their natural environments, which can alter gene expression patterns and diminish environmental interactions that are critical for understanding their function and behavior. In addition, some gene products perform functions that are only relevant in complex natural communities, such as in association with other microorganisms or a host organism. Thus, functional analyses are incomplete and potentially biased when these factors are not considered (Nyholm et al. 2020; O'Donnell et al. 2020; Keating et al. 2024; Reuben and Torres 2025). In this context, meta-omics becomes a relevant option when the research focus is on microbial community dynamics or on microorganisms that are difficult to culture. Meta-omics can provide researchers with the tools to discover community-level interactions of microorganisms and to gain insight into uncultured organisms, both within natural environments and laboratory conditions (Pérez-Cobas et al. 2020).

Microbial omics approaches have significantly improved the research in biotechnology, medicine, environmental sciences, and agriculture. Some examples include pathogen surveillance, antibiotic resistance profiling, investigation of host-microbiome interactions, environmental monitoring, and synthetic biology applications (Baker et al. 2018; Brockhurst et al. 2019). The capacity to investigate microbial genomes, transcriptomes, proteomes, and metabolomes continues to expand with the accelerating pace of technological advances, and consequently, drives new discoveries. This chapter describes microbial omics techniques at both the single-organism and meta-omics levels, along with their potential applications and associated tools.

10.2 Single-Organism Omics

10.2.1 Genomics

Microbial single-organism genomics is the study of microorganism genomes using experimental and computational DNA analysis tools. Today, genomic data are of great importance to microbial taxonomy, systematics, and evolutionary studies (Thompson et al. 2013; Chun and Rainey 2014) as most microorganisms in nature are uncultured and only accessible through genomic methods (Riesenfeld et al. 2004). Modern microbial genomics methods increasingly rely on high-throughput DNA sequencing techniques, which enable detailed investigation of microbial species' characteristics. The experimental workflow for microbial genomics typically includes cell isolation, lysis, library preparation, and sequencing, while bioinformatics analyses vary depending on the research goals and may include genome assembly, annotation, comparative genomics, and functional interpretation.

Among various microbial genomics approaches, whole-genome sequencing (WGS) provides the most comprehensive view of a microorganism's genetic composition. When compared to targeted gene techniques like 16S rRNA gene amplicon sequencing, it enables the identification of both taxonomic and functional characteristics, including species classification, virulence factors, and antibiotic resistance genes (Purushothaman et al. 2022; Iacono et al. 2019; Kwong et al. 2015; Baker et al. 2018). Furthermore, WGS enables researchers to create genotype-to-phenotype correlation maps and thus reveal how certain genetic variants drive features like antibiotic resistance and metabolic plasticity (Aminu et al. 2024).

Over the last decade, WGS of microorganisms has become crucial in the food sector, medicine, and biotechnology (Iacono et al. 2019), and it is now widely utilized in public health, epidemiology, microbial ecology, and synthetic biology studies (Brumfield et al. 2020). The comprehensive datasets generated by WGS support a wide range of downstream analyses, including assembly and annotation, comparative genomics, pangenome analysis, phylogenetic analysis and functional genomics. For each analysis type listed below, we provide a brief description of its general features, highlight potential applications, and outline commonly used tools.

Genome assembly is the reconstruction of the genome sequence of an organism for sequencing reads. Individual, short fragment reads are pieced together to cover the entire genome and determine its sequence. This process is usually carried out by comparing with a reference genome. Unless there are no reference genomes available, in which case, this process is referred to as de novo assembly. The assembly step is essential for fully understanding the genetics and evolutionary inheritance of an organism, as it allows researchers to determine biology and genetic history by providing a complete genome sequence (Li and Durbin 2024). SPAdes is an example tool for genome assembly (Bankevich et al. 2012).

Genome annotation is a crucial step in many genomics analysis methods. It refers to the process of identifying and "annotating" various fragments of a sequence of a genome. Genome annotation starts with raw sequence data and applies various computational analyses to return physiologically relevant information like genes, RNA sections, and functions of their coded proteins (de Sá et al. 2018; Gabriel et al. 2024). Modern annotation algorithms are not only able to use DNA sequences, but also RNA-seq data, protein homology, and ab initio gene predictions. This also allows for a synthesis approach to identify many genetic components like genes, regulatory elements, and functional domains (Holt and Yandell 2011; Brettin et al. 2015). Genome annotation studies can be focused on different levels: nucleotide-level that investigates loci of sequences and components, protein-level that tries to understand gene function and process-level that annotates pathways to reveal interactions (de Sá et al. 2018). Tools such as MicrobeAnnotator (Ruiz-Perez et al. 2021a), Prokka (Seemann et al. 2014), and MEGAnnotator2 (Lugli et al. 2023) enable accurate annotations of microbial genomes using comprehensive databases.

Comparative genomics systematically compares genomes of different microbial species or strains to discover genetic diversity and evolutionary relationships (Armstrong et al. 2019; Nandy et al. 2025). It employs whole-genome alignments in order to detect similarities and differences (Wei et al. 2002), perform comparative annotation of both coding and non-coding regions (Paril et al. 2023), and analyze the structural elements of the genome (Paterson et al. 2000). Comparative genomics has a multitude of applications, including the prediction of gene functions using orthologous relationships (Wei et al. 2002; Brown 2007), elucidation of regulatory networks through conserved, non-coding sequences (Ureta-Vidal et al. 2003), and identification of genes associated with pathogenicity or beneficial traits (Nandy et al. 2025). It also enables researchers to track the evolution of antibiotic resistance, virulence factors, and environmental adaptability (Dewar et al. 2025). As comparative studies become more widespread, tools like whole-genome aligners and phylogenetic frameworks are also becoming increasingly common (Paril et al. 2023). An important extension of comparative genomics is pangenome analysis, which evaluates the entire range of genes shared by all individuals of a microbial population or community. A pangenome is made up of (1) genes shared by all strains, a core genome, and (2) genes found only in some individuals, an accessory genome. This method is able to detect genetic variation that a single reference genome study would not be able to detect (Brockhurst et al. 2019; Golicz et al. 2020; Liao et al. 2023). Graph-based pangenome models are able to carry out high-accuracy alignment and structural variation discovery to effectively describe genomic variety (Hickey et al. 2024; Secomandi et al. 2025). Pangenomic techniques are especially useful for analyzing bacterial populations, as processes like horizontal gene transfer and rapid evolution cause changes in accessory genomes

(Tonkin-Hill et al. 2023). It allows for tracking strain variety in diseases, improving industrial strains for microbial biotechnology, and the development of vaccines through discoveries of conserved potential targets across strains (Zhou et al. 2022a). Comparative genomics relies on whole-genome aligners like progressiveMauve, while pangenome analyses often use frameworks, such as Pagoo and PanGenome Graph Builder (Darling et al. 2010; Ferrés and Iraola 2021; Garrison et al. 2023).

Phylogenetic analysis reconstructs the evolutionary relationships among microorganisms using genetic data on a large scale, such as genomes, metagenomes or marker genes. This method is key to the discovery of evolutionary history, taxonomy and community structure in the context of microbial environments. Whole-genome data or marker genes (e.g., 16S rRNA) are frequently used for taxonomic classification. Phylogenetic analyses also help characterize the composition and structure of microbial communities by revealing ecological or functional roles of taxa as well as the relationships between them. In addition, phylogenetic frameworks allow for the study of microbial evolution, strain-level diversity, and the emergence of traits like antibiotic resistance (Singer et al. 2016; Asnicar et al. 2020; Liu et al. 2024). Phylogenetic analyses utilize marker gene-based approaches through tools like QIIME 2 or conserved protein-coding genes or genome-wide marker approaches with tools like LotuS2 (Estaki et al. 2020; Özkurt et al. 2022).

Functional genomics investigates the ways genes and gene products drive the functions, activities and interactions of microorganisms. It aims to link genetic information to biological functions such as metabolism, biogeochemistry, and responses to the environment by analyzing the presence of genes at a genome-wide scale. This field frequently combines genomics with the study of gene expression (transcriptomics), protein production and function (proteomics), and metabolic activity (metabolomics) to characterize the actions and interactions of main biomolecules (DNA, RNA, proteins, and metabolites, along with their modifications) that link observable characteristics of an organism to the genetic properties. Key methods in functional genomics include high-throughput molecular, genetic, and cellular technologies. These methods range from transcriptome analysis (e.g., RNA sequencing) and proteomics to metabolite profiling and epigenomic mapping. Functional genomics studies utilize software tools like METABOLIC and public databases like NCBI Genome database (Caudai et al. 2021; Zhou et al. 2022b).

Genome mining is the computational approach of analyzing genomic data for the detection and annotation of genes or gene clusters called BGCs (biosynthetic gene clusters) that partake in the biosynthesis of secondary metabolites (Ziemert et al. 2016). These secondary metabolites are called NP's (natural products) in this context and are frequently studied in this manner, correlating with the increase of the amount of public sequencing data (Russel et al. 2020). This approach allows scientist to reveal the biosynthetic potential of microorganisms and is frequently employed for: (1) the discovery of new NP's like antibiotic, anticancer compounds, or other therapeutic molecules, (2) discovering the functions of NP's in their environments, and (3) linking BGC's to their respective NP's in order to better understand the microbial biochemistry (Ziemert et al. 2016; Russel et al. 2020). While tools like Prodigal (Hyatt et al. 2010) are able to run gene predictions from genomic sequences, the actual field of genome mining lies within the potential of predicting metabolites directly from genomes. Genome mining platforms such as PRISM 4 (Skinnider et al. 2020) and antiSMASH (Blin et al. 2023) predict BGCs and their associated metabolite structures, thereby facilitating pathway mapping directly from genomic sequences.

10.2.2 Transcriptomics

Microbial single-organism transcriptomics (hereafter referred to as transcriptomics) is the study of all of the RNA molecules (the transcriptome) present inside microbial cells. It provides a snapshot of the genes that are actively expressed at a given time. Unlike genomics, transcriptomics reveals how microorganisms interact in their population, respond to their environment and carry out vital cell functions through investigating gene activity, rather than just the genetic composition.

Transcriptomics typically involves sequencing and quantifying RNA (mainly mRNA) from microbial cells or populations to determine which genes are turned on or off under specific conditions (Sato 2024; Lowe et al. 2017; Mäder et al. 2011; Van Vliet 2010). These methods are also used for tracking pathogen activities and antibiotic responses (Hébert et al. 2018; Ojala et al. 2023). The experimental procedure typically includes RNA isolation, optional mRNA enrichment or rRNA depletion, reverse transcription, library preparation, and sequencing. RNA sequencing (RNA-seq) is a primary method in which researchers run sequencing on the entire isolated RNA content of a sample. It allows researchers to quantify expression levels across the transcriptome and characterize transcriptional differences both among species and through time to ultimately (Lowe et al. 2017; Zhang et al. 2021a). For each analysis type listed below, we briefly describe its general features, highlight potential applications, and outline commonly used tools.

Gene expression analyses utilize RNA-seq and are designed to measure, model, and interpret the activity of thousands of genes across a species. This allows for the investigation of dynamic expression profiling that can be tracked over time. The initial step used in these applications can utilize pipelines including preprocessing (quality control, normalization, filtering), dimensionality reduction (e.g., PCA, UMAP), clustering, and marker gene identification. Differential gene expression (DGE) analysis is a staple technique in transcriptomics. It is the process of quantifying and analyzing variations in gene expression levels across different conditions. It results in a multi-layered and temporally resolved view of cellular regulation, opening new possibilities in systems biology, regenerative medicine, and precision oncology. However, it requires statistical models to derive gene lists and probability estimates of expression differences and limit false-discovery rates and improve the reliability of resulting comparisons (Nayak and Hasija 2021; Gurung 2023). Clustering and pseudotime analysis are sometimes carried out before DGE (Dar et al. 2021). These findings are often validated through independent methods like qRT-PCR and Northern blotting (Becker et al. 2001). In a 2017 study, Costa-Silva et al. queried the most common and state-of-the-art DGE analysis tools and demonstrated the most balanced tools to be DESeq2 (Love et al. 2014), NOISeq (Tarazona et al. 2011) and limma+voom (Law et al. 2014), when compared to qRT-PCR results (Costa-Silva et al. 2017).

Regulatory RNAs are non-coding RNA molecules that include microRNAs, enhancer RNAs, and antisense RNAs that regulate gene expression via molecular mechanisms including chromatin remodeling, RNA processing, and post-transcriptional regulation. Transcriptomics investigates these RNAs by combining advanced sequencing techniques with specialized computational analyses to identify, quantify, and characterize their regulatory functions. Techniques like capped-small RNA-Seq allow sensitive detection of both stable and unstable regulatory RNA elements, including those with rapid turnover or low abundance, by detecting transcription start sites (Duttke et al. 2019). Network inference methods and integration of regulatory RNA data with epigenomic data can help discover connections between regulatory RNAs and their biological roles and target genes (Xue et al. 2023; Wang et al. 2023). In addition, CRISPR-based perturbation can be combined with single-cell transcriptomics to enable systematic dissection of regulatory RNA function in real time (Xu et al. 2024). In order to carry out rRNA detection, optimized pipelines like GRIL-Seq are powerful tools for identifying direct molecular targets (Han et al. 2016).

Co-expression network construction is a method used to analyze gene expression data by identifying modules: groups of genes that show similar expression patterns across many samples. It is used to identify which genes tend to show coordinated expression patterns across a group of samples. Modules that are highly connected are interpreted as sharing similar biological functions and participating in the same pathways. First, relationships between genes are defined based on the similarity between their expression patterns between them. In this context, genes are represented as nodes, and the connections are represented as edges. The connections reflect the strength of the correlation of expression. Finally, modules are identified using clustering techniques. These modules group genes with similar expression patterns across multiple samples. This approach is used in predicting gene functions, the discovery of novel modules, as well as providing insights into regulatory networks (Dam et al. 2017; Liu et al. 2018). Co-expression network construction is carried out through well-established software that utilize algorithms and databases like WGCNA and DAVID (Langfelder and Horvath 2008; Huang et al. 2009a).

Functional enrichment analysis is a technique usually employed for identifying trends in large-scale biological datasets in the form of gene lists or genome-wide regions of interest. It summarizes regulated gene expression profiles into simplified functional categories, such as signaling or biochemical pathways. These categories are determined by the scientists of the transcriptomics research ecosystem in order to interpret complex data by highlighting which biological functions or pathways are over-represented among a set of genes. It is frequently used in biomedicine to identify diseases and drug mechanisms by highlighting the underlying pathways. This method is capable of interpreting differential expression and comparative analysis (Wu et al. 2021; Wijesooriya et al. 2022). Software packages like clusterProfiler 4.0

allow researchers to interpret transcriptomics data for functional enrichment analysis (Wu et al. 2021).

10.2.3 Proteomics

Microbial single-organism proteomics (hereafter referred to as proteomics) is the large-scale study of all proteins (the proteome) produced by microorganisms, such as bacteria, archaea, fungi, and viruses. It provides direct insight into the actual functions, activities, and interactions of microbes by identifying and quantifying their proteins, revealing what microbes are doing. This process involves identifying and quantifying protein abundance in a population, most often using mass spectrometry–based techniques (Kleiner 2019; Washburn and Yates 2000; Graham et al. 2011).

Proteomics workflows typically involve protein identification, quantification, and subsequent statistical analysis for the biological interpretation. In the widely used bottom-up approach, proteins are enzymatically digested into peptides, most commonly using proteases such as trypsin. The resulting peptides are then analyzed by mass spectrometry, which identifies peptides by measuring their mass-to-charge ratios to detect unique spectral signatures. This process generally includes peptide separation by chromatography or electrophoresis, ionization of peptides, fragmentation of ions to generate spectra, and computational analysis of the spectra to either match them to reference peptide sequences and infer novel proteins. The processed data are subsequently subjected to statistical analysis to characterize the protein composition of a sample or to detect proteome-level changes across experimental conditions (Noor et al. 2021; Jiang et al. 2024).

Differential protein expression analysis involves identifying and quantifying proteins whose abundance significantly changes across different biological conditions, such as healthy versus diseased states or varying environmental contexts. This analysis elucidates how biological systems respond to stimuli, treatments, or environmental perturbations. Differential expression studies typically rely on bottom-up proteomics and mass spectrometry-based quantitative methods, as described previously. The resulting protein abundance data are analyzed to detect true changes while minimizing false positives, often using tools such as MaxQuant (Cox et al. 2009). Statistical analyses are performed using software packages such as DanTE, Perseus, and MSstats 4.0 to assess significance and variability across samples (Polpitiya et al. 2008; Tyanova et al. 2016; Kohler et al. 2023). Web-based platforms like ProteoSign v2 further streamline differential expression analysis, supporting multiple experimental designs and downstream analyses such as gene set enrichment (Theodorakis et al. 2021). Building on these quantitative proteomics approaches, integrative strategies such as proteogenomics further combine protein and genomic data to enhance genome annotation and functional characterization of proteins.

Proteogenomics allows researchers to validate gene predictions, discover unannotated genes, correct errors like reading frame reversals, and identify precise translational start sites, stop codon read-throughs and programmed frameshifts. In addition, mass spectrometry methods can detect events such as multiple initiations, alternative splicing, peptide signal processing, post-translational modifications, constitutive proteolytic processing and sequence polymorphisms. By integrating protein data, researchers can assign functions to hypothetical proteins and improve the accuracy of gene product descriptions. Furthermore, proteogenomics can be applied at the earliest stage of genome projects to provide experimental evidence for gene models in parallel with genome sequencing (Jaffe et al. 2004; Armengaud 2009). Several tools have been developed for this integrative approach, including Pepline, which searches for intron-split peptides (Ferro et al. 2008); GFS (Genomic Feature Scanner), which identifies the corresponding genomic loci for peptides of a given protein (Holmes and Giddings 2008); and GAPP (Genome Annotation and Global Profiling of Posttranslational Modifications in Prokaryotes), which enables large-scale, proteome-wide analysis of post-translational modifications in prokaryotes (Zhang et al. 2016). Beyond refining gene models and protein annotations, proteogenomics generates comprehensive protein datasets that can be further interpreted using bioinformatics approaches such as functional enrichment and pathway analysis to uncover the biological processes and pathways most relevant to the observed proteins.

Functional enrichment analysis (or gene set analysis, GSA) maps genes to biological annotations, such as Gene Ontology (GO) terms or pathways, and tests statistically whether specific terms are overrepresented, highlighting key biological processes. Pathway analysis, a subset of enrichment analysis, focuses on identifying signaling, metabolic, or regulatory pathways significantly represented in a dataset, revealing which biological processes are most affected. Both methods shift the focus from individual genes to functionally related groups, providing a systems-level understanding of molecular data (Huang et al. 2009b; Garcia-Moreno et al. 2022). Gene set analysis is typically divided into three categories. Singular enrichment analysis (SEA) evaluates the statistical significance of each annotation among candidate genes, with tools such as GeneCodis4 performing SEA and DAVID providing a comprehensive annotation database (Garcia-Moreno et al. 2022; Huang et al. 2009a). Gene set enrichment analysis (GSEA) assesses the distribution of genes associated with a given term, ranking them by criteria such as fold changes, using resources like GSEA-P and MSigDB (Subramanian et al. 2005). Modular enrichment analysis (MEA) exploits relationships between annotations to define modules and evaluates their combined significance, with tools such as STRING and DAVID widely used for this purpose (Szklarczyk et al. 2023; Huang et al. 2009a). Functional enrichment and pathway analyses can be complemented by proteomic networks by modeling interactions and co-expression patterns among proteins, enabling the identification of functional modules and key regulatory hubs.

Proteomic networks, often referred to as protein–protein interaction (PPI) networks, model the interactions among proteins, where nodes represent proteins and edges represent physical or functional associations. These networks are generally constructed from protein abundance profiles obtained through mass spectrometry. Co-expression analysis examines how protein abundance patterns correlate across samples, identifying modules of co-expressed proteins that may share functions, regulatory mechanisms, or serve as potential biomarkers. Both PPI and co-expression analyses facilitate the identification of functional modules and hub proteins, providing insights into biological processes and disease mechanisms, with tools such as Cytoscape and WGCNA widely used for these analyses (Gibbs et al. 2013; Vella et al. 2017; Langfelder and Horvath 2008; Saito et al. 2012).

Proteomics is also essential for studying post-translational modifications (PTMs), including phosphorylation, acetylation, and ubiquitination, which regulate signaling and metabolic pathways. Sensitive mass spectrometry, peptide enrichment methods, and advanced algorithms enable the discovery of PTMs even at the single-cell level, revealing their dynamics in heterogeneous populations (Mann and Jensen 2003). Key approaches include data-independent acquisition mass spectrometry (DIA-MS) for high-throughput and accurate PTM identification, often combined with deep learning and spectral libraries (Witze et al. 2007; Yang and Qiao 2023); peptide enrichment techniques, such as immunoenrichment and chemoselective labeling, for detecting low-abundance PTMs (Mann and Jensen 2003); nanopore sensing for direct, single-molecule PTM detection (Nova et al. 2023); and meta-analysis tools like PTM-Shepherd for automated annotation and comparison of PTM datasets (Geiszler et al. 2021).

10.2.4 Metabolomics

Microbial single-organism metabolomics (hereafter referred to as metabolomics) is the study of low-weight metabolites, such as amino acids, sugars, organic acids, and secondary metabolites present in or around microbial cells at a specific time. Unlike transcriptomics and proteomics, which assess the potential gene expression or expressed proteins at a given time, metabolomics provides researchers with a direct snapshot of the physiological and metabolic state of a species. It provides real-time data on microbial activity and reveals the exact mechanisms by which microorganisms respond to environmental conditions, interact with conspecifics or their hosts, and influence health and disease. As the metabolite levels show rapid fluctuation in response to environmental and intracellular signals, this method is particularly valuable in microbial omics studies (Holbrook-Smith et al. 2024; Zhao et al. 2020).

Metabolomics workflows are typically categorized as targeted or untargeted. Targeted metabolomics workflows focus on the accurate quantification of predefined metabolites. This approach is used for validating known biomarkers and studying specific metabolic pathways. While the study scope is limited, these workflows provide high sensitivity and high-accuracy quantification. Untargeted metabolomics workflows, by contrast, aim to detect a wide array of metabolites without prior knowledge. This method is crucial for discovering novel metabolic states or molecules. Since these workflows come with the cost of lower sensitivity and reproducibility, they often require extensive post-acquisition data processing to identify or annotate these newly discovered features. Hybrid workflows that combine both approaches are becoming increasingly common. These workflows allow researchers to discover metabolites in a wide range while also identifying key metabolites of interest (Gao et al. 2018; Zheng et al. 2020; Amer et al. 2023).

Similar to proteomics, mass spectrometry techniques are key technologies for metabolomics, due to their high sensitivity, spatial resolution, and broad molecular coverage. Several techniques are employed to enhance metabolite detection and characterization. Nanospray ionization (nanoESI-MS) provides a simple and consistent approach to increase coverage in spatial metabolomics, addressing a common limitation of conventional studies (Nguyen et al. 2024). Matrix-assisted laser desorption/ionization (MALDI-MS) enables the generation of spatially resolved metabolic maps (Kuik et al. 2024). Nuclear magnetic resonance (NMR) offers quantification and structural elucidation of metabolites without the need for metabolite-specific standards. Although less sensitive than mass spectrometry, NMR remains a valuable complementary method, particularly for validating key findings or analyzing stable isotope-labeled substrates (Muhamadali et al. 2023).

The wide variety of metabolites produced by a microbial species is identified, quantified and interpreted for their function and structure with bioinformatic tools. These tools allow researchers to distinguish microbial metabolites, map them to metabolic pathways, and integrate metabolomics data with other omics layers for a more comprehensive understanding of biological processes. Metabolite identification is performed using tools such as MetOrigin and microbeMASST, which differentiate metabolites by origin through spectral database searches and link them to organism–metabolite interactions while enabling pathway enrichment analysis (Yu et al. 2022; Zuffa et al. 2024). Pathway and functional analysis tools, including MetOrigin, MicrobiomeAnalyst 2.0, and MetaboAnalyst, facilitate mapping of metabolites to metabolic pathways, perform functional enrichment, and support statistical interpretation (Yu et al. 2022; Lu et al. 2023; Marco-Ramell et al. 2018). Combinational approaches, such as integrating NMR with mass spectrometry, further enhance metabolite detection and annotation, enabling identification of key metabolites that might be missed by MS alone and providing more comprehensive pathway coverage (Bhinderwala et al. 2018).

A summary of single-organism omics analyses, their main applications, and representative tools is presented in Table 10.1.

10.3 Meta-Omics

10.3.1 *Metagenomics*

Metagenomics is the study of collective genetic material extracted from an environmental sample (Riesenfeld et al. 2004). This culture-independent approach enables the characterization of entire microbial communities, including organisms that cannot be grown under laboratory conditions, thereby expanding the scope of microbial omics and transforming our understanding of microbial diversity and function (Pérez-Cobas et al. 2020; Zhang et al. 2021b).

Table 10. 1 Omics types, their main applications and representative tools

Omics type	Main applications	Representative tools
Genomics	Genome Annotation	MicrobeAnnotator (Ruiz-Perez et al. 2021a), MEGAnnotator2 (Lugli et al. 2023)
	Comparative Genomics,	progressiveMauve, (Darling et al. 2010)
	Pangenomics	Pagoo (Ferrés and Iraola 2021), PanGenome Graph Builder (Garrison et al. 2023)
	Phylogenetics	QIIME 2 (Estaki et al. 2020), LotuS2 (Özkurt et al. 2022)
	Functional Genomics	METABOLIC (Zhou et al. 2022b)
	Genome Mining	antiSMASH (Blin et al. 2023), PRISM 4 (Skinnider et al. 2020)
Transcriptomics	Gene Expression Profiling	NOISeq (Tarazona et al. 2011), limma+voom (Law et al. 2014), DESeq2 (Love et al. 2014)
	Regulatory RNA Detection	GRIL-Seq, (Han et al. 2016)
	Co-Expression Networks	WGCNA (Langfelder and Horvath, 2008), DAVID (Huang et al. 2009a)
	Functional Enrichment	clusterProfiler (Wu et al. 2021)
Proteomics	Differential Protein Expression	MaxQuant (Cox et al. 2009), Perseus (Tyanova et al. 2016), ProteoSign (Theodorakis et al. 2021)
	Proteogenomics	Pepline, GAPP (Ferro et al. 2008; Zhang et al. 2016)
	Functional Enrichment Analysis	STRING, DAVID (Szklarczyk et al. 2023; Huang et al. 2009a)
	Proteomic Network Analysis	Cytoscape, WGCNA (Langfelder and Horvath 2008; Saito et al. 2012)
	PTM Analysis	PTM-Shepherd, DIA-MS (Geiszler et al. 2021; Yang and Qiao 2023)
Metabolomics	Metabolite Isolation, Quantification	nanoESI-MS (Nguyen et al. 2024), MALDI-MS (Kuik et al. 2024)
	Metabolite Identification	MetOrigin, microbeMASST (Yu et al. 2022; Zuffa et al. 2024)
	Pathway & Functional Mapping	MicrobiomeAnalyst (Lu et al. 2023), MetaboAnalyst (Marco-Ramell et al. 2018)

Metagenomics involves extracting total DNA from environmental or clinical samples using physical (e.g., freeze–thaw) or chemical methods (e.g., protease and SDS) (Gabor et al. 2003). Lysis buffers with detergents and nucleases are commonly used to disrupt membranes and release DNA, which is then purified through chemical steps. Additional treatments may be required to remove contaminants, such as humic acids in soil, which are known to tightly bind DNA (Daniel 2005). The purified DNA is subsequently fragmented, ligated to sequencing adapters, and prepared as libraries for sequencing (Shi et al. 2022; Zhang et al. 2021b). Once high-quality DNA libraries are generated, the next step in metagenomics is bioinformatics analyses such as assembly, binning, comparative metagenomics or metagenome mining.

Metagenomic assembly reconstructs longer contiguous sequences (contigs) from short sequencing reads, which are then grouped (binned) into draft genomes representing individual community members. Assembly can be performed with short-read tools such as metaSPAdes and MetaVelvet (Papudeshi et al. 2017; Zhang et al. 2023), or with long-read assemblers like hifiasm-meta and metaMDBG, which improve contiguity and completeness of metagenome-assembled genomes (MAGs) and can also integrate short- and long-read data (Bickhart et al. 2022; Portik et al. 2024). Binning tools such as MetaBAT2, MaxBin, and CONCOCT cluster contigs based on sequence composition (e.g., k-mer frequencies, GC content) and coverage patterns across samples (Kang et al. 2019; Lin and Liao 2016; Strous et al. 2012). Graph-based methods, including GraphBin and METAMVGL, refine binning using assembly graph data, which is particularly valuable for resolving short or ambiguous contigs (Mallawaarachchi et al. 2020; Zhang and Zhang, 2020). More recently, machine-learning approaches such as VAMB (variational autoencoders) and SemiBin (siamese neural networks) integrate multiple features to enhance binning performance in complex or novel communities (Nissen et al. 2021; Pan et al. 2022; Herazo-Álvarez et al. 2025).

Beyond genome reconstruction, metagenomics enables community profiling at both taxonomic and functional levels. Taxonomic profiling identifies and quantifies microorganisms, often down to the species or strain level, using marker-gene or reference-based methods. Tools such as MetaPhlAn and mOTUs perform taxonomic classification from 16S rRNA genes or reference genomes (Segata et al. 2012; Ruscheweyh et al. 2021). These approaches expand the detectable diversity of uncultured species and novel environments (Beghini et al. 2021; Whelan et al. 2020). Functional profiling, in turn, characterizes the metabolic potential of microbial communities by identifying gene families, pathways, and traits. For example, HUMAnN maps shotgun data to known pathways (Abubucker et al. 2012; Beghini et al. 2021). Together, taxonomic and functional profiling provide a multidimensional perspective, linking taxa to ecological functions, microbial communities to diseases, and revealing ecosystem-level processes such as nutrient cycling, biofilm formation, and virulence (Langille et al. 2013; Beghini et al. 2021).

Comparative metagenomics systematically compares datasets from different environments to assess how community composition and function vary across samples, revealing environment-specific genes, metabolic capabilities, and ecological adaptations. This approach can highlight differences in metabolic pathways and resistance genes (Tringe et al. 2005; Nayfach and Pollard 2016), and by analyzing the abundance and diversity of taxa or genes, researchers can detect habitat-specific signatures and identify enriched functions. Statistical methods are applied to determine differentially abundant genes, taxa, or pathways between groups (Jonsson et al. 2016), with tools such as METAGENassist and IMG/M commonly used for these analyses (Arndt et al. 2012; Chen et al. 2019).

Beyond profiling, metagenome mining enables the prediction of target biomolecules, expanding the discovery space for natural products. Genome mining tools such as BiGSCAPE, CORASON (Navarro-Muñoz et al. 2020), and PRISM 4 (Skinnider et al. 2020) predict biosynthetic diversity directly from metagenomic data. Metagenomics can also support community-scale metabolic modeling, either by using reference-based genomic data or by reconstructing networks from metagenome-assembled genomes (MAGs), with the latter offering improved accuracy and validation when integrated with metabolomic data (Mallick et al. 2019; Majzoub et al. 2024).

10.3.2 Metatranscriptomics

Metatranscriptomics provides a snapshot of which genes are actively being expressed in a given environment by studying all of the RNA transcripts produced by the present microorganisms. In contrast to metagenomics, which identifies the presence of microorganisms and genes, metatranscriptomics elucidates the active genes within microbial communities. This approach enables researchers to understand how communities respond to environmental changes, interact with hosts, and contribute to processes like nutrient cycling, disease, and resistance to antibiotics (Shakya et al. 2019; Zhang et al. 2021a; Ojala et al. 2023; Peng et al. 2024; Zampieri et al. 2023).

Metatranscriptomic workflows begin with RNA extraction, which follows principles similar to DNA extraction but requires more sensitive handling. Depending on the RNA type of interest (e.g., mRNA, microRNA), enrichment steps are applied prior to sequencing. Extracted (and optionally enriched) RNA is then fragmented according to platform requirements and reverse transcribed into cDNA using reverse transcriptase with primers such as random hexamers or oligo(dT). Subsequent steps include adapter ligation, end repair, library cleanup, and cDNA amplification, after which the amplified cDNA is sequenced (Bikel et al. 2015). Processing of metatranscriptomic reads involves quality filtering, assembly, annotation, gene expression profiling, and taxonomic assignment. Taxonomic profiling is achieved with tools such as MetaPhlAn3 (bioBakery 3) and Kraken, which rely on taxon-specific marker genes (Beghini et al. 2021; Wood and Salzberg 2014). Downstream analyses, including differential expression and pathway enrichment, are typically performed with statistical frameworks such as DESeq2 (Love et al. 2014).

Understanding spatial and temporal variation in microbial gene expression is key to linking community function with environmental or host conditions. Spatial metatranscriptomics (SmT) employs spatially barcoded arrays to capture host and microbial transcripts at defined tissue positions, enabling micrometer-scale mapping of microbial "expression hotspots" and host–microbe interactions (Saarenpää et al. 2023; Lyu et al. 2023). Similarly, par-seqFISH, an imaging-based single-cell method, profiles gene expression in situ to reveal microscale heterogeneity within biofilms or tissues (Dar et al. 2021). Temporal metatranscriptomics is addressed by longitudinal sampling, which captures seasonal or conditional transcriptional changes. For example, freshwater lake studies have demonstrated strong seasonal variation in bacterial transcriptional activity, reflecting adaptation to temperature and environmental fluctuations (Gifford et al. 2011; Zhang et al. 2021a; Sadeghi et al. 2023). Coupling time-series sequencing with statistical modeling enables resolution of dynamic shifts in community gene expression, with tools such as RAIN supporting rhythmic or periodic data analysis (Thaben et al. 2014; Linz et al. 2020).

A major application of metatranscriptomics is linking expressed genes with their microbial origin, thereby providing taxon-resolved functional profiles. Pipelines such as MetaFunc (Sulit et al. 2023) and MetaPro (Taj et al. 2023) integrate taxonomic and functional annotations to directly map expressed genes to specific taxa, while resources such as MarFERRet leverage curated databases to improve annotation accuracy (Groussman et al. 2023). In communities lacking reference genomes, de novo assembly approaches can reconstruct transcripts for novel or poorly characterized microbiomes. Tools like Trinity (Grabherr et al. 2011) and rnaSPAdes (Bushmanova et al. 2019) enable functional and taxonomic annotation of assembled transcripts. Some pipelines further enhance interpretation by offering visualization and correlation analyses between expressed genes and microbial taxa (Davids et al. 2016), extending even to host gene expression in host-associated microbiomes (Lyu et al. 2023).

10.3.3 Metaproteomics

Metaproteomics is the large-scale identification and quantification of proteins in entire microbial communities. It offers a snapshot of the protein landscape in microbial communities. Unlike metagenomics (which reveals genetic potential) or metatranscriptomics (which measures gene expression at the RNA level), metaproteomics captures the actual functional proteins to reflect the real-time metabolic and physiological state of the microbial community (Wilmes et al. 2015; Kleiner 2019; Van Den Bossche et al. 2025). Metaproteomic functional profiling is used to study metabolic pathways, identify key enzymes, and link protein expression to specific taxa. It provides further understanding in studying microbial function, interspecies interactions, and ecological functions in various environments. It is also used for biomarker discovery and environmental monitoring, identifying protein markers linked to diseases, environmental stress, or industrial applications, and tracking functional changes in microbial communities (Maron et al. 2007; Armengaud 2022; Pan et al. 2024).

Efficient protein extraction is a critical step in metaproteomics. Chemical lysis commonly employs detergents such as SDS (sodium dodecyl sulfate), which effectively solubilize proteins and yield higher recovery compared to non-ionic detergents or urea. Mechanical disruption methods, including ultrasonication and, less efficiently, bead beating, are often necessary for complex matrices such as soil or biofilms. Protocol choice significantly affects protein yield and depends on sample type and study objectives (Benndorf et al. 2009; Keiblinger et al. 2012; Zhang et al. 2018; García-Durán et al. 2021). Environmental samples may require specific approaches, such as phenol extraction or NaOH treatment for soil and water, and stepwise lysis can selectively enrich proteins from particular microbial taxa, e.g., gram-positive versus gram-negative organisms (Benndorf et al. 2009; Keiblinger et al. 2012; Salvato et al. 2022). Purification of extracted proteins is also essential to remove contaminants that interfere with downstream analyses, with methods including alcohol or acetone precipitation and filter-aided sample preparation (FASP) (Speda et al. 2017; Tanca et al. 2024). Extracted and purified proteins are enzymatically digested and subsequently analyzed using chromatography-coupled mass spectrometry. Metaproteomics data allows both taxonomic and functional analyses.

Metaproteomics data enable both taxonomic and functional analyses, which can be either peptide-centric or protein-centric. In peptide-centric approaches, peptides are annotated immediately after MS identification based on sequence similarity, preserving all potential annotations and providing a comprehensive view of taxa or functions. Protein-centric approaches, in contrast, group peptides into proteins or protein subgroups to reduce redundancy, with annotation applied after aggregation to improve specificity (Van Den Bossche et al. 2025). The choice between these strategies depends on the study's objectives. Additionally, peptide-centric databases, such as DBCGR2, have been shown to enhance the resolution of peptide-specific analyses (Gao et al. 2025).

Taxonomic assignment in metaproteomics can be performed at either the peptide or protein level, and several factors must be considered to ensure accuracy. Peptides unique to a protein provide direct species-level evidence, whereas shared peptides among homologous proteins require careful handling to avoid ambiguous assignments (Tanca et al. 2013; Blakeley-Ruiz and Kleiner 2022). The Last Common Ancestor (LCA) method improves accuracy by assigning each peptide to the most specific taxonomic level common to all matching proteins. Tools such as MetaX (Wu et al. 2025), MetaLab (Cheng et al. 2017), and Unipept (Gurdeep Singh et al. 2018) facilitate genus- or species-level mapping. Employing sample-specific or hybrid databases enhances coverage and precision, particularly for poorly characterized or novel taxa. Optimizing analysis parameters is critical to minimize false positives caused by shared peptides or database redundancy (Tanca et al. 2013; Blakeley-Ruiz and Kleiner 2022; Wu et al. 2024a).

Functional annotation in metaproteomics often integrates peptide identification with curated databases using bioinformatic tools to achieve high accuracy and broad coverage. Peptides detected by mass spectrometry are typically matched to protein sequences from either public repositories or sample-specific metagenomes (Van Den Bossche et al. 2025). The choice of database strongly influences annotation quality and coverage (Tanca et al. 2016; Da Silva Duarte et al. 2024; Gao et al. 2025). Homology-based annotation is commonly performed by aligning peptide sequences against resources such as UniProt (The UniProt Consortium 2023), eggNOG (Hernández-Plaza et al. 2023), and NCBI (Schoch et al. 2020; Schiebenhoefer et al. 2020). Software packages like MetaLab (Cheng et al. 2017) and MetaProteomeAnalyzer (Muth et al. 2015) utilize these databases to streamline functional annotation, while tools such as mPies (Werner et al. 2019) and MetaX (Wu et al. 2025) apply consensus-based approaches that minimize erroneous matches through multiple sequence alignments. Beyond homology-based methods, machine learning has expanded functional annotation capacity. For instance, DeepFRI can predict functions of proteins lacking homologs, thereby improving coverage in metaproteomic datasets (Maranga et al. 2023). Each tool and database used in annotation carries unique strengths in sensitivity, specificity, and taxonomic integration, making their complementary use valuable.

10.3.4 Meta-Metabolomics

Meta-metabolomics is the comprehensive analysis of all metabolites present in microbial communities. It is a key approach for discovering the functional components of microbiomes by directly measuring metabolites, small molecules produced or modified by microorganisms. It reveals environmental or host-associated metabolites through targeted or untargeted profiling methods. Meta-metabolomics generally complements other meta-omics approaches to provide concise results in terms of community metabolic functions and interactions (Addis et al. 2016; Droby and Wisniewski 2018; Zhang et al. 2019; Go et al. 2024).

Community-level metabolite profiling enables researchers to link metabolite patterns to microbial community structure and function. This approach usually utilizes untargeted metabolomics and computational modeling, as well as multi-omics integration (Shaffer et al. 2022; Majzoub et al. 2024). Techniques like LC-MS (liquid chromatography–mass spectrometry) and GC-MS (gas chromatography–mass spectrometry) are used in combination with NMR to identify and quantify metabolites in samples standardized in terms of collection, extraction and analytical methods. The data is then analyzed using various statistical and bioinformatics methods (Wagner et al. 2007).

Researchers use a variety of computational approaches to map metabolites to microbial pathways, revealing their origins and functions within communities (Table 10.2). MetOrigin identifies metabolite origins, performs pathway enrichment, and visualizes metabolite–pathway relationships via Sankey networks, integrating statistical and biological correlations (Yu et al. 2022; Wu et al. 2024b). Likewise, AMON leverages metabolic networks and KEGG data to annotate metabolite origins and generate pathway maps (Shaffer et al. 2019), while microbeMASST links mass spectrometry data to curated microbial monoculture databases to improve annotation confidence (Zuffa et al. 2024). MPA Pathway Tool maps microbial community data, including genes and proteins, to metabolic pathways with customizable taxonomic parameters (Walke et al. 2021). Integrative methods such as metaFISH combine FISH with high-resolution MALDI mass spectrometry imaging to spatially assign metabolites to microbial taxa at micrometer resolution (Geier et al. 2020), and predictive frameworks like PredCMB infer metabolite changes from shotgun metagenomic data by connecting gene abundance patterns

Table 10.2 Meta-omics types, their main applications and representative tools

Omics type	Main applications	Representative tools
Metagenomics	Assembly	metaSPAdes (Papudeshi et al. 2017), MetaVelvet (Zhang et al. 2023)
	Binning	MetaBAT2 (Kang et al. 2019), GraphBin (Mallawaarachchi et al. 2020)
	Taxonomic Profiling	MetaPhlAn 3 (Segata et al. 2012, Beghini et al. 2021), mOTUs (Ruscheweyh et al. 2021)
	Functional Profiling	HUMAnN 3 (Abubucker et al. 2012, Beghini et al. 2021), PICRUSt (Langille et al. 2018).
	Comparative Metagenomics	METAGENassist (Arndt et al. 2012), IMG/M (Chen et al. 2019).
	Metagenome Mining	BiGSCAPE, CORASON (Navarro-Muñoz et al. 2020) and PRISM 4 (Skinnider et al. 2020)
Metatranscriptomics	Gene Expression Profiling	MetaPro (Taj et al. 2023), MetaPhlAn 3, HUMAnN 3 (Beghini et al. 2021), DESeq2 (Love et al. 2014)
	Spatial Transcriptomics	par-seqFISH (Dar et al. 2021)
	Temporal Transcriptomics	RAIN (Thaben et al. 2014)
	Mapping Expression with Taxonomy	MetaFunc (Sulit et al. 2023), MarFERRet (Groussman et al. 2023).
	De novo Assembly	Trinity (Grabherr et al. 2011), rnaSPAdes (Bushmanova et al. 2019)
Metaproteomics	Annotation Databases	UniProt (The UniProt Consortium. 2023), eggNOG (Hernández-Plaza et al. 2023), NCBI (Schoch et al. 2020)
	Functional Annotation	MetaLab (Cheng et al. 2017), mPies (Werner et al. 2019)
	Taxonomic Mapping	MetaX (Wu et al. 2025), Unipept (Gurdeep et al. 2018)
Metametabolomics	Community-Level Profiling	Statistical Analyses on MS and NMR Spectra (Wagner 2007)
	Microbial Pathway Mapping	MetOrigin (Yu et al. 2022), AMON (Shaffer et al. 2019), microbeMASST (Zuffa et al. 2024),
	Spatial Mapping	metaFISH (Geier et al. 2020).
	Gene-To-Metabolite Network Analysis	PredCMB (Ji and Jung 2025).
	Host–Microbiome Interaction	LOCATE (Shtossel et al. 2024), MicrobioLink (Andrighetti et al. 2020)

to metabolite predictions via gene-to-metabolite networks (Ji and Jung 2025). Tools like LOCATE extend this by predicting microbiome-derived metabolites and associating them with host conditions (Shtossel et al. 2024), and computational pipelines such as MicrobioLink integrate predicted microbe–host interactions with molecular networks, enabling mapping of microbial effects on host pathways and gene expression (Andrighetti et al. 2020).

10.4 Multi-Omics Integration and Comparative Analysis

While omics methods offer valuable scientific information, a single type of omics analysis is often considered inconclusive within the modern perspective. Undoubtedly, an investigation of the same biological phenomenon from the different omics levels would yield more accurate and precise results. For this reason, researchers have been utilizing different omics approaches in unison for research. Multi-omics integration is a modern, comprehensive approach that elucidates the structure and functions of microorganisms far better than standalone omics applications. As omics techniques require computational tools, multi-omics integration is accordingly computationally demanding method. Thus, various methods and tools are being developed in the scientific community every day. For example, integrating metagenomic reconstructs from multiple sites with metatranscriptomic datasets allows for comparative analysis of gene expression across locations, such as along different regions of the intestinal tract. This approach enhances the identification of site-specific functional activities, including genes of unknown function (Uehara et al. 2022).

Integration of different omics data is a challenging task because of their heterogeneity and complexity. The input and output data for different omics levels often differ, more so as the study continues, the analyses downstream. This heterogeneous and high amount of data should be leveraged in meaningful, parallel or subsequent analyses. In order to accomplish this, scientists have been developing a wide range of compu tational and statistical tools (Table 10.3). HONMF utilizes matrix factorization (Ma et al. 2023), MintTea (Muller et al. 2024), and sCCA (Deng et al. 2024) use canonical correlation analysis, PALM utilizes dynamic Bayesian networks (Ruiz-Perez et al. 2021b), mmvec uses machine learning approaches through neural networks (Morton et al. 2019). Intermediate integration strategies, such as those implemented in MintTea, take features from multiple omics and combine them into intermediary representations in order to make them easier to interpret and increase the accuracy of predictions. In addition, network-based techniques, such as MetagenoNets (Nagpal et al. 2020) and graph databases, are increasingly utilized to model complex interactions and derive insights from multi-layered data (Kim et al. 2022).

Another important challenge in integrating multi-omics data is standardization of complex analyses. The storage-heavy and complex nature of multi-omics data necessitates streamlined, easy-to-use processes to become standard tools. These workflows need to be scalable for different study ranges and yield reproducible

Table 10.3 Multi-omics applications and representative tools

Multi-Omics application	Description	Representative tools
Statistical and Machine Learning Integration	Employs computational and statistical approaches to account for data heterogeneity and improve analysis quality	mmvec (Morton et al. 2019), PALM (Ruiz-Perez et al. 2021b), MintTea (Muller et al. 2024)
Web-Based Environments	Allow for data aggregation, annotation and comparative, statistical, functional and network analyses.	Qiita (Gonzalez et al. 2018), IMG/M (Chen et al. 2019), MetagenoNets (Nagpal et al. 2020), MicrobiomeAnalyst 2.0 (Lu et al. 2023)
Standardized and Reproducible Pipelines	Provides flexible multi-omics integration and a multitude of reproducible workflows for downstream analyses.	gNOMO2 (Arıkan and Muth 2024), EasyMultiProfiler (Liu et al. 2025), IMP (Narayanasamy et al. 2016)

results for precision. Thus, several bioinformatic platforms and pipelines have been established to carry out reproducible, scalable, and user-friendly multi-omics integration as needed. Web-based environments like Qiita (Gonzalez et al. 2018) and IMG/M (Chen et al. 2019) allow for data aggregation, annotation, and comparative analysis. MicrobiomeAnalyst 2.0 (Lu et al. 2023) and MetagenoNets (Nagpal et al. 2020) enable comprehensive statistical, functional, and network analyses. Meanwhile, modular pipelines like gNOMO2 (Arıkan and Muth 2024) EasyMultiProfiler (Liu et al. 2025), and IMP (Narayanasamy et al. 2016) support flexible integration of various omics types and a multitude of downstream analyses. In addition, complex software platforms like bioBakery 3 offer integrated workflows for taxonomic, functional, and strain-level profiling (Beghini et al. 2021).

10.5 Challenges and Future Directions

Omics approaches offer modern, comprehensive, and high-resolution methods for studying microbial communities. Despite their advantages, several challenges persist. Different omics layers produce diverse data types and formats, often at inconsistent scales and with varying experimental designs. These differences extend to analysis pipelines, complicating integration and making comprehensive biological interpretation difficult (Cocolin et al. 2018; Tarazona et al. 2021; Arıkan and Muth 2023).

Multi-omics datasets are frequently noisy, sparse, and non-normal, which can limit statistical analyses and increase the risk of false discoveries (Mallick et al. 2021; Tarazona et al. 2021). Integration is further hindered by a lack of standardized protocols for sample collection, processing, and analysis, as well as data heterogeneity, compositionality, missing values, and batch effects, reducing reproducibility, especially in cross-study comparisons (Arıkan and Muth 2023; Heintz-Buschart and Westerhuis 2022; Shaffer et al. 2022). Moreover, high-throughput omics generate massive datasets that require advanced computational resources for storage, processing, and analysis. Limited access to such resources, coupled with incomplete development of integration tools, underscores the need for standardization and large public datasets (Chetty and Blekhman 2024). Recent advances in user-friendly, scalable, and comprehensive software aim to address these challenges, enhancing reproducibility and enabling multi-omics integration (Mallick et al. 2021; Arıkan and Muth 2023).

Biological and experimental data do not always align perfectly in multi-omics studies. Linking omics features to specific microbial functions or phenotypes is difficult due to the complexity and diversity of microbial communities (Cocolin et al. 2018; Arıkan and Muth 2023). Many detected proteins and metabolites remain uncharacterized, and ecological interactions are often poorly understood, particularly in the gut microbiome (Daliri et al. 2021). Unculturable or unidentified microorganisms and unknown metabolites further complicate experimental and analytical workflows (Gutleben et al. 2018).

To enhance result reliability, comparative analyses across cohorts and studies are essential. Meta-analyses and cross-cohort studies conducted under standardized protocols facilitate meaningful comparisons across environmental datasets and healthcare studies (Gonzalez et al. 2018; Ning et al. 2023). Standardization of sample processing, data formats, reference databases, and analytical workflows is critical for reproducibility and accurate cross-study comparisons (Heintz-Buschart and Westerhuis 2022; Arıkan and Muth 2023).

10.6 Conclusion

Omics technologies have been central to progress in various fields, and the increasing use of these technologies in microbiology has brought many opportunities. Genomics, transcriptomics, proteomics, and metabolomics provide complementary views of the structure and function of single organisms while meta-omics approaches further illuminate community dynamics, evolutionary relationships, and environmentally driven phenotypes. Despite ongoing challenges and limitations, omics methods remain essential, continually advancing tools for understanding microbial ecosystems.

References

Abubucker S, Segata N, Goll J, Schubert AM, Izard J, Cantarel BL et al (2012) Metabolic reconstruction for metagenomic data and its application to the human microbiome. PLoS Comput Biol 8(6):e1002358

Addis MF, Tanca A, Uzzau S, Oikonomou G, Bicalho RC, Moroni P (2016) The bovine milk microbiota: insights and perspectives from-omics studies. Mol Biosyst 12(8):2359–2372. https://doi.org/10.1039/c6mb00217j

Ali A, Abouleila Y, Shimizu Y, Hiyama E, Emara S, Mashaghi A, Hankemeier T (2019) Single-cell metabolomics by mass spectrometry: advances, challenges, and future applications. TrAC Trends Anal Chem 120:115436

Amer B, Deshpande RR, Bird SS (2023) Simultaneous quantitation and discovery (SQUAD) analysis: combining the best of targeted and untargeted mass spectrometry-based metabolomics. Metabolites 13(5):648. https://doi.org/10.3390/metabo13050648

Aminu S, Ascandari A, Laamarti M, Safdi NEH, El Allali A, Daoud R (2024) Exploring microbial worlds: a review of whole genome sequencing and its application in characterizing the microbial communities. Crit Rev Microbiol 50(5):805–829. https://doi.org/10.1080/1040841X.2023.2282447

Andrighetti T, Bohar B, Lemke N, Sudhakar P, Korcsmaros T (2020) MicrobioLink: an integrated computational pipeline to infer functional effects of microbiome–host interactions. Cells 9(5):1278. https://doi.org/10.3390/cells9051278

Arıkan M, Muth T (2023) Integrated multi-omics analyses of microbial communities: a review of the current state and future directions. Mol Omics. https://doi.org/10.1039/d3mo00089c

Arıkan M, Muth T (2024) gNOMO2: a comprehensive and modular pipeline for integrated multi-omics analyses of microbiomes. GigaScience 13:giae038. https://doi.org/10.1093/gigascience/giae038

Armengaud J (2009) A perfect genome annotation is within reach with the proteomics and genomics alliance. Curr Opmicrobiol 12(3):292–300

Armengaud J (2022) Metaproteomics to understand how microbiota function: the crystal ball predicts a promising future. Environ Microbiol 25:115–125. https://doi.org/10.1111/1462-2920.16238

Armstrong J, Fiddes IT, Diekhans M, Paten B (2019) Whole-genome alignment and comparative annotation. Annu Rev Anim Biosci 7(1):41–64. https://doi.org/10.3410/f.734326805.793570658

Arndt D, Xia J, Liu Y, Zhou Y, Guo AC, Cruz JA et al (2012) METAGENassist: a comprehensive web server for comparative metagenomics. Nucleic Acids Res 40(W1):W88–W95. https://doi.org/10.1093/nar/gks497

Asnicar F, Thomas A, Beghini F, Mengoni C, Manara S, Manghi P, Zhu Q, Bolzan M, Cumbo F, May U, Sanders J, Zolfo M, Kopylova E, Pasolli E, Knight R, Mirarab S, Huttenhower C, Segata N (2020) Precise phylogenetic analysis of microbial isolates and genomes from metagenomes using PhyloPhlAn 3.0. Nat Commun 11. https://doi.org/10.1038/s41467-020-16366-7

Baker S, Thomson N, Weill FX, Holt KE (2018) Genomic insights into the emergence and spread of antimicrobial-resistant bacterial pathogens. Science 360(6390):733–738. https://doi.org/10.1126/science.aar3777

Bankevich A, Nurk S, Antipov D, Gurevich AA, Dvorkin M, Kulikov AS et al (2012) SPAdes: a new genome assembly algorithm and its applications to single-cell sequencing. J Comput Biol 19(5):455–477

Becker P, Hufnagle W, Peters G, Herrmann M (2001) Detection of differential gene expression in biofilm-forming versus planktonic populations of Staphylococcus aureus using micro-representational-difference analysis. Appl Environ Microbiol 67(7):2958–2965. https://doi.org/10.1128/AEM.67.7.2958-2965.2001

Beghini F, McIver LJ, Blanco-Míguez A, Dubois L, Asnicar F, Maharjan S et al (2021) Integrating taxonomic, functional, and strain-level profiling of diverse microbial communities with bioBakery 3. elife 10:e65088. https://doi.org/10.1101/2020.11.19.388223

Benndorf D, Vogt C, Jehmlich N, Schmidt Y, Thomas H, Woffendin G et al (2009) Improving protein extraction and separation methods for investigating the metaproteome of anaerobic benzene communities within sediments. Biodegradation 20(6):737–750. https://doi.org/10.1007/s10532-009-9261-3

Bhardwaj M (2013) The applications of omics technologies and the challenges of ethics in nutritional sciences. In Bioinformatics: Concepts, methodologies, tools, and applications. IGI Global Scientific Publishing. pp 1334–1343. https://doi.org/10.4018/978-1-61692-883-4.ch006

Bhinderwala F, Wase N, DiRusso C, Powers R (2018) Combining mass spectrometry and NMR improves metabolite detection and annotation. J Proteome Res 17(11):4017–4022. https://doi.org/10.1021/acs.jproteome.8b00567

Bickhart DM, Kolmogorov M, Tseng E, Portik DM, Korobeynikov A, Tolstoganov I et al (2022) Generating lineage-resolved, complete metagenome-assembled genomes from complex microbial communities. Nat Biotechnol 40(5):711–719. https://doi.org/10.1038/s41587-021-01130-z

Bikel S, Valdez-Lara A, Cornejo-Granados F, Rico K, Canizales-Quinteros S, Soberón X et al (2015) Combining metagenomics, metatranscriptomics and viromics to explore novel microbial interactions: towards a systems-level understanding of human microbiome. Comput Struct Biotechnol J 13:390–401. https://doi.org/10.1016/j.csbj.2015.06.001

Blainey PC (2013) The future is now: single-cell genomics of bacteria and archaea. FEMS Microbiol Rev 37(3):407–427

Blakeley-Ruiz JA, Kleiner M (2022) Considerations for constructing a protein sequence database for metaproteomics. Comput Struct Biotechnol J 20:937–952. https://doi.org/10.1016/j.csbj.2022.01.018

Blin K, Shaw S, Augustijn HE, Reitz ZL, Biermann F, Alanjary M, Fetter A, Terlouw BR, Metcalf WW, Helfrich EJN, van Wezel GP, Medema MH, Weber T (2023) antiSMASH 7.0: new and improved predictions for detection, regulation, chemical structures and visualisation. Nucleic Acids Res 51:W46–W50. https://doi.org/10.1093/nar/gkad344

Brettin T, Davis JJ, Disz T, Edwards RA, Gerdes S, Olsen GJ et al (2015) RASTtk: a modular and extensible implementation of the RAST algorithm for building custom annotation pipelines and annotating batches of genomes. Sci Rep 5(1):8365. https://doi.org/10.1038/srep08365

Brockhurst MA, Harrison E, Hall JP, Richards T, McNally A, MacLean C (2019) The ecology and evolution of pangenomes. Curr Biol 29(20):R1094–R1103. https://doi.org/10.1016/j.cub.2019.08.012

Brown JR (2007) Comparative genomics: basic and applied research. CRC Press. https://doi.org/10.1201/9781420008876

Brumfield KD, Huq A, Colwell RR, Olds JL, Leddy MB (2020) Microbial resolution of whole genome shotgun and 16S amplicon metagenomic sequencing using publicly available NEON data. PLoS One 15(2):e0228899. https://doi.org/10.1371/journal.pone.0228899

Bushmanova E, Antipov D, Lapidus A, Prjibelski AD (2019) rnaSPAdes: a de novo transcriptome assembler and its application to RNA-Seq data. GigaScience 8(9):giz100

Caudai C, Galizia A, Geraci F, Pera L, Morea V, Salerno E, Via A, Colombo T (2021) AI applications in functional genomics. Comput Struct Biotechnol J 19:5762–5790. https://doi.org/10.1016/j.csbj.2021.10.009

Chen IMA, Chu K, Palaniappan K, Pillay M, Ratner A, Huang J, Kyrpides NC (2019) IMG/M v. 5.0: an integrated data management and comparative analysis system for microbial genomes and microbiomes. Nucleic Acids Res 47(D1):D666–D677. https://doi.org/10.1093/nar/gky901

Cheng K, Ning Z, Zhang X, Li L, Liao B, Mayne J et al (2017) MetaLab: an automated pipeline for metaproteomic data analysis. Microbiome 5(1):157. https://doi.org/10.1186/s40168-017-0375-2

Chetty A, Blekhman R (2024) Multi-omic approaches for host-microbiome data integration. Gut Microbes 16(1):2297860. https://doi.org/10.1080/19490976.2023.2297860

Chun J, Rainey FA (2014) Integrating genomics into the taxonomy and systematics of the Bacteria and Archaea. Int J Syst Evolut Microbiol 64(Pt_2):316–324. https://doi.org/10.1099/ijs.0.054171-0

Cocolin L, Mataragas M, Bourdichon F, Doulgeraki A, Pilet MF, Jagadeesan B et al (2018) Next generation microbiological risk assessment meta-omics: the next need for integration. Int J Food Microbiol 287:10–17. https://doi.org/10.1016/j.ijfoodmicro.2017.11.008

Costa-Silva J, Domingues D, Lopes FM (2017) RNA-Seq differential expression analysis: an extended review and a software tool. PLoS one 12(12):e0190152. https://doi.org/10.1371/journal.pone.0190152

Cox J, Matic I, Hilger M, Nagaraj N, Selbach M, Olsen JV, Mann M (2009) A practical guide to the MaxQuant computational platform for SILAC-based quantitative proteomics. Nat Protoc 4(5):698–705

da Silva Duarte V, Moreira LDPD, Skeie SB, Svalestad F, Øyaas J, Porcellato D (2024) Database selection for shotgun metaproteomic of low-diversity dairy microbiomes. Int J Food Microbiol 418:110706

Dai X, Shen L (2022) Advances and trends in omics technology development. Front Med 9:911861. https://doi.org/10.3389/fmed.2022.911861

Daliri EBM, Ofosu FK, Chelliah R, Lee BH, Oh DH (2021) Challenges and perspective in integrated multi-omics in gut microbiota studies. Biomolecules 11(2):300. https://doi.org/10.3390/biom11020300

Dam S, Võsa U, Van Der Graaf A, Franke L, Magalhães J (2017) Gene co-expression analysis for functional classification and gene–disease predictions. Brief Bioinform 19:575–592. https://doi.org/10.1093/bib/bbw139

Daniel R (2005) The metagenomics of soil. Nat Rev Microbiol 3(6):470–478

Dar D, Dar N, Cai L, Newman DK (2021) Spatial transcriptomics of planktonic and sessile bacterial populations at single-cell resolution. Science 373(6556):eabi4882. https://doi.org/10.1126/science.abi4882

Darling A, Mau B, Perna N (2010) progressiveMauve: multiple genome alignment with gene gain, loss and rearrangement. PLoS One 5. https://doi.org/10.1371/journal.pone.0011147.

Davids M, Hugenholtz F, Martins dos Santos V, Smidt H, Kleerebezem M, Schaap PJ (2016) Functional profiling of unfamiliar microbial communities using a validated de novo assembly metatranscriptome pipeline. PLoS One 11(1):e0146423. https://doi.org/10.1371/journal.pone.0146423

de Sá PH, Guimarães LC, das Graças DA, de Oliveira Veras AA, Barh D et al (2018) Next-generation sequencing and data analysis: strategies, tools, pipelines and protocols. In: Omics technologies and bio-engineering. Academic Press, London, pp. 191–207

Deng L, Tang Y, Zhang X, Chen J (2024) Structure-adaptive canonical correlation analysis for microbiome multi-omics data. Front Genet 15:1489694. https://doi.org/10.3389/fgene.2024.1489694

Dewar AE, Belcher LJ, West SA (2025) A phylogenetic approach to comparative genomics. Nat Rev Genet. 1–11. https://doi.org/10.1038/s41576-024-00803-0

Droby S, Wisniewski M (2018) The fruit microbiome: a new frontier for postharvest biocontrol and postharvest biology. Postharvest Biol Technol 140:107–112. https://doi.org/10.1016/J.POSTHARVBIO.2018.03.004

Duttke SH, Chang MW, Heinz S, Benner C (2019) Identification and dynamic quantification of regulatory elements using total RNA. Genome Res 29(11):1836–1846. https://doi.org/10.1101/gr.253492.119

Estaki M, Jiang L, Bokulich NA, McDonald D, González A, Kosciolek T et al (2020) QIIME 2 enables comprehensive end-to-end analysis of diverse microbiome data and comparative studies with publicly available data. Curr Protoc Bioinform 70(1):e100

Ferrés I, Iraola G (2021) An object-oriented framework for evolutionary pangenome analysis. Cell Rep Method 1. https://doi.org/10.1016/j.crmeth.2021.100085

Ferro M, Tardif M, Reguer E, Cahuzac R, Bruley C, Vermat T, Nugues E, Vigouroux M, Vandenbrouck Y, Garin J, Viari A (2008) PepLine: a software pipeline for high-throughput direct mapping of tandem mass spectrometry data on genomic sequences. J Proteome Res 7(5):1873–1883

Gabor EM, de Vries EJ, Janssen DB (2003) Efficient recovery of environmental DNA for expression cloning by indirect extraction methods. FEMS Microbiol Ecol 44(2):153–163. https://doi.org/10.1016/S0168-6496(02)00462-2

Gabriel L, Brůna T, Hoff KJ, Ebel M, Lomsadze A, Borodovsky M, Stanke M (2024) BRAKER3: fully automated genome annotation using RNA-seq and protein evidence with GeneMark-ETP,

AUGUSTUS, and TSEBRA. Genome Res 34(5):769–777. https://doi.org/10.1101/gr.278090.123

Gaisser KD, Skloss SN, Brettner LM, Paleologu L, Roco CM, Rosenberg AB et al (2024) High-throughput single-cell transcriptomics of bacteria using combinatorial barcoding. Nat Protoc 19(10):3048–3084. https://doi.org/10.1038/s41596-024-01007-w

Gao Y, Chen Y, Yue X, He J, Zhang R, Xu J et al (2018) Development of simultaneous targeted metabolite quantification and untargeted metabolomics strategy using dual-column liquid chromatography coupled with tandem mass spectrometry. Anal Chim Acta 1037:369–379. https://doi.org/10.1016/j.aca.2018.08.042

Gao X, Liang H, Hu T, Zou Y, Xiao L (2025) Cultivated genome references for protein database construction and high-resolution taxonomic annotation in metaproteomics. Microbiol Spectr 13(2). https://doi.org/10.1128/spectrum.01755-24

García-Durán C, Martínez-López R, Zapico I, Pérez E, Romeu E, Arroyo J et al (2021) Distinct human gut microbial taxonomic signatures uncovered with different sample processing and microbial cell disruption methods for metaproteomic analysis. Front Microbiol 12:618566. https://doi.org/10.3389/fmicb.2021.618566

Garcia-Moreno A, López-Domínguez R, Villatoro-García JA, Ramirez-Mena A, Aparicio-Puerta E, Hackenberg M, Carmona-Saez P (2022) Functional enrichment analysis of regulatory elements. Biomedicines 10(3):590

Garrison E, Guarracino A, Heumos S, Villani F, Bao Z, Tattini L, Hagmann J, Vorbrugg S, Marco-Sola S, Kubica C, Ashbrook D, Thorell K, Rusholme-Pilcher R, Liti G, Rudbeck E, Golicz A, Nahnsen S, Yang Z, Mwaniki M, Nobrega F, Wu Y, Chen H, De Ligt J, Sudmant P, Huang S, Weigel D, Soranzo N, Colonna V, Williams R, Prins P (2023) Building pangenome graphs. Preprint at bioRxiv https://doi.org/10.1101/2023.04.05.535718

Geier B, Sogin EM, Michellod D, Janda M, Kompauer M, Spengler B et al (2020) Spatial metabolomics of in situ host–microbe interactions at the micrometre scale. Nat Microbiol 5(3):498–510. https://doi.org/10.1038/s41564-019-0664-6

Geiszler DJ, Kong AT, Avtonomov DM, Yu F, da Veiga Leprevost F, Nesvizhskii AI (2021) PTM-Shepherd: analysis and summarization of post-translational and chemical modifications from open search results. Mol Cell Proteom 20:1-13

Gibbs D, Baratt A, Baric R, Kawaoka Y, Smith R, Orwoll E, Katze M, McWeeney S (2013) Protein co-expression network analysis (ProCoNA). J Clin Bioinform 3:11–11. https://doi.org/10.1186/2043-9113-3-11

Gifford SM, Sharma S, Rinta-Kanto JM, Moran MA (2011) Quantitative analysis of a deeply sequenced marine microbial metatranscriptome. ISME J 5(3):461–472. https://doi.org/10.1038/ismej.2010.141

Go D, Yeon GH, Park SJ, Lee Y, Koh HG, Koo H et al (2024) Integration of metabolomics and other omics: from microbes to microbiome. Appl Microbiol Biotechnol 108(1):538. https://doi.org/10.1007/s00253-024-13384-z

Golicz AA, Bayer PE, Bhalla PL, Batley J, Edwards D (2020) Pangenomics comes of age: from bacteria to plant and animal applications. Trends Genet 36(2):132–145. https://doi.org/10.1016/j.tig.2019.11.006

Gonzalez A, Navas-Molina JA, Kosciolek T, McDonald D, Vázquez-Baeza Y, Ackermann G et al (2018) Qiita: rapid, web-enabled microbiome meta-analysis. Nat Methods 15(10):796–798. https://doi.org/10.1038/s41592-018-0141-9

Grabherr MG, Haas BJ, Yassour M, Levin JZ, Thompson DA, Amit I et al (2011) Trinity: reconstructing a full-length transcriptome without a genome from RNA-Seq data. Nat Biotechnol 29(7):644

Graham C, McMullan G, Graham RL (2011) Proteomics in the microbial sciences. Bioeng Bugs 2(1):17–30. https://doi.org/10.4161/bbug.2.1.14413

Groussman RD, Blaskowski S, Coesel SN, Armbrust EV (2023) MarFERReT, an open-source, version-controlled reference library of marine microbial eukaryote functional genes. Sci Data 10(1):926. https://doi.org/10.1038/s41597-023-02842-4

Gurdeep Singh R, Tanca A, Palomba A, Van der Jeugt F, Verschaffelt P, Uzzau S et al (2018) Unipept 4.0: functional analysis of metaproteome data. J Proteome Res 18(2):606–615
Gurung AB (2023) Human transcriptome profiling: applications in health and disease. In: Transcriptome profiling. Academic Press, London, pp, 373–395
Gutleben J, Chaib De Mares M, Van Elsas JD, Smidt H, Overmann J, Sipkema D (2018) The multi-omics promise in context: from sequence to microbial isolate. Crit Rev Microbiol 44(2):212–229. https://doi.org/10.1080/1040841X.2017.1332003
Han K, Tjaden B, Lory S (2016) GRIL-seq provides a method for identifying direct targets of bacterial small regulatory RNA by in vivo proximity ligation. Nat Microbiol 2. https://doi.org/10.1038/nmicrobiol.2016.239
Hébert FO, Boyle B, Levesque RC (2018) Directin vivo microbial transcriptomics during infection. Trends Microbiol 26(9):732–735. https://doi.org/10.1016/j.tim.2018.07.002
Heintz-Buschart A, Westerhuis JA (2022) A beginner's guide to integrating multi-omics data from microbial communities. Biochemist 44(3):23–29. https://doi.org/10.1042/bio_2022_100
Herazo-Álvarez J, Mora M, Cuadros-Orellana S, Vilches-Ponce K, Hernández-García R (2025) A review of neural networks for metagenomic binning. Brief Bioinform 26(2):bbaf065. https://doi.org/10.1093/bib/bbaf065
Hernández-Plaza A, Szklarczyk D, Botas J, Cantalapiedra CP, Giner-Lamia J, Mende DR et al (2023) eggNOG 6.0: enabling comparative genomics across 12 535 organisms. Nucleic Acids Res 51(D1):D389–D394
Hickey G, Monlong J, Ebler J, Novak AM, Eizenga JM, Gao Y et al (2024) Pangenome graph construction from genome alignments with Minigraph-Cactus. Nat Biotechnol 42(4):663–673. https://doi.org/10.1038/s41587-023-01793-w
Holbrook-Smith D, Trouillon J, Sauer U (2024) Metabolomics and microbial metabolism: toward a systematic understanding. Annu Rev Biophys 53:41-64
Holmes MR, Giddings MC (2008) Using GFS to identify encoding genomic loci from protein mass spectral data. Curr Protoc Bioinform 21(1):13–9
Holt C, Yandell M (2011) MAKER2: an annotation pipeline and genome-database management tool for second-generation genome projects. BMC Bioinform 12(1):491. https://doi.org/10.1186/1471-2105-12-491
Huang DW, Sherman BT, Lempicki RA (2009) Bioinformatics enrichment tools: paths toward the comprehensive functional analysis of large gene lists. Nucleic Acids Res 37(1):1–13
Huang DW, Sherman BT, Lempicki RA (2009) Systematic and integrative analysis of large gene lists using DAVID bioinformatics resources. Nat Protoc 4(1):44–57. https://doi.org/10.1038/nprot.2008.211
Jaffe JD, Berg HC, Church GM (2004) Proteogenomic mapping as a complementary method to perform genome annotation. Proteomics 4(1):59–77. https://doi.org/10.1002/pmic.200300511
Hyatt D, Chen GL, LoCascio PF, Land ML, Larimer FW, Hauser LJ (2010) Prodigal: prokaryotic gene recognition and translation initiation site identification. BMC Bioinform 11(1):119
Iacono G, Massoni-Badosa R, Heyn H (2019) Single-cell transcriptomics unveils gene regulatory network plasticity. Genome Biol 20(1):110
Ji J, Jung S (2025) PredCMB: predicting changes in microbial metabolites based on the gene–metabolite network analysis of shotgun metagenome data. Bioinformatics 41(1):btaf020. https://doi.org/10.1093/bioinformatics/btaf020
Jiang G, Zheng JY, Ren SN, Yin W, Xia X, Li Y, Wang HL (2024) A comprehensive workflow for optimizing RNA-seq data analysis. BMC Genomics 25(1):631. https://doi.org/10.1186/s12864-024-10414-y
Jonsson V, Österlund T, Nerman O, Kristiansson E (2016) Statistical evaluation of methods for identification of differentially abundant genes in comparative metagenomics. BMC Genomics 17(1):78. https://doi.org/10.1186/s12864-016-2386-y
Kang DD, Li F, Kirton E, Thomas A, Egan R, An H, Wang Z (2019) MetaBAT 2: an adaptive binning algorithm for robust and efficient genome reconstruction from metagenome assemblies. PeerJ 7:e7359. https://doi.org/10.7717/PEERJ.7359

Keating C, Fiege K, Diender M, Sousa DZ, Villanueva L (2024) Microbial single-cell applications under anoxic conditions. Appl Environ Microbiol 90(11):e01321–24. https://doi.org/10.1128/aem.01321-24

Keiblinger KM, Wilhartitz IC, Schneider T, Roschitzki B, Schmid E, Eberl L et al (2012) Soil metaproteomics–Comparative evaluation of protein extraction protocols. Soil Biol Biochem 54:14–24. https://doi.org/10.1016/J.SOILBIO.2012.05.014

Kelly RT (2020) Single-cell proteomics: progress and prospects. Mol Cell Proteom 19(11):1739–1748. https://doi.org/10.1074/mcp.R120.002234

Kim DH, Kim YS, Son NI, Kang CK, Kim AR (2017) Recent omics technologies and their emerging applications for personalised medicine. IET Syst Biol 11(3):87–98

Kim S, Thapa I, Ali H (2022) A multi-omics graph database for data integration and knowledge extraction. In: Proceedings of the 13th ACM international conference on bioinformatics, computational biology and health informatics. Association for Computing Machinery (ACM), New York, pp 1–6

Kleiner M (2019) Metaproteomics: much more than measuring gene expression in microbial communities. Msystems 4(3):10–1128. https://doi.org/10.1128/mSystems.00115-19

Kohler D, Staniak M, Tsai TH, Huang T, Shulman N, Bernhardt OM et al (2023) MSstats version 4.0: statistical analyses of quantitative mass spectrometry-based proteomic experiments with chromatography-based quantification at scale. J Proteome Res 22(5):1466–1482

Kuik C, van Hoogstraten SW, Arts JJ, Honing M, Cillero-Pastor B (2024) Matrix-assisted laser desorption/ionization mass spectrometry imaging for quorum sensing. AMB Express 14(1):45. https://doi.org/10.1186/s13568-024-01703-6

Kwong JC, McCallum N, Sintchenko V, Howden BP (2015) Whole genome sequencing in clinical and public health microbiology. Pathology 47(3):199–210

Langfelder P, Horvath S (2008) Wgcna: an r package for weighted correlation network analysis. BMC Bioinform 9: 559

Langille MG, Zaneveld J, Caporaso JG, McDonald D, Knights D, Reyes JA et al (2013) Predictive functional profiling of microbial communities using 16S rRNA marker gene sequences. Nat Biotechnol 31(9):814–821. https://doi.org/10.1038/nbt.2676

Law CW, Chen Y, Shi W, Smyth GK (2014) voom: Precision weights unlock linear model analysis tools for RNA-seq read counts. Genome Biol 15(2):R29

Li H, Durbin R (2024) Genome assembly in the telomere-to-telomere era. Nat Rev Genetics. https://doi.org/10.1038/s41576-024-00718-w

Liao WW, Asri M, Ebler J, Doerr D, Haukness M, Hickey G et al (2023) A draft human pangenome reference. Nature 617(7960):312–324. https://doi.org/10.1038/s41586-023-05896-x

Lin HH, Liao YC (2016) Accurate binning of metagenomic contigs via automated clustering sequences using information of genomic signatures and marker genes. Sci Rep 6(1):24175. https://doi.org/10.1038/srep24175

Linz AM, Aylward FO, Bertilsson S, McMahon KD (2020) Time-series metatranscriptomes reveal conserved patterns between phototrophic and heterotrophic microbes in diverse freshwater systems. Limnol Oceanogr 65:S101–S112

Liu B, Liu Y, Xu S, Wu Q, Wu D, Zhan L et al (2025) EasyMultiProfiler: an efficient multi-omics data integration and analysis workflow for microbiome research. Preprint at bioRxiv https://doi.org/10.1101/2025.04.17.649266

Liu R, Qiao X, Shi Y, Peterson CB, Bush WS, Cominelli F et al (2024) Constructing phylogenetic trees for microbiome data analysis: a mini-review. Comput Struct Biotechnol J 23:3859–3868

Liu W, Li L, Long X, You W, Zhong Y, Wang M, Tao H, Lin S, He H (2018) Construction and analysis of gene co-expression networks in *Escherichia coli*. Cells 7. https://doi.org/10.3390/cells7030019.

Love MI, Huber W, Anders S (2014) Moderated estimation of fold change and dispersion for RNA-seq data with DESeq2. Genome Biol 15(12):550

Lowe R, Shirley N, Bleackley M, Dolan S, Shafee T (2017) Transcriptomics technologies. PLoS Comput Biol 13(5):e1005457. https://doi.org/10.1371/journal.pcbi.1005457

Lu Y, Zhou G, Ewald J, Pang Z, Shiri T, Xia J (2023) MicrobiomeAnalyst 2.0: comprehensive statistical, functional and integrative analysis of microbiome data. Nucleic Acids Res 51(W1):W310–W318. https://doi.org/10.1093/nar/gkad407

Lugli G, Fontana F, Tarracchini C, Milani C, Mancabelli L, Turroni F, Ventura M (2023) MEGAnnotator2: a pipeline for the assembly and annotation of microbial genomes. Microbiome Res Rep 2. https://doi.org/10.20517/mrr.2022.21

Lyu L, Li X, Feng R, Zhou X, Guha TK, Yu X et al (2023) Simultaneous profiling of host expression and microbial abundance by spatial metatranscriptome sequencing. Genome Res 33(3):401–411. https://doi.org/10.1101/gr.277178.122

Ma Y, Liu L, Ma Y, Zhang S (2023) HONMF: integration analysis of multi-omics microbiome data via matrix factorization and hypergraph. Bioinformatics 39(6):btad335

Mäder U, Nicolas P, Richard H, Bessières P, Aymerich S (2011) Comprehensive identification and quantification of microbial transcriptomes by genome-wide unbiased methods. Curr Opin Biotechnol 22(1):32–41. https://doi.org/10.1016/j.copbio.2010.10.003

Majzoub ME, Luu LD, Haifer C, Paramsothy S, Borody TJ, Leong RW et al (2024) Refining microbial community metabolic models derived from metagenomics using reference-based taxonomic profiling. Msystems 9(9):e00746–24. https://doi.org/10.1128/msystems.00746-24

Mallawaarachchi V, Wickramarachchi A, Lin Y (2020) GraphBin: refined binning of metagenomic contigs using assembly graphs. Bioinformatics 36(11):3307–3313. https://doi.org/10.1093/bioinformatics/btaa180

Mallick H, Franzosa EA, Mclver LJ, Banerjee S, Sirota-Madi A, Kostic AD et al (2019) Predictive metabolomic profiling of microbial communities using amplicon or metagenomic sequences. Nat Commun 10(1):3136. https://doi.org/10.1038/s41467-019-10927-1

Mallick H, Rahnavard A, McIver LJ, Ma S, Zhang Y, Nguyen LH et al (2021) Multivariable association discovery in population-scale meta-omics studies. PLoS Comput Biol 17(11):e1009442

Mann M, Jensen ON (2003) Proteomic analysis of post-translational modifications. Nat Biotechnol 21(3):255–261. https://doi.org/10.1038/nbt0303-255

Maranga M, Szczerbiak P, Bezshapkin V, Gligorijevic V, Chandler C, Bonneau R et al (2023) Comprehensive functional annotation of metagenomes and microbial genomes using a deep learning-based method. MSystems 8(2):e01178–22

Marco-Ramell A, Palau-Rodriguez M, Alay A, Tulipani S, Urpi-Sarda M, Sanchez-Pla A, Andres-Lacueva C (2018) Evaluation and comparison of bioinformatic tools for the enrichment analysis of metabolomics data. BMC Bioinform 19(1):1. https://doi.org/10.1186/s12859-017-2006-0

Maron PA, Ranjard L, Mougel C, Lemanceau P (2007) Metaproteomics: a new approach for studying functional microbial ecology. Microb Ecol 53(3):486–493

Morton JT, Aksenov AA, Nothias LF, Foulds JR, Quinn RA, Badri MH et al (2019) Learning representations of microbe–metabolite interactions. Nat Methods 16(12):1306–1314. https://doi.org/10.1038/s41592-019-0616-3

Muhamadali H, Winder CL, Dunn WB, Goodacre R (2023) Unlocking the secrets of the microbiome: exploring the dynamic microbial interplay with humans through metabolomics and their manipulation for synthetic biology applications. Biochem J 480(12):891–908. https://doi.org/10.1042/BCJ20210534

Muller E, Shiryan I, Borenstein E (2024) Multi-omic integration of microbiome data for identifying disease-associated modules. Nat Commun 15(1):2621

Muth T, Behne A, Heyer R, Kohrs F, Benndorf D, Hoffmann M et al (2015) The MetaProteomeAnalyzer: a powerful open-source software suite for metaproteomics data analysis and interpretation. J Proteome Res 14(3):1557–1565

Nagpal S, Singh R, Yadav D, Mande S (2020) MetagenoNets: comprehensive inference and meta-insightsfor microbial correlation networks. Nucleic Acids Res 48:W572–W579. https://doi.org/10.1093/nar/gkaa254

Nandy S, Thapa V, Negi P, Joshi S, Bhatt G, Bhattarai S (2025) Comparative genomics: A modern-day approach for understanding the blue print of life. Int J Zool Appl Biosci. https://doi.org/10.55126/ijzab.2025.v10.i01.003

Narayanasamy S, Jarosz Y, Muller E, Heintz-Buschart A, Herold M, Kaysen A, Laczny C, Pinel N, May P, Wilmes P (2016) IMP: a pipeline for reproducible reference-independent integrated metagenomic and metatranscriptomic analyses. Genome Biol 17. https://doi.org/10.1186/s13059-016-1116-8

Navarro-Muñoz JC, Selem-Mojica N, Mullowney MW, Kautsar SA, Tryon JH, Parkinson EI et al(2020) A computational framework to explore large-scale biosynthetic diversity. Nat Chem Biol 16(1):60–68

Nayak L, Hasija Y (2021) A review on recent advancements in single-cell RNA sequencing. Mol Biol Rep 48(2):1813–1830. https://doi.org/10.1007/s11033-020-06145-1

Nayfach S, Pollard K (2016) Toward accurate and quantitative comparative metagenomics. Cell 166:1103–1116. https://doi.org/10.1016/j.cell.2016.08.007

Nguyen K, Carleton G, Lum JJ, Duncan KD (2024) Expanding spatial metabolomics coverage with lithium-doped nanospray desorption electrospray ionization mass spectrometry imaging. Anal Chem 96(46):18427–18436

Ning L, Zhou Y, Sun H, Zhang Y, Shen C, Wang Z, Xuan B, Zhao YY, Yan Y, Tong T, Huang X, Hu M, Zhu X, Ding J, Zhang Y, Cui Z, Fang J, Chen H, Hong J (2023) Microbiome and metabolome features ininflammatory bowel disease via multi-omics integration analyses across cohorts. Nat Commun 14. https://doi.org/10.1038/s41467-023-42788-0

Nissen J, Johansen J, Allesøe R, Sønderby C, Armenteros J, Grønbech C, Jensen L, Nielsen H, Petersen T, Winther O, Rasmussen S (2021) Improved metagenome binning and assembly using deep variational autoencoders. Nat Biotechnol 39:555–560. https://doi.org/10.1038/s41587-020-00777-4

Noor Z, Ahn SB, Baker MS, Ranganathan S, Mohamedali A (2021) Mass spectrometry–based protein identification in proteomics—a review. Brief Bioinform 22(2):1620–1638

Nova IC, Ritmejeris J, Brinkerhoff H et al (2023) Detection of phosphorylation PTMs along single peptides with nanopores. Nat Biotechnol. https://doi.org/10.1038/s41587-023-01839-z

Nyholm L, Koziol A, Marcos S, Botnen A, Aizpurua O, Gopalakrishnan S, Limborg M, Gilbert M, Alberdi A (2020) Holo-Omics: integrated host-microbiota multi-omics for basic and applied biological research. iScience 23. https://doi.org/10.1016/j.isci.2020.101414

O'Donnell S, O'Donnell S, Ross R, Ross R, Stanton C (2020) The progress of multi-omics technologies: determining function in lactic acid bacteria using a systems level approach. Front Microbiol 10. https://doi.org/10.3389/fmicb.2019.03084

Ojala T, Häkkinen A, Kankuri E, Kankainen M (2023) Current concepts, advances, and challenges in deciphering the human microbiota with metatranscriptomics. Trend Genet. https://doi.org/10.1016/j.tig.2023.05.004

Özkurt E, Fritscher J, Soranzo N, Ng DY, Davey RP, Bahram M, Hildebrand F (2022) LotuS2: an ultrafast and highly accurate tool for amplicon sequencing analysis. Microbiome 10(1):176

Pan S, Zhu C, Zhao X, Coelho L (2022) A deep siamese neural network improves metagenome-assembled genomes in microbiome datasets across different environments. Nat Commun 13. https://doi.org/10.1038/s41467-022-29843-y

Pan H, Wattiez R, Gillan D (2024) Soil metaproteomics for microbial community profiling: methodologies and challenges. Curr Microbiol 81(8):257

Papudeshi B, Papudeshi B, Haggerty J, Doane M, Morris M, Walsh K, Beattie D, Pande D, Zaeri P, Silva G, Thompson F, Edwards R, Dinsdale E (2017) Optimizing and evaluating the reconstruction of metagenome-assembled microbial genomes. BMC Genomics 18. https://doi.org/10.1186/s12864-017-4294-1

Paril J, Zare T, Fournier-Level A (2023) Compare_genomes: a comparative genomics workflow to streamline the analysis of evolutionary divergence across eukaryotic genomes. Curr Protoc 3(8):e876. https://doi.org/10.1002/cpz1.876

Paterson A, Draye X, Bowers J, Elsik C, Jiang C, Katsar C, Burow M, Ming R, Wright R, Lin Y, Lan T (2000) Comparative genomics of plant chromosomes. Plant Cell 12:1523–1539. https://doi.org/10.1105/tpc.12.9.1523

Peng J, Zhou X, Rensing C, Liesack W, Zhu YG (2024) Soil microbial ecology through the lens of metatranscriptomics. Soil Ecol Lett 6(3):230217. https://doi.org/10.1007/s42832-023-0217-z

Pérez-Cobas A, Gomez-Valero L, Buchrieser C (2020) Metagenomic approaches in microbial ecology: an update on whole-genome and marker gene sequencing analyses. Microb Genom 6. https://doi.org/10.1099/mgen.0.000409

Pieters VM, Co IL, Wu NC, McGuigan AP (2021) Applications of omics technologies for three-dimensional in vitro disease models. Tissue Eng Part C Methods 27(3):183–199. https://doi.org/10.1089/ten.TEC.2020.0300

Polpitiya AD, Qian WJ, Jaitly N, Petyuk VA, Adkins JN, Camp DG, Smith RD et al (2008) DAnTE: a statistical tool for quantitative analysis of-omics data. Bioinformatics 24(13):1556–1558

Portik D, Feng X, Benoit G, Nasko D, Auch B, Bryson S, Cano R, Carlin M, Damerum A, Farthing B, Grove J, Islam M, Langford K, Liachko I, Locken K, Mangelson H, Tang S, Zhang S, Quince C, Wilkinson J (2024) Highly accurate metagenome-asomes from human gut microbiota using long-read assembly, binning, and consolidation methods. Preprint at bioRxiv https://doi.org/10.1101/2024.05.10.593587

Purushothaman S, Egli A, Meola M (2022) Combination of whole genome sequencing and metagenomics for microbiological diagnostics. Int J Mol Sci 23. https://doi.org/10.3390/ijms23179834

Reuben R, Torres C (2025) Integrating the milk microbiome signatures in mastitis: milk-omics and functional implications. World J Microbiol Biotechnol 41. https://doi.org/10.1007/s11274-024-04242-1

Riesenfeld C, Schloss P, Handelsman J (2004) Metagenomics: genomic analysis of microbial communities. Annu Rev Genet 38:525–52. https://doi.org/10.1146/ANNUREV.GENET.38.072902.091216

Ruiz-Perez C, Conrad R, Konstantinidis K (2021a) MicrobeAnnotator: a user-friendly, comprehensive functional annotation pipeline for microbial genomes. BMC Bioinform 22. https://doi.org/10.1186/s12859-020-03940-5

Ruiz-Perez D, Lugo-Martinez J, Bourguignon N, Mathee K, Lerner B, Bar-Joseph Z, Narasimhan G (2021) Dynamic bayesian networks for integrating multi-omics time series microbiome data. Msystems 6(2):10–1128. https://doi.org/10.1128/mSystems.01105-20

Ruscheweyh HJ, Milanese A, Paoli L, Sintsova A, Mende DR, Zeller G, Sunagawa S (2021) mOTUs: profiling taxonomic composition, transcriptional activity and strain populations of microbial communities. Curr Protoc 1(8):e218

Russell AH, Truman AW (2020) Genome mining strategies for ribosomally synthesised and post-translationally modified peptides. Comput Struct Biotechnol J 18:1838–1851

Saarenpää S, Shalev O, Ashkenazy H, Carlos V, Lundberg D, Weigel D, Giacomello S (2023) Spatial metatranscriptomics resolves host–bacteria–fungi interactomes. Nat Biotechnol 42:1384–1393. https://doi.org/10.1038/s41587-023-01979-2

Sadeghi J, Shahraki A, Chagant S, Heath D (2023) Functional gene transcription variation in bacterial metatranscriptomes in large freshwater Lake Ecosystems: Implications for ecosystem and human health. Environ Res 116298. https://doi.org/10.1016/j.envres.2023.116298

Saito R, Smoot M, Ono K et al (2012) A travel guide to Cytoscape plugins. Nat Methods 9:1069–1076. https://doi.org/10.1038/nmeth.2212

Salvato F, Vintila S, Finkel OM, Dangl JL, Kleiner M (2022)) Evaluation of protein extraction methods for metaproteomic analyses of root-associated microbes. Mol Plant-Microbe Interact 35(11):977–988

Sato Y (2024) Transcriptomeanalysis: a powerful tool to understand individual microbial behaviors and interactions in ecosystems. Biosci Biotechnol Biochem. https://doi.org/10.1093/bbb/zbae064

Schiebenhoefer H, Schallert K, Renard B, Trappe K, Schmid E, Benndorf D, Riedel K, Muth T, Fuchs S (2020) A complete and flexible workflow for metaproteomics data analysis based on MetaProteomeAnalyzer and prophane. Nat Protoc 15:3212–3239. https://doi.org/10.1038/s41596-020-0368-7

Schoch CL, Ciufo S, Domrachev M, Hotton CL, Kannan S, Khovanskaya R et al (2020) NCBI Taxonomy: a comprehensive update on curation, resources and tools. Database 2020:baaa062

Secomandi S, Gallo GR, Rossi R, Rodríguez Fernandes C, Jarvis ED, Bonisoli-Alquati A et al (2025) Pangenome graphs and their applications in biodiversity genomics. Nat Genet 57(1):13–26. https://doi.org/10.1038/s41588-024-02029-6

Seemann T (2014) Prokka: rapid prokaryotic genome annotation. Bioinformatics 30(14):2068–2069. https://doi.org/10.1093/bioinformatics/btu153

Segata N, Waldron L, Ballarini A, Narasimhan V, Jousson O, Huttenhower C (2012) Metagenomic microbial community profiling using unique clade-specific marker genes. Nat Methods 9(8):811–814

Shaffer J, Nothias L, Thompson L, Sanders J, Salido R, Couvillion S, Brejnrod A, Lejzerowicz F, Haiminen N et al (2022) Standardized multi-omics of Earth's microbiomes reveals microbial and metabolite diversity. Nat Microbiol 7:2128–2150. https://doi.org/10.1038/s41564-022-01266-x

Shaffer M, Thurimella K, Quinn K, Doenges K, Zhang X, Bokatzian S, Reisdorph N, Lozupone C (2019) AMON: annotation of metabolite origins via networks to integrate microbiome and metabolome data. BMC Bioinform 20. https://doi.org/10.1186/s12859-019-3176-8

Shakya M, Lo C, Chain P (2019) Advances and challenges in metatranscriptomic analysis. Front Genet 10. https://doi.org/10.3389/fgene.2019.00904

Shi Y, Wang G, Lau H, Yu J (2022) Metagenomicsequencing for microbial DNA in human samples: emerging technological advances. Int J Mol Sci 23. https://doi.org/10.3390/ijms23042181.

Shtossel O, Koren O, Shai I, Rinott E, Louzoun Y (2024) Gut microbiome-metabolome interactions predict host condition. Microbiome 12. https://doi.org/10.1186/s40168-023-01737-1

Singer E, Bushnell B, Coleman-Derr D, Bowman B, Bowers R, Levy A, Gies E, Cheng J, Copeland A, Klenk H, Hallam S, Hugenholtz P, Tringe S, Woyke T (2016) High-resolution phylogenetic microbial community profiling. ISME J 10:2020–2032. https://doi.org/10.1038/ismej.2015.249

Skinnider MA, Johnston CW, Gunabalasingam M, Merwin NJ, Kieliszek AM, MacLellan RJ et al (2020) Comprehensive prediction of secondary metabolite structure and biological activity from microbial genome sequences. Nat Commun 11(1):6058. https://doi.org/10.1038/s41467-020-19986-1

Speda J, Johansson MA, Carlsson U, Karlsson M (2017) Assessment of sample preparation methods for metaproteomics of extracellular proteins. Anal Biochem 516:23–36. https://doi.org/10.1016/j.ab.2016.10.008

Steen A, Crits-Christoph A, Carini P, DeAngelis K, DeAngelis K, Fierer N, Fierer N, Lloyd K, Thrash J (2019) High proportions of bacteria and archaea across most biomes remain uncultured. ISME J 13:3126–3130. https://doi.org/10.1038/s41396-019-0484-y

Strous M, Kraft B, Bisdorf R, Tegetmeyer HE (2012) The binning of metagenomic contigs for microbial physiology of mixed cultures. Front Microbiol 3:410. https://doi.org/10.3389/fmicb.2012.00410

Subramanian A, Tamayo P, Mootha VK, Mukherjee S, Ebert BL, Gillette MA et al (2005) Gene set enrichment analysis: a knowledge-based approach for interpreting genome-wide expression profiles. Proc Natl Acad Sci USA 102(43):15545–15550

Sulit AK, Kolisnik T, Frizelle FA, Purcell R, Schmeier S (2023) MetaFunc: taxonomic and functional analyses of high throughput sequencing for microbiomes. Gut Microbiome 4:e4. https://doi.org/10.1101/2020.09.02.271098

Szklarczyk D, Kirsch R, Koutrouli M, Nastou K, Mehryary F, Hachilif R et al (2023) The STRING database in 2023: protein–protein association networks and functional enrichment analyses for any sequenced genome of interest. Nucleic Acids Res 51(D1):D638–D646

Taj B, Adeolu M, Xiong X, Ang J, Nursimulu N, Parkinson J (2023) MetaPro: a scalable and reproducible data processing and analysis pipeline for metatranscriptomic investigation of microbial communities. Microbiome 11(1):143. https://doi.org/10.1186/s40168-023-01562-6

Tanca A, Deledda MA, De Diego L, Abbondio M et al (2024) Benchmarking low-and high-throughput protein cleanup and digestion methods for human fecal metaproteomics. MSystems 9(7):e00661–24. https://doi.org/10.1128/msystems.00661-24
Tanca A, Palomba A, Deligios M, Cubeddu T, Fraumene C, Biosa G et al (2013) Evaluating the impact of different sequence databases on metaproteome analysis: insights from a lab-assembled microbial mixture. PLoS One 8(12):e82981. https://doi.org/10.1371/journal.pone.0082981
Tanca A, Palomba A, Fraumene C, Pagnozzi D, Manghina V, Deligios M, Uzzau S (2016) The impact of sequence database choice on metaproteomic results in gut microbiota studies. Microbiome 4(1):51. https://doi.org/10.1186/s40168-016-0196-8
Tarazona S, Arzalluz-Luque Á, Conesa A (2021) Undisclosed, unmet and neglected challenges in multi-omics studies. Nat Comput Sci 1:395–402. https://doi.org/10.1038/s43588-021-00086-z
Tarazona S, García-Alcalde F, Dopazo J, Ferrer A, Conesa A (2011) Differential expression in RNA-seq: a matter of depth. Genome Res 21(12):2213–2223
Thaben PF, Westermark PO (2014) Detecting rhythms in time series with RAIN. J Biol Rhythm 29(6):391–400
The UniProt Consortium (2023) UniProt: theuniversal protein knowledgebase in 2023. Nucleic Acids Res 51(1)):D523–D531. https://doi.org/10.1093/nar/gkac1052
Theodorakis E, Antonakis AN, Baltsavia I, Pavlopoulos GA, Samiotaki M, Amoutzias GD, et al (2021) ProteoSign v2: a faster and evolved user-friendly online tool for statistical analyses of differential proteomics. Nucleic Acids Res 49(W1):W573–W577
Thompson C, Chimetto L, Edwards R, Swings J, Stackebrandt E, Thompson F (2013) Microbial genomic taxonomy. BMC Genomics 14: 913. https://doi.org/10.1186/1471-2164-14-913
Tonkin-Hill G, Corander J, Parkhill J (2023) Challenges in prokaryote pangenomics. Microb Genomics 9(5):001021. https://doi.org/10.1099/mgen.0.001021
Tringe SG, Von Mering C, Kobayashi A, Salamov AA, Chen K, Chang HW et al (2005) Comparative metagenomics of microbial communities. Science 308(5721):554–557. https://doi.org/10.1126/SCIENCE.1107851
Tyanova S, Temu T, Sinitcyn P, Carlson A, Hein MY, Geiger T et al (2016) The Perseus computational platform for comprehensive analysis of (prote) omics data. Nat Methods 13(9):731–740
Uehara M (2024) Decoding host-microbiome interactions through metatranscriptomic analysis. KeiO Associated Repository of Academic resources. Keio University. Tokyo.
Ureta-Vidal A, Ettwiller L, Birney E (2003) Comparative genomics: genome-wide analysis in metazoan eukaryotes. Nat Rev Genet 4(4):251–262. https://doi.org/10.1038/nrg1043
Van Den Bossche T, Armengaud J, Benndorf D, Blakeley-Ruiz J, Brauer M, Cheng K, Creskey M, Figeys D, Grenga L, Griffin T, Henry C, Hettich R, Holstein T, Jagtap P, Jehmlich N, Kim J, Kleiner M, Kunath B, Malliet X, Martens L, Mehta S, Mesuere B, Ning Z, Tanca A, Uzzau S, Verschaffelt P, Wang J, Wilmes P, Zhang X, Zhang X, Li L (2025) The microbiologist's guide to metaproteomics. iMeta 4. https://doi.org/10.1002/imt2.70031
Van Vliet AH (2010) Next generation sequencing of microbial transcriptomes: challenges and opportunities. FEMS Microbiol Lett 302(1):1–7. https://doi.org/10.1111/j.1574-6968.2009.01767.x
Vella D, Zoppis I, Mauri G, Mauri P, Silvestre D (2017) From protein-protein interactions to protein co-expression networks: a new perspective to evaluate large-scale proteomic data. EURASIP J Bioinform Syst Biol. https://doi.org/10.1186/s13637-017-0059-z.
Wagner S, Scholz K, Sieber M, Kellert M, Voelkel W (2007) Tools in metabonomics: an integrated validation approach for LC-MS metabolic profiling of mercapturic acids in human urine. Anal Chem 79(7):2918–2926
Walke D, Schallert K, Ramesh P, Benndorf D, Lange E, Reichl U, Heyer R (2021) MPA_pathway_tool: User-friendly, automatic assignment of microbial community data on metabolic pathways. Int J Mol Sci 22(20):10992
Wang J, Chen Y, Zou Q (2023) Inferring gene regulatory network from single-cell transcriptomes with graph autoencoder model. PLoS Genet 19(9):e1010942. https://doi.org/10.1371/journal.pgen.1010942

Washburn MP, Yates IIIJR (2000) Analysis of the microbial proteome. Curr Opmicrobiol 3(3):292–297. https://doi.org/10.1016/S1369-5274(00)00092-8
Wei L, Liu Y, Dubchak I, Shon J, Park J (2002) Comparative genomics approaches to study organism similarities and differences. J Biomed Inform 35(2):142–150. https://doi.org/10.1016/S1532-0464(02)00506-3
Werner J, Géron A, Kerssemakers J, Matallana-Surget S (2019) mPies: a novel metaproteomics tool for the creation of relevant protein databases and automatized protein annotation. Biol Direct 14(1):21. https://doi.org/10.1186/s13062-019-0253-x
Whelan FJ, Waddell B, Syed SA, Shekarriz S, Rabin HR, Parkins MD, Surette MG (2020) Culture-enriched metagenomic sequencing enables in-depth profiling of the cystic fibrosis lung microbiota. Nat Microbiol 5(2):379–390. https://doi.org/10.1038/s41564-019-0643-y
Wijesooriya K, Jadaan S, Perera K, Kaur T, Ziemann M (2022) Urgent need for consistent standards in functional enrichment analysis. PLoS Comput Biol 18. https://doi.org/10.1371/journal.pcbi.1009935.
Wilmes P, Heintz-Buschart A, Bond P (2015) A decade of metaproteomics: Where we stand and what the future holds. Proteomics 15:3409–3417. https://doi.org/10.1002/pmic.201500183
Witze ES, Old WM, Resing KA, Ahn NG (2007) Mapping protein post-translational modifications with mass spectrometry. Nat Methods 4(10):798–806. https://doi.org/10.1038/nmeth1100
Wood DE, Salzberg SL (2014) Kraken: ultrafast metagenomic sequence classification using exact alignments. Genome Biol 15(3):R46
Wu E, Yang Y, Zhao J, Zheng J, Wang X, Shen C, Qiao L (2024) High-abundance protein-guided hybrid spectral library for data-independent acquisition metaproteomics. Anal Chem 96(3):1029–1037. https://doi.org/10.1021/acs.analchem.3c03255
Wu Q, Ning Z, Zhang A, Zhang X, Sun Z, Figeys D (2025) Operational taxon-function framework in MetaX: unveiling taxonomic and functional associations in metaproteomics. Anal Chem 97(18):9739–9747. https://doi.org/10.1021/acs.analchem.4c06645
Wu S, Zhou H, Chen D, Lu Y, Li Y, Qiao J (2024b) Multi-omic analysis tools for microbial metabolites prediction. Brief Bioinform 25(4). https://doi.org/10.1093/bib/bbae264
Wu T, Hu E, Xu S, Chen M, Guo P, Dai Z, Feng T, Zhou L, Tang W, Zhan L, Fu X, Liu S, Bo X, Yu G (2021) clusterProfiler 4.0: a universal enrichment tool for interpreting omics data. Innovation 2. https://doi.org/10.1016/j.xinn.2021.100141.
Xu Z, Sziraki A, Lee J, Zhou W, Cao J (2024) Dissecting key regulators of transcriptome kinetics through scalable single-cell RNA profiling of pooled CRISPR screens. Nat Biotechnol 42(8):1218–1223. https://doi.org/10.1038/s41587-023-01948-9
Xue L, Wu Y, Lin Y (2023) Dissecting and improving gene regulatory network inference using single-cell transcriptome data. Genome Res 33(9):1609–1621. https://doi.org/10.1101/gr.277488.122
Yang Y, Qiao L (2023) Data-independent acquisition proteomics methods for analyzing post-translational modifications. Proteomics 23(7-8):2200046. https://doi.org/10.1002/pmic.202200046
Yu G, Xu C, Zhang D, Ju F, Ni Y (2022) MetOrigin: discriminating the origins of microbial metabolites for integrative analysis of the gut microbiome and metabolome. IMeta 1(1):e10. https://doi.org/10.1002/imt2.10
Zampieri G, Campanaro S, Angione C, Treu L (2023) Metatranscriptomics-guided genome-scale metabolic modeling of microbial communities. Cell Rep Method 3(1). https://doi.org/10.1016/j.crmeth.2022.100383
Zhang J, Yang MK, Zeng H, Ge F (2016) GAPP: a proteogenomic software for genome annotation and global profiling of posttranslational modifications in prokaryotes. Mol Cell Proteom 15(11):3529–3539
Zhang L, Chen F, Zeng Z, Xu M, Sun F, Yang L, Bi X, Lin Y, Gao Y, Hao H, Yi W, Li M, Xie Y (2021b) Advances in metagenomics and its application in environmental microorganisms. Front Microbiol 12. https://doi.org/10.3389/fmicb.2021.766364.

Zhang X, Li L, Butcher J, Stintzi A, Figeys D (2019) Advancing functional and translational microbiome research using meta-omics approaches. Microbiome 7(1):154. https://doi.org/10.1186/s40168-019-0767-6

Zhang X, Li L, Mayne J, Ning Z, Stintzi A, Figeys D (2018) Assessing the impact of protein extraction methods for human gut metaproteomics. J Proteom 180:120–127. https://doi.org/10.1016/j.jprot.2017.07.001

Zhang Y, Thompson KN, Branck T, Yan Y, Nguyen LH, Franzosa EA, Huttenhower C (2021) Metatranscriptomics for the human microbiome and microbial community functional profiling. Annu Rev Biomed Data Sci 4(1):279–311. https://doi.org/10.1146/ANNUREV-BIODATASCI-031121-103035

Zhang Z, Zhang L (2020) METAMVGL: a multi-view graph-based metagenomic contig binning algorithm by integrating assembly and paired-end graphs. BMC Bioinform 22(Suppl 10):378. https://doi.org/10.1186/s12859-021-04284-4

Zhang Z, Yang C, Veldsman WP, Fang X, Zhang L (2023) Benchmarking genome assembly methods on metagenomic sequencing data. Brief Bioinform 24(2):bbad087. https://doi.org/10.1093/bib/bbad087

Zhao J, Wang G, Chu J, Zhuang Y (2020) Harnessing microbial metabolomics for industrial applications. World J Microbiol Biotechnol 36(1):1. https://doi.org/10.1007/s11274-019-2775-x

Zheng F, Zhao X, Zeng Z, Wang L, Lv W, Wang Q, Xu G (2020) Development of a plasma pseudotargeted metabolomics method based on ultra-high-performance liquid chromatography–mass spectrometry. Nat Protocol 1–19. https://doi.org/10.1038/s41596-020-0341-5

Zhou Y, Zhang Z, Bao Z, Li H, Lyu Y, Zan Y et al (2022) Graph pangenome captures missing heritability and empowers tomato breeding. Nature 606(7914):527–534. https://doi.org/10.1038/s41586-022-04808-9

Zhou Z, Tran PQ, Breister AM, Liu Y, Kieft K, Cowley ES et al (2022) METABOLIC: high-throughput profiling of microbial genomes for functional traits, metabolism, biogeochemistry, and community-scale functional networks. Microbiome 10(1):33. https://doi.org/10.1186/s40168-021-01213-8

Ziemert N, Alanjary M, Weber T (2016) The evolution of genome mining in microbes–a review. Nat Product Rep 33(8):988–1005

Zuffa S, Schmid R, Bauermeister AP, Gomes PW, Caraballo-Rodriguez AM, El Abiead Y et al (2024) microbeMASST: a taxonomically informed mass spectrometry search tool for microbial metabolomics data. Nat Microbiol 9(2):336–345. https://doi.org/10.1038/s41564-023-01575-9

FSC
www.fsc.org
MIX
Papier aus verantwortungsvollen Quellen
Paper from responsible sources
FSC® C105338

If you have any concerns about our products, you can contact us on
ProductSafety@springernature.com

In case Publisher is established outside the EU, the EU authorized representative is:
Springer Nature Customer Service Center GmbH
Europaplatz 3, 69115 Heidelberg, Germany

Printed by Libri Plureos GmbH
in Hamburg, Germany